Hydrology: Beyond the Basics

Hydrology: Beyond the Basics

Edited by Allison Sergeant

SYRAWOOD
PUBLISHING HOUSE
New York

Published by Syrawood Publishing House,
750 Third Avenue, 9th Floor,
New York, NY 10017, USA
www.syrawoodpublishinghouse.com

Hydrology: Beyond the Basics
Edited by Allison Sergeant

International Standard Book Number: 978-1-68286-658-0 (Hardback)

Cataloging-in-Publication Data

Hydrology : beyond the basics / edited by Allison Sergeant.
    p. cm.
Includes bibliographical references and index.
ISBN 978-1-68286-658-0
1. Hydrology. 2. Hydrography. 3. Earth sciences. 4. Water. I. Sergeant, Allison.
GB661.2 .H93 2019
551.48--dc23

TABLE OF CONTENTS

PREFACE

This book attempts to provide information about the key concepts and theories related to the field of hydrology in comprehensive detail. It also discusses the modern principles and practices of this field. Hydrology refers to the study of all aspects related to water such as its movement, quality, distribution, type etc. Some of the prominent applications of hydrology are the estimation of rainfall, construction of dams, facilitation of potable water, development of urban drainage systems, among many others. This book is compiled in such a manner, that it will provide in-depth knowledge about the principles and practice of hydrology, enabling the reader to gain a thorough understanding of the subject. Students, researchers, experts and all associated with the discipline of hydrology will benefit alike from this book.

The information shared in this book is based on empirical researches made by veterans in this field of study. The elaborative information provided in this book will help the readers further their scope of knowledge leading to advancements in this field.

Finally, I would like to thank my fellow researchers who gave constructive feedback and my family members who supported me at every step of my research.

Editor

Groundwater Modeling of Multi-Aquifer Systems Using GMS

Khalaf S* and Abdalla MG

Irrigation and Hydraulics Department, Faculty of Engineering, El-Mansoura University, Egypt

Abstract

The Nubian Sandstone complex in the western desert is a part of the major regional Nubian aquifer system comprising West Africa. The different rock strata deposited in localities forming the units of the aquifer system. Farafra Oasis lies in the northern part of the Dakhla basin. In its southern region, the Nubian Sandstone (deep aquifer) is overlained by Dakhla shale but in the central and northern regions by fissured chalky limestone (shallow aquifer). The two overlaying aquifers in Farafra Oasis represent a typical hydrogeological model of a huge multi-layered artesian basin extending over the territory of Egypt. The Post Nubian aquifer played an essential role in the development of Farafra Oasis for a long time through the many springs issued from this aquifer. The rapid drilling process of deep wells started in 1960s led to stop flowing of many springs and wells plus the depletion in discharges and pressure of many others. Therefore, there is a real danger of either dewatering or increasing the water depths to uneconomic lifting depths for both the shallow and deep aquifers. A two-dimension flow model GMS (Groundwater Modeling System) was used to investigate this problem. Application of the present conditions indicated that drawdowns in the Post Nubian aquifer range from 5 m to about 9 m. The second scenario tries to sustain the groundwater utilities in the Post Nubian aquifer through a group of procedures. Accordingly, drawdowns are expected to range from 5 m to 8.6 m in the Post Nubian aquifer. According to this scenario, 3 m decline in the Nubian Sandstone aquifer followed by declining in the Post Nubian by about 1 m.

Keywords: Groundwater; Multi aquifer; GMS; Calibration; El-Farafra oasis

Introduction

Many aquifers in nature loss or gain water through adjacent confining beds of relatively low permeability. However, such leaky aquifers are often only part of multiple-aquifer systems. When water is withdrawn or recharged in one particular aquifer the head distribution in the entire system will be influenced. In such a system several aquifers can be distinguished, each separated from the others by aquitards (semi pervious layers). These aquifers and aquitards together form a single leaky system when recharge or discharge at one place influences the head distribution in all other parts, but is not felt at its boundaries. In practice, however, the system considered may be restricted by boundaries with a negligible influence during the period of interest. When water is withdrawn from one or more layers of a multiple-aquifer system, the induced drawdown will be dependent on the hydraulic properties of all aquifers and aquitards. Analytical solutions have recently been presented for several types of steady-state flow in systems comprising any number of layers [1]. In an attempt to combine the advantages of an analytical approach with the capability of numerical models to include heterogeneity, Hemker [2] developed a hybrid analytical–numerical solution for transient well flow in vertically heterogeneous aquifers. The radial components of flow are treated analytically, while the finite-difference technique is used to compute the vertical flow components in the horizontally layered aquifer. The resulting drawdown equations in the Laplace domain also account for the effects of a finite diameter pumped well and wellbore storage. To determine transmissivity values for aquifers and hydraulic resistances for aquitards, field investigations have to be conducted which usually involve one or more aquifer tests. As the presented multiple-aquifer well flow solution can only be used for steady state drawdowns, extrapolation of the observed time--drawdown relationship for each piezometer can provide the required information. This implies that for such an analysis a relatively long pumping period should be chosen, especially when aquitard resistances have to be obtained with some accuracy. The availability of analytical solutions for multiple-aquifer transient well flow will provide

improved methods for aquifer test evaluation, probably substantially reducing the required pumping period. These practical implications and the improved possibilities to predict the response of a multiple aquifer system to abstractions have provided a stimulus to further extend the eigenvalue approach in well flow hydraulics. Saafan et al., [3] applied multi-objective genetic algorithm (MOGA) model in deep Nubian Sandstone aquifer in El-Farafra oasis, Egypt to develop the maximum pumping rate and minimum operation cost as well as the prediction of the future changes in both pumping rate and pumping operation cost. They concluded that the optimal pumping rate and the corresponding drawdown range from 190699 to 179423 m³/day and 6.13 to 8.34 m respectively. Moharram et al. presented optimization model based on the combination of the MODFLOW with GA. This model was used in deep Nubian Sandstone aquifer in El-Farafra oasis, Egypt El-Farafra oasis-Egypt to obtain optimal pumping rate during different simulation period.

The Study Area

Farafra Oasis is one of many morphotectonic depressions present in the Western Desert. It lies in the heart of these depressions between latitudes 26° 30\, 27° 30\ and longitudes 27° 30\ and 29°00\ inclosing an area of about 86200 Km² (Figure 1). It forms an irregular triangular shape with an apex to Bahariya Oasis and base towards Dakhla Oasis. Groundwater is the only water source in this remote area. It is supplied

**Corresponding author: Khalaf S, Irrigation and Hydraulics Department, Faculty of Engineering, El-Mansoura University, Egypt
E-mail: Samykhalaf2005@yahoo.com*

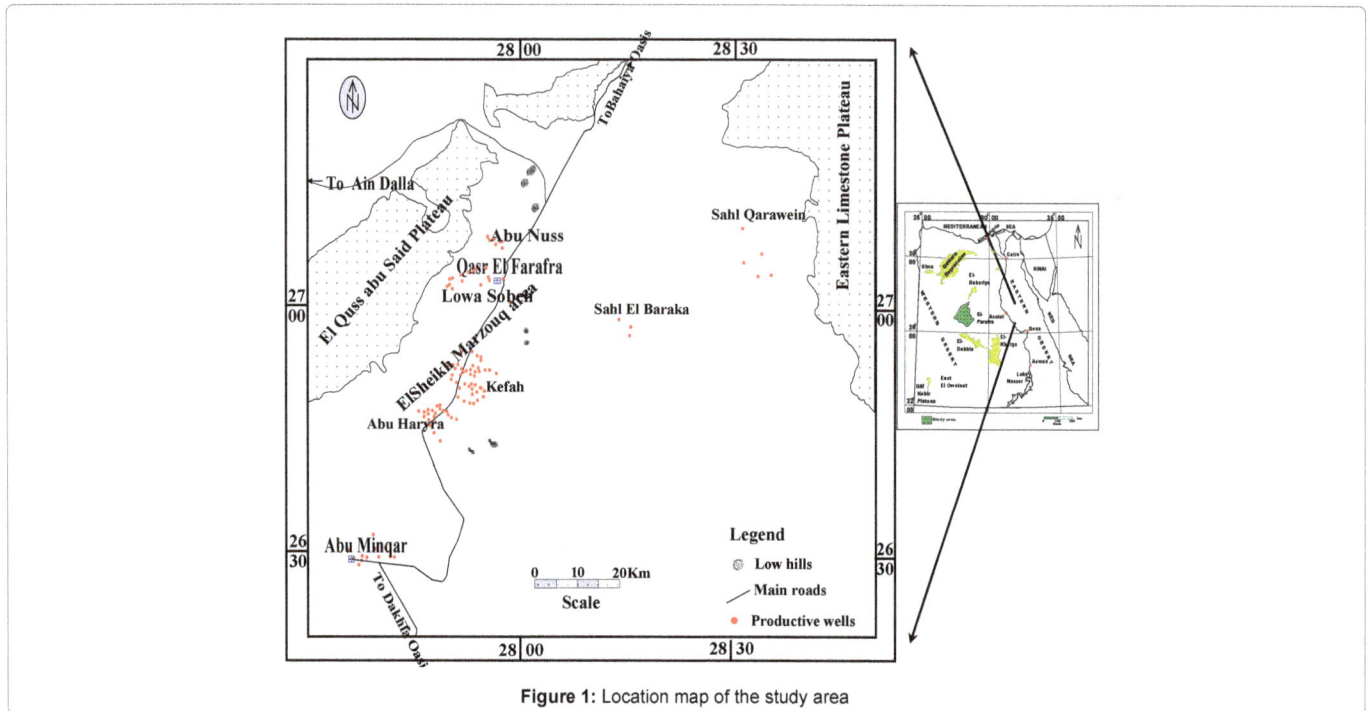

Figure 1: Location map of the study area

from two main aquifers; the shallow Post Nubian aquifer and the deep Nubian Sandstone aquifer.

The Post Nubian aquifer (chalky limestone aquifer) played a vital role in the development of Farafra Oasis for a long time through the many springs issued in this aquifer. These springs formed the only source of water that was used by the inhabitants for domestic and irrigation purposes. Moreover, this water forms the easiest and cheapest water source in the Oasis. Since 1960, rapid drilling process of deep-water wells started to invest the high potentiality of the well-known Nubian Sandstone aquifer. Thus, huge quantities of groundwater are discharging from this aquifer (about 150 million m³/year, Ali, [4] to irrigate the continuously increasing reclaimed land (49,000 feddan) until now. The extensive exploitation of groundwater led to the depletion of the aquifer storage. Therefore, there is a real danger of either dewatering or increasing the water depths of wells to uneconomic lifting depths for both the shallow and deep aquifers. Nowadays, many springs dried out and the remnants are expected to face the same problem in the near future. Accordingly, many social and economic problems appeared (Figure 1).

The present study aims at assessing this problem from the hydrogeological point of view. In this respect, it tries to decrease the rapid head decline and sustains the groundwater supplies of the Post Nubian aquifer through the application of some management plans using mathematical modeling techniques.

Hydrological settings

According to the previous, hydrological studies i.e. [5-7] and the lithology of the old and new drilled wells; Farafra Oasis is distinguished into two distinctive aquifers; the Post Nubian aquifer and the Nubian aquifer (Figure 2).

Post Nubian aquifer

This aquifer is composed of limestone, chalky limestone with shale interbeds. A thin dolomitic bed characterizes the base of this aquifer. The

aquifer thickness shows relatively small variations. It ranges from 90 m in the south at Abu Minqar area to about 170 m in north and west with an average value 130 m. Different oriented and mutually intersecting joints and fissures cut this aquifer. These structures provide the avenue for the upward movement of groundwater from deep horizons under artesian pressure [8]. Groundwater of this aquifer is supplied through number of natural springs and very few wells. Water, of these springs, is utilized by inhabitants for domestic and agricultural purposes.

The number and discharges of these springs were inventoried and compared with the previous data. It was found that the numbers and discharges of these springs had intensively decreased (Figures 3 and 4). Spring numbers were decreased from 67 springs at year 1962 to 11 springs now while the discharge of Ain Tenin for example was decreased from 156 m³/day at year 1962 to about 13 m³/day at present inventory. Although Ain El Balad showed a big anomaly, its discharge was estimated as 2050 m³/day indicating a deep fault structure origin. The high temperature of the spring water (36°C) enhances this result. On the other hand, Most of the dried springs are located at south where the flowing ones are located at extreme north (Figure 5). This situation may be attributed to the topography of the depression surface, which slopes from south to northeast i.e. from +135 m to +50 m.

Based on the groundwater heads and pressure measurements of some non-flowing and flowing springs, groundwater level contour map of the Post Nubian aquifer was available (Figure 6), groundwater levels of the karst aquifer are ranging from 105 m at El Sheikh Marzouq area to 60 m in the extreme north. Groundwater generally flows from south to north and from southeast to northwest. The smooth, regular and spaced contour lines reflect the homogeneity of the chalky limestone aquifer (Figures 3-6).

Nubian Sandstone aquifer

This aquifer consists of thick alternating sequence of coarse classics sediments of sandstone and clay beds. Accordingly, three successive water-bearing formations are well defined; Taref, Sabaya and Six Hills Formations, named Zone A, Zone B and Zone C (Figure 2). The clay

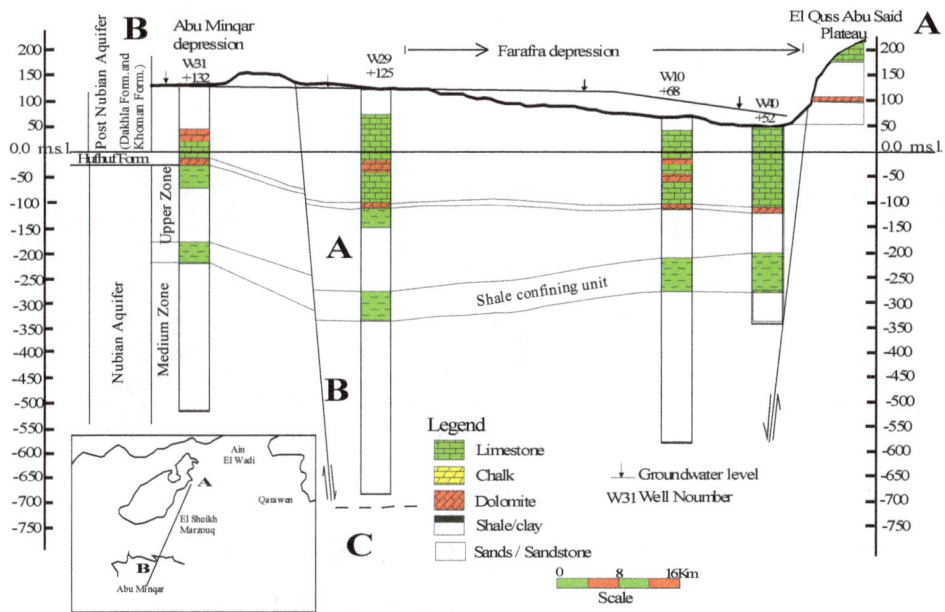

Figure 2: Hydrogeological section B-A in Farafra Oasis (modified after Salem, 2002)

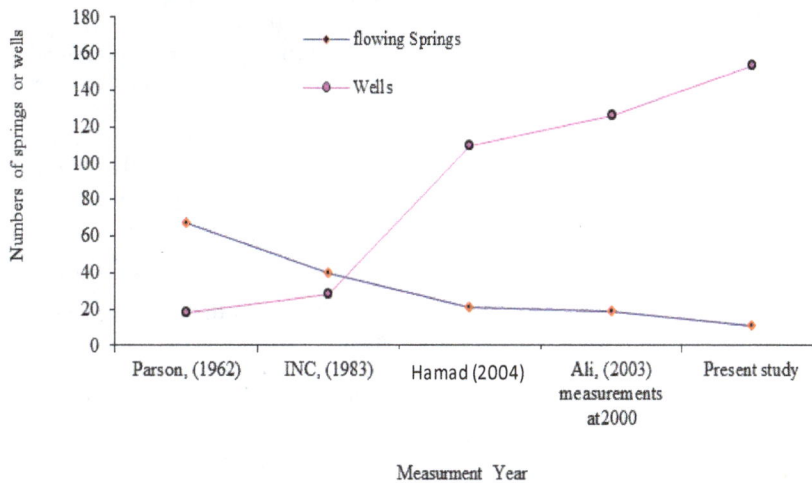

Figure 3: Relation between number of deep wells and flowing springs

beds that form aquicludes slow down the vertical movement of the groundwater through the aquifer layers. As a result, the sandstone beds form single aquifers in the regional aquifer complex. The bottom level of the Nubian aquifer in the study area varies from -1700 m to about -2300 m [9]. The number of wells tapping this aquifer increased from 18 wells in 1960s to about 140 wells in the present time. Hence, the pumping from the aquifer was increased in the last decade to reach about 145 million m³/year (Ali, 2003). This figure is expected to increase in the near future due to continuous increment in drilling of wells.

The aquifer is generally a high potential aquifer due to the large saturated and prevalence of sand facies in the aquifer sediments. According to Hamad, [10], transmissivity of Zone A recorded 148.6 m²/day with an average permeability 1.3 m/day while zone B recorded 1613 m²/day and 5.7 m/day. On the other hand, the third productive zone (zone C) recorded the highest transmissivity and permeability

values of the aquifer (1642 m²/day and 7 m/day) reflecting the thick sandy aquifer section.

Based on the pressure measurements of some flowing wells and the depth to water in some non-flowing wells, a groundwater level contour map was constructed (Figure 7). From this map, groundwater generally flows from south to north with the direction of head decline i.e. from +125 m at Abu Minqar area to about + 88 m at north Qasr El Farafra area. Two cones of depression were detected El Sheikh Marzouq and north Qasr El Farafra area indicated by the depression cones formed in such areas. This situation reflects the extensive exploitation of groundwater in those areas. These depression cones form local groundwater flow directions to the center of these cones. Sahl El Baraka and Sahl Qarawen areas showed small drawdown, no depression cones were formed indicated by regular contour lines reflecting the small well number and recent pumping in that areas (Figure 7).

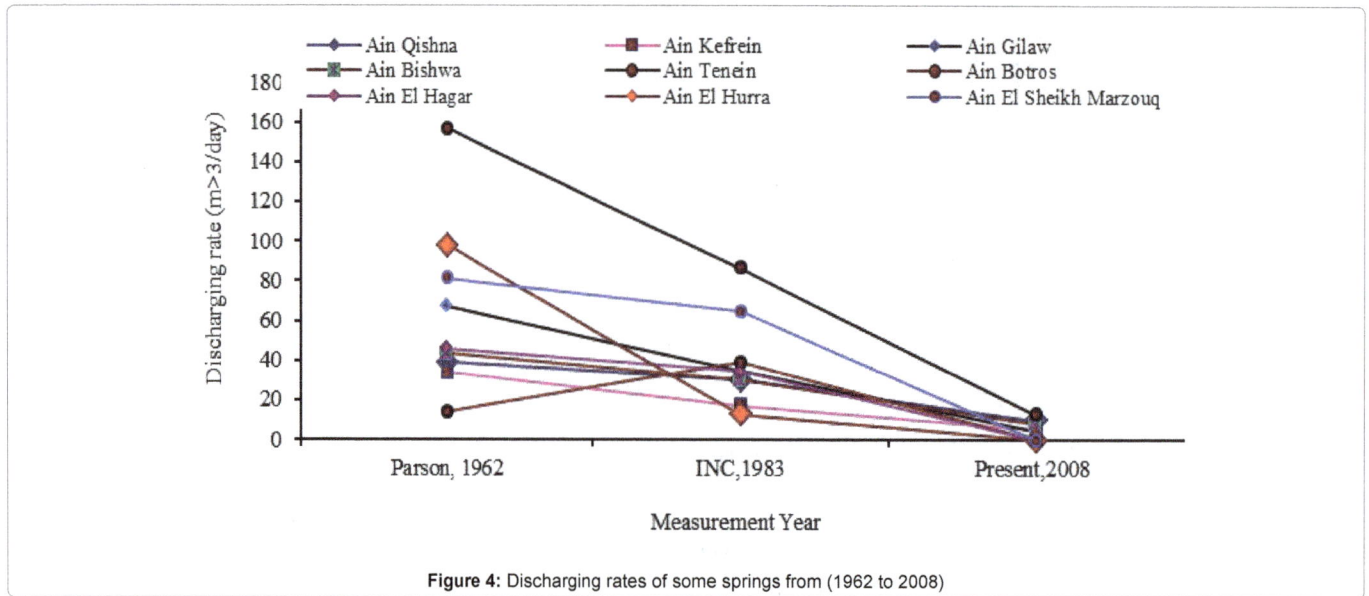

Figure 4: Discharging rates of some springs from (1962 to 2008)

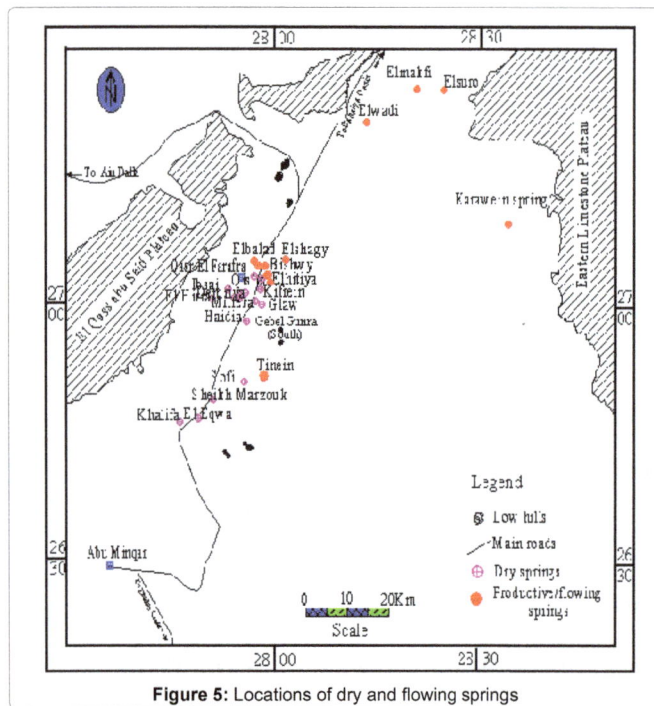

Figure 5: Locations of dry and flowing springs

Material and Methods

Modeling of Groundwater

$$\frac{\partial}{\partial x}\left(K_{XX}\frac{\partial h}{\partial x}\right)+\frac{\partial}{\partial y}\left(K_{YY}\frac{\partial h}{\partial y}\right)+\frac{\partial}{\partial Z}\left(K_{ZZ}\frac{\partial h}{\partial Z}\right)-W=S_s\frac{\partial h}{\partial t}$$

The model describes groundwater flow of constant density under non-equilibrium conditions in a heterogeneous and anisotropic medium according to the following equation (Bear [11]).

K_{xx}, K_{yy} and Kzz are values of hydraulic conductivity (L T^{-1}); along the x, y, and z coordinate axes; h is the potentiometric head (L); W is the volumetric flux per unit volume and represents sources and/or sinks of water (T^{-1}); Ss is the specific storage of the porous material (L^{-1});

and t is time (T). The model was used the finite difference approach to solve the groundwater flow equation.

Three dimensional, numerical time dependent flow model of finite difference GMS program was used to investigate the hydrodynamic impacts of the present discharging rates of the deep Nubian and Post Nubian aquifers on the groundwater levels in Farafra Oasis. Furthermore, it was used to study the proposed investigated plans to sustain the groundwater supplies from the Post Nubian aquifer, which faces acute depletion problem. Our approach involves the development of a two-layer groundwater flow model through the establishing of the boundary conditions, initial conditions and calibration of every layer in the model.

Boundary and initial conditions

According to the hydrogeological setting of the study area, two aquifer systems are present. Each one has its individual characteristics. The first (Post Nubian aquifer) is an unconfined aquifer of one homogenous layer. Its top is the ground surface and its bottom is the level of the marked dolomitic limestone bed that defines the border between the Nubian and Post Nubian aquifer. On the other hand, the Nubian Sandstone aquifer (the second layer) is a confined aquifer type. It was considered as one complex layer of vertical and horizontal variations in sediments and hydraulic parameters. The aquifer top is represented by the previously mentioned dolomitic limestone bed while the aquifer bottom is defined by the bottom of the third productive zone (Zone C) where the deepest wells were immersed (-1200 m). A grid of 3600 cell (60 columns and 60 row) was constructed to cover the model domain (Figures 8 and 9). This grid is refined in the area of springs for the first layer and wells for the second layer where detailed information on the aquifer properties is available. The groundwater system was built for the two layers by assigning the hydraulic conductivity, transmissivity, porosity, leakance and storativity to the grid cells using all information of the previous and present work.

Referring to the boundary conditions (Figures 8 and 9), no-flow boundary prescribed the eastern and western sides of the modeled area since it matches the flow directions and traces the major faults between the scarps and the depression area. General Head boundaries (GHB) are applied to the southern and northern boundaries of the model domain. The hydraulic heads and hydraulic conductance of both layers were

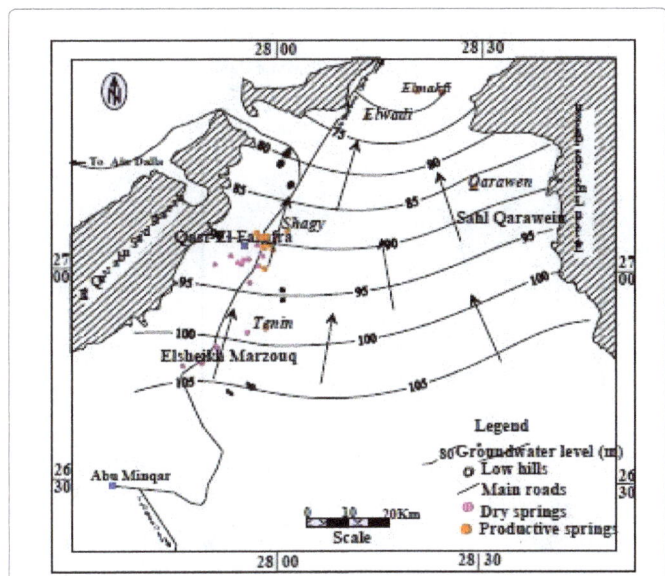

Figure 6: Head distribution map of the Post Nubian aquifer (April 2008)

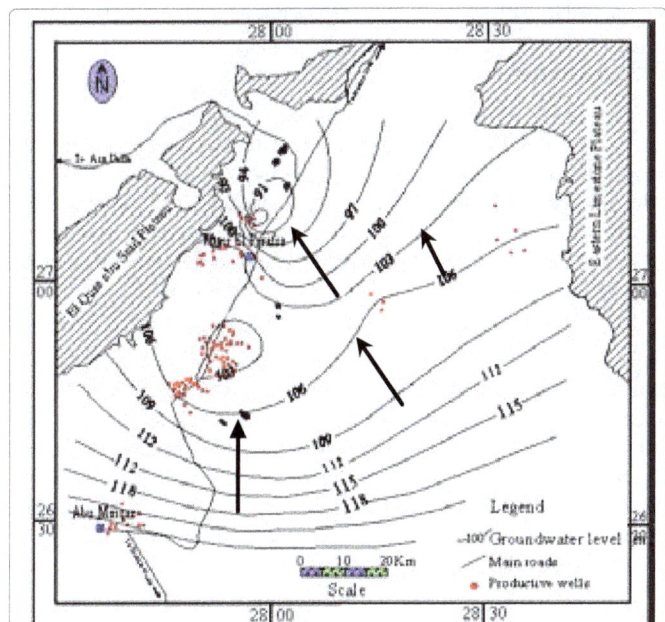

Figure 7: Piezometric head contour map of the Nubian Sandstone aquifer (April 2008)

assigned to boundary cells. For the upper layer (Post Nubian aquifer), the southern boundary was applied as 120 m head in the boundary and 520 m²/day as hydraulic conductance while the northern boundary was simulated as 50 m as head in the boundary and 520 m²/day as hydraulic conductance. On the other hand the southern heads of the lower layer (Nubian aquifer) was applied as 130 m as head on the boundary and 1250 m²/day as hydraulic conductance while they were assigned as 60 m and 750 m²/day at the northern boundary. These heads varies during the simulation process according to different stresses applied on the modeled area (Figures 8 and 9).

Model calibration

The model was firstly run under steady state conditions till the year 2004 then under transient conditions from 2004 to 2008. The calculated heads for both the first and second layer were compared by the measured ones at 2008. Big differences between the calculated and the measured heads are present, so the calibration of the model is critical and important. Calibration was carried out by a trial–and-error method through modifying the hydraulic parameters of the aquifer until a satisfied difference between the calculated and measured heads reached (Figure 10). The difference was lowered to 0.05 m for the first layer and to 0.65 m for the second layer. As a result, head distribution maps resulted from the model are closely related to the actually measured maps (Figures 6 and 7).

The calibrated hydraulic conductivity values of the first layer range from 9 m/day to about 25 m/day. These values reflect the intensive secondary porosity characterize the Post Nubian aquifer and clarify

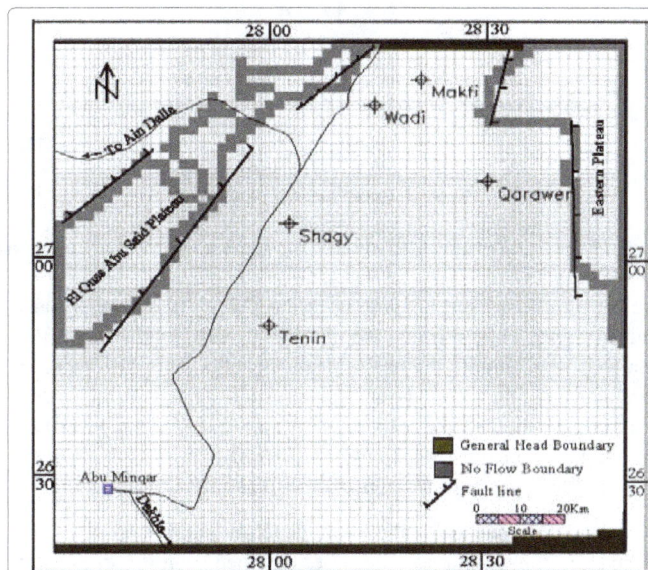

Figure 8: Grid and boundary conditions of the modeled area (Layer 1)

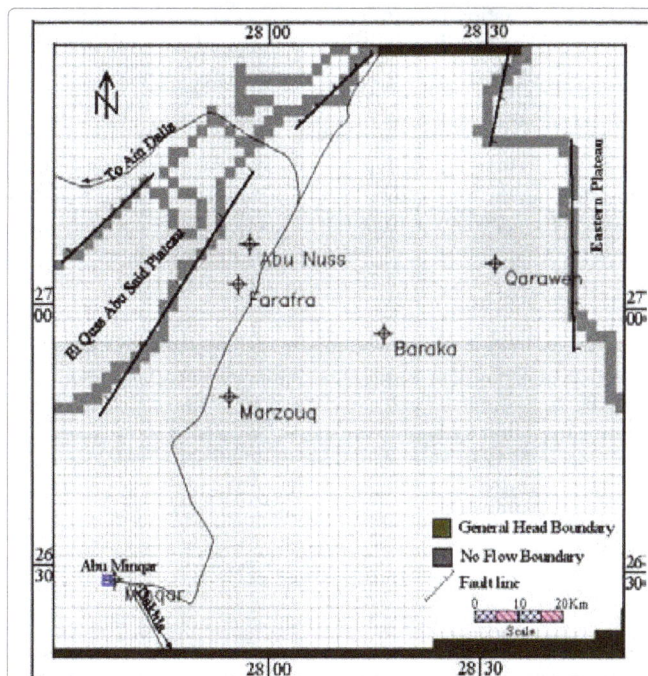

Figure 9: Grid and boundary conditions of the modeled area (Layer 2)

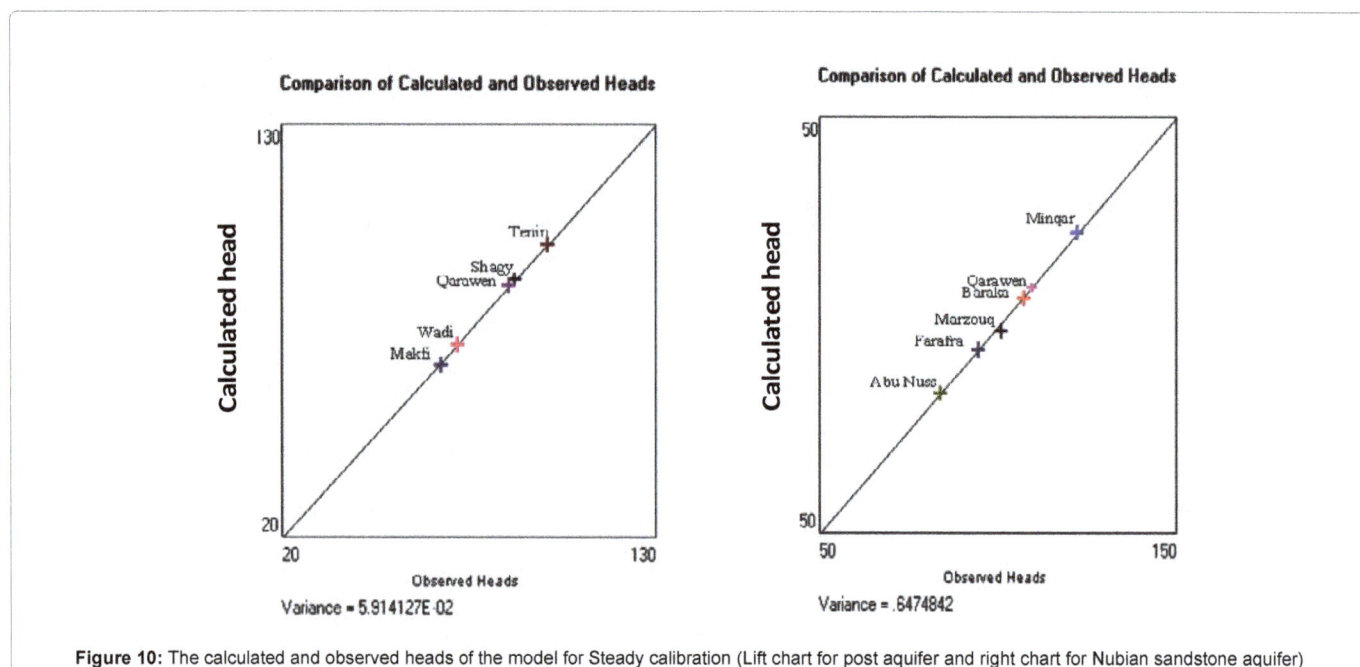

Figure 10: The calculated and observed heads of the model for Steady calibration (Lift chart for post aquifer and right chart for Nubian sandstone aquifer)

the big number of springs present in the Oasis. On the other hand, transmissivity ranges from 205 m²/day to 320 m²/day signifying relatively small aquifer thickness. Meanwhile, the calibrated hydraulic conductivity of the second layer (Nubian aquifer) ranges from 2 m/day to 14 m/day reflecting the coarse clastic nature and primary porosity of the Nubian sediments while transmissivity ranges from 1320 m²/day to 1750 m²/day reflecting very large aquifer thickness.

Successful transient calibration depends mainly on the good estimation of hydraulic conductivities obtained from the steady state condition. Generally, specific storage for confined aquifer is the main parameter that is changed during the transient calibration. In the process of calibration of transient state, specific storage values were modified on a trial and errors basis, until a good match between the observed heads of years 2004 and 2008 and the calculated heads were achieved. The range of the resulted specific yield for post aquifer after the final calibration of the transient state was found to be varying from 0.18 to 0.26. The range of the resulted specific storage for Nubian sandstone after the final calibration of the transient state was found to be varying from 0.00001 to 0.0001. It can be seen that in general, there is good agreement between the observed and simulated head.

Model application

After calibrating the model under steady state, the model is converted into transient conditions (time dependent conditions) by using the time parameter in the parameter list of the model. The heads resulting from the steady state simulation are used as starting heads in the transient analysis. New parameters such as storage coefficient and wells were assigned to the model grid. Simulation period of 20 years was applied to predict the aquifer heads changes under different plans. It is subdivided into 20-time step. In this way, heads at every year in the period 2009 to 2029 can be calculated and mapped.

Under transient conditions, two management plans were applied to investigate the probable head changes in both the Post Nubian and Nubian aquifers and their impact on the sustainable development of the Oasis.

First Scenario

This scenario investigates the impact of the present discharging rates (490,000 m³/day) from about 140 water well, drilled in Nubian Sandstone aquifer and about 3000 m³/day from about 25 spring (Flowing, pumped springs and wells) engrossed in the Post Nubian aquifer. The model was run under these stresses and heads were estimated for every time step (next 20 years) to the upper and lower layer by the result extractor tool of the GMS program.

The head distribution maps of the Post Nubian aquifer (First layer) showed gradual decline in the groundwater levels indicated by the gradual migration of contour lines southward (Figure 11). Contour line 85 passes through Ain Qarawen and immediately south Ain Shagy at the first year and it slowly moves southward to about 12 Km south Ain Shagy at the end of the simulation time (20 years). The smooth and displaced contour lines indicate aquifer homogeneity. No changes in the groundwater flow directions were recorded due to the regional head decline in the whole aquifer.

Drawdown calculations in the Post Nubian aquifer (Table 1) indicate that the drawdowns range from 5 m to about 9 m. The maximum drawdowns are expected in Ain Tenin and Ain Shagy reflecting the direct impact of the intensive pumping of water from the Nubian Sandstone aquifer in these areas (Farafra and Elsheikh Marzouq). On the other hand, the minimum drawdowns are expected in the extreme north in the area of Ain El Makfi where no pumping occurs. The drawdowns also indicated that the springs that are still flowing in the northern part are expected to sand up within the next 3 years where the water level of these springs are about 2 m above the ground surface (Figure 11).

On the other hand, head distribution maps (Figure 12) showed considerable decline in groundwater levels of the second layer (Nubian Sandstone aquifer). These drawdowns are expected to take place in the areas of productive wells manifested by widening the depression cones already formed now. The two depression cones formed at Abu Nuss and Marzouq areas are expected to form one large depression cone surrounded by contour line 85 at the end of simulation time (year 2029)

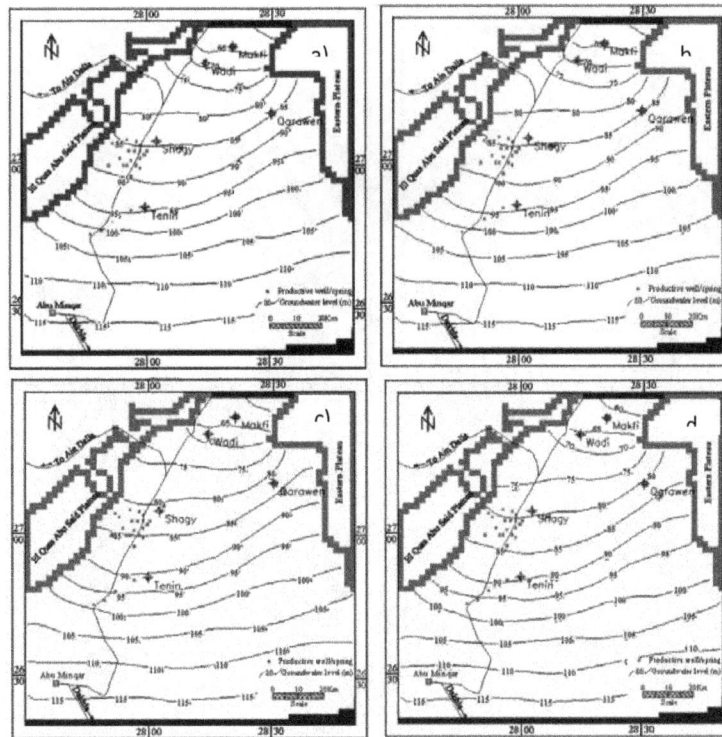

Figure 11: Predicted head distribution map of the Post Nubian aquifer (1st scenario) a) year 2009   b) year 2013   c) year 2018   d) year 2028

Years	Predicted Post – Nubian aquifer drawdowns (m)						Predicted Nubian Sandstone aquifer drawdowns (m)						
	El Makfi	El Wadi	Shag-y	Teni-n	Qar-wen	Average	Abu Nus	Farafra	Marzouq	Minqar	Baraka	Qara-wen	average
2009	1.2	1.1	1.6	2.1	1.2	1.4	29.5	17.5	15.9	0.8	2.0	1.9	11.2
2010	1.5	1.6	1.9	2.9	1.5	1.9	42.5	24.9	22.8	1.5	3.1	2.7	16.3
2011	1.9	2.0	3.0	3.5	1.8	2.4	50.5	30.0	26.7	2.3	4.2	3.4	19.5
2012	2.1	2.3	3.5	4.0	2.1	2.8	56.0	33.7	29.2	2.9	5.2	4.1	21.9
2013	2.4	2.6	4.2	4.6	2.5	3.3	60.1	36.5	31.0	3.5	6.1	4.8	23.7
2014	2.7	3.0	4.8	5.2	3.0	3.8	63.2	38.7	32.4	4.1	7.0	5.5	25.1
2015	3.0	3.4	5.4	5.8	3.4	4.2	65.6	40.4	33.6	4.5	7.7	6.1	26.3
2016	3.3	3.7	5.9	6.3	3.8	4.6	67.6	41.8	34.5	4.9	8.4	6.7	27.3
2017	3.6	4.0	6.4	6.7	4.2	5.0	69.2	43.0	35.3	5.2	9.0	7.3	28.2
2018	3.8	4.3	6.8	7.1	4.6	5.3	70.6	44.0	36.0	5.5	9.6	7.7	28.9
2019	4.0	4.6	7.2	7.4	4.9	5.6	71.7	44.8	36.6	5.8	10.0	8.2	29.5
2020	4.2	4.8	7.5	7.7	5.1	5.9	72.6	45.5	37.1	6.0	10.5	8.6	30.0
2021	4.3	4.9	7.8	8.0	5.4	6.1	73.4	46.0	37.5	6.1	10.8	9.0	30.5
2022	4.4	5.1	8.0	8.2	5.6	6.3	74.1	46.5	37.8	6.3	11.2	9.3	30.9
2023	4.6	5.3	8.2	8.4	5.8	6.4	74.7	47.0	38.2	6.4	11.5	9.6	31.2
2024	4.7	5.4	8.4	8.6	5.9	6.6	75.2	47.3	38.4	6.5	11.7	9.8	31.5
2025	4.8	5.5	8.6	8.7	6.1	6.7	75.6	47.7	38.7	6.6	11.9	10.0	31.8
2026	4.8	5.6	8.7	8.9	6.2	6.8	76.0	47.9	38.9	6.7	12.1	10.2	32.0
2027	4.9	5.7	8.8	9.0	6.3	6.9	76.3	48.2	39.1	6.8	12.3	10.4	32.2
2028	5.0	5.7	8.9	9.1	6.4	7.0	76.6	48.4	39.2	6.8	12.5	10.5	32.3

Table 1: Expected drawdowns in Farafra Oasis (First Scenario)

[12-16]. Drawdowns range from 6.8 m to 76.6 m (Table 1). These large variations may be attributed to the high discharging rates and large number of wells in such areas. i.e. in Abu Nuss area (the maximum decline), the number of wells are about 11 wells with discharging rates of about 14,000 m³/day/well while in Abu Minqar, most of wells stop flowing, the number of operating wells are 9 well with discharging rate attains 2000 m³/day/well. Sahl Baraka and Sahl Qarawen areas showed

reasonable variations (12.5 m and 10.5 m) (Figure 12).

This may be due to the few wells number and the large aquifer thickness distinguishing these areas. The relation between the predicted drawdowns in Post Nubian and Nubian Sandstone aquifer showed rational relationship (Figure 13). Head decline in the Nubian Sandstone aquifer by about 13.8 m is expected to follow by decline in the Post Nubian aquifer by about 1 m. When the declines in the Nubian

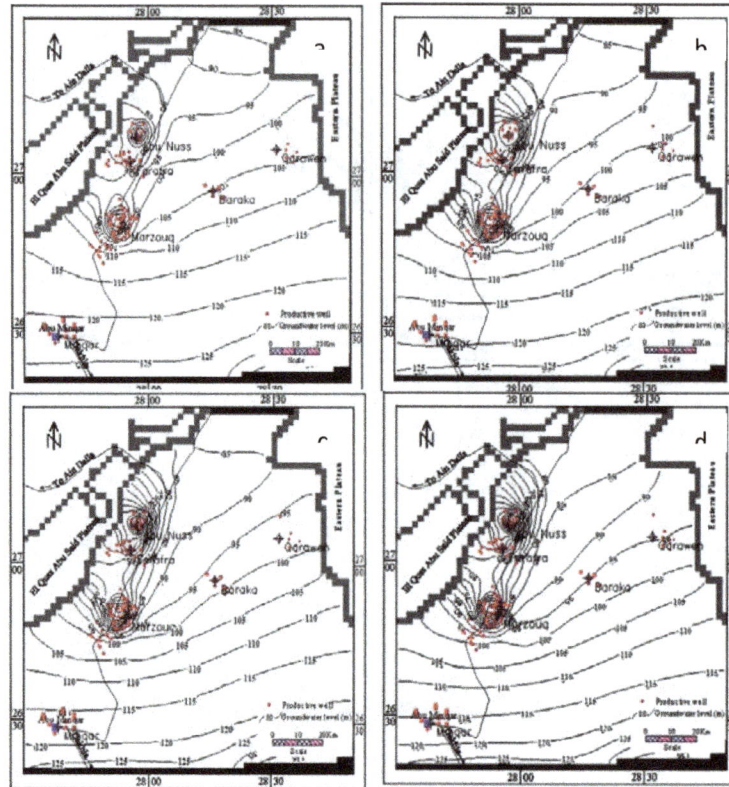

Figure 12: Predicted head distribution map of the Nubian Sandstone aquifer (1st scenario) a) year 2009   b) year 2013   c) year 2018   d) year 2028

aquifer exceed 25 m, a sharp decline in the Post Nubian aquifer may occur indicated by the abrupt change in the curve direction. The small drawdowns of the Post Nubian aquifer compared to the drawdowns in the Nubian aquifer indicate that potentialities of the Post Nubian aquifer are not well utilized (about 3000 m³/day only) due to the dryness of many flowing springs. Therefore, many wells are recommended to be drilled in the aquifer that will be discussed in the second scenario (Figure 13).

Second Scenario

This scenario tries to sustain the groundwater utilities in the Post Nubian aquifer by substituting the drying springs by shallow wells (19 well with depths 150 – 200 m and discharge rate 1000 m³/day/well) and cleaning up the flowing ones to enhance their productivities. Furthermore, keeping the present discharging rates at 490,000 m³/day with rearranging the wells immersed in the Nubian aquifer by substituting the stop flowing wells by new deep wells (depths >1000 m , Zone C) in the eastern and southeastern parts (Baraka and Qarawen areas) where the aquifer thickness and characteristics are reliable. No more wells immersed in the first productive zone since it forms the main recharging source of the Post Nubian aquifer. In addition, reducing the discharging rates of some high productive wells to about 5000 m³/day/well especially in Abu Nuss area.

The model was run under these new stresses and heads were estimated for the next 20 years for the upper and lower layer. The head distribution maps of the Post Nubian aquifer (upper layer) showed relatively gradual decrease in the groundwater levels indicated by the slow migration of contour lines southward (Figure 14), i.e. contour line 85 slowly moves southward to about 10 Km south Ain Shagy at the end

of the simulation time (20 years). This behavior may be attributed to the new wells imposed in the aquifer that maintains the drawdown rates closely related to that of the first scenario. Drawdowns are expected to range from 5 m to 8.6 m (Table 2). The maximum drawdowns were recorded in Ain shaggy and Tenin reflecting the new pumping activities in such areas. Groundwater flow directions were not changed through this scenario (Figure 14).

On the other hand, head distribution maps of the Nubian Sandstone aquifer in this scenario (Layer 2) showed relatively small drawdowns compared with the declines occurred in the first scenario (Figure 15). The maximum drawdowns were recorded at Abu Nuss area as 35.8 m while the minimum drawdowns were recorded at Abu Minqar area as 5.1 m (Table 2). Depression cones formed at Abu Nuss and Marzouq areas enlarged with time progress but with smaller rate than the first scenario. These drawdowns are direct result to the number and rate of well discharges at these areas. The drawdowns in Baraka and Qarawen areas showed small difference compared with the first scenario in spite of the increase of the well numbers. This may be attributed to the control of the well discharges and good well distribution. Drawdowns recorded about 12 m at the end of the simulation time.

The drawdowns in Post Nubian and Nubian Sandstone aquifers showed continuous relationship (Figure 16). Three meters head decline in the Nubian Sandstone aquifer is followed by decline in the Post Nubian by about 1 m. The drawdown rate in the Post Nubian aquifer through the simulation time (20 years) not exceed 7 m and this is an acceptable value regarding the proposed extraction quantity (19000 m³/day) which represent about 4% of the total water exploited from the Oasis (Figures 15 and 16).

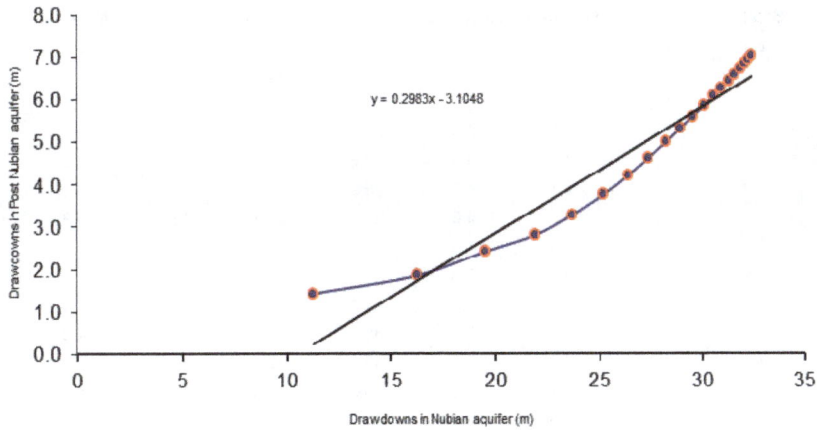

Figure 13: Relation between mean drawdowns in Post Nubian and Nubian Sandstone aquifer (First Scenario)

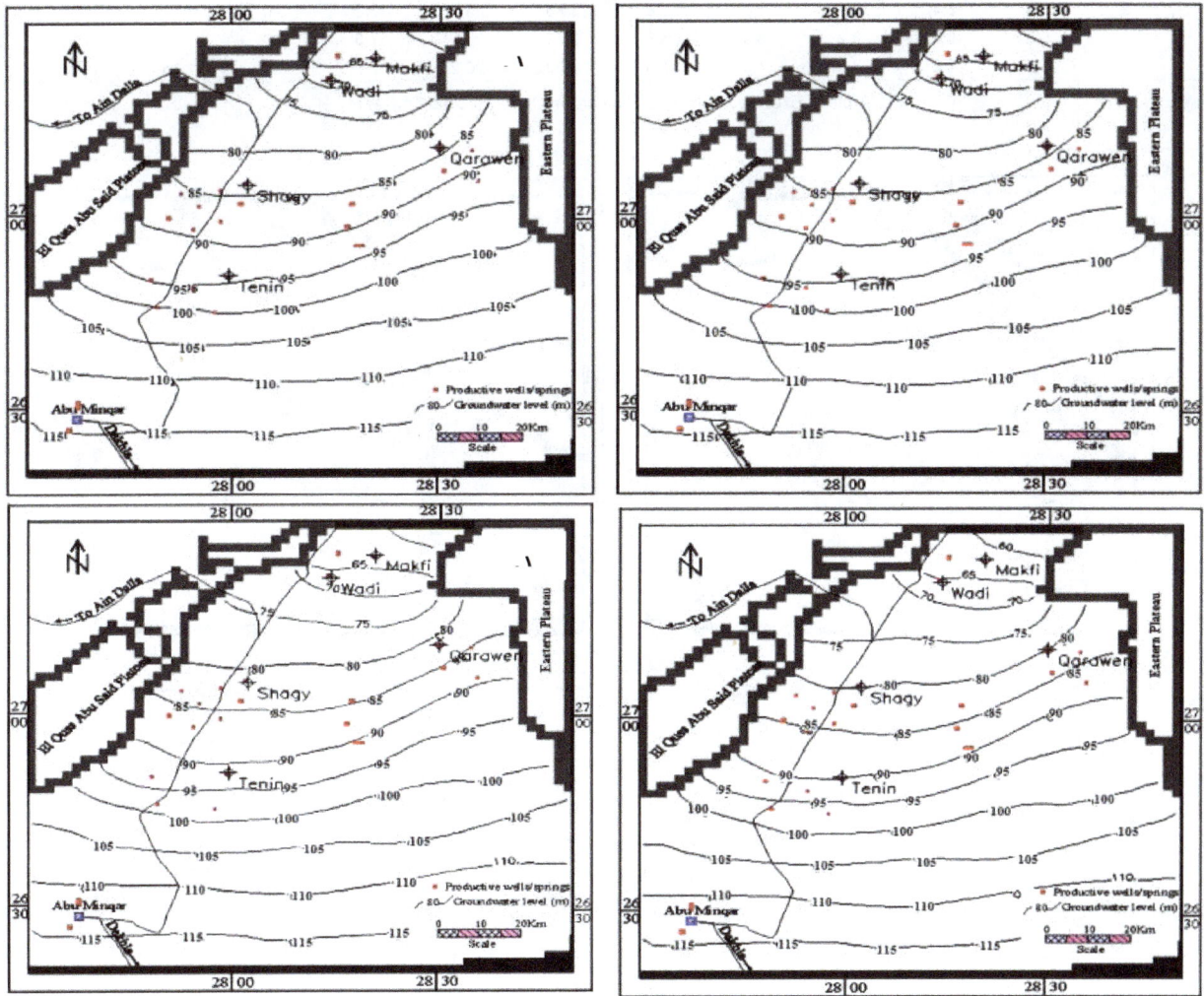

Figure 14: Predicted head distribution map of the Post Nubian aquifer (2nd scenario) a) year 2009　b) year 2013　c) year 2018　d) year 2028

Conclusion and recommendations

According to the study, a real danger of either dewatering or increasing the water depths to uneconomic lifting depths faces both the shallow and deep aquifers in Farafra Oasis. Many springs dried out and the discharging rate of the flowing ones intensively decreased, i.e Ain Tenin discharging rate decreased from 156 m³/day in 1962 to about 13 m³/day in 2008. Groundwater levels of the Post Nubian aquifer

Year	Post – Nubian aquifer drawdowns (m)						Nubian Sandstone aquifer drawdowns (m)						
	El Makfi	El Wadi	Shag-y	Teni-n	Qar-wen	Aver-age	Abu Nus	Farafra	Marzouq	Minqa-r	Baraka	Qara-wen	Average
2009	2.0	1.9	2.5	2.0	1.5	2.0	7.8	6.4	4.7	0.7	2.7	2.5	4.1
2010	2.3	2.2	3.0	3.3	2.2	2.6	13.7	10.5	7.0	1.2	4.1	3.8	6.7
2011	2.4	2.5	3.8	3.8	2.9	3.1	18.0	13.4	8.8	1.8	5.2	4.8	8.7
2012	2.6	2.8	4.3	4.2	3.3	3.4	21.3	15.7	10.2	2.2	6.2	5.6	10.2
2013	2.8	3.1	4.8	4.7	3.7	3.8	23.8	17.5	11.4	2.6	7.0	6.4	11.5
2014	3.1	3.4	5.3	5.2	4.1	4.2	25.9	18.9	12.3	3.0	7.8	7.1	12.5
2015	3.4	3.8	5.8	5.6	4.5	4.6	27.5	20.1	13.1	3.3	8.5	7.7	13.4
2016	3.6	4.0	6.2	6.0	4.8	4.9	28.9	21.1	13.8	3.6	9.1	8.3	14.1
2017	3.8	4.3	6.6	6.4	5.2	5.3	30.1	22.0	14.4	3.8	9.6	8.8	14.8
2018	4.0	4.5	6.9	6.7	5.5	5.5	31.1	22.7	14.9	4.0	10.1	9.2	15.3
2019	4.2	4.7	7.2	6.9	5.7	5.7	31.9	23.3	15.4	4.2	10.5	9.6	15.8
2020	4.3	4.9	7.4	7.2	6.0	6.0	32.6	23.8	15.7	4.4	10.8	10.0	16.2
2021	4.5	5.0	7.7	7.4	6.2	6.2	33.2	24.3	16.1	4.5	11.2	10.3	16.6
2022	4.6	5.2	7.9	7.6	6.3	6.3	33.8	24.7	16.4	4.6	11.4	10.6	16.9
2023	4.7	5.3	8.0	7.7	6.5	6.4	34.2	25.0	16.6	4.7	11.7	10.8	17.2
2024	4.7	5.4	8.2	7.9	6.6	6.6	34.6	25.3	16.8	4.8	11.9	11.0	17.4
2025	4.8	5.5	8.3	8.0	6.7	6.7	35.0	25.6	17.0	4.9	12.1	11.2	17.6
2026	4.9	5.6	8.4	8.1	6.8	6.8	35.3	25.8	17.2	5.0	12.2	11.4	17.8
2027	4.9	5.6	8.5	8.2	6.9	6.8	35.5	26.0	17.3	5.0	12.4	11.5	18.0
2028	5.0	5.7	8.6	8.3	7.0	6.9	35.8	26.2	17.5	5.1	12.5	11.6	18.1

Table 2: Expected drawdowns in Farafra Oasis (Second Scenario)

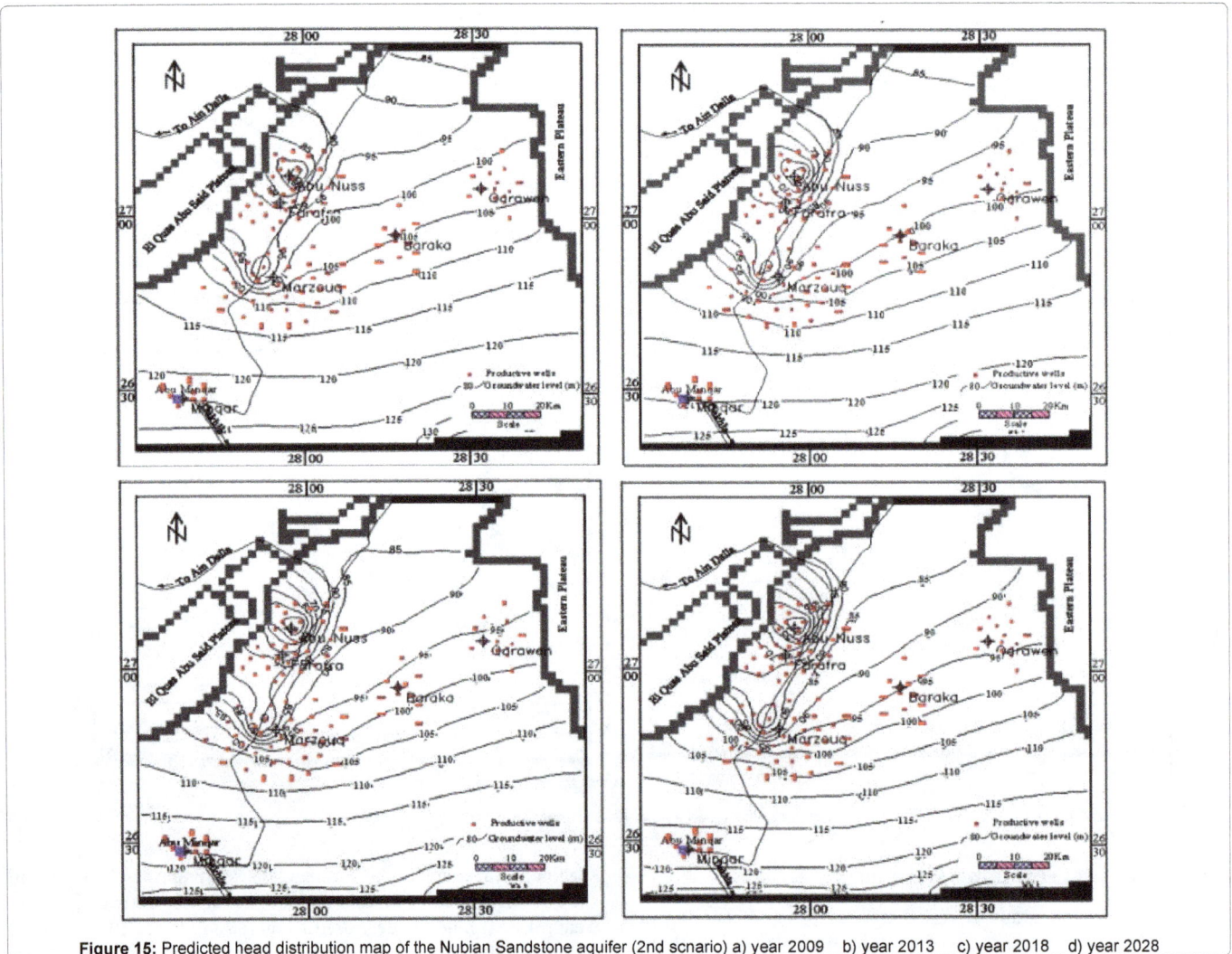

Figure 15: Predicted head distribution map of the Nubian Sandstone aquifer (2nd scnario) a) year 2009   b) year 2013   c) year 2018   d) year 2028

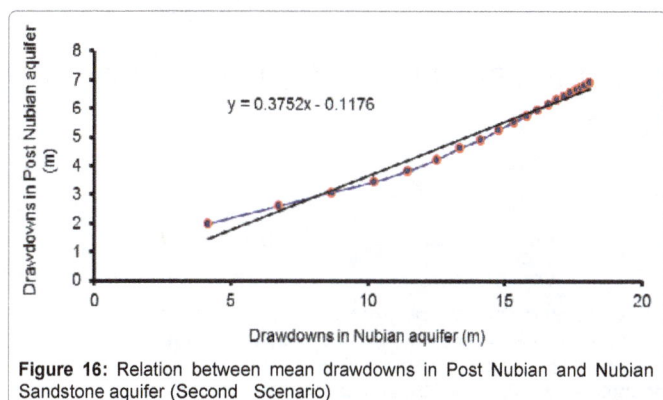

Figure 16: Relation between mean drawdowns in Post Nubian and Nubian Sandstone aquifer (Second Scenario)

range from 105 m at El Sheikh Marzouq area to about 60 m in the extreme north while in the Nubian aquifer; it ranges from +125 m at Abu Minqar area to about +88 m at north Qasr El Farafra area. The application of the model reveals that the present discharging rates in the Oasis causes maximum drawdowns attain 9 m in the Post Nubian aquifer and 76.6 m in the Nubian aquifer. The second scenario reveals maximum drawdowns as 8.6 m in the Post Nubian aquifer and 35.8 m in the Nubian aquifer with the same discharging rate. The first and the second scenarios expected the drying up of all springs within the next three years. The second scenario is recommended to sustain the groundwater supplies of the Post Nubian aquifer through a group of procedures:

- Substituting the drying springs by shallow wells (19 well by depths about 150 – 200 m and discharge rate 1000 m³/day/well).

- Cleaning up the flowing springs to enhance their productivities.

- Keeping the present discharging rates at 490,000 m³/day with rearranging the wells drilled in the Nubian aquifer by substituting the stop flowing wells by new deep ones (depths >1000 m, Zone C) in the eastern and southeastern parts (Baraka and Qarawen areas) where the aquifer thickness and characteristics are reliable.

- No more drilling of wells in the first productive zone since it forms the main recharging source of the Post Nubian aquifer.

- Reducing the discharging rates of some high productive wells to about 5000 m³/day/well especially in Abu Nuss area.

References

1. Hemker CJ (1984) Steady groundwater flow in leaky multiple aquifer systems. J Hydrol 72: 355-374.

2. Hemker CJ (1999) Transient well flow in vertically heterogeneous aquifers. J Hydrol 225: 1-18.

3. Saafan TA, Moharram SH, Gad MI, Khalaf S (2011) A multiobjective optimization approach to groundwater management using genetic algorithm. J Water Resour Environ Eng 3: 139-149.

4. Ali MT (2004). Evaluation of groundwater resources of El Sheikh Marzouq area at Farafra Oasis in the Desert of Egypt (Doctoral dissertation, PH. D. Thesis, Geol. Dep. Fac. of Sci. Menoufiya Univ., Egypt).

5. Himida IH (1970) The Nubian artesian basin, its regional hydrogeology aspects and palaeohydrogeological reconstruction. Journal of Hydrology 9: 89-116.

6. Shata AA (1982) "Hydrogeology of the great Nubian Sandstone basin, Egypt"Jornal of Eng. Geol., London 15: 127-133.

7. Ebraheem AM, Riad S, Wycisk P, Seif El-Nasr AM (2002) Simulation of impact of present and future groundwater extraction from the non-replenished Nubian Sandstone aquifer in southeast Egypt. Journal of Env Geology 43: 188-196.

8. Idris H (1996)" Springs in Egypt" Jornal of Env Geology 27: 99-104.

9. Thorweihe U (1990) "Das Groundwsser der Ostsahara" Die Geowissenschaten 8: 211-219.

10. Hamad MHM (2004) "Subsurface geological hydrological and hydrochemical studies on the Inter-stratal waters in Farafra Oasis, Western Desert, Egypt. M. Sc. Thesis, Geology Dep., Fac. of Sci., Cairo Univ, Egypt, pp. 277.

11. Bear J (1979) "Hydraulics of groundwater" (New York: Mc-Graw Hill).

12. Anderson MP, Woessner WW (1992) "Applied groundwater modeling: Simulation of flow and Advection transport" Academic Press, San Diego, New York Boston pp. 318.

13. Il Novo Castro (INC) (1985) "Technical and economic feasibility study for the reclamation of 50,000 feddans in Farafra oasis, Report submitted to the General Authority for Reclamation Projects and Agriculture Development (GARPAD) 4: 250.

14. Moharram SH, Gad MI, Saafan TA and Khalaf S (2011) "Optimal Groundwater Management Using Genetic Algorithm in El-Farafra Oasis, Western Desert, Egypt" Water Resour Manage pp. 9865-3.

15. Parson M (1962) "Bahariya and Farafra area, New Valley Project, Western Desert, Egypt, Final Report" A report submitted to GDDO, Parson Eng. C, California, USA pp. 150.

16. Salem AAA (2002) "Hydrogeological studies on the Nubian sandstone aquifer in Bahariya and Farafra depressions, Western Desert Egypt" PH. D. Thesis, Geol Dep Fac of Sci. Ain Shams Univ Egypt, pp. 157.

2

Assessment of the Influence of Rainfall and River Discharge on Sediment Yield in the Upper Tana Catchment in Kenya

Imelda N Njogu[1]* and Johnson U Kitheka[1]

[1]School of Water Resources Science and Technology, South Eastern Kenya University, Kitui, Kenya

Abstract

Sediment yield in the Upper Tana Basin in Kenya has implications on the sustainability of Hydro-Electric Power (HEP) dams and water resources development projects. Therefore, a study was undertaken in the basin to establish the extent to which rainfall and river discharges influence the sediment yield in the catchment. The study was based on hydrological data obtained from the Water Resources Management Authority (WARMA) and Kenya Meteorological Department (KMD). The river discharge data was obtained from three RGS Maragua (4BE01), Gura (4AD01) and Tana Sagana (4BC02) and rainfall data was obtained from Sagana Fish Farm and Nyeri Ministry of Works for the of period 1960-2013.The study also applied the Soil Water Assessment Tool (SWAT) Model to determine the extent to which the model can be used to simulate streamflow and sediment yield in the basin. The results of the study showed that there is a significant variability in streamflow and sediment yield in the Upper Tana Basin. In the period between 1960 and 2015, the mean total annual river discharge of Tana Sagana was 128 m3s-1, and the maximum and minimum river discharges were 29.94 m3s-1 and 3.15 m3s-1, respectively. There was an indication of increasing trend in rainfall and subsequently sediment yield in the basin, which may be attributed to alteration of land use and climatic change. The results showed that SWAT model was quite good in simulating the variability of river discharge. The analysis revealed a poor relationship between sediment yield and rainfall. However, the relationship between rainfall and stream flow was strong with r value of 0.9 which is significant at p=0.05. Relationship between simulated and observed river discharge had a R2 of 0.442, r of 0.665 and NSE of -89.43. The relationship between simulated and observed sediment yield had a R2 of 0.733, r of 0.86 and NSE of 0.69. The results of this study showed that SWAT model can be used to predict sediment yield in the Upper Tana catchment. The model had good performance when daily rainfall, stream flow and sediment yield data were used. Thus, the model can be used to establish the relationship between rainfall, discharge, and sediment yield in a highly human-impacted tropical catchment area. The study puts also forward various recommendations on land and water resources management in the basin.

Keywords: SWAT model; Upper tana; Sediment yield; Rainfall; Discharge

Introduction

Sediment yield in the Upper Tana Basin in Kenya is influenced by many factors among them landuse, vegetation cover, river discharge and rainfall variability [1]. The influences of rainfall and river discharge are particularly important but these are usually complicated by changes in land use and vegetation cover, both seasonally and inter-annually. As a result, the amount of sediment delivered by the river to the hydro-electric power dams located downstream is highly variable. While the influence of land use and vegetation cover change has received a lot of attention in past studies, this cannot be said to be so in case of the influence of rainfall and river discharge variability. In this regard, the assessment of the relationship between river discharge and rainfall and sediment yield is important since it is these two variables that shows significant variability as a result of climate change and whose future trends will have major implications on the dams located in the Upper Dam basin. There is still gap in our understanding of the factors influencing the observed sediment yield in the Upper Tana Basin. In particular, the extent to which rainfall variability and hence climate change influences sediment yields in the basin has not been fully established, as is true in many other similar tropical river basins in Africa. This has been attributed to lack of data. Therefore, to fill the above gaps, we carried out an analysis of the relationship between rainfall, river discharge and sediment yield in some of the key streams draining the Upper Tana Basin in Kenya. This study compliments other studies on sediments yield in the Tana Basin [2-8].

Previous studies carried out on sediment yield in the upper Tana basin have not addressed the relationship between rainfall and discharge and sediment yield. Maingi, et al. [7] examined the hydrologic impacts following construction of dams along the Tana River and Dunne, et al. [3] came up with an approach for estimating the sedimentation in the Upper Tana catchment. Mango, et al. [1] used a calibrated model to explore the potential impacts of continued land use and future climate change in the upper Tana catchment. Dunne T [9] study noted that land use was the main factor that influenced sediment yield. Archer D [10] study in Nairobi region concluded that the sedimentation rates in the reservoirs was high in areas with rainfall amounts ranging between 1000 and 1600 mm and runoff ranging between 350 and 700 mm. Due to geographical differences, land use, climatic and socio-economic differences, Archer D [10] generalization for Nairobi region further to the south cannot be applied to the Upper Tana catchment. This study on the Upper Tana catchment is important since it provides information on measures that can be used to mitigate the effects of landuse change in specific areas of the basin which are considered to be the main sources of sediments in the Tana river [11]. By providing an analysis on sediment yield in the basin, the study can contribute in the

*Corresponding author: Imelda N Njogu, School of Water Resources Science and Technology, South Eastern Kenya University, Kitui, Kenya
Email: imeldan@seku.ac.ke

sustainable management of the Upper Tana Basin and the dams that have been constructed downstream e.g Masinga, Kamburu, Gitaru, Kiambere and the planned Grand Falls dam [8].

Description of the upper Tana basin

The Upper Tana catchment in Central Kenya, is the main catchment area for the Tana river-the largest river in Kenya (Figure 1). The sub-basin covers a surface area of about 12,500 km² with elevation ranging from 400 to 5,199 m above sea level. The main rivers in the catchment are Sagana, Thiba, Maragua, Mathioya, Chania, Nyamindi, Chania, Rupingazi, Ena, Tunga and Gura [12]. Five major reservoirs are found in the lower reaches of the sub-basin (Figure 2), namely Kindaruma (completed in 1968), Kamburu, Gitaru, Masinga and Kiambere. These HEP reservoirs combined provide approximately three quarters of electricity in Kenya. Recent studies have shown that they have modified the flow of the Tana river, particularly in downstream reaches of the river up to the Tana Delta [8]. The Upper Tana catchment experiences two rainy seasons every year as a result of the migration of the Inter-Tropical Convergence Zone (ITCZ). The long rain season lasts from March to June, and the short rains from September to December. There is a great variance in the rainfall patterns in Central Kenya highlands. The average annual rainfall ranges between 400 mm and 2300 mm [13]. The basin is one of the main water sources of the Tana River basin-the largest and most important river system in Kenya covering basin area of 5,950 km². The Tana river basin covering 17% of the total land area in Kenya contributes to 27% of the total mean discharge in Kenya.

The Upper Tana sub-basin vegetation consists of (i) the forests in the Aberdares conservation area, (ii) the middle zones consisting of farming areas and (iii) the lower drier grazing zone. In the conservation area, vegetation is determined by rainfall distribution and temperature. The forest vegetation in the Aberdares is divided into four categories, namely: the wet evergreen forests; dry evergreen forests; *Juniperus podocarpus*/Olive forests; and low altitude shrubs. The 10 most common species of trees in the three forest reserves of the Aberdares Conservation Area are *Nuxia congesta, Juniperus procera, Olea europaea, Podocarpus latifolius*, and *Neboutonia macrocalyx* [14,15].

The land use in the Upper Tana catchment can be divided into three main classes, namely (i) natural vegetation (forest, grassland, and wetlands), (ii) rain-fed and irrigated agriculture (tea, coffee, maize and cereals) and (iii) Rangeland. The catchment also includes different agro-ecological zones, which corresponds to the different land use types [16]. The population in the Upper Tana catchment is approximately 3.1 million people. The largest urban centres are: Thika, Sagana, Karatina, Murang'a and Nyeri. The population density which is as high as 300 people/km² declines with elevation partly due to decreasing rainfall and soil fertility [17]. Most of the people in the basin rely heavily on farming as well as the associated agro-industries which are sources of employment opportunities. The main crops grown include coffee, tea, potatoes, pyrethrum, maize, rice, and bananas.

The soils in the sub-basin are dominated by the humic nitisols which have formed as result of the volcanic deposits on the high-altitude zones. The Nitisol soils are however highly vulnerable to erosion where soil conservation measures are not applied [18]. The other soil types in the catchment include vertisols, cambisols, andosols ferralsols and leptisols. The geology of the area (Figure 3) is characterized by the volcanic rocks of the Cainozoic era, and metamorphic rocks of the Mozambique belt [13]. Mt. Kenya, an extinct volcano formed between 100-4000 million years ago is located in the west of the catchment and is a source of most of the rivers in the Tana basin [19].

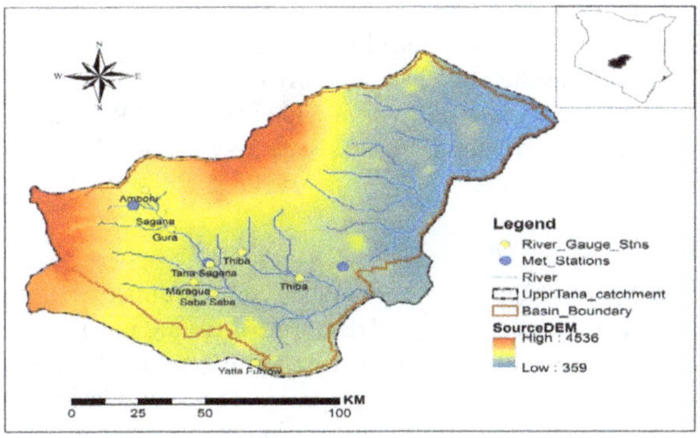

Figure 1: location of the River Gauging Stations (RGS) and rainfall stations in the Upper Tana Basin.

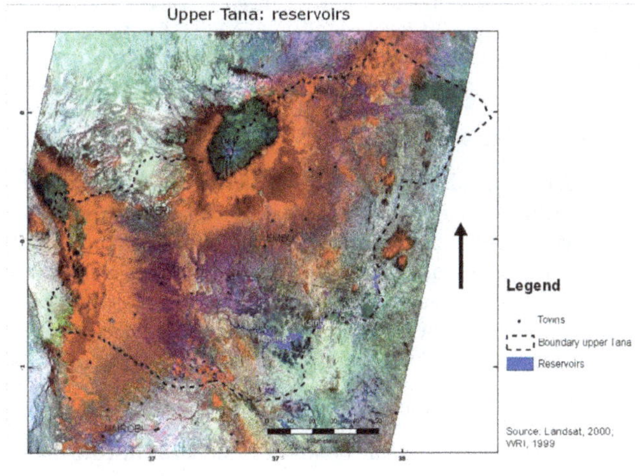

Figure 2: Reservoirs in the Upper Tana catchment.

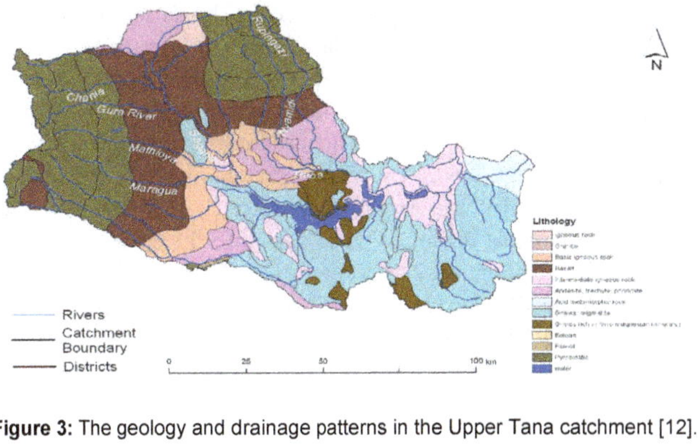

Figure 3: The geology and drainage patterns in the Upper Tana catchment [12].

Methodology

Streamflow data

Secondary data on the rainfall, river discharge and sediment yield were obtained from the Kenya's Water Resources Management Authority (WRMA) at Embu. River discharge data were obtained from 8 River Gauging stations (RGS), namely Amboni (4AB05), Sagana (4AC03), Gura (4AD01), Tana Sagana (4BC02), Maragua (4BE01), Thiba 1 (4DD02), Saba Saba (4BF02) and Thiba 2 (4DA10) (Figure 1).

Rainfall data

The rainfall data was obtained from Sagana Fish Farm and Nyeri Ministry of Works Stations that had more consistent records without gaps. Other stations which data was provided by the WARMA included Kiritiri Chiefs Camp (Embu), Sagana State Lodge, Nyeri Met Station, and Meru Forest Station.

Sediment data

There is a huge gap in terms of availability of sediment data in the Upper Tana basin. Sediment yield data were therefore not available in most of the River Gauging Stations with the exception of the Sagana River Gauging station where continuous data was available for the period 1957-1980. Theother sediment yield data for this station were obtained from WRMA for the following RGS- Tana Sagana (4AC03), Gura (4AD01), Maragua (4BE01) and Sagana- Grandfalls. The Sagana-Grandfalls station had continous data for the period 1957 to 1980. The data on the total suspended sediment concentrations (TSSC) was only available for the period between 2010 and 2011 for Maragua (RGS 4BE01), Mathioya (RGS 4BDNEW), Saba Saba (RGS 4BF01) and Sagana (RGS 4BE10).

Data processing and analysis

The data obtained for this study was subjected to time-series analysis to show the trends in rainfall, sediment yield and river discharge in the period between 1957 and 1990 and in some instances, up to 2015.The flow duration curves and mass curves were plotted for rainfall and river discharge data for Sagana fish farm and Gura RGS, respectively. This was done in order to check the accuracy of the SWAT model and also for detecting shifts in rainfall and river discharges in the basin.

The soil water assessment tool (SWAT) model

The Soil Water Assessment Tool (SWAT) Model was used to simulate flow rates of the Sagana river which is the main branch emanating from the Upper Tana Basin. The SWAT model has been used in other parts of the world [20-23]. The model dealt with two parameters in the input stage rainfall and discharge of the river. The output of the model was plotted in Excel to show the relationship between measured data and simulated model data on rainfall, sediment yield and discharge. Various quantitative methods used included the measures of central tendency (mean, range), regression analysis, correlation and coefficient of determination and Nash Sutcliffe Efficiency (NSE). The SWAT model was built using the Arc-Map interface (i.e Arc SWAT) which provided the suitable means to enter data into the SWAT code. The SWAT Model deals with soil and water parameters, with primary data inputs being the Digital Elevation Model (DEM), land use and soils shape files. Modelling allowed quantification of soil erosion processes at non-gauged areas and during periods when measurements were absent [24]. The Model was calibrated using data for the period between 1981 and 2010. Model validation was done using datasets from the Gura river gauging station (1980-1985). For modeling purposes, the Upper Tana watershed was partitioned into a number of sub-basins. Input information for each sub-basin was then grouped or organized into the following categories: climate and hydrologic response units (HRUs). The first step in creating a SWAT model involved the delineation of the sub-watersheds in the Upper Tana basin for which each of them is treated as individual units. The sub-basins were further divided into hydrologic response units (HRUs) which had homogenous land use practices, soil type and management practices. The Nash-Sutcliffe efficiency (NSE) was used to as a normalized statistic to determine the relative magnitude of the residual variance compared to the measured data variance [25]. NSE indicates how well the plot of observed versus simulated data fits the 1:1 line. Servat, et al. [26] noted that NSE provides the best objective function for reflecting the overall fit of a hydrograph. The NSE equation used is presented in equation 1:

$$NSE = 1 - \left[\sum_{i=1}^{n} \frac{\left(y_i^{obs} - y_i^{sim}\right)^2}{\left(y_i^{obs} - y_i^{mean}\right)^2} \right] \tag{1}$$

Where,

Y_i^{obs}-i^{th}=observation for the constituent being evaluated

Y_i^{sim}-i^{th}=simulated value for the constituent being evaluated

Y^{mean}=mean of the observed data for the constituent being evaluated

N=total number of observations.

The Nash Sutcliffe Efficiency ranges from $-\infty$ to +1, where the acceptable levels of performance are the values greater or equal to 0.0 to 1. If the NSE number is less than 0.0 it indicates that the mean observed value is a better predictor than the simulated value, which indicates unacceptable performance of the model.

The long-term calibration was validated with inflow data from the Gura River Gauging station that was found to have the least data gaps. A monthly dataset was available from 1981-2010.The erosion rates and sediment yields were calibrated using the model values that were compared with the model predictions to calibrate the soil erosion and sediment routing parameters of the model. The mode fine-tuning was done using the data available from the various RGS and meteorological stations to improve the accuracy of the model parameters and the output. The long-term calibration spurned a 30-year period in which the basin has changed considerably in terms of land use and in terms of infrastructure roads, small-scale hydraulic works, and diversions.

Statistical data analysis

Multiple regression models presented in equations 2 and 3 were used in this study, respectively.

$$Y = ax_1^2 + bx_2 + c \tag{2}$$

$$Y = a + bX + e \tag{3}$$

In above equations,

Y=Sediment Yield (tons. month^{-1})

a and b=the coefficients value of X variables

X_1=River discharge (m^3 month^{-1})

X_2=Rainfall (mm.month^{-1})

c=Constant.

The regression equation was deemed to be significant when the p-value is less or equal to 0.05. The correlation analysis was based on the Pearson's correlation coefficient (r) that was used to describe the degree of collinearity between simulated and measured data according to equation 4.

$$r = \frac{n\left(\sum xy\right) - \left(\sum x\right)\left(\sum y\right)}{\sqrt{\left[n\sum x^2 - \left(\sum x\right)^2\right]\left[n\sum y^2 - \left(\sum y\right)^2\right]}} \tag{4}$$

In above equation,

r=Correlation coefficient

y=Dependent variable (e.g. Sediment Yield)

x=Independent variable (e.g. River Discharge)

n=Total number of values.

The coefficient of determination R^2 describing the proportion of the variance in measured data explained by the model was computed using equation 5 [27].

$R^2=\sqrt{r}$ (5)

Where R^2=coefficient of determination

r=Pearson's correlation coefficient

R^2 ranges from 0 to 1, with higher values indicating less error variance, and typically, values greater than 0.5 are considered acceptable [28,29]. Analysis of Variance (ANOVA) was used to determine whether there exist significant differences between two or more variables at a selected probability level. One-way analysis of variance was used in testing hypothesis at 95% confidence level. The null hypothesis was rejected when the critical F>F and when the p-value (x)>0.05.

Results

Seasonal and inter-annual variability of river discharges

The stream flow in the rivers draining the Upper Tana Basin showed significant seasonal and inter-annual variabilities. The high stream flows are experienced during the two rainy seasons, namely long rain season (March, April, and May) and short rain season (October, November, and December). The low stream flows are experienced during the dry seasons (June, July, August, September and January, February). The streamflows also show significant inter-annual variabilities. The maximum river discharges for Amboni (RGS 4AB05), Maragua (RGS 4BE01), Gura (RGS 4AD01) and Tana Sagana (RGS 4BC02) were 4.84 m^{3}s^{-1}, 33.25 m^{3}s^{-1}, 29.94 m^{3}s^{-1} and 59.26 m^{3}s^{-1}, respectively. The mean river discharges for Amboni (RGS 4AB05), Maragua (RGS 4BE01), Gura (RGS 4AD01) and Tana Sagana (RGS 4BC02) were 1.5 m^{3}s^{-1}, 14.3 m^{3}s^{-1}, 19.7 m^{3}s^{-1} and 21.2 m^{3}s^{-1}, respectively. Tana Sagana had relatively higher discharge since it represents the main tributary of the Tana draining from the northwest Upper Tana Basin. The flow duration curves of the rivers showed that the high magnitude flows >80 m^{3}s^{-1} are experienced in <10% of the time, while river discharges <10 m^{3}s^{-1} occur in 80% of the time (Figure 4).

Seasonal and inter-annual variability of sediment yield

The data on the total suspended sediment concentrations (TSSC) for four RGS for the period between 2010 and 2011 showed existence of significant seasonal and inter-annual variability of TSSC. Maragua river (RGS 4BE01) had a minimum, maximum and mean TSSC of 3 mgl^{-1}, 0.517 gl^{-1} and 0.065gl^{-1}, respectively. Mathioya (4BDNEW) had a minimum, maximum and mean of 3 mgl^{-1}, 0.258 gl^{-1} and 0.043 gl^{-1} respectively. Saba Saba (4BF01) had a minimum, maximum and mean of 0.010 gl^{-1}, 1.433 gl^{-1} and 0.243 gl^{-1} respectively. Sagana (4BE10) had a minimum, maximum and mean of 0.03 gl^{-1}, 0.50 gl^{-1} and 0.057 gl^{-1}, respectively. The TSSC ranged from 0.03 to 0.50 gl^{-1} in Sagana river in the period between 2010-2011. The highest TSSC were recorded in the Sagana river which is the main branch of the Tana draining from the northwest Upper Tana Basin. As with TSSC, Sediment yield in the Upper Tana Basin showed both seasonal and inter-annual variabilities. The data for the period between 1955 and 1995, showed that the variability of sediment yield is related to the variability of rainfall and subsequently river discharges (Figure 5).

The computation of sediment yield provided a maximum yield of 39,050 ton.yr^{-1} in 1961 and >50,000 ton.yr^{-1} in 2015. The 1961 sediment yield data are much lower than the 2015 yield, probably pointing to an increase in sediment production in the basin. There is thus an indication of increased sediment yield in the basin. The sediment yields for Maragua, Mathioya, Saba Saba and Sagana were 1,365,603 ton.yr^{-1}, 798,365 ton.yr^{-1}, 216,210 ton.yr^{-1} and 2,782,547 ton.yr^{-1}, respectively. Previous studies in the basin by Dunne, et al. [3] reported sediment yields ranging from 883,000 ton.yr^{-1} and 2,302,000 ton.yr^{-1}, respectively. From the results of sediment yield from different sub-basins, it seems that Maragua and Mathioya are the main sources of sediment in the Upper Tana basin.

Seasonal and inter-annual variations of rainfall

The rainfall trend shows normal year-to year variability although there is an indication of declining rainfall trend. As with rainfall, the river discharges show normal year to year variations that are related to the variability in rainfall. The sediment yields also showed significant inter-annual variations. Rainfall in the Upper Tana Basin based on data obtained at Nyeri Public Works station and Sagana Fish Farm station showed significant inter-annual variations with an indication of declining trend in the period between 1980 and 2010 (Figure 5). There is also an indication of an increase in the degree of variability indicating an increase in the frequency and magnitude of extreme rainfall events. It is possible that the declining trend in rainfall observed in the Upper Tana Basin is a result of climate variability or land use change. It is however difficult at this stage to determine exactly what is the main driver. While rainfall shows a declining trend, the degree of variability seems to have increased after 1990.

The relationship between rainfall and streamflow in the basin was significant with correlation coefficient r of 0.99 and coefficient of determination R^2 of 0.98. This indicates that variations in rainfall explains 98% of the variations in river discharges in the Upper Tana Basin. This would also indicate rapid basin response with limited baseflow contributions to streamflow. This could be a result of land degradation occasioned by increased destruction of catchment areas for agriculture and settlements (Figure 6).

Land-use change in the upper Tana basin

There has been significant change in land-use in the Upper Tana Basin when one compares the situation during the pre-colonial and post-colonial period. Major significant changes in land-use begun in 1920s following introductions of agriculture in Central Kenya Highlands by European settlers [30]. In the last 100 years, there has

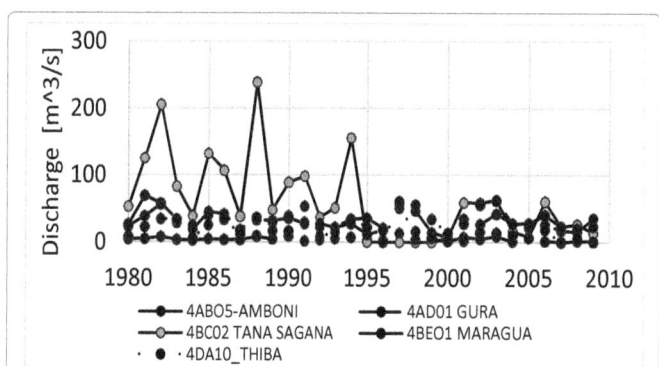

Figure 4: Variability of river discharge at Amboni and Gura River Gauging Stations (RGS) in the Upper Tana Basin in the period between 1980 and 2009.

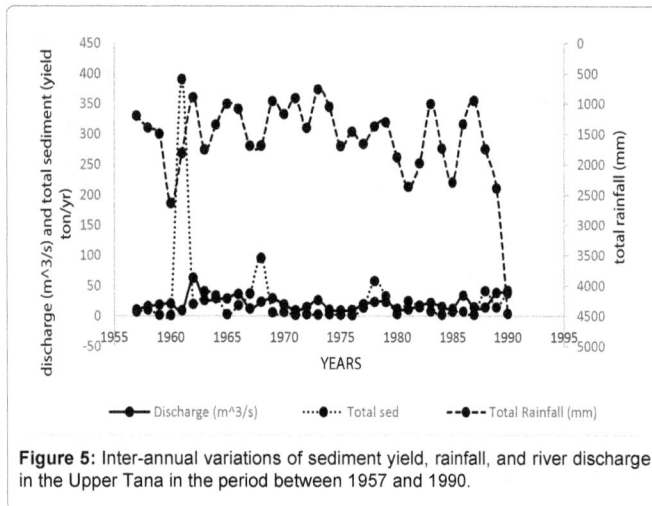

Figure 5: Inter-annual variations of sediment yield, rainfall, and river discharge in the Upper Tana in the period between 1957 and 1990.

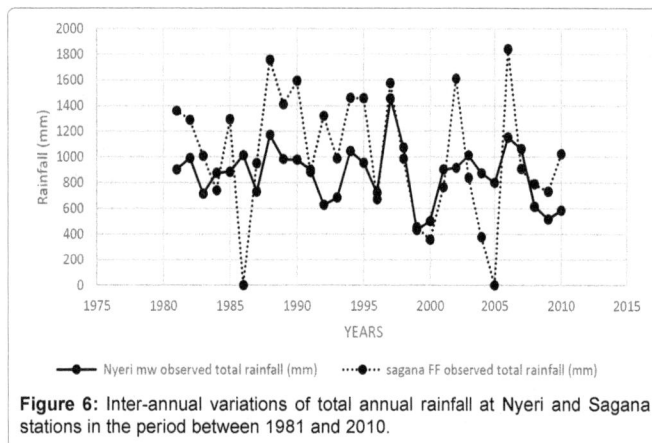

Figure 6: Inter-annual variations of total annual rainfall at Nyeri and Sagana stations in the period between 1981 and 2010.

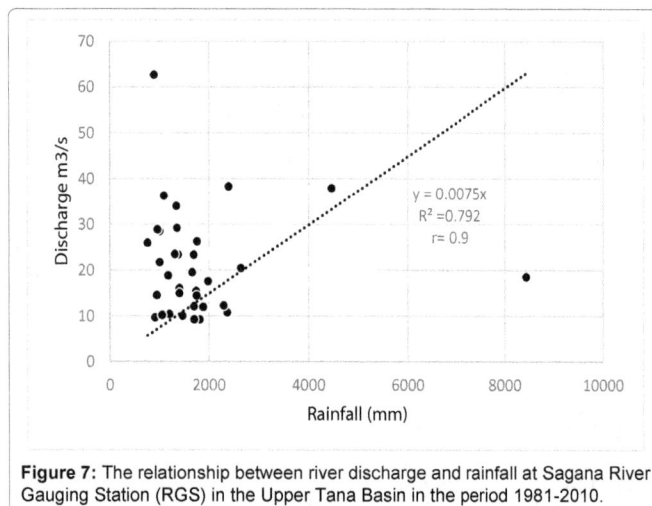

Figure 7: The relationship between river discharge and rainfall at Sagana River Gauging Station (RGS) in the Upper Tana Basin in the period 1981-2010.

been major changes in land use and vegetation cover in the basin. The analysis of satellite images and aerial photographs showed that the forest cover has reduced by more than 50%. Agriculture and settlements have increased by more than 100%. The population in the Upper Tana Basin has increased from <250,000 people in 1969 to >2 million people in 2015 [30]. High population density and intensive cropping in the region leave the soils susceptible to soil erosion. Most of the cultivation

is done without the application of appropriate soil erosion protection measures and the cultivation on marginal lands having steep slopes or high erodibility have increased (Figure 7).

Simulation of stream flow and sediment yield

The streamflow in the Upper Tana was modeled using SWAT model using data obtained at Gura River Station for the period between 1980 and 2010. The results of simulation of river discharges are shown in Figure 8. The relationship between simulated and observed river discharge yielded a correlation coefficient r of 0.67 with R^2 value of 0.44. The NSE index for the relationship was 89 showing the model simulated 89% of the streamflows (Figure 9 and 10).

The sediment yields were simulated using data for the period between 1980 and 2010 at Sagana River Gauging station. The comparison between simulated and observed sediment yield showed that the relationship was weak with r value of 0.02 with R^2 value<0.01. Thus, the SWAT model did not simulate the measured sediment yield satisfactorily (Figure 11 and 12).

An attempt was made to establish whether there is a significant relationship between rainfall and measured sediment yield at Sagana RGS in the Upper Tana Basin, for the period 1980 and 2010. The results showed a relatively weak relationship with a correlation coefficient (r) of 0.16 and a coefficient of determination (R^2) of 0.026. These results imply that there is no significant relationship between total annual rainfall and total annual sediment yield in the Upper Tana Basin. This perhaps points to other factors that are playing an important role in determining the annual sediment yield in the basin. It could also be due to problems with the quality of data or due to the fact that actual rainfall was not used to establish the link with the actual sediment yield.

The suspended sediment concentrations in the Upper Tana vary widely with as low as 0.03 gl^{-1} and values greater than 1.40 gl^{-1}. Maragua river (RGS 4BE01) had a minimum, maximum and mean of 0.03 gl^{-1}, 0.517 gl^{-1} and 0.065 gl^{-1}, respectively. Mathioya river (4BDNEW) had a minimum, maximum and mean of 0.03 gl^{-1}, 0.258 gl^{-1} and 0.043 gl^{-1}, respectively. Saba Saba river (4BF01) had a minimum, maximum and mean of 0.010 gl^{-1}, 1.433 gl^{-1} and 0.243 gl^{-1}, respectively. Sagana river (4BE10) had a minimum, maximum and mean of 0.03 gl^{-1}, 0.500 gl^{-1} and 0,057 gl^{-1}, respectively. These river gauging stations Maragua, Mathioya, Saba Saba and Sagana had sediment yields of 1,365,603 ton.yr^{-1}, 798,365 ton.yr^{-1}, 216,210 ton.yr^{-1} and 2,782,547 ton.yr^{-1}, respectively.

The results of regression analysis showed that there is no significant relationship between rainfall and river discharge (p=0.9). The same can be said to be true for the relationship between river discharge and sediment concentration, and that between river discharge and sediment yield. In other words, there is no indication that an increase in rainfall will cause a corresponding increase in stream flow. The weak relationship between river discharge and sediment yield and TSSC could be attributed to the time lag between the variables.

Discussion

Stream flow and rainfall

The discharge of the main rivers found in the Upper Tana Basin showed an increase in the range of low and high flows, indicative of large variation in stream flow. For instance, in Tana Sagana river, the high flows reached 239 m^{3}s^{-1} while the base flows were of the order

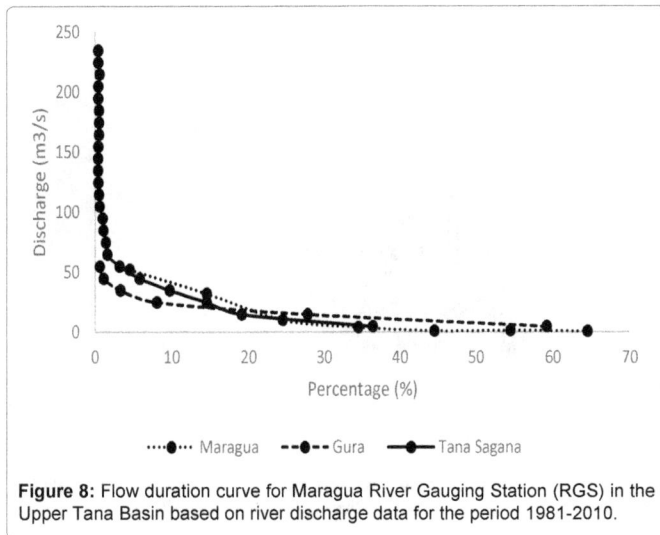

Figure 8: Flow duration curve for Maragua River Gauging Station (RGS) in the Upper Tana Basin based on river discharge data for the period 1981-2010.

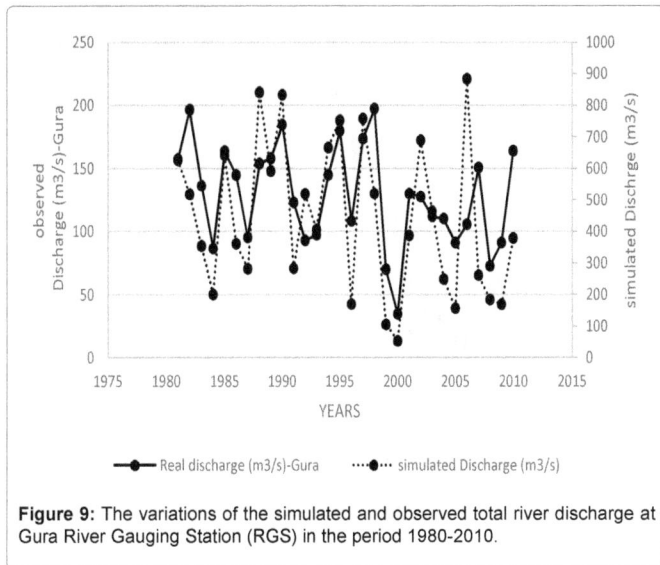

Figure 9: The variations of the simulated and observed total river discharge at Gura River Gauging Station (RGS) in the period 1980-2010.

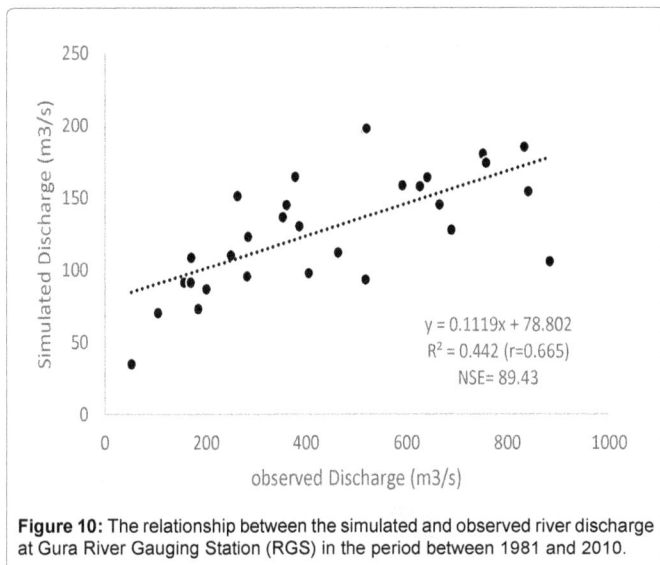

Figure 10: The relationship between the simulated and observed river discharge at Gura River Gauging Station (RGS) in the period between 1981 and 2010.

0.12 m³s⁻¹. There is also an indication of the reduction of stream flow which can be attributed to changes in land-use practices (e.g. increased cultivation, increased settled areas), increased deforestation of the catchment areas and reduced glacial cover on Mt. Kenya as a result of climate change. The variability of rainfall in the basin which has showed great variability could also be a factor in the increased variability of stream flow. However, the decreased flow could be due to increased abstraction of water (for irrigation, domestic and municipal water supply). Some of the rivers in the Upper Tana catchment have been tapped for the supply of water of water to Ndakaini Dam that supplies water to the City of Nairobi. It is expected that the tapping of the high flows >80 m³s⁻¹ under the Northern Collector Tunnel will further cause a reduction in the total stream flow of the Tana Sagana [31].

Sediment yield and rainfall and discharge

Sediment yield in the Upper Tana Basin is a consequence of poor land-use practices that causes high rates of soil erosion. The period 1960-1961 was characterized by high sediment yield which can be attributed to high rainfall events during the same period. There is an indication of a lag between the occurrence of high rainfall event in 1960 and the corresponding high sediment transport rate in early 1961. However, there is no clear indication of an increase in sediment transport rate since 1960, despite high variabilities that are clearly evident in our data. It is not certain whether reforestation and soil conservation programmes in the basin could explain this scenario.

The rainfall in the Upper Tana Basin exhibits significant spatial variations. The maximum total rainfall at Sagana is 1,839 mm while at Nyeri it is 1,456.3 mm. The minimum values for Sagana and Nyeri were 357 mm and 501.3 mm, respectively. The results of simulation of rainfall showed that variabilities are largely due to the difference in the location of the stations. Both Sagana and Nyeri stations showed a negative gradient (-12.49 and -4.93) meaning the amount of rainfall is reducing in the region. This is in agreement with findings of other studies done in the region [8].

Simulation of river discharges and sediment yield in the upper Tana basin

The results of simulation of river discharges in the Upper Tana are based on the results for Gura river. The results had (r) value of 0.665, a R^2 value of 0.442 and NSE of -89.43. These results show a good relationship between the two datasets, although the R^2 value is low. This can be explained by the gaps in the available data. The R^2 value showed a small error in the variance of the two datasets. The NSE of -89.43 shows that the mean observed river discharge for the Gura RGS is a better predictor than the mean simulated value and hence an unacceptable performance of the model. The high values for the Gura RGS were over-predicted by the model while in some instances they were under-predicted. The low river discharge values were generally under-predicted which explains the low efficiency in the simulation.

The results obtained for the observed and simulated sediment yield provided a (r) value of 0.02, a R^2 value of 0.0005 and NSE of -0.6. The r and R^2 are low because of relying on the model data. There was no data to input for the model to run as the available data was provided annually and the model required data in daily form. The NSE clearly shows that the mean observed value is a better predictor than the mean simulated value, this means that this is an unacceptable level of performance for the model to simulate sediment yield, and we would rather rely on the observed data as opposed to simulated data. The high values for the sediment yield were highly over-predicted with the low values being highly under-predicted.

Factors influencing sediment yield

The main objective of this study was to determine how rainfall influences sediment yield in the upper Tana catchment. According to the results of SWAT model, the amount of rainfall had a low significance influence on the amount of sediment yield in the Upper Tana catchment (Figure 12), with a r value of -0.25 and R^2 value of -0.062. The simulated data on rainfall was used for the plotting because the model produced a good simulation between the observed and simulated data for the Sagana fish culture farm. The results obtained for the relationship between rainfall and sediment yield is quite low on the r and R^2 which can be attributed to low performance of the model to simulate sediment yield. The data available was in annual form while the model required data in daily format. It would be expected from the graph that an increase in rainfall would result to an increase in the amount of sediment yield but this is not the case in many years. This shows that there are other factors apart from rainfall that influence the amount of sediment yield in the Upper Tana catchment. According to the results provided by the model, the amount of rainfall and sediment yield has no threat to the future of the Masinga dam reservoir. However, land use and the type of soil are the major determinant of the sediment supply to the reservoir. This is in agreement with other studies carried out by Kitheka, et.al. [8], that found a correlation of 0. 67. Mutiso (1980) found a correlation coefficient r of 0.8 which is much higher than the one reported in this study which is 0.5. The difference is attributed to the approaches used to determine the relationship between the parameters and the difference in geographical locations. This can also be attributed to the use of different data sets. The multiple regression analysis on the extent to which sediment yield is influenced by stream flow and rainfall variations yielded r value of 0.973 and R^2 of 0.94 at 95% confidence level (p=0.05). This shows the combined influences of rainfall and river discharge influence sediment yield. This study established that there is a good relationship between river discharge and sediment yield in the Upper Tana basin. This is expected since increased river discharge is associated with the increased capacity to transport the detached sediments [32].

Variability of sediment yield

The production of sediments in the Upper Tana Basin is a consequence of soil erosion processes in the sub-basin. Soil erosion is influenced by numerous factors among them rainfall, runoff, land-use, vegetation cover, and soil erodibility. This study has examined the influence of river runoff and rainfall. The extent to which land-use change and soil erodibility influences sediment production in the basin has not been examined in detail due to lack of data. However, we based our arguments on their contributions from the results of other studies.

The computation of sediment yield in the basin was done using the river discharge data and suspended sediment concentrations (TSSC) data for Sagana river. This was based on the mean, maximum and minimum TSSC values of 0.384 gl^{-1}, 0.032 gl^{-1} and 0.0042 gl^{-1} respectively and the corresponding mean, maximum and minimum river discharges of 20. 53 m^{3}s^{-1}, 62.56 m^{3}s^{-1} and 9.23 m^{3}s^{-1}, respectively. These data yielded the mean, maximum and minimum sediment yield of 31,062,960 ton.km^2/yr, 781,777,440 ton.km^2/yr and 157,680 ton. km^2/yr respectively.

The Upper Tana catchment has changed considerably in terms of land use with increased extent of Cultivated areas due to increased population, increased livestock keeping and in terms of infrastructure roads, small-scale hydraulic works, and diversions [9,33]. There exists a strong relationship between the land-use and sediment yield in the

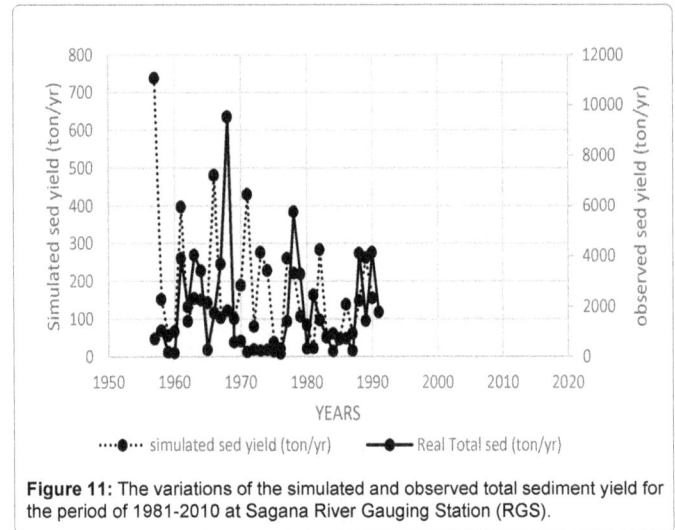

Figure 11: The variations of the simulated and observed total sediment yield for the period of 1981-2010 at Sagana River Gauging Station (RGS).

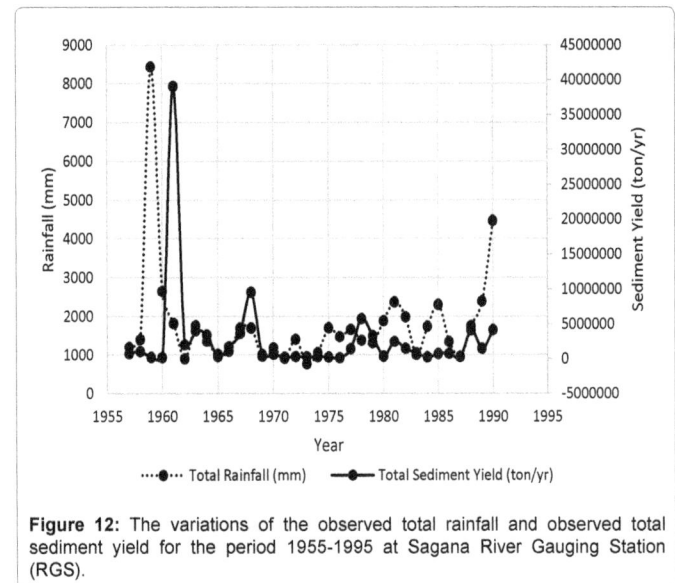

Figure 12: The variations of the observed total rainfall and observed total sediment yield for the period 1955-1995 at Sagana River Gauging Station (RGS).

Upper Tana catchment. The consequence of high sediment yield is the siltation of the Masinga dam reservoir [4,5,7-9]. The primary concern of high sediment discharge in the Masinga reservoir is the loss of storage capacity which affects the economic life of the reservoir. Loss of storage capacity hampers a reservoir's ability to fulfill its main functions which includes; store water, which impacts on other functions, e.g. power generation, irrigation supply, flood control, navigation, and domestic water supply. Due to the nonlinear relationship between water yield and reservoir capacity, even a small loss of reservoir capacity can result in substantial reductions in reservoir yield [34].

Sustainable land-use and protection of reservoirs in the upper Tana basin

Sustainable land use practices are seen as one of the best approaches of reducing sediment yield in the Upper Tana Basin. This can be done through application of various management practices which have the potential to benefit both upstream as downstream stakeholders. These include strip cropping, gully erosion control [9], and reforestation. There is a need for a comprehensive reforestation programme in the

Upper Tana catchment. There is also a need for eco-hydrological studies to establish the relationship and feedbacks between ecological and the hydrological processes [35]. In addition, there is need for comprehensive monitoring of river discharges, rainfall and sediment yield by the WRMA. Data collection programme should be consistent to avoid gaps in hydrological and climatological data [36-41].

References

1. Mango LM, Melesse AM, McClain ME, Gann D, Setegn SG (2011) Land use and climate change impacts on the hydrology of the upper Mara River Basin, Kenya: results of a modeling study to support better resource management. Hydrology and Earth System Sciences 15: 2245-2258.

2. Kitheka JU, Nthenge P, Obiero M (2003) Dynamics of sediment transport and exchange in the Tana estuary in Kenya. School of Water Resources Science and Technology.

3. Dunne T, Ongweny GS (1976) A new estimate of sedimentation rates on the upper Tana River. The Kenyan Geographer 2: 109-126.

4. Ongwenyi GSO (1978) Erosion and sediment transport in the Upper Tana catchment with special reference to the Thiba basin. Unpublished PhD thesis, University of Nairobi.

5. Ongwenyi GSO (1979) Patterns of sediment production in the Upper Tana basin in Eastern Kenya. In: Hydrology of Areas of Low Precipitation. Proceedings of Canberra Symposium, pp: 447-457.

6. Maingi SM (1991) Sedimentation in Masinga Reservoir. Unpublished MSc Thesis, Appropriate Technology Centre, Kenyatta University, Nairobi.

7. Maingi JK, Marsh SE (2002) Quantifying hydrologic impacts following dam construction along the Tana River, Kenya. Journal of Arid Environments 50: 53-79.

8. Kitheka JU (2014) Assessment of modification of the Tana River runoff due to developments in the Upper Tana Basin. In: The Proceedings of the 2nd Hydrological Society of Kenya (HSK) Workshop: "Hydrology in Water Cooperation and Security for Sustainable Economic Development". Sub-Theme: Hydrology and Sustainable Development, Nairobi, Kenya, pp: 178-191.

9. Dunne T (1979) Sediment yield and land use in tropical catchments. Journal of hydrology 42: 281-300.

10. Archer D (1996) Suspended sediment yields in the Nairobi area of Kenya and environmental controls. IAHS Publications-Series of Proceedings and Reports-Intern Assoc Hydrological Sciences 236: 37-48.

11. Ongwenyi GS (1983) Development of water resources. In the Kenyan Geographer-special issue on the Proceedings of the Nairobi workshop: Strategies for developing the resources of the semi-arid areas of Kenya, pp. 36-47.

12. Geertsma R, Wilschut L, Kauffman JH (2010) Review for the Green Water Credits Pilot Operations in Kenya. Green Water Credits Report 8 / ISRIC Report 2010/02, ISRIC- World Soil Information, Wageningen.

13. Notter B, MacMillan L, Viviroli D, Weingartner R, Liniger HP (2007) Impacts of environmental change on water resources in the Mt. Kenya region. Journal of Hydrology 343: 266-278.

14. Mutangah JG, Mwangangi O, Mwaura PK, Shem OM, Abiero L, et al. (1992) Kenya Indigenous Forest Conservation Programme.

15. Lamond G (2007) Local knowledge of biodiversity and ecosystem services in smallholder coffee farms in Central Province, Kenya (Doctoral dissertation, Master Thesis, Bangor. School of the Environment and Natural Resources, University of Wales, Bangor).

16. Jaetzold R, Helmut SH (1983) Farm Management Handbook of Kenya: Natural Conditions and Farm Management Information, Central Kenya: (Rift Valley and Central Provinces)/ Ministry of Agriculture, Kenya, pp: 3-29.

17. Vahabi J, Nikkami D (2008) Assessing dominant factors affecting soil erosion using a portable rainfall simulator. International Journal of Sediment Research 23: 376-386.

18. Nachtergaele FO, van Velthuizen H, Verelst L, Batjes NH, Dijkshoorn JA, et al. (2008). Harmonized world soil database (Version 1.0). Food and Agric Organization of the UN (FAO); International Inst. for Applied Systems Analysis (IIASA); ISRIC-World Soil Information; Inst of Soil Science-Chinese Acad of Sciences (ISS-CAS); EC-Joint Research Centre (JRC).

19. Baker BH (1967) Geology of the Mount Kenya area. Geological Survey of Kenya, pp: 78-79.

20. Tripathi MP, Panda RK, Raghuwanshi NS (2003) Identification and prioritisation of critical sub-watersheds for soil conservation management using the SWAT model. Biosystems Engineering 85: 365-379.

21. Parajuli PB, Mankin KR, Barnes PL (2008) Applicability of targeting vegetative filter strips to abate fecal bacteria and sediment yield using SWAT. Agricultural water management 95: 1189-1200.

22. Levesque E, Anctil F, Van Griensven ANN, Beauchamp N (2008) Evaluation of streamflow simulation by SWAT model for two small watersheds under snowmelt and rainfall. Hydrological sciences journal 53: 961-976.

23. Rostamian R, Jaleh A, Afyuni M, Mousavi SF, Heidarpour M, et al. (2008) Application of a SWAT model for estimating runoff and sediment in two mountainous basins in central Iran. Hydrological Sciences Journal 53: 977-988.

24. Neitsch SL, Arnold JG, Srinivasan R (2002) Pesticides fate and transport predicted by the soil and water assessment tool (SWAT). Final report submitted to Office of Pesticide Programs. Washington, DC: USEPA.

25. Nash JE, Sutcliffe JV (1970) River flow forecasting through conceptual models part I- A discussion of principles. Journal of hydrology 10: 282-290.

26. Servat E, Dezetter A (1991) Selection of calibration objective functions in the context of rainfall-runoff modelling in a Sudanese savannah area. Hydrological Sciences Journal 36: 307-330.

27. Legates DR, McCabe GJ (1999) Evaluating the use of "goodness-of-fit" measures in hydrologic and hydroclimatic model validation. Water resources research 35: 233-241.

28. Santhi C, Arnold JG, Williams JR, Dugas WA, Srinivasan R, et al. (2001) Validation of the SWAT model on a large water river basin with point and nonpoint sources. JAWRA Journal of the American Water Resources Association, 37: 1169-1188.

29. Van Liew MW, Arnold JG, Garbrecht JD (2003) Hydrologic simulation on agricultural watersheds: Choosing between two models. Transactions of the ASAE, 46: 1539-1551.

30. Jacobs J, Angerer J, Vitale J, Srinivasen R, Kaitho R, et al. (2004) Exploring the potential impact of reforestation on the hydrology of the upper Tana River catchment and the Masinga Dam, Kenya. Impact Assessment Group, Cent. For Natural Resource Information Technol., Texas A and M Univ., College Station, pp: 11-13.

31. Runo KS (2015) Evaluation of geological conditions along the proposed northern collector tunnel using electrical resistivity tomography method in Gatanga, Kenya, Doctoral dissertation, University of Nairobi.

32. Kitheka JU, Mavuti KM, Nthenge P, Obiero M (2014) The dynamics of the turbidity maximum zone in a tropical Sabaki estuary in Kenya.

33. McCully P (1996) Rivers no more: the environmental effects of dams, Zed Books, pp: 29-64.

34. Annandale GW (1987) Reservoir sedimentation. Elsevier 99-116.

35. Zalewski M (2002) Eco-hydrology-The use of cological and hydrological processes for sustainable management of water resources. Hydrological Sciences Journal 47: 823-832.

36. Palmieri A, Shah F, Dinar A (2001) Economics of reservoir sedimentation and sustainable management of dams. Journal of Environmental Management 61: 149-163.

37. Mati BM, Morgan RP, Gichuki FN, Quinton JN, Brewer TR, et al. (2000) Assessment of erosion hazard with the USLE and GIS: A case study of the Upper EwasoNg'iro North basin of Kenya. International Journal of Applied Earth Observation and Geoinformation 2: 78-86.

38. Kitheka JU, Obiero M, Nthenge P (2005) River discharge, sediment transport and exchange in the Tana Estuary, Kenya. Estuarine, Coastal and Shelf Science 63: 455-468.

39. Kitheka JU, Mavuti KM (2016) Tana Delta and Sabaki Estuaries of Kenya: Freshwater and Sediment Input, Upstream Threats and Management Challenges. In Estuaries: A Lifeline of Ecosystem Services in the Western Indian Ocean. Springer International Publishing, pp: 89-109.

40. Baker TJ, Miller SN (2013) Using the Soil and Water Assessment Tool (SWAT) to assess land use impact on water resources in an East African watershed. Journal of Hydrology 486: 100-111.

41. Fao I (2002) WFP (2002) Reducing poverty and hunger: the critical role of financing for food, agriculture and rural development, pp: 18-22.

Tide and Mixing Characteristics in Sundarbans Estuarine River System

Goutam KS[1]*, Tanaya D[1], Anwesha S[2], Sharanya C[1] and Meenakshi C[3]

[1]*School of Oceanographic Studies, Jadavpur University, Kolkata 700032, West Bengal, India*
[2]*Department of Mathematics, Jadavpur University, Kolkata 700032, West Bengal, India*
[3]*Basanti Devi College, Kolkata 700029, West Bengal, India*

Abstract

Sundarbans Estuarine System (SES, 21.25°-22.5° N and 88.25°-89.5° E), comprising the southernmost part of the Indian portion of the Ganga-Bramhaputra delta bordering the Bay of Bengal, is India's largest monsoonal, macro-tidal delta front system. Sundarbans Estuarine Programme (SEP), the first comprehensive observational programme to study tidal as well as salinity features was conducted during 18-21 March, 2011 (Equinoctial spring phase). The main objective of this program was to monitor tides and salinity characteristics within the SES. Out of 30 observation stations, spread over more than 3,600 sq km covering seven inner estuaries, we have chosen river Jagaddal, which is connected with Saptamukhi East Gulley (SEG) in the West and river Thakuran in the East, due to the fact that the station Indrapur situated on this river at location very close to Bay of Bengal represents the condition at the mouth of all seven estuaries.

Tidal elevation, salinity, bathymetry and vertical profile of salinity using CTD were measured during the observation period. Observed current data collected from different sources have been used for comparison with computed tidal current. Finally, the estuarine current, bottom drag coefficient and gradient Richardson number have also been computed. Computed values of these parameters have analyzed for interpreting variations for tidal, current and mixing feature prevailing in the estuary.

Keywords: Tidal current; Estuarine circulation; Salinity; Mixing; Sundarban; Estuary; Bottom drag coefficient

Introduction

Sundarbans Estuarine System (SES, the region lying between 21.25°-22.5° N and 88.25°-89.5° E) is India's largest monsoonal, macro-tidal delta-front estuarine system in the eastern coastal state of West Bengal. River Hoogly, the first deltaic offshoot of the river Ganga forms the western boundary of the SES. River Raimangal, a tributary of river Ichhamati, an easterly distributary of the Ganga forms the eastern boundary of the SES. Dense natural mangrove forests of Sundarbans in the southern fringes of the Bay of Bengal forms the southern limit. An imaginary line named Dampier-Hodges line based on 1829-1832 survey is the northern boundary of SES. North-south flowing rivers: the Saptamukhi, Thakuran, Matla, Bidya, Gomdi, Gosaba, Gona, Harinbhanga and Raimangal form the principal estuaries of SES on the east of the Hoogly as a complex network of numerous west-east interlinked channels, canals and creeks. In addition to the tidal effects at their mouths, during floods these inner estuaries accumulate considerable amount of Gangetic fresh water in their upstream along with summer monsoon rainfall which are the major driving factors in retraining the estuarine character of the SES. The rivers, water bodies, intertidal mudflats, creeks, saline swamps, sandy shoals in SES are remarkable not only for the astounding biodiversity including a wide variety of benthic and pelagic fauna (such as fish and crustaceans), spawning zone of reptiles (such as crocodiles), fisheries but also for their important role as a natural filter to pollutants released from the human settlements and industrial zone in northern reaches and major pathways for nutrient recycling. Moreover impenetrable mangrove forests acts like a protective barrier to the densely populated metropolitan city of Kolkata to its north by absorbing direct impact of cyclonic storms and surges from the Bay of Bengal to a great extent [1].

All the principal estuaries in the SES are funnel-shaped, widths converge rapidly northwards, having very wide mouths at confluences with the Bay of Bengal and following meandering courses with sharp bends. Over this distance the characteristics of the SES has shaped in these different environments (variable depth, width and stretches) by the interplay of a variety of physical processes. This investigation is a part of a series of intensive on-going study of the tidal elevations, salinity and temperature at 30 locations situated on the principal inner estuaries of the SES, conducted by Sundarbans Estuarine Programme (SEP) during 18-21 March 2011(Pre-monsoon period), 2013 and one month phase of study in August, 2015 (Monsoon period).

In the first observation programme (during 18-21 March 2011, Pre-monsoon period, equinoctial spring phase) in SES it was reported that estuaries have shown a great diversity of tidal forcing characteristics with different depth, width, size and shape. In general tidal range has increased from mouth to head (northwards) in all estuaries due to convergent channel geometry (funnel-shaped) and frictional effects with the variable degree and rates of amplification over the various estuarine stretches as a result of flood-dominant and ebb-dominant tidal asymmetries [1]. In such estuary system tidal currents, salinity variation and river discharge predominates in estuarine flow pattern more than meteorological forcing (wind stress). So, the study of tidal flows and salinity structure are essential to assess a broad range of estuarine phenomena including dispersion of salinity, nutrients, pollutants, sediments and their transportation into and out of the estuary and to know how quickly the estuary responds to different

*Corresponding author: Goutam KS, School of Oceanographic Studies, Jadavpur University, Kolkata 700032, West Bengal, India
E-mail: gksju@yahoo.co.in

forcing such as tidal mixing and river discharge which affect the ecological health and the water quality [1,2-8].

In this paper we are intended to study the flow pattern in the narrow tidal channel of River Jagaddal, a part of SES. Our attempt is to quantify the indicators such as tidal amplitude, tidal current, residual current (estuarine circulation), bottom drag coefficient and level of stratification which are essential to characterize the flow pattern in different channel geometry along a single stretch of a river.

Study Area and Relevant Measurements

River Jagaddal is flowing parallel to the Saptamukhi East Gullay (SEG) in the west, the stronger branch of the river Saptamukhi debouncing into the Bay of Bengal having links with the adjoining estuary of River Thakuran in the east through Dhanchi Khaal (cannel). In the northern limit (head) river Jagaddal has linked with river Barchara, river Kalchara and Pakhi Nala (Figure 1). River Barchara and

river Kalchara are influenced by fresh water discharge of Hoogly river whereas Pakhi Nala is connected with water flow of river Thakuran.

In this study the data required for tidal elevation, salinity are collected from the field observation of four stations on Jagaddal: Indrapur (mouth), Dhanchi and Ramganga/Pakhirala (head).Observed current data collected from different sources have been used. These stations are chosen for their significant importance along the channel. The station Indrapur (S1) on river Jagaddal is closest (about 8 Km north of Bay of Bengal) to the Bay of Bengal, representing the conditions at the mouth. The station Dhanchi (S2) is at the junction of river Jagaddal and river Thakuran through Dhanchi Cannel. The station Ramganga (S6) is at the junction of three different linkages of river Barchara, river Kalchara (connected to Hoogly River) and Pakhi Nala (connected to Thakuran). The station Pakhirala (S7) is in the bank of Pakhi Nala, serving as the connecting junction of river Jagaddal to river Thakuran. Both Ramganga and Pakhirala serve as the head of this channel.

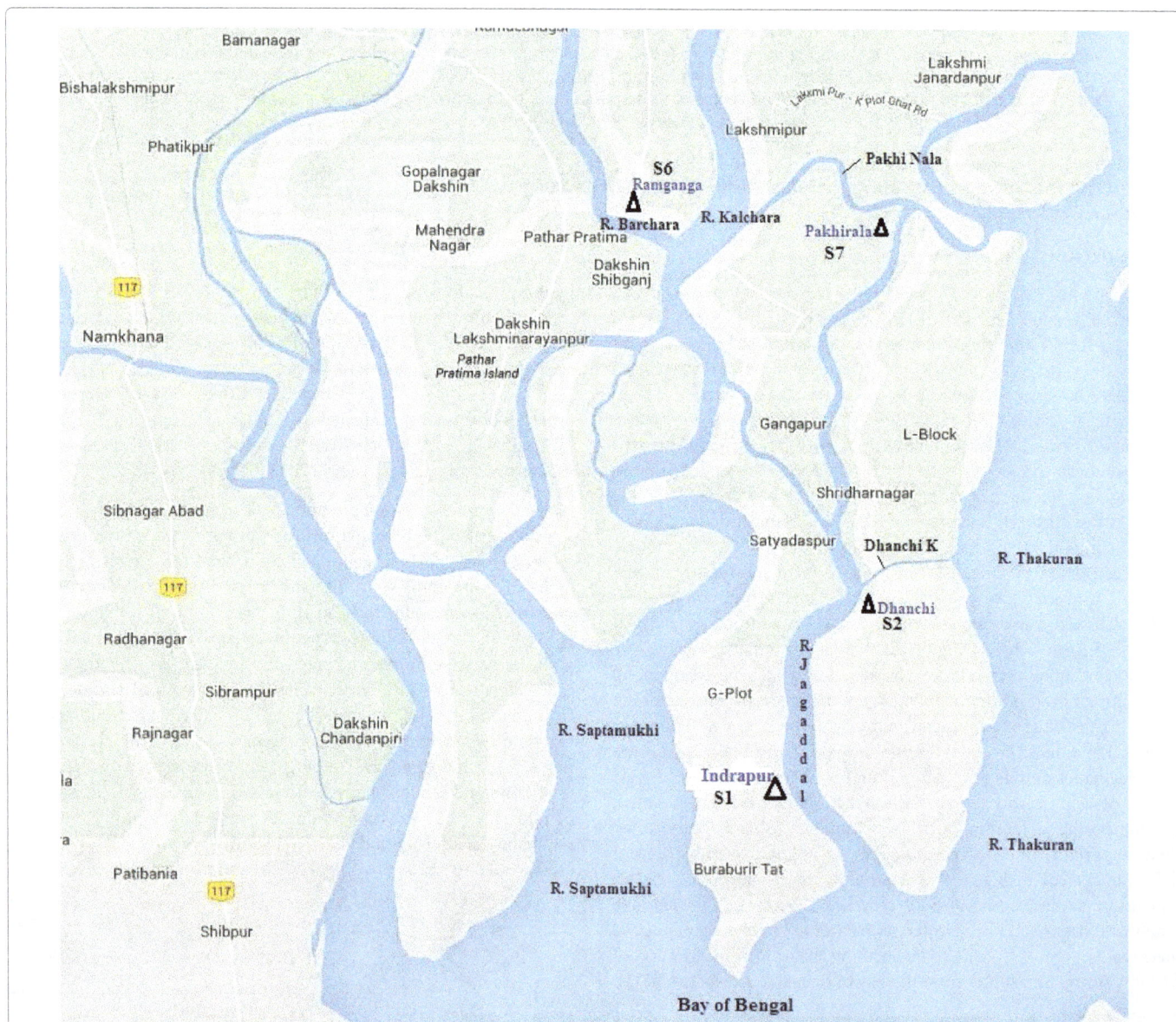

Figure 1: Study Area: Jagaddal River Estuary. Triangles show the observation stations along the river; Abbreviation R. is used to refer "River".

Methodology

The methodology follows five steps:-

(i) The first step solves a numerical model with a given spatial distribution of friction coefficient in a channel with variable along (-x) and cross (-y) channel depth variation and along channel width variation to compute tidal amplitudes and velocity components.

(ii) The second step uses the velocity field computed in the previous step and applies the phase-matching method to recover the drag coefficient.

(iii) Then by using the drag coefficient estimated from the phase-matching method as a function of along-channel position in the numerical model, we have estimated the tidal current and tidal amplitude and compare with the observed amplitude. Step (i) to (iii) are repeated if the computed tidal amplitude are not close to the observed data.

(iv) Using the values of bottom drag coefficient at different depth the magnitude of residual current (estuarine circulation) is calculated.

(v) At last, the level of stratification along the channel is parameterized as the gradient Richardson number using the values of tidal current and residual current or estuarine circulation.

First step: Computing depth functions

We have assumed bathymetric functions for computational convenience, by fitting bathymetry data along cross-channel directions. Several such bathymetry profiles have been used for different segment of the river to examine the effect of complicated bathymetry. The depth function is assumed to vary in both along-(x) and cross-channel (y) directions as needed having the following form:

$$h = b_0 + b_1 * \exp\left(-\frac{\left(y - \frac{D}{2}\right)^2}{b_2}\right) \quad (1)$$

where D is the transverse coordinate, the estuary width, and b_0, b_1, b_2 are the parameters that adjust the depth profile and vary with x. The width D is allowed to vary from 0 (at mouth) to maximum value for each of the transect starting from mouth to head of the channel.

Second step: Solving numerical model

We have used the depth-averaged hydrodynamic model for shallow tidal estuaries (maximum water depth is on the order of a few tens of meters) to compute tidal current and amplitude as described by Li et al. [9,10] to study the variation of intratidal flow field in James river estuary. This method is a modification of the adjoint variational method, the full-scale assimilative numerical models to compute tidal current and amplitude with variable bottom drag coefficient in Chesapeake Bay by Ulman and Wilson et al., Spitz and Klinck et al. [6,11].

We have assumed the estuarine geometry as a curved channel with variable width and an arbitrary lateral depth variation. By following approach similar to Li et al. [7], due to anisotropy of width and depth variation, we have to opt for numerical approach. The x axis is taken to lie along the river boundary and points toward the head of the estuary. The y axis lies along the open boundary at x=0.

The depth-averaged, shallow water momentum and continuity equations [Li and Levinson] [9] are numerically solved to obtain the tidal amplitude and tidal current:-

$$\frac{\partial u}{\partial t} + u\frac{\partial u}{\partial x} + v\frac{\partial u}{\partial y} - fv = -g\frac{\partial \varsigma}{\partial x} - \beta\frac{u}{h} + \frac{\beta}{h^2}u\varsigma$$

$$\frac{\partial v}{\partial t} + u\frac{\partial v}{\partial x} + v\frac{\partial v}{\partial y} + fu = -g\frac{\partial \varsigma}{\partial y} - \beta\frac{v}{h} + \frac{\beta}{h^2}v\varsigma \quad (2)$$

$$\frac{\partial \varsigma}{\partial t} + \frac{\partial (h+\varsigma)u}{\partial x} + \frac{\partial (h+\varsigma)v}{\partial y} = 0$$

Where u:- longitudinal velocity, v: lateral velocity, ς : elevation, h: water depth, x: longitudinal coordinate, y: lateral coordinate, t: time, β : friction coefficient, f: coriolis parameter, g: gravitational acceleration.

The friction coefficient β is defined by Proudman and Parker [12,13]:-

$$\beta = \frac{8C_D U_O}{3\pi} \quad (3)$$

where C_D and U_0 are bottom drag coefficient and magnitude of longitudinal velocity respectively. The friction coefficient $\hat{a}$, a function of x and y, is dependent on the velocity amplitude and the bottom drag coefficient. Spatial distribution of C_D for every segment of the channel and also the interpolated longitudinal velocity U_0 along the channel from observed current data are used to provide initial values for C_D and then β to solve equation (2) numerically.

Recovering bottom drag coefficient

In depth-averaged hydrodynamic model, the intensity of the overall tidal energy dissipation is related to mainly bottom drag coefficient for the major tidal frequency component [9,10]. Again we have assumed that the bathymetry effect alone will be able to produce significant spatial variation in bottom drag coefficient with an effect on tidal current and amplitude in a tidal channel. Hence we have chosen arbitrary along and across channel depth variations with the emphasis on the effect of channel-shoal configuration and bottom slopes.

Using the computed results for tidal current and amplitude from numerical simulation of depth-averaged, shallow water momentum and continuity equations (2), spatial variation of C_D in tidal channels with significant lateral variation of depth is estimated by applying the phase-matching method [10]. The calculation involves the following steps:

(a) The amplitude and phase of the computed tidal current (φ_U) are obtained by harmonic decomposition of horizontal velocity.

(b) The phase of the tidal current is then fitted to a quadratic function of the water depth to obtain the phase of the pressure gradient φ_{Ax}.

(c) From the phase relationship of the longitudinal momentum equation, the drag coefficient is calculated as a function of the transverse position using the following formula:

$$C_D = \frac{3\pi\sigma h}{8U_{o\tan(\varphi_{Ax-\varphi_U})}} \quad (4)$$

Where σ is the angular frequency of the semi-diurnal tide (M2 tide for this study), h is the undisturbed water depth, U_0 is the amplitude of the tidal current, φ_U is the phase of the computed tidal current and φ_{AX} is the phase of the pressure gradient.

Estimation of residual or estuarine circulation

Pritchard [14] first introduced the concept of "residual" or estuarine circulation driven by horizontal salinity gradient, a key dynamical variable that makes estuaries different from any other marine environment and plays a key role in maintaining salinity stratification in estuary which inhibits vertical mixing and leads to one way of classifying estuaries based on strength of stratification: well mixed, partially mixed, highly stratified and salt edge. He pointed out that if the vertically varying horizontal currents are measured through the course of the tidal cycle, and then averaged, the estuarine circulation would be revealed, though the tidal currents are typically much stronger the estuarine circulation.

Using tidal current U_T we have calculated the magnitude of estuarine circulation, U_e expressed as (Geyer et al.,) [15]:

$$U_e = a_0 \frac{\beta_s h^2 g \frac{\partial s}{\partial x}}{C_D U_T} \quad (5)$$

Where a_0 is a dimensionless constant related to turbulent momentum flux, β_s is the coefficient of saline contraction, g is the acceleration of gravity, h is the water depth, U_T is the tidal velocity.

Estimation of level of mixing

Numerous studies have undertaken to analyze importance of salinity gradient, stratification and tidal currents in estuarine circulation. Both river discharge and mixing are crucial for the different tidal flushing process and tide-induced transport of materials in flood and ebb conditions. During the flood tide when the more saline sea water is coming into estuary with a faster advancement at the surface, the heavier salt water on the surface promotes vertical mixing to reach a more stable state. During ebb tide when the fresher less saline estuarine water is going out of the estuary with a faster advancement at the surface on the heavier more saline water at bottom, a more stable water column is formed suppressing vertical mixing. The sensitive dependence of stratification on tide also has been revealed in a number of studies of the spring-neap variation of tidal cycle with stratification

with maximal tidal mixing and minimum stratification during spring tides [7,15].

Geyer et al. [15] obtained the top-to-bottom salinity difference, Δs is estimated from $\frac{\partial s}{\partial x}$, the horizontal salinity gradient obtained from observed salinity data in Hudson estuary,

$$\Delta s = a_1 \frac{\frac{\partial s}{\partial x} h U_e}{C_D U_T} \quad (6)$$

Where a_1 is a constant that depends on the shape of salinity and velocity profile, $\frac{\partial s}{\partial x}$ is the horizontal salinity gradient and C_D is the bottom drag coefficient.

To quantify the importance of stratification on mixing, the gradient Richardson number R_{iT} is estimated which is defined as the ratio between vertical salinity difference and tidal velocity,

$$R_{iT} \frac{\beta_s g \Delta s h}{U_T^2} \quad (7)$$

On the basis of the estimated value of gradient Richardson number we have differentiated the level of mixing along the channel.

Results and Discussions

During observation period of 18-21 March 2011 (Pre-monsoon period) an increased trend in tidal range from mouth (Indrapur) to head (Ramganga, Pakhirala) was observed along Jagaddal. But interestingly, a decrease in tidal range was observed between Indrapur and Dhanchi (Figure 2: near the junction of Jagaddal and Dhanchi Canal (Chatterjee et al., [1]. Some striking features regarding salinity variation was also observed during observation where the salinity decreases from mouth to head upto Ramganga, but it increases from Ramganga to Pakhirala (Figure 3).

From Indrapur towards Dhanchi, at the junction of Jagaddal and Dhanchi canal, the depth has decreased suddenly (Figure 4). In this shallow region, the stronger tidal current has observed. Here the width of the channel has increased due to linking channel with two rivers

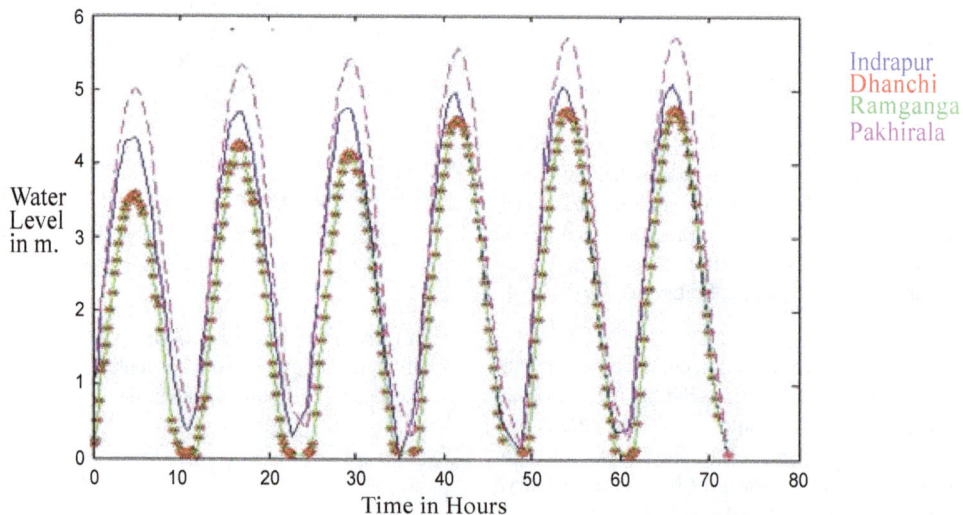

Figure 2: Observed water level variation in four observation stations.

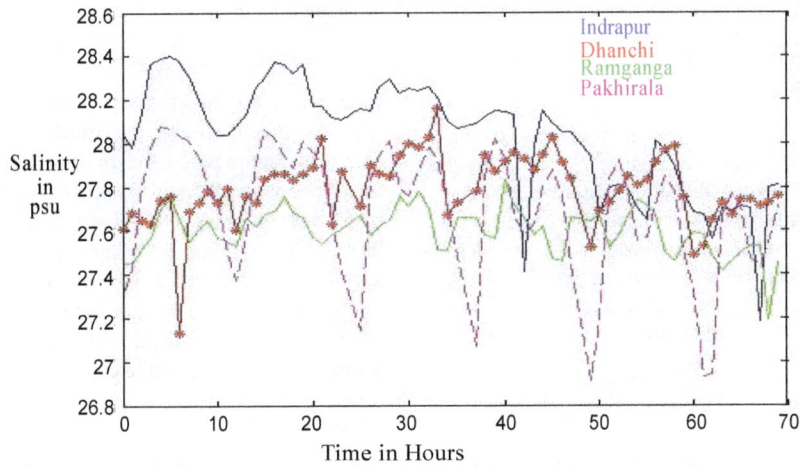

Figure 3: Observed salinity variation in four observation stations.

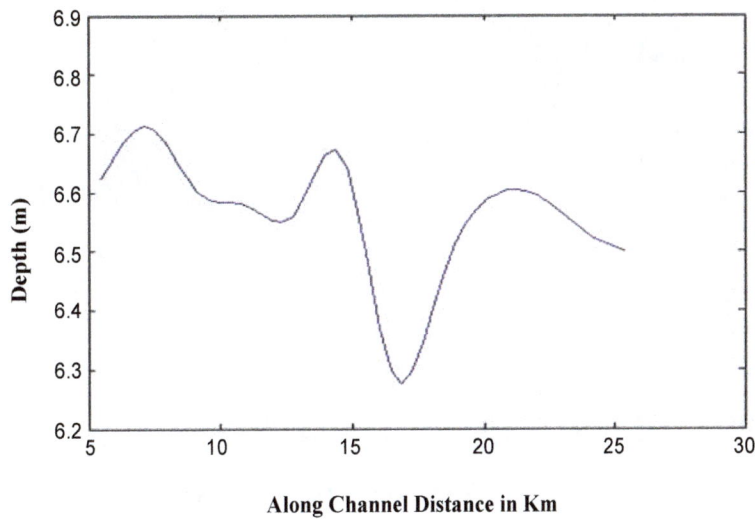

Along Channel Distance in Km

Figure 4: Depth distribution along the channel.

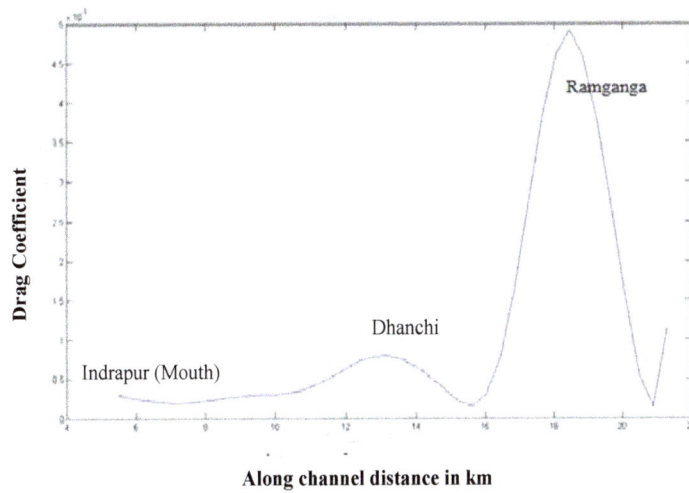

Along channel distance in km

Figure 5: Drag coefficient variation along the channel.

(Saptamukhi, in the west) and Thakuran, in the east connected through Dhanchi canal. In addition, the bottom drag coefficient (Figure 5) has increased suddenly after this junction. With the combined effect of increased C_D and river width, the water level (Figure 2) and tidal current (Figure 6) has dropped after this point.

Near the west bank of Jagaddal, the cross channel current is stronger and generates stronger residual circulation (Figure 7). The estuarine current has started to increase from Dhanchi to Indrapur. Vertical salinity gradient has increased due to landward advection of denser and more saline seawater at bottom and seaward advection of fresh riverine water on the top layer. As a result, an increase in gradient Richardson number (>0.25, Figure 8) from Dhanchi to Indrapur (towards the sea) has found which indicates that the stratification has already started.

Along the channel from Dhanchi to Ramganga, there are three adjoining rivers: Kalchara, Barchara and Pakhi Nala which draw large amount of less saline water from Hoogly. Depth has increasing trend here and tidal current tends to take a sharp rise. From CTD cast it is observed that vertical gradient of salinity over this stretch of the river is negligible. The gradient Richardson number at this region initiates strong mixing by making stratification negligible (the channel in this area acts as a well-mixed estuary) with an increased bottom drag coefficient along this portion of the river. The increased river inflow (through linked rivers or channels) increases the tidal current and amplitude suppressing the role of drag coefficient (opposite to the observations in Dhanchi). Similar type of phenomenon was reported by Li et al. in James river estuary where increased river discharge has limited tidal mixing, lowering the bottom drag coefficient. For this reason, decrease in tidal current and amplitude due to large value of bottom drag coefficient have not observed here in contradiction with Dhanchi. A small peak in estuarine circulation is found due to this limited effect of drag coefficient.

Conclusion

The main intension of this study is to interpret the cause about the

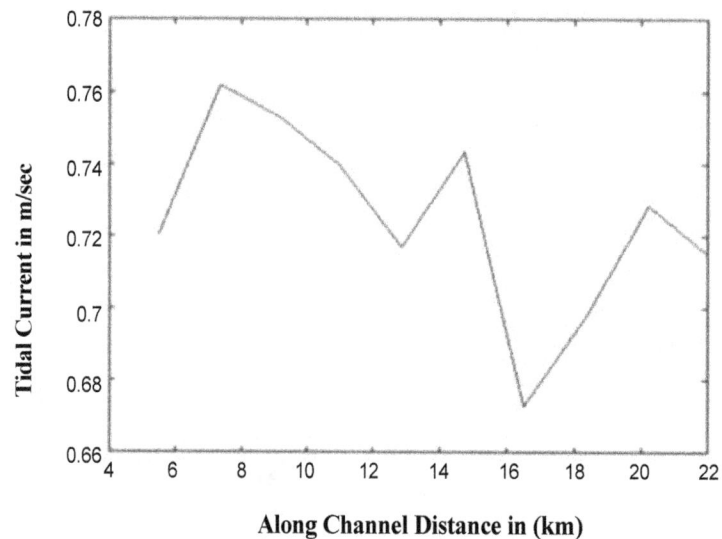

Figure 6: Tidal current variation along the channel.

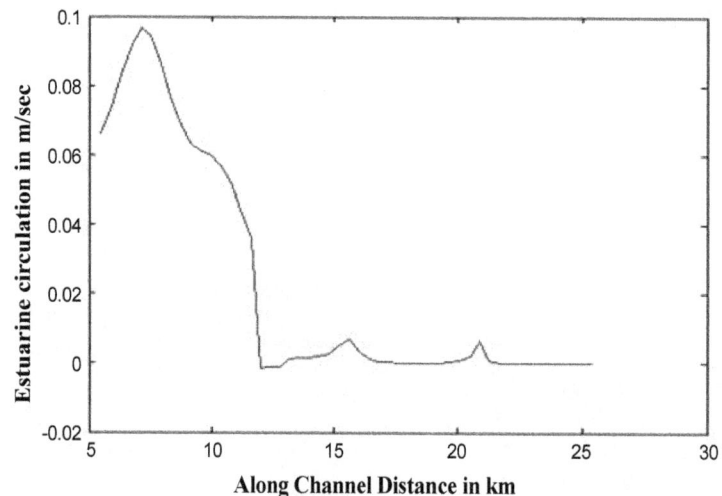

Figure 7: Estuarine current variation along the channel.

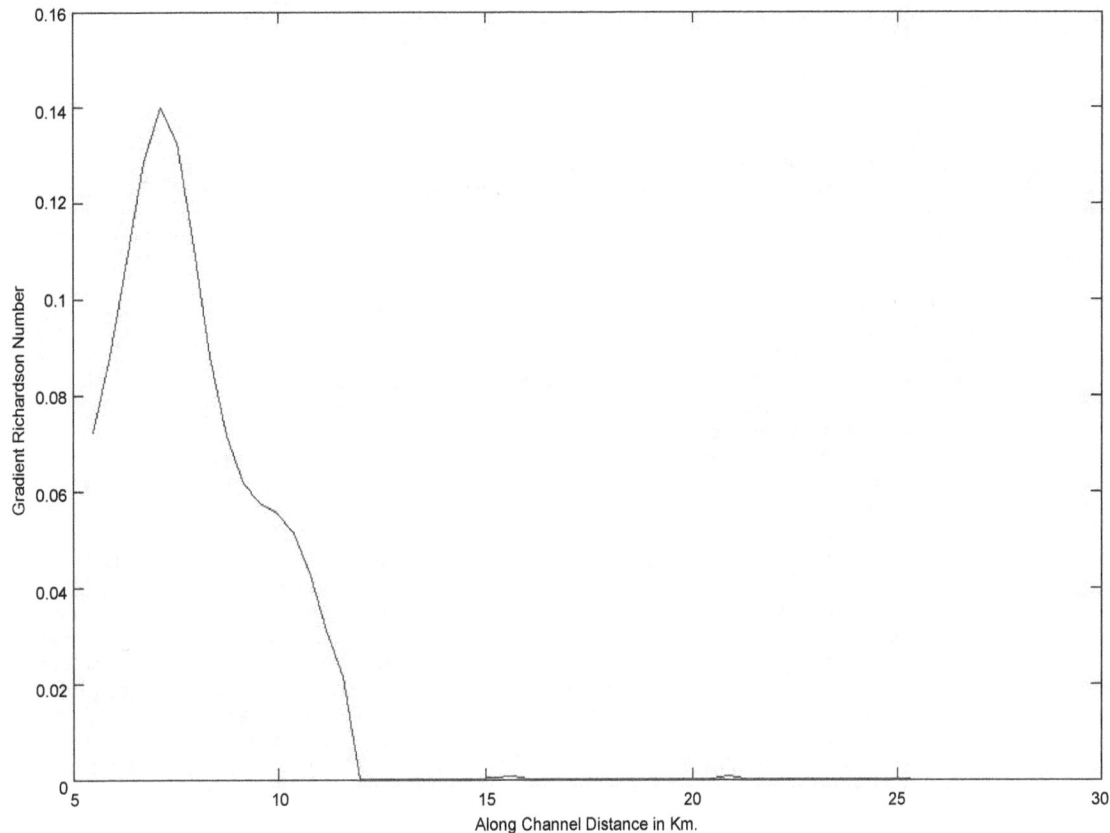

Figure 8: Gradient Richardson number variation along the channel.

variation of water elevation at several locations on Jagaddal. Cross-channel flow (west-east) is a dominant feature in this region (though rigorous measurement on this aspect is not available except some cross channel current measurement near river junction). Variational features of bottom drag coefficient and estuarine current, due to inflow through two intervening channels near Dhanchi, one connected with Thakuran and other connected with Saptamukhi, contribute to tidal amplitude and current variation. River Jagaddal (as observed from, gradient Richardson number variation) also shows stratification characteristics from Indrapur to Dhanchi. But beyond Dhanchi upto the head (further north) it exhibits a zone of strong mixing. This zone of strong mixing is indicative of the feature of considerable fresh water flow from river Hoogly. All the salient findings corroborated in above mentioned statements have been observed with considerable elaboration in our discussion part.

Acknowledgements

This work has been supported by the Sundarbans Estuarine Programme (SEP) funded by Indian national Centre for Ocean Information Services (INCOIS), Ministry of Earth Sciences, Government of India. The authors are grateful to Dr. Meenakdhi Chatterjee, Project Coordinator of Sundarbans Estuarine Programme (SEP) for appreciable assistance and for providing data and other necessary information.

References

1. Chatterjee M, Shankar D, Sen GK, Sanyal P, Sundar D (2013) Tidal variation in the Sundarbans Estuarine System, India. J Earth Syst Sci 122(4): 899-933.

2. Jay DA, Smith JD (1990) Residual circulation in shallow estuaries: 1. Highly stratified, narrow estuaries J Geophys Res 95: 711-731.

3. Jay DA, Smith JD (1990) Residual circulation in shallow estuaries. 2. Weakly stratified and partially mixed, narrow estuaries. J Geophys Res 95: 733-748.

4. Valle-Levinson A, Li C, Royer TC, Atkinson LP (1998) Continental Shelf Research 18: 1157-1177.

5. Mac Cready P (1999) Estuarine adjustment to changes in river flow and tidal mixing. J Phys Oceanogr 29: 708-726.

6. Spitz YH, Klinck JM (1998) Estimate of bottom and surface stress during a spring-neap tide cycle by dynamical assimilation of tide gauge observations in the Chesapeake Bay. J Geophys Res 103: 12761-12782.

7. Li C (2006) Modeling of Bathymetry Locked Residual Eddies in Well mixed Tidal Channels with Arbitrary Depth Variations, J Phys Oceanogr 36: 1974-1993.

8. Li C, Swenson E, Weeks E, White JR (2009) Asymmetric tidal straining across an inlet: Lateral invertsion and variability over a tidal cycle, Estuarine, Coastal and Shelf Science 85: 651-660.

9. Li Chunyan, Valle-Levinson A (1999) A two-dimensional analytic tidal model for a narrow estuary of arbitrary lateral depth variation: The intratidal motion. J Geophys Res 104: 23525-23543.

10. Li C, Valle-LA, Atkinson, LP, Wong K, Lwiza K (2004) Estimation of drag coefficient in James River estuary using tidal velocity data from a vessel-towed ADCP, J Geophys Res 109: C3.

11. Ullman D, Wilson RE (1998) Model parameter estimation from data assimilation modeling: Temporal and spatial variability of the bottom drag coefficient. J Geophys Res 103: 5531-5549.

12. Proudman J (1953) Dynamical Oceanography, Methuen, New York, p. 409.

13. Parker BB (1984) Frictional effects on the tidal dynamics of shallow estuary, (Ph.D. dissertation) Johns Hopkins Univ., Baltimore, pp. 291.

14. Pritchard DW (1956) The dynamic structure of a coastal plain estuary. J Mar Res 15: 33-42.

15. Geyer WR (2001) Estuarine salinity structure and stratification.

4

Modelling Phosphorus Losses from Tropical Agricultural Soils in Gilgel Gibe Watershed, Ethiopia

Yalemsew Adela[1]* and Christian Behn[2]

[1]Jimma University, Institute of Technology, School of Civil and Environmental Engineering, Ethiopia
[2]Rostock University, Faculty of Agricultural and Environmental Sciences, Resources Management and Soil Physics, Rostock, Germany

Abstract

Phosphorus (P) is a vital nutrient for plants, however its excess loss from agricultural lands cause eutrophication on aquatic environment. The Gilgel gibe reservoir located in the southwest part of Ethiopia is exposed to this phenomenon whereby the water quality has been classified as mesotrophic with P concentration of 0.86 mg/l. The objectives of this study were to identify the operating P loss mechanisms from agricultural lands, quantify the amount of P exported and evaluate the factors for P loss using the best management practices (BMPs) concepts. Therefore, experimental data and the annual phosphorus loss estimation (APLE) model were used to study the underlying processes. Catenas surrounding the reservoir, used as arable and pastureland, were investigated. Topsoil samples were taken and analyzed at three slope positions. The soils are mainly nitisol and a smaller portion of vertisol where the parent materials are basalt and rhyolite. The APLE model was set using soil data from the arable lands with nitisol soil and hydrologic records. The phosphorus loss was simulated from 2001 to 2010. Besides, the experimental P sorption data were used to check the logical consistency of the model output. On average 12.66 ± 0.7 kg P ha^{-1} yr^{-1} is lost in the form of particulate and total dissolved P. Generally, 56% of P is lost in the form of particulate P due to erosion, and 44% as soils dissolved and direct fertilizer runoff P. A significant variation observed between the sediment and soil dissolved P loss (p-value= 0.000) which is attributed to the soil chemical and physical properties that control the phosphorus dynamics. Obviously the dominant P transfer from agricultural lands into the Gilgel Gibe River and reservoir is particulate P loss. An evaluation of causing factors using BMPs indicated that a reduction of sediment by 5-20% resulted to retain P from 2-9%. Similarly, a reduction of soil P content reduces the P loss from 2-8.5%. However, a reduction of fertilizer quantity applied on the fields within the same percent range is hardly reducing P loss relative to the earlier factors. Therefore, attention should be given to the application of precision agriculture to avoid such problems.

Keywords: Particulate loss, P loss, P sorption, Soil dissolved P, Tropical soil

Introduction

Phosphorus (P) is a limiting nutrient in the aquatic environment where its excess presence impairs water quality significantly by accelerating the growth of algae and aquatic plants causing eutrophication [1,2]. Previously the impact of phosphorus to cause a problem as a non-point source pollutant was underestimated in comparison with nitrogen due to its hardly mobility in the agricultural landscape. Currently there is a consensus that phosphorus is a major non-point source pollutant that causes eutrophication in surface waters. The intrinsic property that affects the mobility of phosphorus is its sorption affinity for soil particles. However, in the presence of high soil and stream bank erosion, an increased amount of phosphorus is transported in surface water. Additionally, the contribution of dissolved phosphorus is also more significant than previously thought [3].The Gilgel gibe catchment can be referred as agricultural catchment due to the intensive agricultural practice carried out. Despite the catchment comprise a very sensitive development infrastructure, there had not been a mechanism to evaluate and manage the problem associated with nutrient transport and loading. The impact of phosphorus in this catchment has also been continuing on the reservoir water quality. Scientifically researched and published information on the overall aquatic environment of the reservoir is hardly found. The Gilgel gibe reservoir had high turbidity and elevated phosphate concentrations of 0.3-0.7 mg/L, which is much higher than the permissible limit of 0.025 mg/L for lakes and reservoir [4]. To address the problem of phosphorus, the transport process needs critical attention. Several fields and laboratory techniques developed to understand the movement of

phosphorus and its fate. Besides this, transport and loading models are available that are taking special attention recently. To understand the problem in this catchment, appropriate model development or utilization would help a lot. Models developed to evaluate P loss under various conservation practices give emphasis both on steady state and dynamic process, which attributes complexity of the model features [5]. Most commonly, US Environmental Protection Authority uses Spreadsheet Tool for the Estimation of Pollutant Load (STEPL) and P load tools to evaluate conservation practices [1]. Beyond these, simple field-scale models, phosphorus indices are used for conservation planning and regulation under the application of animal manure in the US and Europe [6]. A phosphorus index is typically a tool that yields a categorical rating of phosphorus loss from a single field. These indices are valuable tools to assess the potential risk of phosphorus leaving a site and travelling toward a water body, but they were not initially developed to be quantitative predictors of phosphorus loss [1,7]. A tool that reliably quantifies field-scale phosphorus loss is an alternative to qualitative P Indexes and process- based models and remains easy

*Corresponding author: Yalemsew Adela, Jimma University, Institute of Technology, School of Civil and Environmental engineering, Ethiopia
E-mail: yalemsew.adela@ju.edu.et

to use with its feature that requires only readily obtainable inputs. A field-scale P loss quantification tool offers attractive characteristics for P loss reduction planning. This is due to the ease in designing realistic assumptions for the relative effect of different management practices on P loss and validation with measured data, which eventually give a clear picture of P loss forms [6]. Though process based models like the Soil and Water Assessment Tool (SWAT) can be used to identify management strategies and gives larger picture at the watershed scale for water quality assessment, the requirement of large data makes it difficult for immediate consumption [8]. In addition, watershed-scale predictions of P loadings to fresh water bodies are not reliable unless extensive, site-specific calibration is used. In Gilgel gibe watershed there are no gauging stations for in-stream phosphorus concentration and so many other soil data which are required to assess the transport of P from land into water bodies. Therefore, it is better to use the field scale phosphorus model which gives a clear picture to visualize the other part of the catchment heuristically. Annual Phosphorus Loss Estimator (APLE) as a field-scale P loss quantification tool possesses a remarkable feature. The soil P algorithms developed make the model applicable for a wide variety of soil types, climates, and management conditions [8]. Therefore, the objectives of this study were to identify the dominant P losses mechanism, quantify the amount of P loss from agricultural fields and evaluate the factors for P loss using BMPs concept.

Methods and Materials

Description of the catchment

Gilgel gibe catchment is located in the south-western part of Ethiopia in Jimma zone, Oromia region situated within Omo-Gibe basin. The whole Gilgel gibe sub catchment which sheds water to the reservoir lies between latitude of $7°21'$ to $7°58'$N and longitude of $3631'$ to $37°26'$E covering an area of about 4300 km². The area is flat plateau about 1,650 m a.s.l and consists of a series of gentle sloping low hills and broad plains surrounded by hills or mountains. The Gilgel gibe River, which flows from southwest to northeast, is a tributary of the Great Gibe River and is extremely variable in the course and gradient. The Gilgel gibe is the main river, which creates the reservoir that covers an area of 55 km². The annual rainfall of the catchments area varies from a minimum of 1,300 mm near the confluence with the Great Gibe River, to a maximum of about 1,800 mm on the mountains. Rainfall decreases throughout the catchments with a decrease in elevation. The average annual rainfall over the whole Gilgel gibe basin where it joins the Great Gibe River is 1,527 mm. It appears that 60 per cent of the total amount of annual rainfall occurs between June and September 30 per cent from February to May and only 10 per cent between Octobers to January [9]. The gentle slopes and the central plains of the foothills of the ridge is intensely cultivated and densely populated area. The agricultural sector uses various type of fertilizers (i.e. superphosphate (P_2O_5), Diammonum phosphate (DAP) and Urea) to enhance the productivity. However, the management of these agricultural inputs is traditional and unsupported with soil fertility test examination before application.

Description of the arable lands investigated for model set up

Soil property: The red balloon on figure 1(a) and (b) represents the six catenas where soil samples were taken around the reservoir. Six catenas surrounding the reservoir (120 to 440 m long), used as arable and pastureland, extended on the flatter edge of the catchment were investigated. Topsoil samples were taken at three slope positions. Catena 5 and 6 are uniquely vertisol at each slope position, however, the other four catenas possess a nitisol soil at least at one position along the catena. Therefore, as it is also reported in several literatures, nitisol soils are common for this watershed. Consequently, this study focused on the P loss from the fields which are nitisol by type. For P loss estimation the land which grows teff is considered for both types of the soils. The P loss from the vertisol lands was used for comparative reasoning and to check the consistency of the model. Detailed laboratory analysis was

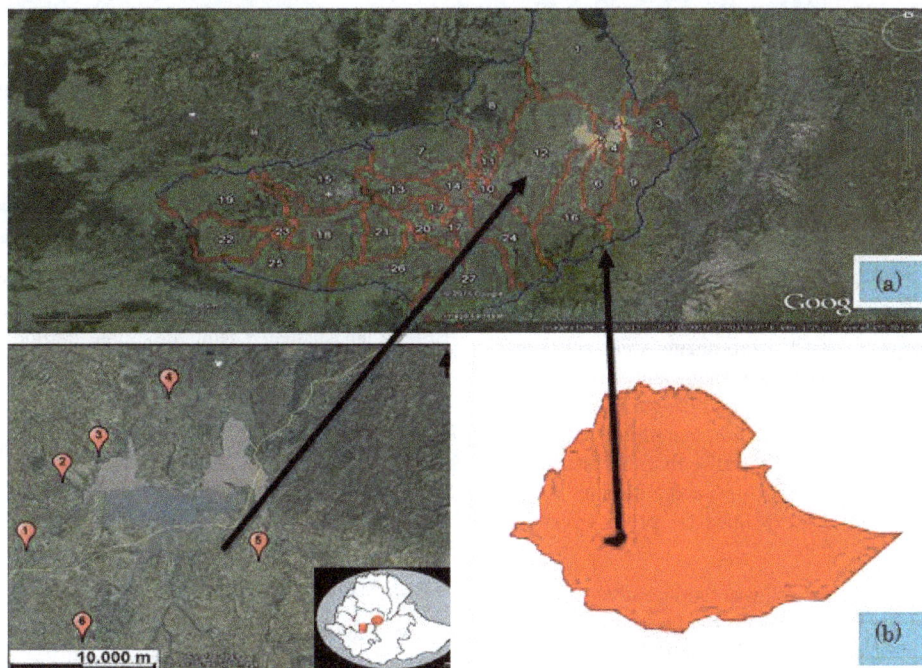

Figure 1: (a) Delineated KML map of Gilgel Gibe catchment and (b) Location of catenas around the Gilgel Gibe reservoir and location of Gilgel Gibe catchment in Ethiopia.

Catena	Slope	Soil	Landuse	Spot	Texture
1	7.6	Vertisol	Fallow	1	Silty clay loam
				2	Clay
				3	Clay
2	2.6	Nitisol	Grazing	1	Clay
				2	Clay
				3	Clay
3	8.2	Nitisol	Deforested	1	Clay
				2	Clay
				3	Clay
4	4	Nitisol	Teff	1	Clay
				2	Clay
				3	Clay
5	4.6	Vertisol	Grazing	1	Clay
				2	Clay
				3	Clay
6	2.3	Vertisol	Teff	1	Clay
				2	Clay
				3	Clay

Table 1: Soil properties for each catena.

Soil type	Soil test P (ppm)	Soil clay %	SHC, ks, cm /day	Soil OM %	Al(g/kg)	Fe(g/kg)	C %	N %	BS (%)	Unstable aggregates (%)	Fertilizer applied kg/ha/yr
Nitisol	510	70	19.2	2.8	64.8	88.5	2.6	0.25	96.1	23.2	92
Vertisol	290	54	36.7	2.2	42.4	54.3	2.1	0.19	99.2	41.3	92

Table 2: General characteristics of the arable lands (BS: Base Saturation, OM: Organic Matter, SHC: Saturated Hydraulic Conductivity).

Freundlich Isotherm parameters				Langmuir Isotherm parameters			
Soil Type	P(mg kg^{-1})	k	1/n	r^2	Soil Type	P(mg kg^{-1})	b	k	r^2
Nitisol	469	943	0.17	0.94	Nitisol	469	2931	0.08	0.92
Vertisol	273	513	0.18	0.95	Vertisol	273	1629	0.07	0.9

Table 3: P-sorption data as explained by Freundlich and Langmuir sorption isotherms.

carried out for several variables where the model requires. Standard laboratory methods were employed for all experiments. A summary of the soil characteristics and other relevant information about the fields are indicated in table 1and 2. To run the model, an average crop P uptake of 7.6 kg/ha was considered based on literatures. Even though pastureland is available in this catena, only croplands were considered for simulating the P loss due to the assumption that the inorganic fertilizer from the croplands causes the problem. However, pasturelands with higher cattle density could contribute a considerable amount of phosphorus from dung. In this catchment, cattle dung is used for energy purpose by the residents so that picking the dung from the field is common.

Soil Phosphorus sorption: Batch experiments were conducted with 7 P concentrations ranging from 0 to 500 mg/l, and the adsorption isotherms were evaluated using Freundlich and Langmuir models. As it is shown in table 3 and figure 2, P sorption fit to both the Freundlich and Langmuir isotherms. The result is indicative that the data best fit to the Freundlich isotherm (Vertisol, $r^2 = 0.95$ and Nitisol, $r^2 = 0.94$) though the Langmuir isotherm also explained the process very well (Vertisol, $r^2 = 0.9$ and Nitisol, $r^2 = 0.92$).

The sorption maximum of nitisol soil was significantly higher than the vertisol. This implied that P sorbed into the two soils also varies by which the higher sorption tendency of P to nitisol soils observed. This phenomenon was associated to soil chemical characteristics (clay content, Al and Fe concentrations). Accordingly, results and discussion plausibly conclude that nitisol soils had shown a higher tendency to

fix P. Similar results are also found for such a soil type [10, 11 and 12].

Annual rainfall and runoff: While studying the transfer of phosphorus from soils into the nearby aquatic environment, it is very important to have rainfall-runoff data which determines the major P loss dynamics. Figure 3 shows the average annual rainfall-runoff data used in this study. Gilgel gibe is well known for receiving higher precipitation relative to other part of the country. The dominant clay texture of the area, gentle steepness, and lower land cover cause a moderate to higher runoff annually causing large amount of soil erosion.

The runoff coefficient is determined using curve number method based on the lands feature which is approximated 0.25. The annual runoff then calculated using the simple method where it is calculated as a product of annual runoff volume, and a runoff coefficient (Rv). Runoff volume is calculated as:

$$R = P * Pj * Rv \hspace{3cm} \text{Equation 1}$$

Where: R = Annual runoff (mm), P = Annual rainfall (mm), Pj = Fraction of annual rainfall events that produce runoff (usually 0.9), Rv = Runoff coefficient.

APLE Model and Modelling Approach

Model Description: The APLE model is an empirical model, with process-based equations based only on experimental data and not spatially explicit model that runs on an annual time step. The model simulates sediment bound and dissolved P loss in surface runoff. It

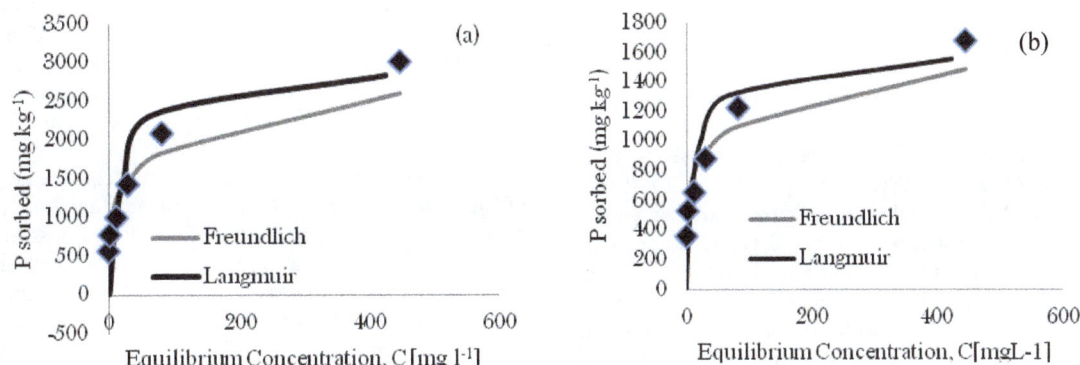

Figure 2: Sorption isotherms for (a) nitisol and (b) vertisol soils.

does not consider subsurface loss of P through leaching to groundwater or artificial drainage networks. It simulates edge-of-field P loss for uniform fields of several hectares in size, or smaller. APLE does not simulate P loss through grassed waterways or buffers that may occur beyond the field edge. The model considers different kinds of animal manure, applied either by machine or by grazing beef or dairy cattle, but considers only highly soluble commercial fertilizers such as superphosphate, triple superphosphate, or mono- and di-ammonium phosphate. APLE model calculates annual total surface P loss from agricultural fields as:

$$P_{tot} = DP_{man} + DP_{fert} + DP_{soil} + P_{sed} \qquad \text{Equation 2}$$

Where P_{tot} is the total annual P loss from surface runoff (kg ha^{-1}),

DP_{man} is annual dissolved P loss in runoff from applied manure (kg ha^{-1}),

DP_{fert} is annual dissolved P loss in runoff from applied fertilizer (kg ha^{-1}),

DP_{soil} is annual dissolved P loss in runoff from soil (kg ha^{-1}), and

P_{sed} is annual sediment P loss from eroded soil (kg ha^{-1}).

The model gives a detailed information about fertilizer dissolved P loss in runoff, sediment bound and dissolved phosphorus runoff from soil, soil phosphorus processes, soil Mixing between topsoil layers and phosphorus leaching from topsoil layers. APLE mixes P between the two topsoil layers based on the user-defined degree of soil mixing based on tillage or natural mixing processes, such as mixing by earthworms or freeze-thaw actions. It also estimates a concentration of dissolved P (mg L^{-1}) in the soil leachate based on a phosphorus sorption isotherm, which relates the amount of P sorbed on the soil and the amount dissolved in the soil water. A detail of the model can be obtained from the APLE theoretical documentation [13].

APLE model set up: APLE requires soil, rainfall, runoff, erosion, and fertilizer application, soil mixing method and depth and annual crop P uptake data. Soil data include the soil test P, clay content, organic matter and depth of the top two soil layers. In order to set up the model, the data were obtained from regional agricultural offices, farmers' interview, field measurements, Ethiopian Institute of Agricultural Research (Jimma Branch). Soil properties (chemical and physical) and soil test P were obtained from laboratory analysis. Rainfall data were collected from the nearby Asendabo and Jimma metrology stations. Sediment data were obtained from Vlamse Interuniversiteit Raad-Flemish Inter University Cooperation (VLIR-IUC) projects soil degradability study team. The model was set for the period of 10

years from 2001 to 2010 on arable land. It is assumed that P due to manure application was set to nil because of no application of manure. Moreover, the role of cattle dung during grazing is also negligible due to the collection practice (researcher observation) for energy purpose by the residents and weak practice of composting the farmlands. Therefore, the outputs of this study fundamentally consider the phosphorus from inorganic fertilizers and soils. Because of the nature of the model, calibration is not required. However validating the model output is essential and P sorption data were used to check whether the model output is realistic or not. This was done by comparing the P lost from two different lands with different soil, i.e. nitisol and vertisol. The limitation of this study design could be associated with the data quality for runoff and sediment. This is due to the assumption made for runoff computation and the sediment data assumed to be similar with the actual field data which was obtained from the nearby fields. The simulated model output data analyses were dispatched using Minitab version 16.00 statistical packages. Descriptive and non-parametric inferential (Mann-Whitney Test, ANOVA, Correlation) statistics were done for different variables. In order to show the presence of water quality problem due to P, random water samples were collected by grab sampling technique from the river and reservoir to evaluate the concentration of Orthophosphates phosphorus, the analyses were done in the laboratory according to standard methods.

Results and Discussion

Water quality

The fate of most diffuse source pollutants in agricultural catchment is the aquatic environment found in proximity. Particularly in agricultural catchments, the main nutrients that diffuse are nitrogen and phosphorus. In this catchment the intensive agricultural practice with the presence of the rugged topography, makes the movement of nutrients facilitated. In addition to these, there is no strategy to test the nutrient status of the soil before applying fertilizers. These and others made the situation very serious and attention seeking. As shown in table 4, the average concentration of phosphate is thirty-four times greater than the permissible value of 0.025 mg/L for lakes and reservoir [14]. This finding confirms that the reservoir is highly influenced by the intrusion of phosphorus from diffuse sources that cause severe water quality and aquatic ecosystem perturbations. Similarly Gilgel gibe reservoir exhibited higher likelihood of having a rapid nutrient enrichment [15]. Vividly all these facts are indicative of the presence of phosphorus enrichment. The following results and discussion below are taking this finding as a footstep to understand the transport phenomena of phosphorus in this catchment.

Phosphorus loss

Total P loss: Phosphorus loss from agricultural lands is commonly controlled by the hydrologic events, such as surface runoff. The runoff can transport P as sediment bound (particulate) or dissolved form. In this study, an average of 12.66 ± 0.7 kg ha^{-1} yr^{-1} phosphorus is lost from the nitisol dominated lands. This value was based on yearly soil erosion rate of 2.75 t ha^{-1} and runoff coefficient of 0.25 independent of soil type. The total P loss is the sum of the sediment bound, soil dissolved and fertilizer runoff loss. A field scale quantitative nutrient flow analysis reported that 7-11 kg ha^{-1}yr^{-1} of P is lost from cropland at the lowlands of this catchment where exactly the investigated fields for this study are found [16]. This showed that the finding of this study has shown a strong agreement with the reported one. Positive correlation was found between runoff and total P loss (r^2= 0.89, p-value = 0.001). These variables are also correlated positively and strongly in other catchments [17].

Particulate P loss: Phosphorus is well understood to its affinity to bind with the soil that affects its movement on surface and subsurface water. Lands with a higher concentration of aluminum and iron oxides, potentially bind P where its movement depends on this property. From this study field, particulate P contributes the largest portion (7.1 ± 0.3 kg ha^{-1}) of its total loss. Analysis of variance within the nitisol group revealed that significant variation observed between the sediment and soil dissolved P loss (p-value= 0.000) which is attributed to the soil chemical and physical properties that control the phosphorus dynamics. The higher clay content, aluminum and iron oxide in nitisol soils favor high adsorption of P (p-value= 0.001) [10-12,18]. Consequently, the prevailing hydrological events, such as runoff and erosion transport P at large. Henceforth, the high P sorption affinity to soil in the presence of surface runoff would result in a magnificent sediment P loss.

Total dissolved P loss: The total dissolved P (soil dissolved and fertilizer runoff P) loss (4.5 ± 0.6 kg ha^{-1}) was also found as considerable amount on the overall phosphorus loss. It was found strongly correlated with run off (r^2 = 0.91, p-value= 0.000) which is also discussed in other literatures [17,19]. In conclusion, this study found that the major phosphorus loss mechanism was particulate P (56%). The remaining fraction of P transported as soil dissolved P and direct fertilizer runoff (36% of P is lost as soil dissolved P and 8% as fertilizer runoff). Research output indicate that that up to 86% of the total P load per annum is contributed in the form of sediment bound P, which supports the present finding in this catchment [20].

P pools: While understanding all the facts and operating processes on the P loss at agricultural fields, it is important to identify the pools whereby P is stored. Therefore, major phosphorus pools that contribute to the studied transport dynamics, the model had simulated the three inorganic P pools (Labile, Active, and Stable) and one Organic P pool.

Looking into figure 4 the largest portion of P lost through erosion is contributed from the soil stable P pools due to the sorption process taking place in this pool; followed by active and labile P pools. On the contrary, the lowest contributor is the organic P pool where this tendency is related to the lower soil organic matter. This implies that the high iron and aluminum oxides in the soil mainly affected the P behavior where sorption occurs. Consequently the prevailing erosion would transport P mainly for the overall transfer from lands into rivers.

Model sensitivity and Uncertainty analysis

This section examines the effect of model input variables in APLE model and identifies the degree of sensitivity observed and uncertainties associated with the model predictors. Sensitivity and uncertainty analysis done for APLE model using First Order Approximation (FOA) and Monte Carol Simulation (MCS) showed that the MCS was found better for its overall uncertainty analysis [21]. Therefore, in this study the MCS technique was employed after reducing the DP$_{man}$ component of equation 2 due to the absence of manure application on the study area. There are nine parameters which affect P loss in APLE model. A sensitivity analysis was done on five model predictors, which are significantly determining the P loss in this catchment. Consequently, the sediment, runoff coefficient, runoff, labile P and fertilizer applied parameters were found to be the most sensitive parameters. The input parameter values were obtained from independent measurements and records. To make the simulated output credible and the model prediction valid, the uncertainty analyses were performed using Monte Carlo simulation with an adequate number of simulations (i.e. 200000) by assuming a triangular distribution of uncertainty which gives better distribution of data sets [21]. The model prediction uncertainties were calculated using the same input data sets as of sensitivity analysis (Table 5). As it is shown in table 6, high uncertainty value is associated with annual runoff, erosion and runoff coefficient parameters. Relatively lower uncertainties can occur due to the remaining parameters. Uncertainties in measured P loss data are a function of errors introduced when measuring runoff, erosion, and concentration of P in solution and attached to sediment [21]. After Monte Carlo simulation, the predicted value of P loss to 95% CI [10.59, 10.94] was 10.77 ± 0.17 kg ha^{-1}. Comparing with the initial APLE model output for total P loss value (12.67 ± 0.7 kg ha^{-1}), and nearly 1.9 ± 1 kg ha^{-1} variation was observed. This implied that there is a comparable result between the initial simulated P loss and P loss after MCS. Though uncertainty analysis was incorporated, the incorporation of uncertainties with P model predictions is still not standard practice [5]. Even though these arguments have been in place, in this study the need for uncertainty analysis will play a great role to make the output result credible and robust model prediction.

Comparison of P loss from nitisol and vertisol soils

The comparison of P loss from these two different soils aspires to check the logical consistency of the model output to that of the experienced fact. The logic behind this argument was that, these two soils have different clay and mineralogical content so that their phosphorus sorption capacity would also be different. This prompts to examine different response from the two soils after running the model. Since APLE simulate P loss in the form of sediment and dissolution via runoff, the P sorbed to soils would prefer the sediment transport and the dissolved P follow the dissolved transport mechanism. A statistical analysis showed that on average 12.66 ± 0.7 kg ha^{-1} yr^{-1}

Parameters	Mean	St Dev	Min	Max
Temperature	21.3	2.08	19	23
pH	7.33	0.25	7.1	7.6
Turbidity(FTU)	100	11.14	90	112
Conductivity(μS/cm)	92.67	7.64	86	101
DO(mg/L)	6.03	0.68	5.5	6.8
BOD(mg/L)	2.47	0.4	2.1	2.9
Nitrate(mg N/L)	1.66	0.99	0.62	2.6
Phosphate(mg P/L)	0.86	0.32	0.5	1.1

Table 4: Water quality status of the Gilgel Gibe reservoir for three different seasons in 2013 and in river P concentration (Min: Minimum, Max: Maximum, BOD$_5$: 5-day Biological Oxygen Demand, DO: Dissolved Oxygen).

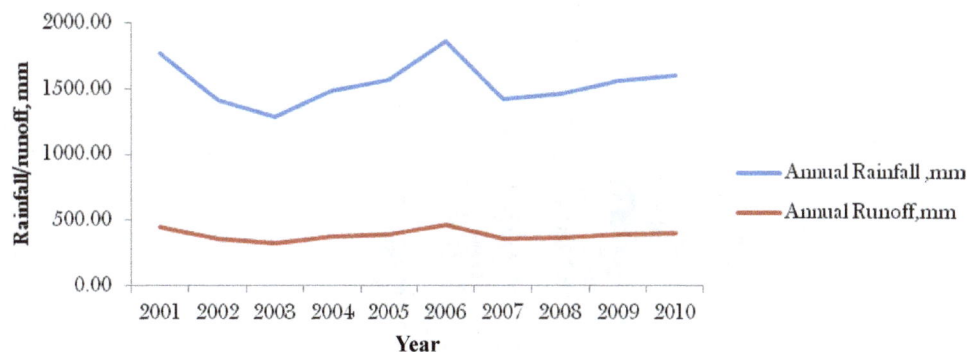

Figure 3: Average annual rainfall-runoff curve.

Sensitivity Rank	Model Variable	Input value	Variable range	Lower and Upper boundary
1	SED, kg ha^{-1}	2790.4	0.47 to 1.34×10^5	±15
2	RO/PT%	25	Dec-40	±25
3	RO, mm	385.8	0-720	±25
4	LP, mg kg^{-1}	173.4	5.5-500	±15
5	FERT$_{TP}$	80.75	0-150	±15

FERT$_{TP}$-total phosphorus applied, LP-labile phosphorus, RO annual runoff, RO/PT- runoff/precipitation; SED- erosion.

Table 5: Input data set, variable range and boundaries.

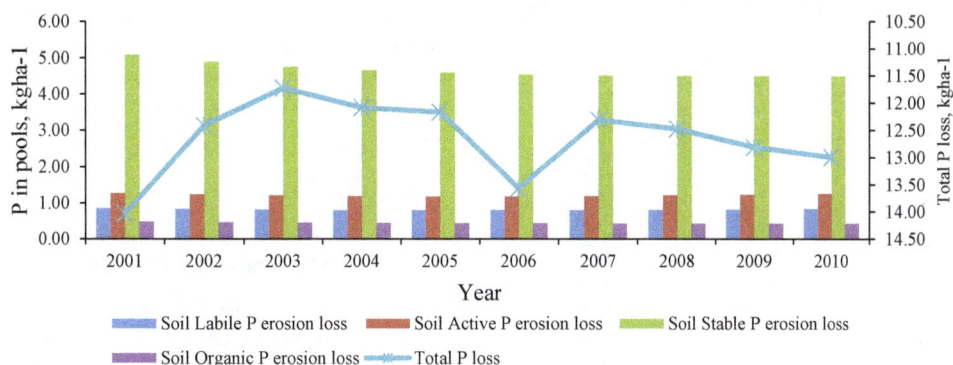

Figure 4: P pools versus total P loss.

phosphorus is lost from the nitisol based fields and 9.37 ± 0.5 kg ha^{-1} yr^{-1} phosphorus from vertisol lands which differ significantly(p-value= 0.0002). Accordingly, particulate P loss from nitisol and vertisol is 7.1 ± 0.3 kg ha^{-1} and 5.7 ± 0.2 kg ha^{-1} respectively (p-value = 0.0001). This variation can be attributed to the high sorption tendency of the nitisol than the vertisol so that in the presence of runoff higher particulate P could be lost from nitisol fields. The higher clay content, aluminum and iron oxide in nitisol soils favor high adsorption of P (p-value= 0.001) [10-12,18]. In both soils, the P loss follows the dominant mode of sediment transport to the nearby receiving pools. Similar findings are reported in literatures [17,22,23]. Together with other hydrology and soil properties of the area, the movement of P is highly facilitated. For instance, the lower water permeability of the clay-dominated soils with the prevailing precipitation would ease P transport process via runoff. The APLE model that simulates the movement of P through surface runoff contains modules for sorption phenomenon based on the soil texture of the land considered. Consequently, the nitisols of the catchment should contribute a larger amount of P in comparison with the vertisol. Therefore, the nitisol soils with a higher sorption power

were found to loss higher amount of particulate P that validates the logical output of the APLE model. However, to give the full picture of validation (both predictive and structural validity) of the model output, the P loss data are a mandatory, which is not available in the case of this study.

Evaluation of factors for P loss using best management practices (BMPs) principles

Though P is essential nutrient for productive crop and livestock agriculture, its loss causes eutrophication of receiving surface waters. The best management practices (BMPs) to mitigate P export to surface water include soil and water conservation practices, other management techniques, and social actions appropriate for specific agronomic, environmental, and socioeconomic conditions. Source BMPs are designed to minimize P available to runoff, managing fertilizer to lower its soluble P content and reduce farm P imports. Transport BMPs are designed to limit runoff, erosion, and leaching as important pathways of P loss. These include such practices as conservation tillage, terracing, and stream buffers. Because source and transport BMPs do not address

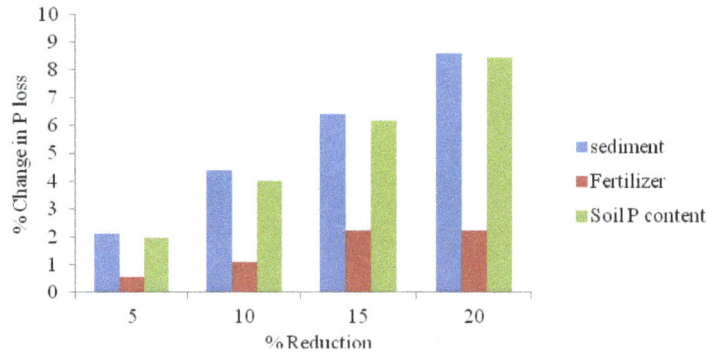

Figure 5: Change in P loss upon reduction of causing factors.

Parameters	Uncertainties	Relative contribution
Soil labile P	5.7×10^{-2} kg/ha	0.08
Annual runoff	0.23 kg/ha	0.34
Eroded sediment	0.21 kg/ha	0.31
Fertilizer P	6.7×10^{-2} kg/ha	0.09
RO/PT	0.11 kg/ha	0.16

Relative contribution of 1.00 is the sum of all the contributions by each parameter.

Table 6: Uncertainties of input parameters in prediction of P loss.

Sampling dates	Phosphate, C(mg/L)	Discharge, Q(m³/s)	Flux, (mg/l*m³/s)
Mar/9/2008	0.021	3.98	0.084
Mar/22/2008	0.18	4.22	0.76
Oct/13/2008	0.09	31.45	2.83
Nov/6/2009	0.05	9.03	0.45
Nov/19/2009	0.03	11.41	0.32
Jun/5/2010	0.4	55.32	22.13
Jun/15/2010	0.06	62.15	3.98
Jun/21/2010	0.02	64.13	1.22
Jun/27/2010	0.02	72.31	1.45
Oct/4/2010	0.12	52.64	6.32
Oct/19/2010	0.08	24.35	1.95
Jul/6/2011	0.09	81.68	7.35
Jul/22/2011	0.09	142.34	12.81
Sep/11/2012	0.11	2.44	0.27
Dec/9/2012	0.07	5.55	0.39
Mar/12/2013	0.23	8.49	1.95
Aug/24/2013	0.55	184.22	101.32
Jan/8/2014	0.13	10.65	1.38
Jul/3/2014	0.34	180.12	61.24

Table 7: Measured orthophosphate and discharge data from Gilgel gibe river downstream of Asendabo Bridge.

the main problem of farm and regional P surpluses, long-term solutions must extend beyond the farm gate [24]. As indicated in figure 5, while reducing the sediment by 5%, 10%, 15% and 20%, the simulated P loss reduction is 2.12%, 4.4%, 6.4% and 8.6% respectively. With similar percent reduction for fertilizer, a total P loss reduction ranges between 0.5 to 2.2%. The other factor, which has great implication, i.e. soil P content was also reduced from 5% to 20% and the simulated P loss reduction ranges from two to 8.5 percent. This evaluation assumes two stages to be followed. First applying different BMPs techniques and analyze the observed change in each variable; second is evaluating the retained P that can be transferred from the land. Therefore, the values are computed after the first assumption is met. This implies

that, the effort made to mitigate the P transfer into the nearby aquatic environment should give BMPs for sediment erosion and soluble P in the soil using standard practices such as conservation tillage, terracing, and stream buffers.

Phosphorus load estimation in Gilgel gibe river

To estimate phosphorus loading, it is necessary to sum the flux, which is expressed as mass per unit time, over the period of interest. Since the flux varies with time, this summing process can be expressed in integral form as shown in equation 3. Flux is often computed as the product of concentration and flow. Thus the three basic steps for estimating P load are: measuring water discharge, measuring P concentration, and calculating P loads (multiplying discharge times concentration over the time frame of interest).

The phosphorus load is the integral over time of the flux:

$$Pload = k \int_{t_1}^{t} flux(t)dt \qquad \text{Equation 3}$$

Where $flux(t) = C(t) \cdot Q(t)$, and k = is a constant for converting units (34.7 used for ton/year), and t is time.

As it is shown in table 7, there are P concentration and discharge data which gives the flux. Using equation 3 the P load is estimated from the beginning of the sampling date (March, 2008) until July 2014. The time interval between these periods is 77 months (6.4 years). Taking the average of both the concentration and discharge within the time interval ($t_1 = 0$ to t = 6.4 years), the average annual load is approximately 1660.6 ton. The cumulative load is determined by adding the calculated fluxes over all sampling intervals. Consequently, the cumulative load is nearly 7758.7 ton. Therefore, the Gilgel gibe reservoirs found to the downstream of the sampling station receive such an amount of P where the non-point source diffusion of P predominately contributes. The limitation to this approximation is the data inadequacy between the integral time intervals.

Conclusion

Gaining a detailed understanding of the operating processes in agricultural fields is important to explain how non-point sources of pollution affect the water quality of the aquatic environment. Phosphorus has been known to its potential damage to aquatic environment once found in excess. For decades there are several research conducted to study the transport and effect of P on the water bodies. However, still miles remain to develop efficient method that explicitly put the phosphorus fate. Therefore, studying the phosphorus transport has got an attention worldwide. In this study similar efforts were made to scrutinize the dominant transport mechanisms and amount transported through surface runoff. Accordingly, the result clearly indicated that, the main transport mechanism of P from the agricultural land is particulate mode of transport. However, a significant amount of P is also lost via dissolution. Consequently, the findings of this study conclude that the agricultural activities in conjunction with the rugged topography of the catchment rapidly affect the water quality of the reservoir. However, it found that applying the BMPs could also reduce substantial amount of P export. Particularly an effort made on sediment, soil P content and runoff generation reduction could bring positive impact to retain P. In conclusion, the management practices designed to control the P transport should give emphasis to control erosion and available soil P. The result of this finding also suggests that most tropical soils in Sub-Sahara countries may follow similar P loss pattern. Application timing and quantity of inorganic fertilizers could play a great role to achieve remarkable P transfer control using BMPs. Generally, attention should be given to the application of precision agriculture to avoid such problems.

Acknowledgment

We are highly thankful to Deutschen Akademischen Austauschdienst (DAAD) and Jimma University for financial support and Ethiopian National Metrology Agency. We would also thank Dr. Seid Tiku for the phosphorus concentration data.

References

1. Michael J, White a, Daniel E. Storm b, Philip R, et al (2010) A quantitative phosphorus loss assessment tool for agricultural fields. Environmental Modelling and Software. 25: 1121-1129.

2. Vadas PA, Owens LB, Sharpley AN (2008) An empirical model for dissolved phosphorus in runoff from surface-applied fertilizers. Agriculture, Ecosystems and Environment.

3. Zaimes GN, Schultz RC (2002) Phosphorus in Agricultural Watersheds. A Literature Review. 5-33.

4. Ambelu A, Koen Lock, Goethals PLM (2013) Hydrological and anthropogenic influence in the Gilgel Gibe I reservoir (Ethiopia) on macro invertebrate assemblages. Lake and Reservoir Management. 29: 143-150.

5. Radcliffe DE, Freer J and Schoumans O (2009) Diffuse Phosphorus Models in the United States and Europe: Their Usages, Scales, and Uncertainties. J Environ Qual 38: 1956-1967.

6. Vadas PA, Joern BC, Moore PA Jr (2012) Simulating Soil Phosphorus Dynamics for a Phosphorus Loss Quantification Tool. J Environ 41: 1750-1757.

7. Lemunyon JL, Gilbert RG (1993) The concept and need for a phosphorus assessment tool. Journal of Production Agriculture. 6: 483-486.

8. Veith T, Sharpley A, Weld J, Gburek W (2005) Comparison of measured and simulated phosphorus losses with indexed site vulnerability. Transactions of the ASAE. 48: 557-565.

9. [EELPA]-Ministry of Mines and Energy Ethiopian Electric Light and Power Authority (1997) Gilgel Gibe hydroelectric project. Environmental assessment main report. Ethiopia (Addis Ababa) 15-20.

10. Hadgu F, Gebrekidan H, Kibret K, Yitaferu B (2014) Study of phosphorus adsorption and its relationship with soil properties, analysed with Langmuir and Freundlich models. Agriculture, Forestry and Fisheries. 3: 40-51.

11. Behn C, Janssen M, Yalemsew GA, Lennartz B (2013) Phosphorus contents and phosphorous sorption in soils of the Gilgel Gibe catchment, SW Ethiopia. Geophysical Research Abstracts. 15: 8308.

12. Chimdi A, Gebrekidan H, Tadesse A, Kibret K (2013) Phosphorus Sorption Patterns of Soils from Different Land. American-Eurasian. Journal of Scientific Research 8: 109-116.

13. Vadas PA (2013) Annual Phosphorus Loss Estimator (1st edn.) Madison, Wisconsin, USA: U.S. Dairy Forage Research Center.

14. US EPA]-US Environmental Protection Agency (1986) Quality Criteria for Water. EPA 440/5, 86-001. Washington (DC).

15. Devi R, Tesfahune E, Legesse W, Deboch B. (2008). Assessment of siltation and nutrient enrichment of Gilgel Gibe dam, Southwestern Ethiopia. Bioresour Technol. 99: 975-979.

16. Nebiyu A (2011) The Sustainable Use of Soil Resources of Gilgel Gibe Dam Catchment. Proceeding of the national workshop in integrated Watershed management on gibe-Omo Basin. Prepared by Jimma University and PHE Ethiopia Consortium. 39-42.

17. Eghball B, Gilley JE (2001) Phosphorus risk assessment index evaluation using runoff measurement. Journal of Soil and Water Conservation. 56: 202-206.

18. McDowell RW, Sharpley AN, Condron LM, Haygarth PM, Brookes PC (2001) Processes controlling soil phosphorus release to runoff and implications for agricultural management. Nutrient Cycling in Agro ecosystems. 59: 269-284.

19. DeLaune PB, Moore PA Jr (2001) Predicting Annual Phosphorus Losses from Fields Using the Phosphorus Index for Pastures. Better Crops. 85: 16-23.

20. Prairie YT, Kalff J (1986) Effect of catchment size on phosphorus export. Water Resour. Bull. 22: 465-470.

21. Bolster CH, Vedas PA (2013) Sensitivity and Uncertainty Analysis for the Annual Phosphorus Loss Estimator Model. J Environ Qual 42: 1109-1118.

22. David MB, Gentry LE (2000) Anthropogenic inputs of nitrogen and phosphorus and riverine export for Illinois, USA. J Environ Qual 29: 494-508.

23. Vaithiyanathan P, Correll DL (1992) The Rhode River Watershed: Phosphorus distribution and export in forest and agricultural soils. J Environ Qual 21: 280-288.

24. Sharpley AN, Daniel T, Gibson G, Bundy L, Cabrera M, et al. (2006) Best Management Practices to Minimize Agricultural Phosphorus Impacts on Water Quality. U.S. Department of Agriculture, Agricultural Research Service, ARS 163: 50.

5

Watershed Response to Bias-Corrected Improved Skilled Precipitation and Temperature under Future Climate - A Case Study on Spencer Creek Watershed, Ontario, Canada

Sadik Ahmed and Ioannis Tsanis*

Department of Civil Engineering, McMaster University, Hamilton, Ontario L8S 4L7, Canada

Abstract

It is widely acknowledged that the statistical properties of precipitation and temperature will change under the future climate condition, and this will cause a significant impact on water resources and its management at watershed scale. This study investigated the hydrological response to climate change for Spencer Creek watershed located in Southern Ontario, Canada. The precipitation and temperature projection used in this study were obtained from the North American Regional Climate Change Assessment Program (NARCCAP) climate simulations. NARCCAP climate projections were bias- corrected for meteorological stations representative of the watershed. The bias-corrected NARCCAP climate projections were used as input in a calibrated hydrological model Hydrologiska Byråns Vattenbalans-avdelning (HBV) to simulate flows at the outlet of the watershed. The improvement of bias-corrected NARCCAP precipitation and temperature is revealed by Brier and Rank Probability Skill Score (BSS and RPSS, respectively). The comparison of current and future simulated flow results reveals an increase in winter daily average flows and decrease in other seasons, and approximately 13% increase in annual evapotranspiration under future climate condition. An increase in high flows and decrease in low flows under future climate is revealed by flow-duration analysis.

Keywords: Climate change; Bias correction; Hydrology; Watershed; Canada

Introduction

The recent Intergovernmental Panel on Climate Change (IPCC) Assessment Report [1] indicates that our climate is undergoing substantial warming, and it is likely that an increasing trend of extreme precipitation will continue. The watershed hydrology will be affected by climate change in many ways because the hydrological cycle is linked with changes in atmospheric temperature and radiative fluxes [2]. The changes in temperature will have a significant effect on the hydrological processes that involve precipitation, snowmelt, evapotranspiration, soil moisture and flow. The prediction of the forthcoming climate change on hydrological processes is vital in water resources management and planning. In this study, climate change impact on hydrological processes has been performed by forcing climate model output to a hydrological model in order to evaluate changes in future flow in the Spencer Creek watershed located in Southern Ontario.

In the last decade, researchers as well as users have shown particular interest in the hydrological impact of climate change. Past research on climate change impact assessment revealed that the hydrological regime of different watersheds could be significantly modified due to the anticipated changes in temperature and precipitation under future climate during the present century [3-5]. The assessment results of climate change impact on hydrology at the watershed scale vary significantly with the climate model projections, greenhouse gas emission scenarios, data downscaling/ correction techniques, and hydrologic models. Grillakis et al. [3] examined the climate change impact on future hydrology of Spencer Creek watershed. The study revealed inter-annual trends for precipitation and temperature both in the past data and future simulation. The analysis shows an annual average precipitation increase by approximately 10% to 15% and temperature increase by approximately +2.2°C and +2.3°C at Hamilton Airport and Hamilton RBG. The study also shows that the yearly average flow at Spencer Creek at Dundas increases by about 12% when future projected flows are

compared with the observed flow. Sultana and Coulibaly [4] assessed the climate change impact on hydrological processes of this watershed using a distributed coupled MIKE SHE/MIKE 11 hydrologic model and the projected daily precipitation and temperature from Canadian global climate model (CGCM 3.1). The downscaled GCM predictions show a 14-17% increase in the annual mean precipitation and 2-3°C increase in annual mean maximum temperature. The coupled hydrologic model predicted about 1-5% annual decrease in snow storage, 1-10% increase in annual ET, 0.5-6% decrease in the annual groundwater recharge, 10-25% increase in annual stream flows for all sites for the 2050s when downscaled GCM scenarios were used. Boyer et al. [5] assessed the impact of climate change on the hydrology of St. Lawrence tributaries (Quebec, Canada) located about 650 km northeast of Spencer Creek. The hydrological model HSAMI was used to produce flow in the future by inputting GCM projections for three 30 year horizons (2010-2039, 2040-2069 and 2070-2099, respectively referred to as 2020s, 2050s and 2080s). The future daily climate (precipitation and temperature) for three 30 year horizons were produced by adding anomalies (monthly mean difference between GCMs in the future and the reference period 1961-1990) to the observed temperature and precipitation during the reference period. The study results indicate that the regime will gradually shift from snow to rain. Most of the future flow simulations show an increase in winter discharge and a decrease in spring discharge.

*Corresponding author: Ioannis Tsanis, Department of Civil Engineering, McMaster University, 1280 Main Street West, Hamilton, Ontario L8S 4L7, Canada
E-mail: tsanis@mcmaster.ca

The study results also show that the center volume date for the winter/spring period is expected to be in advance 22-34 days depending on the location of the watershed.

The most widely used approach to predict climate change impact on hydrological processes is done by inputting climate model simulations into hydrological models. The climate model (GCM or RCM) provides gridded data, and the climate projected from it is not the same as the climate coming from the observations. Therefore, modelers use different techniques for establishing relationship between climate model outputs and observations for correcting the climate model projections both for current and future period to get more realistic results from the hydrological model. A number of dynamical and statistical downscaling methods are available to downscale climate model gridded data at the target points where the meteorological or rainfall stations are located [6-10]. Sharma et al. [11] examined the necessity of correction of raw RCM data by using a statistical downscaling method (SDSM) and a data-driven technique called a time-lagged feedforward network (TLFN) on raw CRCM4.2 data. They revealed that the downscaling did improve raw RCM precipitation, and consequently, the downscaled CRCM4.2 data improves the HBV hydrologic model ability to simulate streamflow accurately as compared to the use of the raw CRCM4.2 data. Although the statistical downscaling methods have been used in many studies, the application and calibration of this method are complex and highly dependent on expert judgment [12]. The regional climate models (RCMs), generated from dynamical downscaling methods, provide climate projections at much finer scale that is largely used in hydrological impact studies in many watersheds around the world. However, recent studies [11,13] revealed that there are systematic differences between the raw RCMs output and the observations, and the bias-correction methods alternative to statistical and dynamical downscaling method has shown effectiveness in removing the bias between raw RCMs output and the observations [13,14]. The bias correction methods used by Ines and Hansen [15] and Samuel et al. [13] have been used in this study for correcting the NARCCAP climate model output. One of the novelty of this study is that two probabilistic verification measures, namely the Brier skill score (BSS) and the rank probability skill score (RPSS) have been used in this study to assess the improvement of NARCCAP precipitation and temperature data when bias correction method was applied.

The availability of higher spatial and temporal resolution climate data, provided by the NARCCAP created from multiple GCMs and RCMs, has facilitated the climate change impact studies. Using ensemble climate model data will provide multiple possible estimations of flow regime, which assists the water manager towards a sustainable planning and design. Because of the high uncertainty in the climate model projections, Mearns et al. [16] emphasized on the use of ensemble climate model projections for climate change impact study by using climate model simulations. NARCCAP provides both precipitation and temperature time series for both current and future period for eight RCM+GCM pairs at same spatial scale. All the available climate model data have been used in this study. A number of hydrological models have been used by the researchers for climate change impact studies in different countries. In this study, a semi-distributed conceptual model, HBV, was chosen for hydrologic simulation using bias-corrected NARCCAP projections. The motivation of choosing this particular model is that the model was used in previous studies [3,17,18] on Canadian watersheds and showed a good performance.

These recent studies on hydrological impact analysis indicate an overall increasing trend in the mean annual flow in Canadian watersheds. However, further investigation of extreme events such as high and low flow analyses is required. This study focused on the investigation of climate change impact on high and low flows using a number of climate model simulations. The overall objective of this study is to investigate the climate change impact on hydrological processes by using bias-corrected NARCCAP climate model projections for Spencer Creek watershed located in Southern Ontario, Canada. The objective was achieved by correcting the bias of raw NARCCAP precipitation and temperature time series, assessment of improvement in bias-corrected NARCCAP projections, performing hydrologic simulation and assessment of flow regime under current and future climate conditions.

Study Area and Data

Study area

The case study area for this study is Spencer Creek watershed located in the Southern Ontario, Canada and is shown in Figure 1. The watershed has an area of 160.4 km². The surface runoff in the watershed is collected by an extensive network of rivers and stream and discharged into Cootes Paradise at the western end of Lake Ontario. The land-use of the study area can be also characterized by agricultural land use, forest area, wetlands and the urban and paved area in the lower part of the watershed. The watershed is complex because of its extensive river and stream network, heterogeneous soil property and diverse land use [19].

Observed hydro-meteorological data

The observed daily precipitation (total precipitation in the form of liquid and snow, measured in mm) and temperature (in °C) data were obtained from meteorological stations; namely the Hamilton Airport, Hamilton RGB, Hamilton RBG CS meteorological station. The meteorological data for 1971-2014 at the stations were collected

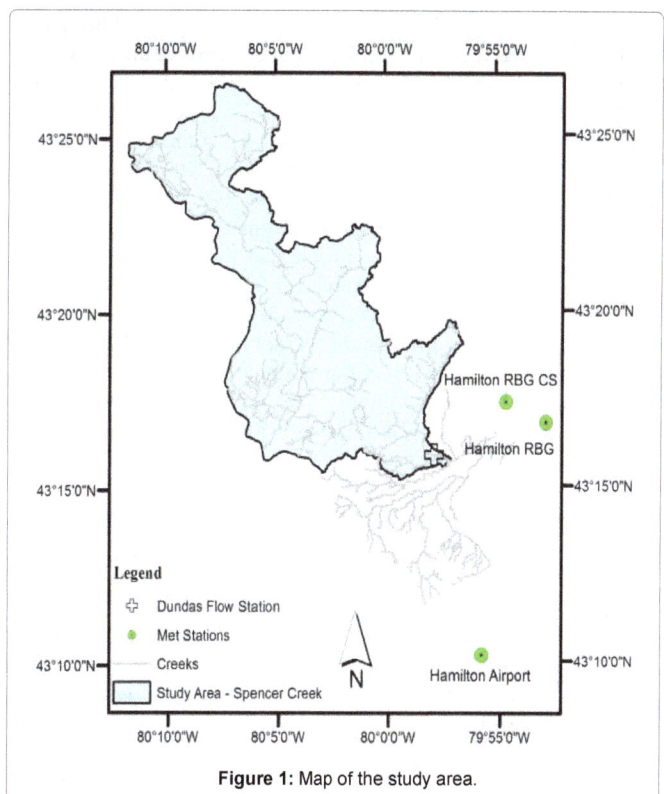

Figure 1: Map of the study area.

from Environment Canada. The observed daily flow data for 30 years, from 1985 to 2014, were obtained for a hydrometric station namely Spencer Creek at Dundas (station ID 02HB007) located at latitude and longitude of 43.27°N and 79.96°W, respectively. The daily flow data were collected from Water Survey Canada. The climate of the study area is humid-continental. Based on the meteorological data from 1971 to 2014 at Hamilton Airport, the daily average maximum and minimum temperatures are 13.4°C and 4°C, and extreme maximum 37.4°C and extreme minimum temperature -30 °C were observed on 7 July, 1988 and 16 January, 2004, respectively. The yearly average precipitation is 893.2 mm based on data from 1971 to 2014 at Hamilton Airport, and the maximum daily rainfall and precipitation 107 mm were observed on 26 July, 1989. The yearly average flow is 2.02 m³/s with highest and lowest monthly average of 4.14 m³/s and 0.59 m³/s on March and August, respectively and the maximum daily average flow 32.4 m³/s was observed on 14 March 2010. These values were obtained based on the available daily time series data from 1985 to 2014 observed at hydrometric station namely Spencer Creek at Dundas.

NARCCAP climate data

The North American Regional Climate Change Assessment Program (NARCCAP) [20,21] is an international program that serves the high resolution climate scenario for the United States, Canada, and Northern Mexico. It provides the data sets in order to investigate uncertainties in regional scale projections of future climate and generate climate change scenarios for use in impacts research. All the NARCCAP future simulations are driven by a GCM that follows greenhouse gas and aerosol concentration based on A2 emission scenario described in the Special Report on Emissions Scenarios (SRES) [22]. NARCCAP provides data produced by several RCM+GCM pairs, and this study used eight RCM+GCM pairs simulated precipitation and temperature time series. The names of the RCMs and GCMs/drivers produced the data, used in this study, are listed in Table 1.

The NARCCAP output data are provided at a gridded horizontal resolution of 50 km, and the precipitation and temperature (maximum and minimum) are provided for three hourly and daily temporal resolutions, respectively. The NARCCAP experimental output spans for two time periods of 33 years – the first time span is for the current/historical period spanning from 1968 to 2000, and the second time span is for the future span from 2038-2070. These two periods permit assessment of mid twenty-first century changes relative to late twentieth century climate. It is notable that the first three years, the spin-up periods [23], of both current and future simulation have been discarded in this study. NARCCAP data are stored in the NetCDF files in 2D arrays. The array dimensions (yc, xc) for the Hamilton Airport, Hamilton RBG/Hamilton RBG CS are found from the grid cell maps for each RCM. The array dimensions (yc, xc) of nearest point of Hamilton Airport for CRCM, HRM3, RCM3 and WRFG are (51,100), (57, 105), (44, 94) and (48, 93), respectively, and array dimensions (yc, xc) of

nearest point of Hamilton RBG/Hamilton RBG CS for CRCM, HRM3 and RCM3, WRFG are (51,100), (58, 105), (45, 93) and (48, 93).

Methodology

The procedure followed in this study involves (1) bias correction of NARCCAP precipitation and temperature time series data and analysis of skill score; (2) transforming bias-corrected precipitation and temperature into flows and evapotranspiration using a hydrological model, and (3) comparing hydrologic regime under current and future climate.

Bias correction

The NARCCAP temperature and precipitation data are gridded areal average, and not point estimates. Bias correction method is used to remove bias between climate model simulated data and observation at a point location to get more accurate results from the hydrological model when NARCCAP data are inputted.

The bias-correction method presented by Ines and Hansen [15] was used to correct the frequency and the intensity of daily precipitation of NARCCAP. This two-step procedure corrects the frequency of daily precipitation at first, and then it corrects the intensity for each of 12 calendar months. The mean precipitation $\bar{X}_{(m)}$ (mmd⁻¹) in calendar month m is the product of mean intensity, μ_1 (mm wd⁻¹)(wd⁻¹) (is wet day, for a threshold 0.1 mm) and relative frequency, π (wd d⁻¹). Therefore, the correction of any bias of these two components also corrects the monthly total precipitation. In this study, this bias-correction method was applied to remove the bias between the daily precipitation data from NARCCAP and observations at Hamilton Airport and Hamilton RBG meteorological stations.

In the first step in order to correct the frequency of precipitation, the empirical distribution of the raw NARCCAP precipitation was truncated above the $\bar{x}_{NARCCAP}$ threshold value, in such a way that the mean frequency of precipitation above the threshold matches the observed mean precipitation frequency. The threshold value $\bar{x}_{NARCCAP}$ are calculated from the observed and NARCCAP precipitation distributions as show in the following equation,

$$\bar{x}_{NARCCAP} = F_{NARCCAP}^{-1}\left(F_{obs}\left(\bar{x}\right)\right) \tag{1}$$

Where F (.) and F⁻¹(.) denotes the cumulative distribution function (CDF) and its inverse, and subscripts indicate NARCCAP precipitation forecasts or observed_daily precipitation. The threshold observed precipitation amount (x) of a day was set to 0.1 mm to define wet day.

In the second step to correct the intensity of precipitation, a two-parameter gamma distribution as shown in Equation 2 was used to fit the truncated daily NARCCAP and observed precipitation data, and then CDF of the truncated daily NARCCAP precipitation data are mapped to the CDF of the observed data as shown in Equation 3.

RCM+GCM Pairs	RCM	GCM/Drivers
CRCM+CCSM	Canadian Regional Climate Model [24]	Community Climate System Model [29]
CRCM+CGCM3	Canadian Regional Climate Model [24]	Third Generation Coupled Global Climate Model [30]
HRM3+GFDL	Hadley Regional Model 3 [25]	Geophysical Fluid Dynamics Laboratory GCM [31]
HRM3+HADCM3	Hadley Regional Model 3 [25]	Hadley Centre Coupled Model, version 3 [32,33]
RCM3+CGCM3	Regional Climate Model version 3 [26,27]	Third Generation Coupled Global Climate Model [30]
RCM3+GFDL	Regional Climate Model version 3 [26,27]	Geophysical Fluid Dynamics Laboratory GCM [31]
WRFG+CCSM	Weather Research Forecasting Model Grell [28]	Community Climate System Model [29]
WRFG+CGCM3	Weather Research Forecasting Model Grell [28]	Third Generation Coupled Global Climate Model [30]

Table 1: List of RCM+GCM data pairs used in this study.

$$F_G(x,\alpha,\beta) = \frac{1}{\beta^\alpha \tilde{A}(\alpha)} \int_{\bar{x}} x^{\alpha-1} exp\left(-\frac{x}{\beta}\right); x \geq \bar{x} \qquad (2)$$

$$F_G(x,\alpha,\beta) = \int_{\bar{x}} f(t)\,dt \qquad (3)$$

Where the shape parameter (α) and the scale parameter (β) of the gamma distribution are determined by Maximum Likelihood Estimation. The corrected NARCCAP precipitation amount x' on day i is calculated by substituting the fitted gamma CDFs into the following equation:

$$x_i' = \begin{cases} F_{I,obs}^{-1}\left(F_{I,NARCCAP}(x_i)\right) & x_i \geq \bar{x} \\ 0 & x_i < \bar{x} \end{cases} \qquad (4)$$

The bias in the NARCCAP temperature series was corrected using a method presented by Samuel et al. [13]. The distribution of the daily NARCCAP temperature was mapped onto the distribution of observed temperature for each of the 12 calendar months. In the case of temperature, correction of frequency distribution and truncation of the empirical distribution of the raw daily NARCCAP temperature data was not performed by using a normal distribution used in this bias correction method to map the temperature distribution. The CDF of the normal temperature distribution was calculated by using Equation 5. The CDF of the daily NARCCAP temperature are mapped to the CDF of the observed data using equation 6. The corrected NARCCAP temperature y' on day i is calculated by Equation 7:

$$F(y;\mu,\alpha,) = \frac{1}{2}\left[1 + erf\left(\frac{y-\mu}{\sqrt{2\alpha^2}}\right)\right]; y \in \Re \qquad (5)$$

$$F_G(y;\mu,\alpha) = \int_0^\Re f(t)\,dt \qquad (6)$$

$$y_i' = F_{T,OBS}^{-1}\left(F_{T,NARCCAP}(y_i)\right) \qquad (7)$$

Description of skill scores

Two probabilistic verification measures, namely the Brier skill score (BSS) and the rank probability skill score (RPSS), mostly used in the assessment of meteorological forecasts [34-36], were used in this study to assess the quality of bias-corrected climate model simulated precipitation and temperature time series. The BSS and RPSS are based on the Brier score (BS) and the rank probability score (RPS), respectively.

The Brier score [37], which is essentially the mean-square error of probabilistic forecasts, is the most commonly used scalar measure for probability forecasts. It is widely used for dichotomous predictands [35]. This score is also applied to continuous-valued forecast [38]. The continuous valued forecasts are converted into a binary event using a threshold filter which can either be exceeded or not [38,39]. In this study, for comparison purpose and consistency, 0.1 mm/day (threshold to define wet day) for precipitation, and the means of the daily mean temperature of each month for temperature are used as BS thresholds. The Brier score BS is calculated by the equation (8):

$$BS = \frac{1}{n}\sum_{k=1}^{n}(y_k - o_k)^2$$

Where n represents the number of days, k is the number of the n simulation/event pair, y_k is the simulation probability and o_k is the observed probability (occurrence and non-occurrence of the event being simulated). y_k is derived by the relative frequency of the ensemble members exceeding the chosen threshold. The observations o_k are translated similar to the simulated values, i.e., the observation $o_k=1$ if the event occurs (if the threshold is exceeded) and $o_k=0$ if the event does not occur. The Brier score ranges between 0 and 1 because the

observation and probability simulations are bounded by 0 and 1, a perfect simulation exhibiting BS=0 and less accurate forecasts receive higher Brier score. The Brier skill score (BSS) is computed using equation (9) in order to make comparison between a simulation relative to reference simulation:

$$BSS = \frac{BSS_{ref} - BSS}{BSS_{ref}} \qquad 9)$$

The RPS [35] is a score derived from the Brier score to the multi-category [40]. The RPS is calculated by equation (10):

$$RPS = \sum_{m=1}^{j}(Y_m - O_m)^2 \qquad (10)$$

Where, Y_m is the cumulative probability of the simulation for category m and O_m is the cumulative probability of the observation for category m. For a group of n forecasts, the RPS is the average ($\overline{RPS}$) of the n RPSs:

$$\overline{RPS} = \frac{1}{n}\sum_{k=1}^{n}RPS_k \qquad (11)$$

In this study, the procedure presented by Clark and Hay [41] and Gangopadhyay et al. [42] was used to calculate RPS: At first, the observed time series data are used to differentiate 10 (j) possible categories (i.e., the minimum value to the 10th percentile, the 10th percentile to the 20th percentile, the 20th percentile to the 30th percentile up to the 90th percentile to the maximum value). These categories were determined separately for each month. In the next step, the number of ensemble member simulation in each category is determined (out of 8 members), and their cumulative probabilities were computed for each simulation-observation pair. Then, in the same way, the observation's cumulative probabilities were computed. All categories below the observation's position are assigned '0', and all categories equal to and above the observation's position are assigned '1'. The RPS was determined as the squared difference between cumulative probabilities of the observations and simulation, and the summation of squared differences over 10 categories. RPS is zero for a perfect simulation and positive otherwise. The ranked probability skill score (RPSS) was calculated in order to make comparison between a simulation relative to a reference simulation:

$$RPSS = \frac{\overline{RPS_{ref}} - \overline{RPS}}{\overline{RPS_{ref}}} \qquad (12)$$

In this study, NARCCAP simulated raw data was used as the reference simulation to calculate BSS and RPSS. Here, the calculated BSS and RPSS show the percentage improvement of bias-corrected NARCCAP precipitation and daily mean temperature data over the NARCCAP simulated data.

Hydrologic modeling

HBV hydrologic model: Although hydrological models have been around for quite some time, there is yet to be one exclusive model that can stand apart from the rest and be declared best at modeling all aspects of the hydrologic system' [43]. A hydrologic model HBV [44] was chosen to simulate flows for current and future period at the outlet of the Spencer Creek Watershed. The model was developed at the Swedish Meteorological and Hydrological Institute (SMHI) and its first application dates back to the early 1970s [45]. The HBV model which includes conceptual numerical descriptions of hydrological processes at the catchment scale is best characterized as a semi-distributed conceptual hydrologic model. The model is usually run on the daily values of precipitation, temperature and estimates of potential evapotranspiration. Flow observations are used for calibration and validation of the model. For most of the applications, the model is run on a daily time step, but it is possible to use shorter time steps. The evapotranspiration values can be used as monthly averaged or daily values. The potential evapotranspiration is calculated using air temperature. The model contains routines for snow accumulation and melt, soil moisture accounting, runoff generation and a routing procedure. The snowmelt routine of the HBV model is a degree-day approach. It is based on air temperature, with a water holding capacity of snow which delays runoff. The soil moisture routine of the model controls runoff formation, accounts for soil field capacity and change in soil moisture storage due to rainfall/snowmelt and evapotranspiration. The excess water from the soil moisture zone transforms to runoff in the response routing. The response function of the model consists of two reservoir-one upper nonlinear, one lower linear, and one transform function. The runoff is computed by adding the contribution from the upper and lower reservoir, and the generated runoff is routed through a transformation function in order to get a proper shape of the hydrograph at the outlet of the watershed.

Model calibration and validation: The process of optimization of model parameters to minimize the difference between model output and observed data is referred to as calibration. A calibrated model needs to be verified for ensuring that the optimized parameters are a good representation of the physical behavior of the catchment. The parameters of the HBV model need to be calibrated in order to provide model output that closely resembles observed data as it is a conceptual model. The HBV manual [46] recommends using at least 10 years of data for the calibration period. It is also recommended to use 75% of total data for model calibration and 25% of data for model validation.

The first 22 years of data (from 1985 to 2006) were used to calibrate the hydrologic model and the last 8 years of data (2007-2014) were used to validate the model. The calibration and validation of the hydrologic model were carried out using the observed and simulated flow hydrograph of daily time step at the outlet (Spencer Creek at Dundas hydrometric station) of the watershed. Following the recommendation of the HBV manual during calibration, the evaluation of the results was mainly done by comparing the explained variance/ Nash and Sutcliffe coefficient R^2 [47], and visually inspecting and comparing the simulated and observed hydrographs. The Nash and Sutcliffe coefficient is the variance around the mean explained by the model. The optimum value of the Nash and Sutcliffe coefficient is one (1), and a value less than 0.7 represent poor performance [48]. The model calibration and validation results of the Spencer Creek watershed model show a good performance according to the Nash and Sutcliffe coefficient of 0.76 for the calibration period and 0.75 for the validation period. The equation used to calculate the Nash and Sutcliffe coefficient (R^2) is as follows:

$$NASH(R^2) = 1 - \frac{\sum_{i}^{N}(y_i - y'_i)^2}{\sum_{i=1}^{N}(y_i - y_{mean})^2} \qquad (13)$$

where, y_i is the observed streamflow at time step i, y'_i is simulated streamflow at time step i, y_{mean} is the mean of observed streamflow, and N is the number of data points.

Figures 2 and 3 demonstrate observed flow and simulated flow from the hydrologic model for two years of both calibration and validation period, respectively.

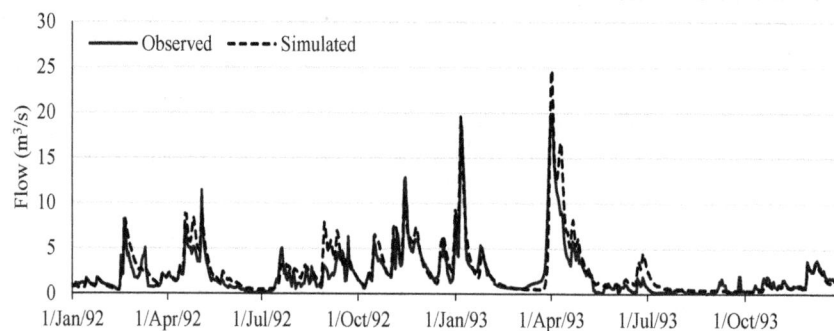

Figure 2: Observed and simulated flows at Dundas station in the calibration period of 1992-1993.

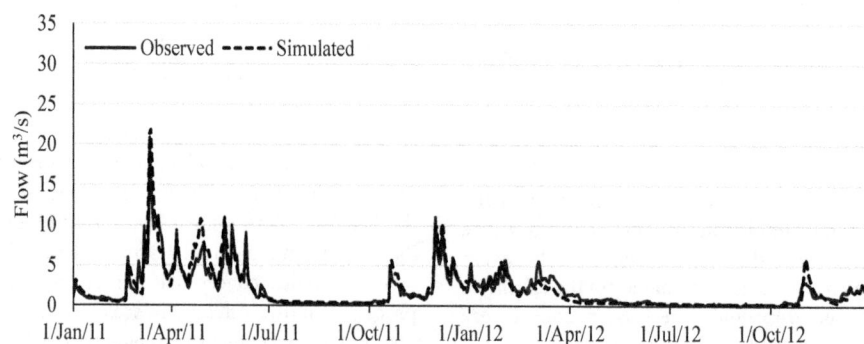

Figure 3: Observed and simulated flows at Dundas station in the validation period of 2011-2012.

Hydrologic simulation: The calibrated HBV model was used to simulate flows at the outlet and evapotranspiration from the watershed at a daily time step for both current (1971-2000) and future (2041-2070) period. The bias-corrected daily total precipitation and daily mean temperature from eight RCM+GCM pairs for current (1971-2000) and future (2041-2070) period were used as input in the watershed model for hydrologic simulation.

Results and Discussion

Evaluation of bias-corrected data

Bias correction was applied to raw NARCCAP daily precipitation and mean temperature (calculated from NARCCAP daily maximum and minimum temperature) both for current (1971-2000) and future (2041-2070) period using the method described in Bias Correction Section. The improvement in bias-corrected NARCCAP projections was assessed using skill score BSS and RPSS described in Description of Skill Scores section, and the BSS and RPSS show the percentage improvement in this study. The BS and RPS were calculated for raw and bias-corrected NARCCAP eight RCM+GCM pair's data for the current period. The scores and skill scores calculated for two meteorological stations namely Hamilton Airport and Hamilton RBG are presented in Tables 2-5. The skill scores represent the improvement of bias-corrected NARCCAP data over raw NARCCAP data produced by eight RCM+GCM pairs. The calculated skill scores revealed an overall improvement in both precipitation and temperature at both stations when bias correction is made. In the case

Month	BS	BS	BSS (%)	RPS	RPS	RPSS (%)
	Raw	Bias-Cor		Raw	Bias-Cor	
Jan	0.29	0.28	3.4	1.1	1.09	0.9
Feb	0.32	0.29	9.4	1.14	1.12	1.8
Mar	0.32	0.29	9.4	0.81	0.8	1.2
Apr	0.31	0.28	9.7	0.87	0.82	5.7
May	0.34	0.28	17.6	0.57	0.54	5.3
Jun	0.33	0.27	18.2	0.57	0.53	7
Jul	0.32	0.26	18.8	0.58	0.53	8.6
Aug	0.3	0.26	13.3	0.55	0.53	3.6
Sep	0.29	0.27	6.9	0.54	0.54	0
Oct	0.31	0.28	9.7	0.86	0.84	2.3
Nov	0.32	0.29	9.4	0.83	0.81	2.4
Dec	0.33	0.29	12.1	1.15	1.13	1.7

Table 2: Skill score of bias-corrected precipitation at Hamilton Airport.

Month	BS	BS	BSS (%)	RPS	RPS	RPSS (%)
	Raw	Bias-Cor		Raw	Bias-Cor	
Jan	0.32	0.29	9.4	0.84	0.83	1.2
Feb	0.33	0.29	12.1	0.85	0.82	3.5
Mar	0.34	0.28	17.6	0.55	0.55	0
Apr	0.32	0.28	12.5	0.89	0.83	6.7
May	0.34	0.27	20.6	0.55	0.53	3.6
Jun	0.33	0.26	21.2	0.56	0.52	7.1
Jul	0.32	0.26	18.8	0.56	0.53	5.4
Aug	0.3	0.27	10	0.54	0.53	1.9
Sep	0.3	0.28	6.7	0.55	0.55	0
Oct	0.31	0.27	12.9	0.57	0.55	3.5
Nov	0.33	0.29	12.1	0.84	0.83	1.2
Dec	0.34	0.29	14.7	0.85	0.84	1.2

Table 3: Skill score of bias-corrected precipitation at Hamilton RBG.

Month	BS	BS	BSS (%)	RPS	RPS	RPSS (%)
	Raw	Bias-Cor		Raw	Bias-Cor	
Jan	0.31	0.3	3.2	1.97	1.96	0.5
Feb	0.29	0.28	3.4	1.91	1.89	1
Mar	0.26	0.25	3.8	1.76	1.73	1.7
Apr	0.26	0.25	3.8	1.75	1.73	1.1
May	0.25	0.24	4	1.68	1.62	3.6
Jun	0.28	0.25	10.7	1.87	1.7	9.1
Jul	0.29	0.28	3.4	2.03	1.93	4.9
Aug	0.29	0.28	3.4	2.03	1.92	5.4
Sep	0.27	0.26	3.7	1.85	1.76	4.9
Oct	0.27	0.27	0	1.86	1.81	2.7
Nov	0.27	0.27	0	1.84	1.8	2.2
Dec	0.3	0.3	0	1.91	1.9	0.5

Table 4: Skill score of bias-corrected temperature at Hamilton Airport.

Month	BS	BS	BSS (%)	RPS	RPS	RPSS (%)
	Raw	Bias-Cor		Raw	Bias-Cor	
Jan	0.33	0.3	9.1	2.11	1.96	7.1
Feb	0.3	0.28	6.7	2.09	1.88	10
Mar	0.28	0.26	7.1	1.84	1.76	4.3
Apr	0.26	0.25	3.8	1.8	1.74	3.3
May	0.25	0.25	0	1.81	1.65	8.8
Jun	0.31	0.26	16.1	2.06	1.72	16.5
Jul	0.32	0.3	6.3	2.29	1.96	14.4
Aug	0.3	0.28	6.7	2.12	1.9	10.4
Sep	0.29	0.26	10.3	2.03	1.77	12.8
Oct	0.27	0.26	3.7	1.95	1.78	8.7
Nov	0.3	0.27	10	2.07	1.81	12.6
Dec	0.33	0.29	12.1	2.12	1.92	9.4

Table 5: Skill score of bias-corrected temperature at Hamilton RBG.

of precipitation, the BS values do not show a seasonal pattern in skill, but the RPS values show that overall skill in other seasons is better than the skill in winter months. Both BSS and RPSS results shown in Table 2 indicate that improvement is higher in the late spring and summer months than others months, and the highest improvement is shown in the month of July with BSS and RPSS values of 18.8% and 8.6%, respectively. A similar seasonal pattern in the improvement of bias-corrected precipitation at Hamilton RBG is shown in Table 3, and it also shows that the highest improvement is in the month of June with BSS and RPSS values of 21.2% and 7.1%, respectively. Results in Tables 2 and 3 show that the improvement presented by RPSS is higher when the RPS of raw NARCCAP precipitation is lower in general. For example, RPS values for raw NARCCAP precipitation at Hamilton Airport are 0.57 and 1.15 in the month of June and December, respectively, and the corresponding RPSS values are 7% and 1.7%. Both BSS and RPSS values shown in Tables 4 and 5 indicate that there is a significant improvement in quality in bias-corrected daily mean temperature for both meteorological stations, and the improvement is slightly better for Hamilton RBG station than Hamilton Airport station. Both BSS and RPSS values also show that the improvement in the quality of bias-corrected daily mean temperature is highest in the month of June for both stations. The BSS and RPSS values are 10.7% and 9.1% in the month of June for Hamilton Airport, and these values are 16.1% and 16.5% for Hamilton RBG station. The RPSS values show that the overall improvement in the quality of bias-corrected temperature is better in the summer months than other seasons.

Monthly average changes in climate variables and flows

The bias-corrected NARCCAP precipitation and daily mean temperature time series over thirty years for both current (1971-2000) and future (2041-2070) periods were analyzed to show the changes under future climate condition. The hydrologic model simulated actual evapotranspiration for the same periods was also analyzed to show any changes. The monthly average values for these variables were calculated to get insight about how the changes are distributed seasonally. Here, the monthly average values were calculated from the average of eight RCM+GCM pairs. The calculated monthly average precipitation and daily mean temperature for Hamilton Airport and Hamilton RBG stations are presented in Figures 4-7. Figure 4 and 5 shows that precipitation increases significantly for the most part of the year except the summer months including September. The increase in precipitation under future climate at Hamilton Airport station varies between 3% and 17%, and the lowest and highest increase are in a fall month, October and a winter month, January, respectively. A similar increase in future precipitation is shown at Hamilton RBG as the lowest increase of 6% in a fall month-November and the higher increase of 16-17% in two winter months, December and January. The increase in precipitation during March, April and May are similar as shown in Figures 4 and 5. The decrease in precipitation will be highest (10-11%) in the summer month of July for both Hamilton and Hamilton RBG station, and the decrease of precipitation in June is insignificant for both meteorological stations. Figures 4 and 5 also show that the higher

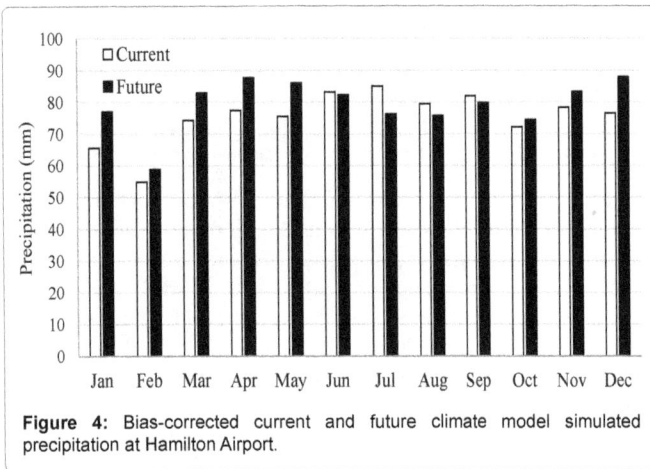

Figure 4: Bias-corrected current and future climate model simulated precipitation at Hamilton Airport.

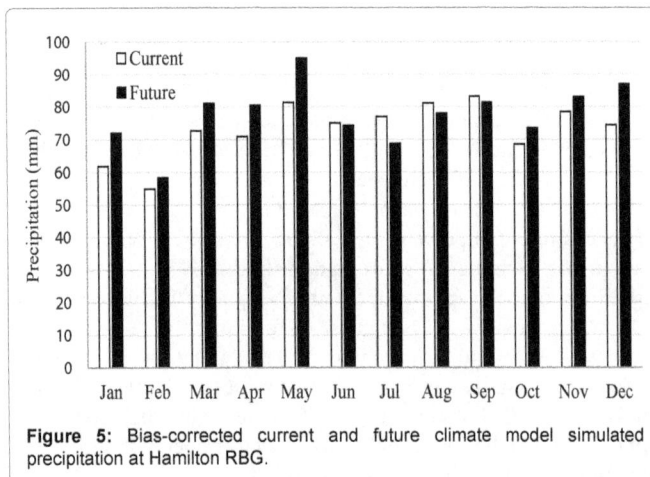

Figure 5: Bias-corrected current and future climate model simulated precipitation at Hamilton RBG.

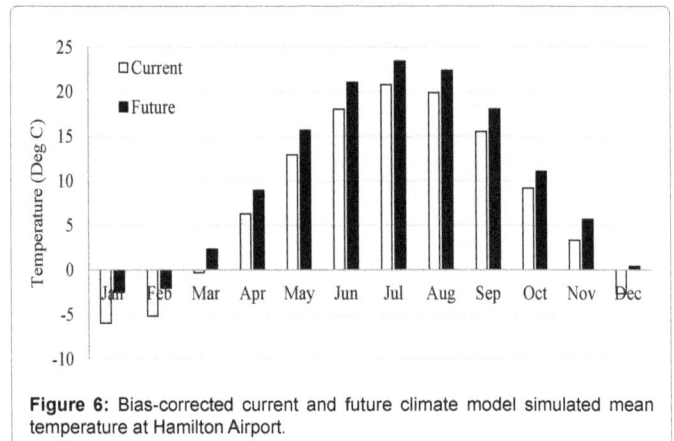

Figure 6: Bias-corrected current and future climate model simulated mean temperature at Hamilton Airport.

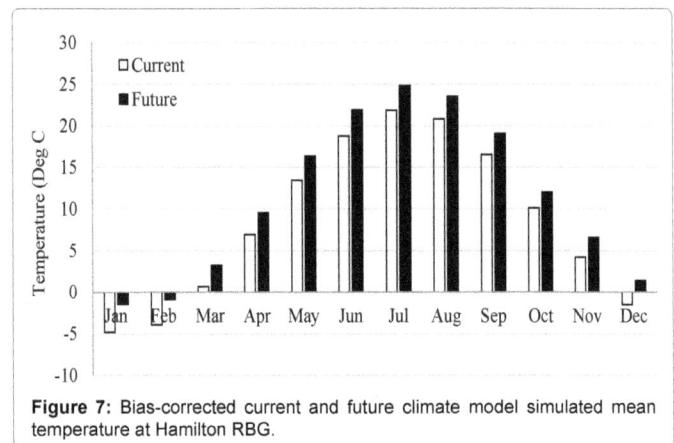

Figure 7: Bias-corrected current and future climate model simulated mean temperature at Hamilton RBG.

amounts of monthly average precipitation in the future are in April and December at Hamilton Airport, and in May and December at Hamilton RBG station. The precipitation projection of average RCMs shows a clear signal of seasonal distribution of change in the precipitation regime. From Figures 6 and 7, it appears that the daily mean temperature will increase in all months at both meteorological stations. Figures 6 and 7 show that the highest and lowest daily mean temperatures are in July and January, respectively at both stations. The increase in temperature under future climate varies between 1.91°C and 3.44°C at Hamilton Airport station and 1.9°C and 3.37°C at Hamilton RBG. The lowest and highest increases are in October and January, respectively. These increases of temperature are close to the increases revealed by Sultana and Coulibaly [4]. A higher increase in the summer month of June than other months in spring, summer and fall are also shown in the figures. Overall, the increase in daily mean temperature in all winter months is higher than other seasons. The temperature projection of average RCMs shows a clear signal of seasonal distribution of change in the temperature. The actual evapotranspiration on daily time step was simulated by the hydrologic model for current and future periods, and the monthly average values of the average of eight RCM+GCM pairs are presented in Figure 8. It can be seen from Figure 8 that the actual evapotranspiration in the future is higher than the current period in all months except in July and August. Increase in evapotranspiration in July is insignificant because of an insignificant decrease in monthly average precipitation, although there is a significant increase in temperature (2.97°C and 3.16°C at Hamilton Airport and Hamilton RBG) in future. The higher decrease in precipitation and lower increase in temperature in other summer months, July and August than in June resulted in about 10% decrease in evapotranspiration. Figure 8 also shows a small amount of evapotranspiration in the winter. Although the actual evapotranspiration is very low in the winter months, the percentage increase is higher in the winter than in other seasons because

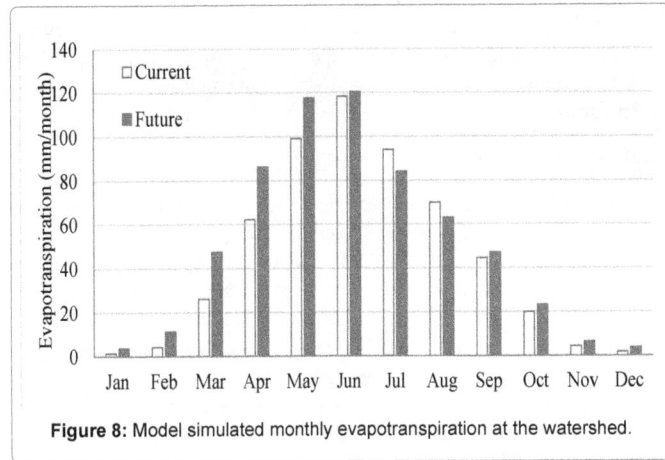

Figure 8: Model simulated monthly evapotranspiration at the watershed.

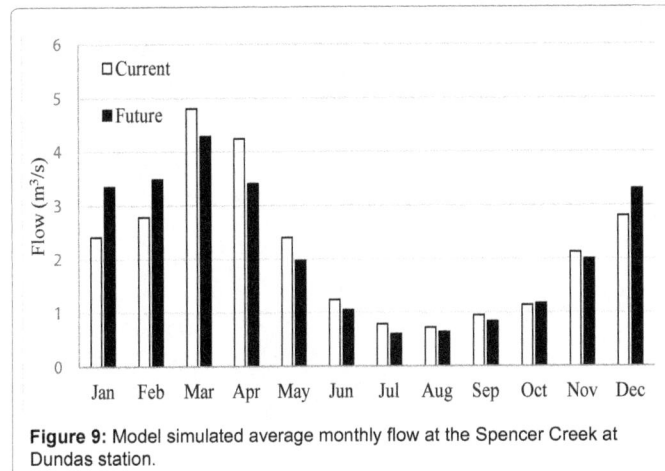

Figure 9: Model simulated average monthly flow at the Spencer Creek at Dundas station.

of a higher increase in temperature and increase in precipitation. In can be seen from Figure 8 that the total increase in evapotranspiration in each month (21 mm, 24 mm and 18 mm in March, April and May, respectively) of spring is much higher than in other seasons.

Figure 9 presents the monthly average flows of the average of eight RCM+GCM pairs for both current and future periods. It can be seen from Figure 9 that the monthly flows increase in the winter and decrease in other seasons except an insignificant increase (0.03 m³/s) in October. The seasonal distribution of future flow is similar to the findings presented by Grillakis et al. [3]. The increase in flow in December, January and February are 17.9% (or 0.50 m³/s), 38.5% (or 0.93 m³/s) and 25.3% (0.70 m³/s), respectively. The decrease in flows in future varies between 6% (or 0.13 m³/s) and 24.1% (or 0.19 m³/s), and the lowest and the highest decrease are in November and in July, however the highest decrease in terms of flow magnitude is in April, where the change is 0.83 m³/s (or 19.7%). The effect of a change in precipitation, temperature, and evapotranspiration for the resultant change in future flow is complex. The effect of a change in evapotranspiration on flow in the winter months is small as the monthly average evapotranspiration is very small in these months. The higher increase in flows in winter months could be attributed to the increase in both winter month's precipitation and temperature in future. The warmer winter temperature in future will increase the winter snowmelt, and will result in the decrease of snowpack for annual basis and termination of snowmelt in the earlier in the spring. Despite the increase in the future precipitation in the spring, the flow will be decreased in the spring because thinner snowpack left to be melted and high evapotranspiration increase in this season. The decrease in flows in the summer is caused by the decrease in precipitation in future, and a comparatively small decrease in the fall could be attributed to the increase of evapotranspiration.

Yearly average changes in climate variables and flows

The difference between the current and future climate variables and flows were analysed, and the annual average of precipitation, daily mean temperature, and the hydrologic model simulated actual evapotranspiration and flows are presented in Table 6. It can be seen from Table 6 that six RCM+GCM pairs out of eight pairs projected

Station/ Watershed	Element	Period	Crcm Ccsm	Crcm Cgcm3	Hrm3 Gfdl	Hrm3 Hadcm3	Rcm3 Cgcm3	Rcm3 Gfdl	Wrfg Ccsm	Wrfg Cgcm3	Average
Hamilton	P (mm/day)	Current	2.44	2.46	2.48	2.48	2.46	2.45	2.52	2.51	2.47
Airport	P (mm/day)	Future	2.47	2.67	2.35	2.74	2.64	2.69	2.45	2.84	2.57
	%	Change	1	8.7	-5.2	10.6	7.3	9.6	-2.8	13	4.1
	T (°C)	Current	7.7	7.7	7.7	7.7	7.7	7.7	7.7	7.7	7.7
	T (°C)	Future	10.61	10.81	11.11	10.31	10.28	10.25	9.98	9.99	10.48
	°C	Change	2.9	3.11	3.4	2.61	2.58	2.54	2.27	2.28	2.77
Hamilton	P (mm/day)	Current	2.37	2.38	2.43	2.46	2.37	2.37	2.45	2.43	2.4
RBG	P (mm/day)	Future	2.39	2.6	2.34	2.62	2.62	2.68	2.4	2.77	2.52
	%	Change	0.8	9.2	-3.7	6.7	10.3	13	-1.8	14	4.9
	T (°C)	Current	8.65	8.65	8.65	8.65	8.65	8.65	8.65	8.65	8.65
	T (°C)	Future	11.53	11.73	12.25	11.36	11.21	11.21	10.92	10.92	11.46
	°C	Change	2.87	3.08	3.59	2.71	2.56	2.56	2.27	2.26	2.8
Spencer	E (mm/day)	Current	1.6	1.74	1.59	1.74	1.71	1.73	1.63	1.71	1.68
	E (mm/day)	Future	1.47	1.48	1.52	1.51	1.5	1.51	1.49	1.47	1.49
	%	Change	8.7	17.6	4.8	14.7	14.4	14.2	9.6	16.3	12.51
Dundas	Q (m³/s)	Current	2	2.19	1.87	2.28	2.2	2.32	1.96	2.51	1.92
	Q (m³/s)	Future	2.16	2.18	2.19	2.23	2.13	2.11	2.28	2.28	1.91
	%	Change	-7.3	0.1	-14.7	1.9	3.3	10	-13.8	10	0.5

Table 6: Changes in annual average precipitation, daily mean temperature, evapotranspiration and flow.

an increase in precipitation with the highest increase projected by WRFG+CGCM3 model at 13% and 14% for Hamilton Airport and Hamilton RBG, respectively. It is worth mentioning that all the RCM+GCM pairs show an increase in the annual average of daily mean temperature. The increase in temperature in future varies between 2.27°C and 3.4°C at Hamilton Airport and between 2.26°C and 3.59°C at Hamilton RBG. The greatest change in terms of temperature increase is projected by HRM3+GFDL models and the most conservative change is projected by WRFG+CCSM and WRFG+CGCM3 models for Hamilton Airport and Hamilton RBG, respectively. The CRCM model projected temperature change is higher than other models except the HRM3+GFDL, and the WRFG model projected temperature increase is lower than the other models. It is notable that the annual average evapotranspiration under future climate compared to current climate will increase for all the RCM+GCM pairs. Table 6 shows that the annual average flows at the Spencer Creek at Dundas hydrometric station will be increased in the case of five RCM+GCM pairs out of eight pairs, and the higher increase (10%) is exhibited by RCM3+GFDL and WRFG+CGCM3 model while the highest decrease (14.7%) is exhibited by HRM3+GFDL model. Overall, the decrease in flows is also shown by one GCM (CCSM) with two RCMs. Averages of eight RCM+GCM

pairs show an increase for all climate variables and a small increase of annual average flow in future. In the case of annual average flows, a difference from the study done by Grillakis et al. [3] is noticed, and the difference resulted due to the use of a different period of data for bias correction and flow comparison for current and future period, and the use of a different number of RCM+GCM pairs. Taking into account the future increase in annual average flow, analysis of flow duration were performed to get insight into how the flow regime will be changed under future climate condition.

High and low flows

Figures 10 and 11 present the flow duration curves created using simulated flows at the Spencer Creek at Dundas hydrometric station for the current (1971-2000) and future period (2041-2070). The simulated current and future flows were obtained by inputting the bias-corrected NARCCAP's eight RCM+GCM pair's precipitation and temperatures into a calibrated hydrologic model. Figure 10 presents the flow duration curves for four RCM+GCM pairs, and Figure 11 presents the flow duration curve for the other four RCM+GCM pairs. For better visualization of the difference between flow duration curves, the maximum value on the ordinates is set to 6 m³/s, and the eight

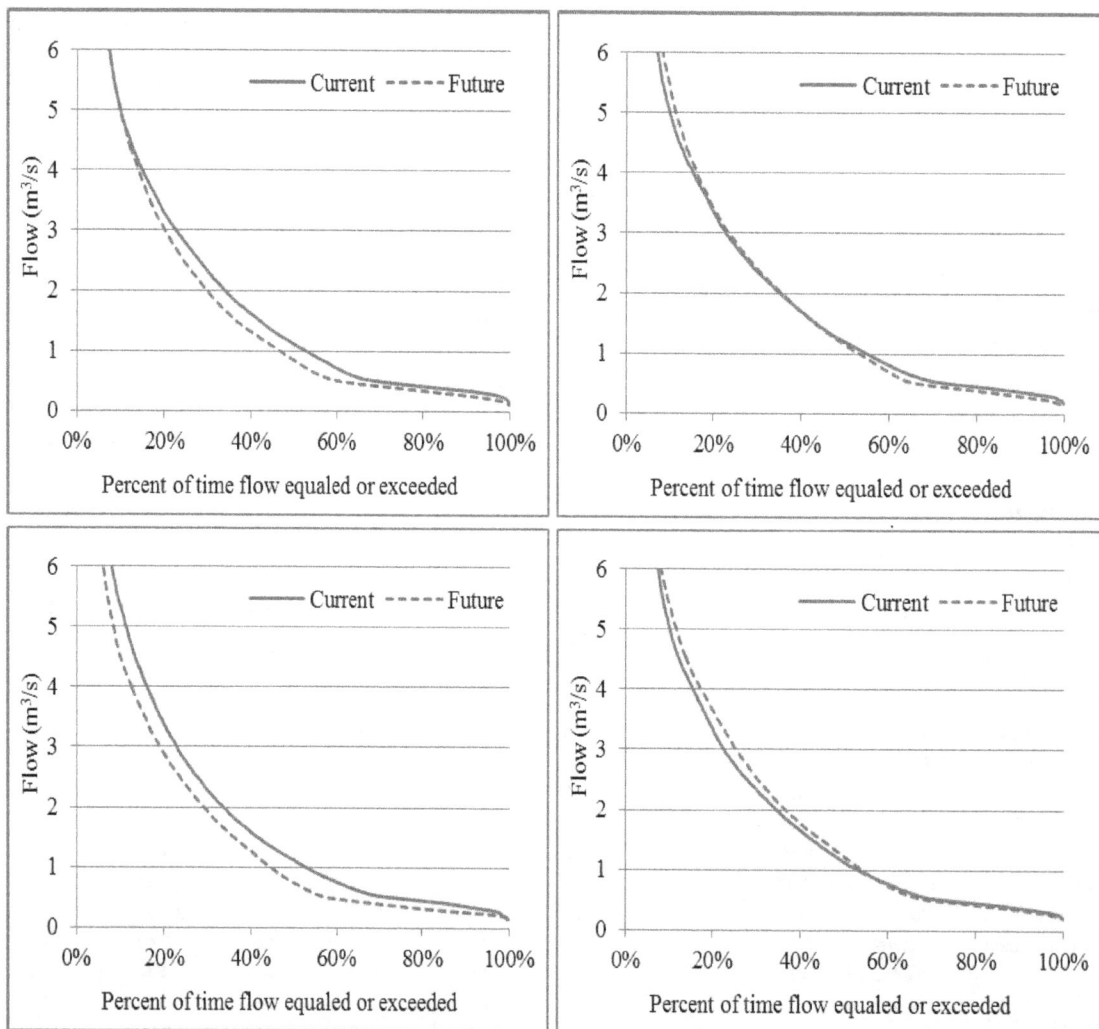

Figure 10: Flow duration curves for the Spencer Creek watershed (top left, top right, bottom left and bottom right represent CrcmCcsm, CrcmCgcm3, Hrm3Gfdl and Hrm3Hadcm3, respectively).

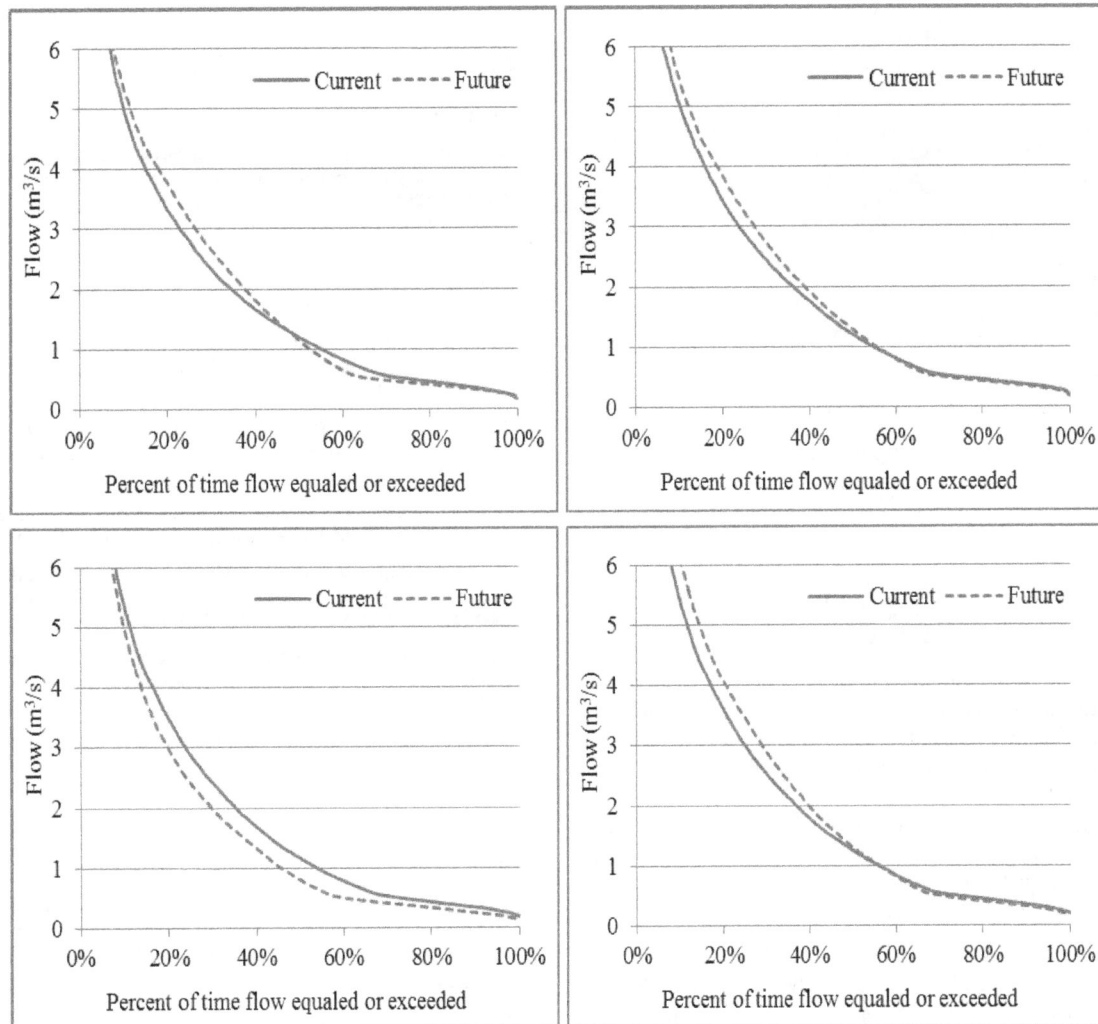

Figure 11: Flow duration curves for the Spencer Creek watershed (top left, top right, bottom left and bottom right represent Rcm3Cgcm3, Rcm3Gfdl, WrfgCcsm and WrfgCgcm3, respectively).

RCM+GCM results are presented in two Figures. For every time series of daily flow data, the exceedance probability of each flow was calculated, and flow duration curves were produced by plotting discharge on the ordinate and exceedance probability on the abscissa. The large difference between the highest and the lowest flow values as shown in the flow duration curves in Figures 10 and 11 reveals that the watershed has a relatively low flow during the dry periods, but responds to extreme precipitation event with a relatively high flow. Two flow statistics, Q_{95} and Q_{10}, were used to compare the flow duration curves for current and future periods. Q_{95} and Q_{10} are the flow values that are equaled or exceeded 95% and 10% of the time, respectively. Q_{95} and Q_{10} are used for analysis of low flow and high flow, respectively [49]. As illustrated in Figures 10 and 11, the low flow decreased for all RCM+GCM pairs with the highest decrease (28.6%) exhibited by WRFG+CCSM model and lowest decrease (3.4%) exhibited by RCM3+CGCM3, and the calculated average low flow value for eight models indicates a decrease by 16.7% under future climate condition. It can be seen from Figures 10 and 11 that the high flow will increase for five RCM+GCM pairs (8.2%, 7.3%, 6.9%, 8.1% and 13.2% for CRCM+CGCM3, HRM3+HADCM3, RCM3+CGCM3, RCM3+GFDL and WRFG+CGCM3, respectively), and decrease for other three RCM+GCM pairs (0.4%, 16.1% and 7.1%

for CRCM+CCSM, HRM3+GFDL and WRFG+CCSM, respectively), and the calculated average high flow value for eight models indicates an increase by 2.4% under future climate condition. The maximum increase in high flow is obtained for WRFG+CGCM3 model, for which the increase of annual average precipitation is the highest, and a maximum decrease of high flow is obtained for HRM3+GFDL model, for which the decrease of annual average precipitation and increase of mean temperature is the highest. Taking into account that the increase in precipitation is consistent, and the high flow is mainly attributed to the precipitation amount, flow duration curves were constructed for the average of five RCM+GCM pairs (CRCM+CGCM3, HRM3+HADCM3, RCM3+CGCM3, RCM3+GFDL and WRFG+CGCM3), and shown in Figure 12. It is shown in the figure that the high flow will increase, and low flow will decrease significantly. The Q_{95} is obtained as 0.31 m^3/s and 0.27 m^3/s for current and future climate that resulted in 12.9% decrease in low flow, and Q_{10} is obtained as 5.11 m^3/s and 5.56 m^3/s for current and future climate that resulted in 8.8% increase in high flow.

Conclusions

The potential impact of climate change on the hydrology of Spencer Creek watershed was analyzed based on the NARCCAP provided

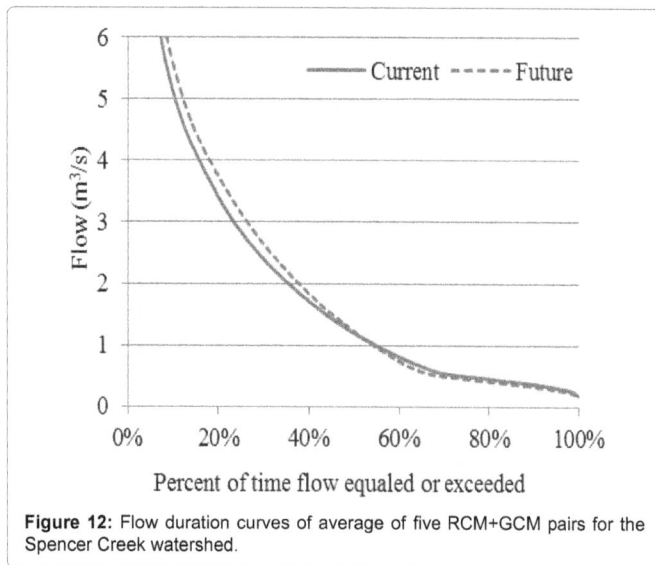

Figure 12: Flow duration curves of average of five RCM+GCM pairs for the Spencer Creek watershed.

eight RCM+GCM pair's precipitation and temperature projections and simulations by using HBV hydrologic model for the current (1971-2000) and future (2041-2070) period.

The NARCCAP meteorological projections were bias corrected to get more realistic simulations from the hydrologic model. An overall improvement for the quality of NARCCAP precipitation and temperature simulations at both Hamilton Airport and Hamilton RBG meteorological stations was achieved when bias correction was applied. Both BSS and RPSS indicate that improvement is high in the late spring and summer months in the case of precipitation. The overall improvement in the quality of bias-corrected temperature is the best for summer months as revealed by RPSS with the highest improvement obtained in June as revealed by both BSS and RPSS.

The climate variables and flow were analyzed on monthly and annually to get insight into the seasonal and overall change under future climate compared to the current climate. Finally, high and low flow were analyzed by using the flow duration curves for eight RCM+GCM pair's data and average of the RCM+GCM pair's data. The RCM+GCM average shows that the precipitation will increase in the fall, winter and spring and decrease in the summer, the temperature will increase in all months, actual evapotranspiration will increase in all months except July and August, and the flow will increase in the winter and decrease in the other seasons. The RCM+GCM averages also show a significant increase in all climate variables and a small increase in annual average flow. The small increase in annual average flow could be attributed to the very high decrease in low flow despite an increase in high flow. The WRFG+CGCM3 model projected the greatest increase in high flow by 13.2% and the WRFG+CCSM model projected the highest decrease in low flow by 28.6%. The averages of eight RCM+GCM pairs show an increase of high flow by 2.4% and a decrease of low flow by 16.7% and the average of five RCM+GCM pairs (precipitation projected by this model are consistent) revealed an increase of high flow by 8.8% and a decrease of low flow by 12.9%. The changes in winter and spring flow will influence the water management at watershed scale. The authorities have to adopt new strategies to manage higher winter and lower spring flow and higher uncertainty in flows for the watershed management infrastructures.

Acknowledgements

We wish to thank the North American Regional Climate Change Assessment Program (NARCCAP) for providing the data used in this paper. The authors gratefully acknowledged Hamilton Region Conservation Authority, Water Survey Canada and Environment Canada for providing data for the study area. This study was supported by the Natural Sciences and Engineering Research Council (NSERC) of Canada in the form of discovery grant (RGPIN04808-14).

References

1. IPCC (2014) Climate Change 2014: Synthesis Report. Contribution of Working Groups I, II and III to the Fifth Assessment Report of the Intergovernmental Panel on Climate Change. Core Writing Team, Pachauri PK, Meyer LA (eds.) IPCC, Geneva, Switzerland, p: 151.

2. Islam Z, Gan TY (2014) Effect of climate change on the surface-water management of South Saskatchewan river basin. Journal of Water Resources Planning and Management 140: 332-342.

3. Grillakis MG, Koutroulis AG, Tsanis IK (2011) Climate change impact on the hydrology of Spencer creek watershed in Southern Ontario, Canada. Journal of Hydrology 409: 1-19.

4. Sultana Z, Coulibaly P (2011) Distributed modeling of future changes in hydrological processes of Spencer Creek watershed. Hydrological Processes 25 (8): 1254-1270. [DOI:10.1002/hyp.7891].

5. Boyer C, Chaumont D, Chartier I, Roy AG (2010) Impact of climate change on the hydrology of St. Lawrence tributaries. Journal of Hydrology 384: 65-83.

6. Dibike YB, Coulibaly P (2005) Hydrologic impact of climate change in the Saguenay watershed: comparison of downscaling methods and hydrologic models. Journal of Hydrology 307 (1-4): 145-163.

7. Fowler HJ, Blenkinsop S, Tebaldi C (2007) Linking climate change modeling to impacts studies: Recent advances in downscaling techniques for hydrological modeling. Int J Climatol 27 (12): 1547-1578.

8. Praskievicz S, Chang HJ (2009) A review of hydrological modeling of basin-scale climate change and urban development impacts. Progress in Physical Geography 33 (5): 650-671.

9. Prudhomme C, Reynard N, Crooks S (2002) Downscaling of global climate models for floodfrequency analysis: Where are we now? Hydrol. Processes 16 (6): 1137-1150.

10. Kalra A, Ahmad S (2011) Evaluating changes and estimating seasonal precipitation for Colorado River Basin using stochastic nonparametric disaggregation technique. Water Resour Res 47: W05555.

11. Sharma M, Coulibaly P, Dibike Y (2011) Assessing the need for downscaling RCM data for hydrologic impact study. Journal of Hydrologic Engineering 16: 534-539. [DOI:10.1061/(ASCE)HE.1943-5584.0000349].

12. Mpelasoka FS, Chiew FHS (2009) Influence of rainfall scenario construction methods on runoff projections. J Hydrometeor 10: 1168-1183. [DOI:10.1175/2009JHM1045].

13. Samuel J, Coulibaly P, Metcalfe RA (2012) Evaluation of future flow variability in ungagged basins: Validation of combined methods. Advances in Water Resources 35: 121-140.

14. Payne JT, Wood AW, Hamlet AF, Palmer RN, Lettenmaier DP (2004) Mitigating the effect of climate change on the water resources of the Coumbia river basin, Clim Change 62: 233-256.

15. Ines AVM, Hansen JW (2006) Bias correction of daily GCM rainfall for crop simulation studies. Agricultural and Forest Meteorology 138: 44-53.

16. Mearns LO, Gutowski WJ, Jones R, Leung LY, McGinnis S, et al. (2009) A regional climate change assessment program for North America. EOS 90: 311-312.

17. Ahmed S, Coulibaly P, Tsanis I (2015) Improved spring peak-flow forecasting using ensemble meteorological predictions. Journal of Hydrologic Engineering 20 (2): 04014044 [DOI:10.1061/(ASCE)HE.1943-5584.0001014].

18. Liu X, Coulibaly P (2011) Downscaling ensemble weather predictions for improved week-2 hydrologic forecasts. Journal of Hydrometeorology 12 (6): 1564-1580.

19. HRCA (1990) Hamilton Region Conservation Authority - MacLaren Plansearch Lavalin, Technical Report: Canada/Ontario Flood Damage Reduction Program.

20. NARCCAP (2013) North American Regional Climate Change Assessment Program. <http://www.narccap.ucar.edu/> (January26, 2013).

21. Mearns LO et al. (2007) updated 2012. The North American Regional Climate Change Assessment Program dataset, National Center for Atmospheric Research Earth System Grid data portal, Boulder, CO. Data downloaded 2014-07-07. [DOI:10.5065/D6RN35ST].

22. Nakicenvoic N, Davidson O, Davis G, Grübler A, Kram T, et al. (2000) Special Report on Emissions Scenarios. A Special Report of Working Group III of the Intergovernmental Panel on Climate Change. Cambridge University Press, Cambridge, p: 599.

23. Mailhot A, Beauregard I, Talbot G, Caya D, Biner S (2012) Future changes in intense precipitation over Canada assessed from multi-model NARCCAP ensemble simulations. Int J Climato 32: 1151-1163.

24. Music B, Caya D (2007) Evaluation of the hydrological cycle over the Mississippi River Basin as simulated by the Canadian regional climate model (CRCM). J Hydrometeor 8: 969-988.

25. Jones R, Noguer M, Hassell D, Hudson D, Wilson S, et al. (2004) Generating high resolution climate change scenarios using PRECIS. Met Office Hadley Center. Exter, p: 40.

26. Elguindi N, Bi X, Giorgi F, Nagarajan B, Pal J, et al. (2007) RegCM Version 3.1 User's Guide, Trieste, Italy <https://users.ictp.it/RegCNET/regcm.pdf> (July 9, 2015).

27. Giorgi F, Marinucci MR, Bates GT (1993) Development of second generation regional climate model (RegCM2) I: boundary layer and radiative transfer processes. Mon Weather Rev 121: 2794-2813.

28. Skamarock WC, Klemp JB, Dudhia J, Gill DO, Barjer DM, Wang W, Powers JG (2005) A description of the advanced research WRF Version 2, NCAR Tech Notes-468+STR <http://www.mmm.ucar.edu/wrf/users/docs/arw_v2.pdf> (January 25, 2016).

29. Collins WD, Bitz CM, Blackmon ML, Bonan GB, Bretherton CS, et al. (2006) The community climate system model version 3 (CCSM3). J Climate 19: 2122-2143.

30. Flato GM (2005) The Third Generation Coupled Global Climate Model (CGCM3) <http://www.ec.gc.ca/ccmac-cccma/default.asp?n=1299529F-1> (August 9, 2015).

31. Anderson JL, Balaji V, Broccoli AJ, Cooke WF (2004) The New GFDL global atmospheric and land model AM2-LM2: Evaluation with prescribed SST simulations. J Climate 17: 4641-4673.

32. Gordon C, Cooper C, Senior CA, Banks H, Gregory JM, et al. (2000) The simulation of SST, sea ice extents and ocean heat transports in a version of the Hadley Centre coupled model without fluxadjustments. Climate Dynamics 16: 147-168.

33. Pope VD, Gallani ML, Rowntree PR, Stratton RA (2000) The Impact of new physical parameterizations in the Hadley Centre climate model-HadAM3. Climate Dynamics 16: 123-146.

34. Murphy AH, Brown BG, Chen Y (1989) Diagnostic verification of temperature forecasts. Weather Forecasting 4 (4): 485-501.

35. Wilks DS (2000) Diagnostic verification of the Climate Prediction Center long-lead outlooks, 1995-1998. Journal of Climate 13 (13): 2389-2403.

36. Hartmann HC, Pagano TC, Bales R, Sorooshian S (2002) Confidence builders: Evaluating seasonal climate forecasts from user perspectives. Bull. American Meteorological Society 83: 683-698.

37. Brier GW (1950) Verification of forecasts expressed in terms of probabilities. Monthly Weather Review 78: 1-3.

38. Renner M, Werner MG, Rademacher S, Sprokkereef E (2009) Verification of ensemble flow forecasts for the river Rhine. Journal of Hydrology 376 (3-4): 463-475.

39. Roulin E (2007) Skill and relative economic value of medium-range hydrological ensemble predictions. Hydrology and Earth System Sciences 11 (2): 725-737.

40. Mullen SL, Buizza R (2011) Quantitative precipitation forecasts over the United States by the ECMWF ensemble prediction system. Mon Weather Rev 129 (4): 638-663.

41. Clark MP, Hay LE (2004) Use of medium-range numerical weather prediction model output to produce forecasts of streamflow. Journal of Hydrometeorology 5 (1): 15-32.

42. Gangopadhyay S, Clark M, Brandon D, Wener K, Rajagopalan B (2004) Effects of spatial and temporal aggregation on the accuracy of statistically downscaled precipitation estimates in the upper Colorado river basin. Journal of Hydrometeorology 5 (6): 1192-1206.

43. Sharma M (2009) Comparison of downscaled RCM and GCM data for hydrologic impact assessment. MASc Thesis, Department of Civil Engineering, McMaster University, Hamilton, Ontario, Canada.

44. Bergstrom S (1991) Principles and confidence in hydrological modelling. Nordic Hydrology 22: 123-136.

45. Lindstrom G, Johansson B, Magnus P, Gardelin M, Bergstrom S (1997) Development and test of the distributed HBV-96 hydrological model. Journal of Hydrology 201: 272-288.

46. Swedish Meteorological and Hydrological Institute (SMHI) (2012) Integrated hydrological modelling system manual, version 6.3. Norrköping, Sweden.

47. Nash JE, Sutcliffe JV (1970) River flow forecasting through conceptual models. Part I – a discussion of Principals. Journal of Hydrology 10: 282-290.

48. Coulibaly P, Anctil F, Bobee B (2001) Multivariate reservoir inflow forecasting using temporal neural networks. Journal of Hydrologic Engineering 6: 367-376.

49. Zahmatkesh Z, Karamouz M, Goharian E, Burian SJ (2015) Analysis of the effect of climate change on urban storm water runoff using statistical downscaled precipitation data and a change factor approach. Journal of Hydrologic Engineering 20: 05014022. [DOI: 10.1061/(ASCE)HE.1943-5584.0001064].

6

Severity Classification and Characterization of Waterlogged Irrigation Fields in the Fincha' a Valley Sugar Estate, Nile Basin of Western Ethiopia

Getahun Kitila[1], Gizachew Kabite[1]* and Tena Alamirew[2]

[1]College of Natural and Computational Science, Wollega University, 395, Ethiopia

[2]Haramaya University, Institute of Technology, 138, Ethiopia

Abstract

Waterlogging is becoming the major threat to the sustainability of irrigated agricultural lands in Fincha'a Valley Sugar Estate (FVSE). In the present study timely and accurate detection of waterlogged areas through piezometer monitoring and remote sensing indicators, along with their characterization and severity classification has been made. Accordingly, spatial maps of groundwater table (GWT) depth were produced in a Geographic information system (GIS) (ArcGIS 10.2) environment from 40 groundwater monitoring piezometer data. Results of the study revealed that FVSE, after nearly 20-25 years of irrigation, is experiencing a serious waterlogging problem. About 324.4 km^2 (75.5%) of the delineated plantation fields are severely waterlogged and 105 km^2 (24.5%) are critically waterlogged. The study also revealed that the GWT depth for all selected irrigation fields is very shallow in winter compared to spring, autumn and summer seasons. The seasonal fluctuation and spatial variability of groundwater table in the irrigated fields is owing to excess irrigation water application, nature of the soil, topography and high seepage from water bodies and poor drainage system; hence are the main causes for waterlogging (GWT rise) problem in the study area. The groundwater depth is extremely shallow (<1 m below ground) in most of the piezometer sites (about 94.7% of the study area) throughout the entire season and showed great spatio-seasonal variability. The rate of annual increment of groundwater rise, coupled with seasonal fluctuation, has obvious repercussions and grave consequences for the sustainability of Fincha'a Valley Sugar Estate. The serious problem of the rising groundwater table can be tackled by adopting improved irrigation water management practices, designing drainage system and further geological investigations. Therefore, it is highly suggested to critically study the causes, consequences and solutions of the waterlogging problem (GWT rise) in a concerted and integrated manner to get out of this vicious problem.

Keywords: Waterlogging; Groundwater table; GIS; Piezometer; Drainage; Topography

Introduction

In agricultural terms, the soil should be considered waterlogged when the water table is within such a distance from the surface of the ground that it reduces the crop production below its normal yield that would be expected from the soil type of that area DIRU (Department of Irrigation, Uttar Pradesh) (2011). In physical context, an area is said to be waterlogged when the water table rises to an extent that the soil pores in the root zone of a crop become saturated, resulting in restriction of the normal circulation of air and decline in the level of oxygen that further increases the level of carbon dioxide. The actual depth of water table, which is considered to be harmful, would depend upon the type of crop, the type of soil and the quality of water and the period for which the water table remains high. The actual depth of water table when it starts affecting the yield of the crop adversity may vary over wide range from zero for the paddy to about 1.5 m for the other crops. The crops, which otherwise, would have grown in the wet season cannot be grown then due to high water table.

For sugarcane crop, the groundwater contribution increases as a function of increment in GWT depth Kahlown and Azam [1,2] recommended the critical depth resulting in a decrease of sugarcane yield is 1.5 m below the ground. Harshika [3] reported that the yield of sugarcane crop suffered when the water table depth is less than 1 m. Furthermore, the shallow groundwater table, in agricultural fields, can cause crops to perish and fields become inaccessible for machinery and harvesting operations [4].

Waterlogging is often compounded by soil compaction. However, reduced tillage and permanent bed systems may alleviate soil compaction and the severity of waterlogging. Cloudy weather associated with wet seasons may enhance the waterlogging effect as well as the incidence of some cotton diseases. Low rates of evaporation and reduced radiation (sunshine) may encourage waterlogging and reduction in yield. Monitoring, diagnosis and mapping of waterlogged area in irrigated agriculture is a perquisite for management of valuable land resources. Groundwater table monitoring can indicate whether the groundwater table depth is rising, falling or remaining static and hence, used to identify the areas at risk of soil salinization DNRE [5]. A rising trend of GWT under irrigated agriculture can provide an early indicator of an increased risk of soil salinity and vice versa [3]. A rise in groundwater results when irrigation induced recharge is greater than the natural discharge. Groundwater rise has subsequently led to waterlogging and the related salinity problems in many irrigated lands around the world, which has happened where the pace of drainage development is not in balance with irrigation development, or where maintenance of drainage has largely been neglected [6].

The main factor challenging the sustainability of the sugar estate is

*Corresponding author: Mostafa Said Barseem, Geophysical exploration department, Desert Research Center, Matahaf El Matariya, Cairo, Ethiopia
E-mail: Barseem2002@hotmail.com

the rise of GWT depth to the crop root zone. The major cause for the rise of GWT depth in the area is an intensive use of furrow irrigation system for long periods of time, coupled with poor drainage systems [7]. Groundwater table rising to the crop root zone is one of the most unfavorable effects of irrigation projects, which occur slowly, and its problem tends to emerge over years [1,2,7]. The adverse impacts of shallow GWT depth to human health, environment, and crop production are well documented by different (local and international) studies [2,7-9].

Geographic information system (GIS) offers an excellent alternative to conventional techniques in monitoring and assessing the extent of waterlogged and saline areas. In the past, several studies have demonstrated the usefulness of remote sensing and GIS techniques in detecting and monitoring waterlogged areas and saline/alkaline soils. Some scientists have used visual interpretation technique for the mapping of waterlogged areas and salt affected soils in IGNP Command areas [10].

Mothikumar and Bhagwat [11] studied the salt affected land using Landsat at 1:50,000 scale by visual interpretation. According to FAO/UNEP [12] guide lines groundwater table depth <2 m are critically waterlogged areas; groundwater table depth which ranges from 2-3 m are considered to be potentially waterlogged, whereas groundwater table >3 m is considered to be deep and hence, safe from waterlogging [7,13]. High temperatures tend to exacerbate the negative effects of waterlogging. The GWT depth <3 m is expected to contribute to the crop evapotranspiration [1] and its effect is maximum when the depth is less than 1 m.

In waterlogged fields, sucrose inversion may result, which affects the sugar quality and quantity. Waterlogging also affects the nutrient and water uptake of roots by restricting root development, which is limited by moisture, aeration and temperature. Plant roots are susceptible for lack of oxygen for respiratory processes when drainage is inadequate (under anaerobic condition) or soils are heavily compacted [14]. Estimates of the global extent of irrigation-induced soil salinity vary, but there is widespread agreement that the twin menaces of waterlogging and salinization represent serious threats to the sustainability of irrigated agriculture in many arid and semi-arid regions [15]. Therefore, this study was carried out with the objective to investigate waterlogged areas, along with their characterization and severity classification in the FVSE.

Materials and Methods

Geographical environment of the study area

Fincha'a Valley Sugar Estate (FVSE) is located in the western highlands of Ethiopia, within the Nile basin, Ethiopia and bounded by the Amhara National Regional State in the north, Guduru District in the South and east, Horro District in the west and Jarte and Amuru District in the North West (Figure 1). It lies between 1055000 m and 1109500 m N and 302000 and 338000 m E. The elevation in the watershed varies from 892 to 2520 meters above sea level (masl). The littoral and alluvial deposits of recent sediments underlie the area Fincha'a River originates from the Chomen and Fincha'a swamps on the highlands and divides the scheme into west and east banks and joins the Nile River of Western Ethiopia. Many streams join the Fincha'a River, the main tributaries being Agamsa, Korke, Fakaree, and Boye from the western side and Sargo-Gobana, Aware, Sombo, and Andode from the eastern side.

The thirty two years (1979-2011) climatic data from the FVSE Meteorological Station recorded a yearly average rainfall of 1316 mm

which is characterized by unimodal rainfall pattern. About 80% of the annual rain falls between May to September. Its mean annual maximum and minimum temperatures are 31 and 15ºC, respectively (Figure 2). The average annual relative humidity is about 84% The FVSE has alternate wet (during May to October) and dry (during the rest of the months) seasons. Wind speed in the FVSE is low as the surrounding escarpments hinder wind movement. However, wind speed is high between the months of March to June [16,17]. The soils in the FVSE are made of alluvia land and colluvial materials from the surrounding escarpments. Six major soil types were identified in the FVSE areas of which Luvisols and Vertisols are predominant. These soils account for more than 95% of the cultivated and irrigated lands.

As indicated in the Figure 3 above, maximum rainfall in the area is obtained in July while minimum rainfall is on January. Furthermore, the rainy season in the area is summer while winter is the dries season.

Data source and analysis

A total of 28 piezometer tubes (F=80 mm and Length=3 m) were installed in November 2010 to characterize the seasonal behavior and spatial variability of GWT depth of the study area. The piezometers are all PVC tubes and fairly distributed in the area. Different sources of water like (irrigation canals, streams and drainage canals), slope and soil type were taken into consideration for the selection of piezometer sites. The PVC tubes were installed manually using auger tubes. The locations (latitude, longitude and elevation) of each piezometer (Table 1) were registered using hand held GPS. Monthly bimonthly monitoring of groundwater depth monitoring commenced as of January 2010, until December l 2012; with the monitoring frequency of two readings per month. Water levels were monitored using a graded contact gauge that provides sound and light signals when it touches water in the tube.

Care was taken to collect the GW levels in all tubes within a

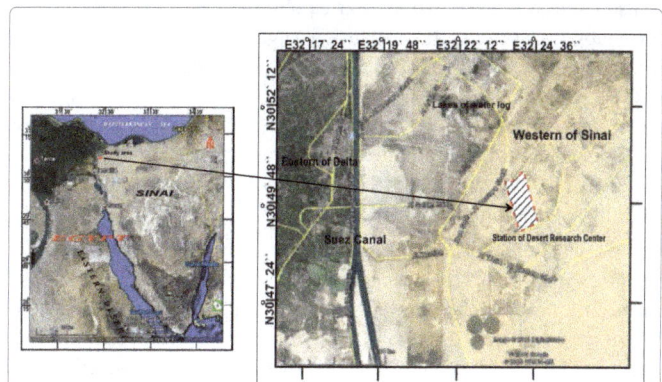

Figure 1: Location map of the study area.

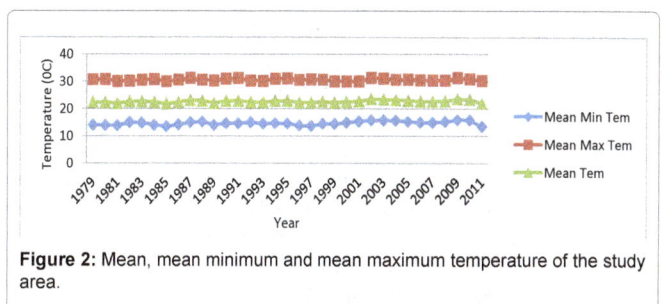

Figure 2: Mean, mean minimum and mean maximum temperature of the study area.

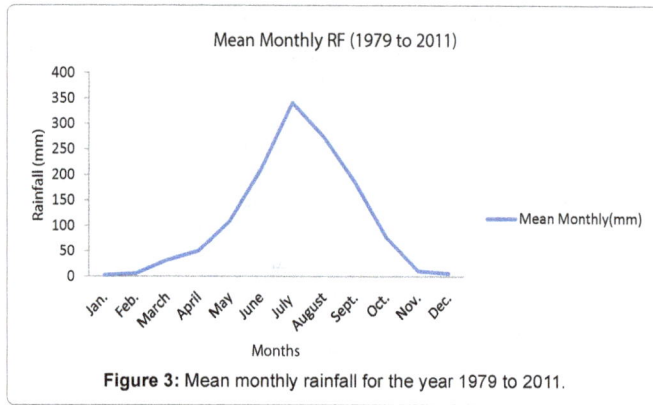

Figure 3: Mean monthly rainfall for the year 1979 to 2011.

WT depth (m)	Area (km²)	Area (%)	Waterlogging condition
-0.2 to 0.2	8.7	2.1	Severe
0.2 to 0.5	67.5	15.7	Severe
0.5 to 0.8	248.2	57.7	Severe
0.8 to 1	82.4	19.2	Critical
1 to 1.6	22.6	5.3	Critical

Table 1: Severity classification of waterlogging of sugarcane irrigation fields in the FVSE.

minimum possible time. The previous GW records (2000-2009) were obtained from the database of FVSE. The Digital Elevation Model (DEM) (30 m resolution) was downloaded from Shuttle Radar Topography Mission (SRTM). Ground water table data of the year (2011-2012) were collected from the readings of the pre-installed piezometers and groundwater monitoring commenced which are spread all over the study area (Figure 1). Topographic maps with sufficient accuracy to determine the expansion of Fincha'a Sugar Estate are not available for the past decades. Therefore an attempt was made to use Landsat imagery, which started observation in the early 1970's, for mapping current and expansion sugarcane fields. The selected images were all cloud free and cover Fincha'a Sugar Estate. A Digital CAD format Plantation (base) map showing all the roads, irrigation and drainage networks, field plots, Fincha'a river was collected from the Department of Civil Engineering of the Fincha'a Sugar Estate and the 1980 Fincha'a top sheet (scale 1:50,000) was purchased from the Ethiopian Mapping Agency (EMA).

The DEM was processed in ArcGIS environment for the study and the surrounding area, assisted by topographic and plantation base maps. The piezometer readings were analyzed in an excel spreadsheet to monthly, seasonal, and annual values for each piezometer. Any missing data were filled by regression analysis. Then, the extent of waterlogging was mapped from point-monitored data showing the piezometric surfaces. The spatio-seasonal maps of GWT depth were produced in ArcGIS 10.2 using the Inverse Distance Weight (IDW) interpolation technique. With the help of these maps, detailed explanations were provided regarding the waterlogged condition of the area for each of the four Ethiopian seasons (winter, autumn, summer, and spring) represented by four months (January, April, July, and October), respectively. Furthermore, water table depth ranges, area coverage and waterlogging condition of the study area were analyzed (Table 1). The DWDG is seamless groundwater table representation and reclassified into four distinct classes viz. most critical/severe (GWT<1 m), critical (1 m<GWT<2 m), less critical (2<GWT<3 m) and not critical/moderate (GWT>3 m), following the FAO/UNEP [12] guidelines [13].

Results and Discussion

Characterization and Severity Classification of Waterlogging in the FVSE

Waterlogged plantation fields in the FVSE are delineated, the status of GWT depth and sensitive irrigation fields to waterlogging are shown in Figure 4. The study revealed that groundwater table of the study area ranges from -0.2 to 1.6 m. In general, the results shows that GWT depth of the field is categorized as very shallow (<2 m) and hence, varied from severe with GWT depth <1 m (94.7%) to critical with GWT depth from 1 to 1.6 m (4.3%) waterlogging condition (Table 1). According to Kahlown and Hutchinson [18-20] water table rises as a consequence of poor drainage design, poor water management and is expected to contribute to the crop evapotranspiration. The GWT depth for most of observed plantation fields in the FVSE was <1 m and this can affect the yield of sugarcane [3] reported the importance of shallow GWT for soil salinization for other similar areas. It seems that the shallow perch GWT leads to high capillary movement of water in such areas and increases the risk of salinization and land degradation provided the water is saline. Kahlown and Azam [2] reported the GWT depth (1.5 m) below the ground is the critical level recommended for sugarcane crop and shallower GWT will result in a decrease of sugarcane yield. This coincided with current situation in the FVSE.

The delineated plantation critically and/or severely waterlogged fields have low topography and the soils are heavy textured (Vertisols) with very high available water holding capacity and slow infiltration rate. This was implication of the effect of topography on GWT rise. The study also revealed that all of the delineated fields had the GWT

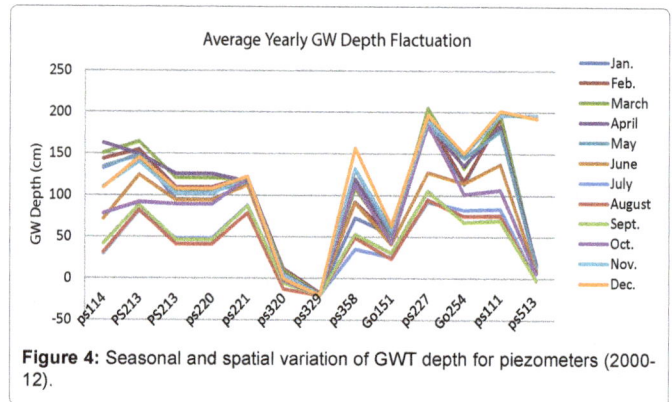

Figure 4: Seasonal and spatial variation of GWT depth for piezometers (2000-12).

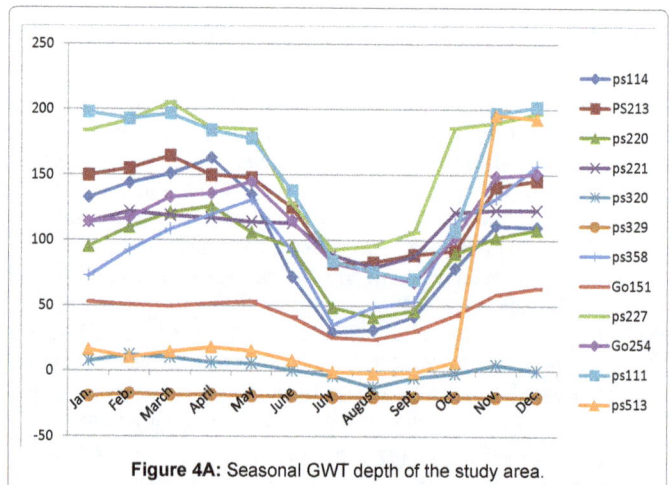

Figure 4A: Seasonal GWT depth of the study area.

depth value <2 m below the ground. This condition of GWT depth will affect the plant available water and water productivity. As a principle, the lower the GWT depth impels water is available at shallow depth while the higher GWT depth value means water is at deeper level which shows available water in the study area is found at shallow depth.

In general, GWT rise correlates negatively with the slope of the area. Furthermore, intensive grazing, over population, intensive cultivation, the vegetation cover change, deforestation and land use/land cover changes, of the upstream area of the FVSE might be affected hap hazardously. This may trigger the agricultural runoff and soil erosion in the upstream areas. The agricultural runoff from the upstream may significantly worsen the downstream areas by polluting the quality of groundwater and rising water table of the plantation fields. The upstream farmers might use different inputs to cultivate various crops. These inputs such as fertilizers, herbicides, and insecticides might increase the ionization of the ionic constituents of the surface as well as GWT rise in the downstream.

Spatial and seasonal variability of groundwater table depth for piezometer sites

Temporal (seasonal) and spatial variability of GWT depth for selected irrigated fields in the FVSE is displayed in Figures 4 and 5. Both on irrigation and off irrigation seasons were considered to identify the seasonal (monthly) fluctuations of GWT depth for these fields. The study revealed that the GWT depth for all selected irrigation fields is very shallow in winter compared to spring, autumn and summer seasons (Figure 4 and summary of Table 2). The GWT depth varies from 0.5 m in the summer to 1.09 m in the autumn. During summer, almost all of the piezometers have very shallow GWT depth (<1.0 m) below the ground while, Ps-513 and Ps-329 have shallower depth above the surface (-0.06 and -0.02 m), respectively in almost all seasons and all of the selected piezometers sites have GWT depth below the critical (1.5 m) recommended depth for sugarcane yield crop reduction.

In general, the GWT depth of the area is very shallow and showed great seasonal and spatial variability. This implies that the selected irrigation fields fall under severe to critical waterlogging condition and exposed to sugarcane yield reduction. The average rainfall pattern of the study area (Figure 3) shows that winter is the driest season or has the lowest average monthly rainfall as compared to the other three seasons. The rising of GWT during this season may be an indicator of the change of GW flows from the upper stream, recharge from irrigation, phreatic GW, poor irrigation water management, nature of basement materials and drainage condition in the study area.

The above table shows the spatial and seasonal variability of some piezometer sites which have groundwater table depth from the year 2000 to 2012. The result of the analysis shows that the average GWT depth from March to May, June to August, September to November and December to February is about 1 m, 0.5 m, 0.8 m and 1 m respectively. The result also reveals that GWT depth is very shallow in summer while relatively deep in both winter and spring. However, generally, the average GWT depth of the study area is less than 1 m which characterizes the area as area with shallow water table.

Piezometers are labeled by the field number of the plantation. Ps and Go refers to the Pump station and gravity off take, respectively for the piezometers installed in the FVSE.

That means the contribution of salinity to the crop root zone is significant as the GWT depth is being becoming shallower and shallower. The sugarcane growth was observed to be stunted and out of production. Significant irrigated fields were abandoning, almost all Black cotton (Vertisols) because of their severe to critical waterlogging

condition and GWT was rising to the surface in some irrigated fields (Table 1 and Figure 5).

During winter season the GWT depth varies from -0.2 to 2 m, with average value of 0.50 m (Figure 4C). About 62% of the selected piezometer sites have GWT depth <1.0 m and have relatively had very shallow GW depths (Table 2), which is in line with the topographic feature of the area. From July to January, the GWT depth value in some selected piezometer sites or irrigated fields showed a slight increment (Table 2) may be due to higher rainfall. Winter season (January, February and December) is the dry period in the study area, characterized by lowest rainfall (Figure 3). Hence, the plantation area is not under the influence of direct rainfall and runoff. Based up on this fact, a significant reduction in GW levels has been inevitable during this period, which is actually not the case. However, GWT depth of the study area, even in this dry season, is shallow (<2 m) and much less in some piezometer sites (ps-329 and ps-513) as revealed in Table 2. The possible sources for GW recharge during this season might be due to excess irrigation water, seepage from the irrigation and drainage canals and nature of the basement as identified This was also mostly due to the poor performance of irrigation and drainage systems of the area. However, it is important to note that the minimum GWT depth significantly increased (from 0.47 m to 0.86 m), with the exception of (Ps-329, Ps-375, Ps-513, Ps-412, Ps-320, Go-219, Go-3110, Go-266) with very shallow GWT depth during the main rainy season and remained shallower owing to their lower altitude (Figure 4 and Table 2).

In autumn season, the water table continued to decrease, but its magnitude is similar to that of the winter season. The GWT depth varied from -0.2 to 1.84 m (Figure 4 and Table 2), with average value of 1.16 m below the ground. GWT depth is relatively less shallow in this season compared with all the other seasons. About 69% of the plantation fields have GWT depth >1 m and 24% of the plantation had GWT depth >1.5 m and found to be safe from significant yield reduction. The irrigated fields (Ps-114, Ps-213, Ps-220, Go-251 and Ps-111) have relatively deep GWT depths; whereas the irrigated fields (Ps-221, Ps-320, Ps- 329, Ps-154) have relatively shallow GWT depth (Figures 4A and 4B), which is in line with the topography of the area. During this season, irrigation fields (Ps-154, Ps- 227, Ps-251, Ps-111 and Ps-513), showed significant reduction in GWT depth (Figure 4C) compared to the preceding January (winter) season. This period, like spring and winter, is irrigation season, just prior to the summer (main rainy) season. The Sugar Estate had difficulty in getting enough water for irrigation during this season; thus, sugarcane plant usually showed signs of wilting (moisture stress). The peak crop water demand and significant reduction of Fincha'a reservoir are the main reasons for the shortage of water at FVSE during this season; the later one is a major concern for sustainability of irrigation development in the Fincha'a valley, in general. The water shortage in the Fincha'a Reservoir, during this season was due to the harsh climatic conditions (high temperature, high ET, more sunshine hours, high humidity, low rainfall) and the peak water demand in the area in particular and most of the areas in the western region of the country.

Climate change, ecological change and deforestation are expected to reduce water availability in the Fincha'a Reservoir in the near future and have negative effect on the sustainability of the irrigation development. Although this season is characterized by a small but occasionally appreciable amount of rainfall, it is compensated by the peak crop water demand, high evapotranspiration and reduced water application rates due to water shortage due to minimum effects of direct rainfall and the incoming runoff on GWT depth fluctuation. Furthermore, these areas are known to receive high magnitude of runoff coming from the

Depth in (cm)		Coordinate			Months											
Year	Field	X	Y	Z	Jan.	Feb.	March	April	May	June	July	August	Sept	Oct.	Nov.	Dec
2000	ps114	327278	1069859	1534	74	70	78	60	60.5	42	3	4	6	4	54	5
2001	"				74	70	81	67.5	60.5	51	3.5	30	40	69.5	96.5	90
2002	"				198	155	147	257	141	116	27	28	72	129	221	224
2003	"				135	255	260	266	185	153	29	26	47	153	49	50
2004	"				44	92	254	266	177	61	17	16	26	65	90	92
2005	"				194	216	194	178	95	56	15	45	46	69	77	80
2006	"				107	19	18	17	35	65	6	4	0	145	212	220
2007	"				222	250	250	260	110	52	30	20	46	32	93	95
2008	"				128	130	152	147	150	107	62	24	28	37	79	80
2009	"				170	258	160	90	88	82	80	90	106	108	118	120
2010	"				169	171	164	233	254	80	80	77	70	79	42	45
2011	"				40	85	108	140	194	36	16	22	26	60	152	160
2012	"				174	101	97	135	200	40	20	24	30	70	160	168
Average					133	144	151	163	135	72	29.9	31.538	41.8	78.5	111	110
2000	PS213	327239	1072310	1546	42	90	158	154	133	57	48	76	33	14	145	150
2001	"				50	98	150	153	153	54	17	14	82	46	153	155
2002	"				184	177	130	193	130	193	108	105	102	108	204	206
2003	"				243	240	242	140	158	150	35	34	60	80	126	130
2004	"				245	245	245	245	245	245	108	105	102	108	204	209
2005	"				243	240	242	143	158	108	28	62	70	105	145	150
2006	"				145	145	187	180	195	200	193	129	80	215	187	190
2007	"				214	229	231	231	225	219	218	174	185	201	210	225
2008	"				132	141	195	206	208	183	183	177	154	159	171	178
2009	"				195	193	235	229	231	133	53	64	71	73	104	110
2010	"				103	74	71	59	63	63	65	62	59	64	121	126
2011	"				126	109	20	11	13	6	3	38	79	12	29	30
2012	"				25	28	33	12	14	10	6	40	80	15	30	40
Average					150	155	164.5	150	148	125	81.9	83.077	89	92.3	141	146
2001	ps220	322044	108433	1486	30	56	96	118	112	76	21	21	34	100	110	115
2002	"				91	82	80	98	90	28	19	18	24	118	65	70
2003	"				71	40	100	180	97	12	17	10	19	104	116	118
2004	"				112	130	122	99	55	130	19	27	25	110	84	92
2005	"				119	140	143	140	109	107	6	31	30	88	75	80
2006	"				114	110	173	170	131	126	8	6	6	95	75	78
2007	"				84	140	160	165	120	114	49	85	88	80	100	112
2008	"				92	102	120	110	113	105	39	23	26	39	60	65
Depth in (cm)		Coordinate			Months											
Year	Field	X	Y	Z	Jan.	Feb.	March	April	May	June	July	August	Sept	Oct.	Nov.	Dec
2009	"				80	104	98	88	76	45	11	20	32	33	96	98
2010	"				114	105	110	106	108	155	154	10	32	80	106	109
2011	"				98	158	144	138	157	121	121	120	120	112	168	178
2012	"				138	155	107	105	105	120	120	125	122	120	170	180
Average					95.3	110	121.1	126	106	94.9	48.7	41.333	46.5	89.9	102	108
2000	ps221	326804	1072774	1558	53	73	67	93	111	109	120	94	133	173	103	105
2001	"				59	60	61	121	117	110	153	84	107	179	119	120
2002	"				103	115	118	109	121	125	112	73	105	133	163	165
2003	"				127	103	115	101	94	71	77	82	103	150	107	109
2004	"				146	94	100	99	94	71	40	60	87	149	104	105
2005	"				142	162	151	132	163	120	56	104	63	84	127	130
2006	"				120	145	140	103	100	89	78	69	52	53	108	110
2007	"				113	154	183	149	93	105	81	42	31	100	89	90
2008	"				93	98	127	134	107	127	60	44	62	37	61	65
2009	"				95	98	139	129	97	158	97	82	136	138	153	156
2010	"				119	123	118	130	124	121	117	116	110	137	133	143

Year	Field	X	Y	Z	Jan.	Feb.	March	April	May	June	July	August	Sept	Oct.	Nov.	Dec.
2011	"				190	256	118	118	131	139	83	89	54	118	163	135
2012	"				128	108	109	104	125	130	80	90	100	120	165	163
Average					114	122	118.9	117	114	113	88.8	79.154	87.9	121	123	123
2000	ps320	327382	1074827	1508	-40	-11	-5	-10	-15	-40	-13	-30	-20	-13	-23	-20
2001	"				-8	-13	-4	-6	-14	-13	-28	-32	16	-5	-21	-24
2002	"				-10	-2	-6	-9	-1	-12	-9	-20	-16	9	6	-6
2003	"				12	24	40	41	40	-2	9	-7	6	21	9	10
2004	"				4	10	22	41	42	37	0	-9	-1	13	7	12
2005					15	-3	0	9	21	-1	-13	-8	-14	-6	14	18
2006	"				7	4	-11	-4	-6	-4	0.3	-10	-2	-9	9	10
2007	"				36	40	33	10	2	26	1	-3	-5	-1	6	8
2008	"				17	11	24	18	18	32	32	-4	-5	-8	-2	-1
2009	"				17	18	14	10	7	11	0	-2	5	8	8	10
2010	"				35	18	11	9	8	6	4	3	0	-10	-10	-8
2011	"				-13	40	-8	-10	-12	-14	-20	-18	-16	-12	29	-4
2012	"				29	24	26	-15	-15	-20	-12	-20	-16	-12	29	-4
Average					7.77	12.3	10.46	6.46	5.77	0.46	-3.7	-12.31	-5.23	-1.9	4.69	0.08
2000	ps329	327183	1076187	1493	-23	-20	-20	-20	-20	-20	-20	-20	-20	-20	-20	-20
2001					-20	-21	-21	-21	-21	-21	-21	-21	-21	-21	-21	-21
2002					-20	-20	-20	-20	-20	-20	-20	-20	-20	-20	-20	-20
2003					-20	-20	-20	-20	-20	-20	-20	-20	-20	-20	-20	-20

Depth in (cm)		Coordinate			Months											
Year	Field	X	Y	Z	Jan.	Feb.	March	April	May	June	July	August	Sept	Oct.	Nov.	Dec.
2004					-20	-20	-20	-20	-11	-12	-20	-20	-20	-20	-20	-20
2005					-20	-20	-20	-20	-20	-20	-20	-20	-20	-20	-20	-20
2006					-20	-20	-20	-20	-20	-20	-20	-20	-20	-20	-20	-20
2007					-20	-20	-20	-20	-20	-20	-20	-20	-20	-20	-20	-20
2008					-20	-4	-16	-20	-20	-20	-20	-20	-20	-20	-20	-20
2009					-20	-20	-20	-20	-20	-20	-20	-20	-20	-20	-20	-20
2010					0.5	0.4	0.8	0.7	-13	-15	-16	-18	-18	-20	-20	-20
2011					-20	-20	-20	-20	-20	-20	-20	-20	-20	-20	-20	-20
2012					-20	-20	-20	-20	-20	-20	-20	-20	-20	-20	-20	-20
Average					-19	-17	-18.2	-18	-19	-19	-20	-19.92	-19.9	-20	-20	-20
2000	ps358	327152	1078545	1462	0	33	122	165	192	183	8	6	47	145	242	245
2001	" "				19	46	147	180	169	25	16	10	118	162	252	258
2002	" "				18	80	141	130	177	70	22	12	24	170	17	20
2003	" "				110	198	235	41	108	32	21	9	32	89	18	24
2004	" "				37	57	112	130	164	158	31	16	32	110	162	165
2005	" "				11	25	55	25	14	8	6	42	10	9	212	220
2006	" "				230	240	231	230	224	12	18	12	8	87	68	70
2007	" "				165	200	26	130	68	54	11	13	15	59	118	120
2008	" "				23	26	54	33	22	28	12	10	9	16	170	180
2009	" "				183	34	69	160	165	115	28	61	129	131	150	160
2010	" "				123	130	148	180	216	126	129	121	109	153	45	50
2011	" "				32	30	73	80	190	192	160	167	160	171	266	267
2012	" "				120	106	79	78	180	190	145	160	160	171	266	267
Average					73.2	92.7	108.7	120	131	91.8	35.5	49.154	53.3	113	132	157
2000	Go151	3227019	1081228	1454	52	49	57	63	58	55	18	20	34	48	46	50
2001	" "				50	56	18	36	55	38	26	24	45	50	53	55
2002	" "				50	48	48	50	72	43	24	23	24	55	98	100
2003	" "				74	60	85	74	50	19	39	17	38	62	55	58
2004	" "				71	67	74	65	72	27	10	26	32	52	50	55
2005	" "				46	52	60	80	46	57	16	18	40	20	46	50
2006	" "				46	24	47	56	60	28	16	15	16	42	46	54
2007	" "				69	42	51	49	45	45	42	46	51	69	151	160
2008	" "				42	40	44	48	25	28	12	11	9	18	41	45
2009	" "				40	46	40	68	66	51	30	26	24	23	32	35
2010	" "				40	41	38	48	42	51	36	30	28	38	41	46

Year	Field	X	Y	Z	Jan.	Feb.	March	April	May	June	July	August	Sept	Oct.	Nov.	Dec
2011	" "				50	54	30	35	48	49	32	27	30	41	52	58
2012	" "				60	80	54		52	50	34	30	30	41	52	58
Average					53.1	50.7	49.69	51.7	53.2	41.6	25.8	24.077	30.8	43	58.7	63.4

Depth in (cm)		Coordinates			Months											
Year	Field	X	Y	Z	Jan.	Feb.	March	April	May	June	July	August	Sept	Oct.	Nov.	Dec
2000	ps227	326094	1084349	1418	193	204	215	235	235	195	114	155	170	214	220	240
2001	" "				220	204	230	238	225	163	146	160	210	225	222	230
2002	" "				235	234	210	232	160	178	138	131	134	240	244	245
2003	" "				240	224	212	232	240	166	208	138	153	220	210	220
2004	" "				188	212	214	185	226	205	168	170	172	210	268	270
2005	" "				212	227	243	228	214	70	14	130	98	105	96	100
2006	" "				210	206	190	17	13	19	16	8	8	208	215	220
2007	" "				195	160	190	200	220	200	80	97	92	110	176	180
2008	" "				185	193	230	213	106	93	56	41	34	150	172	178
2009	" "				-2	130	142	151	192	140	66	108	160	190	192	200
2010	" "				158	154	159	156	156	150	143	140	125	172	220	242
2011	" "				180	172	216	160	200	40	28	-20	9	184	115	117
2012	" "				175	182	216	165	214	50	32	-15	9	184	115	117
Average					184	192	205.2	186	185	128	93	95.615	106	186	190	197
2000	Go254	323156	1086264	1437	48	46	141	33	113	90	15	9	14	33	108	109
2001	" "				88	90	148	34	103	118	36	14	10	8	116	118
2002	" "				88	89	140	180	188	17	34	9	13	27	110	112
2003	" "				43	24	21	162	180	13	42	10	15	83	161	165
2004	" "				79	69	55	154	73	165	24	42	92	135	137	140
2005	" "				146	175	210	132	201	121	86	131	20	112	156	156
2006	" "				113	155	160	178	190	200	180	90	96	127	135	135
2007	" "				148	100	90	120	90	36	100	85	82	120	207	204
2008	" "				200	196	160	116	99	114	77	143	82	110	134	140
2009	" "				90	140	144	161	150	110	113	92	128	130	156	156
2010	" "				154	158	160	165	162	176	82	110	102	117	177	178
2011	" "				141	142	154	162	175	163	147	115	118	166	178	180
2012	" "				140	138	150	165	160	178	142	134	112	156	167	170
Average					114	117	133.3	136	145	115	82.9	75.692	68	102	149	151
2000	ps111	327065	1068758	1458	133	154	148	189	158	155	68	33	18	-8	123	124
2001					182	123	169	165	167	134	24	134	127	245	218	220
2002					230	215	200	220	273	278	250	136	124	84	218	241
2003					240	278	274	229	270	170	50	40	65	208	270	278
2004					226	266	172	167	123	93	53	24	22	100	260	270
2005					270	276	280	201	200	160	46	47	19	80	249	250
2006					278	280	280	172	115	50	48	46	30	67	228	229
2007					260	262	280	280	280	170	50	31	50	81	111	112
2008					140	156	162	164	151	47	64	51	63	71	138	140

Depth in (cm)		Coordinate			Months											
Year	Field	X	Y	Z	Jan.	Feb.	March	April	May	June	July	August	Sept	Oct.	Nov.	Dec
2009					138	170	176	145	143	143	134	126	123	132	140	142
2010					176	179	260	250	250	250	232	228	198	161	152	160
2011					94	31	34	76	80	77	32	36	38	90	227	228
2012					210	120	127	130	98	72	45	57	36	80	225	227
Average					198	193	197.1	184	178	138	84.3	76.007	70.2	107	197	202
2000	ps513	325645	1085546	1445	0	0	-3	42	-1	-1	-2	-3.5	-3.5	42.5	220	220
2001					39	-1.5	-3	-5	-4	-7.5	-9.5	-9	-1.5	2.5	222	231
2002					55.5	19.5	47.5	71	60.5	22	13	13.5	21.5	20	244	245
2003					27.5	26.5	28.5	24	34	30.5	15	13	15.5	16	210	220
2004					16.5	23	17.5	17.5	13.5	8	3	4.5	5.5	14.5	268	270
2005					16.5	23	18.5	18.5	16.5	8	3	4.5	5.5	14.5	96	100
2006					15.5	6	8.5	6	4	2.5	-1.5	-2	-1	3	215	220
2007					17.5	14.5	11.5	9	6.5	4.5	2	-3	-4	-0.5	176	180
2008					2	5.5	19	35	70.5	36.5	18	10	8	10	172	172

2009			10	10	10	7	5	40	0.3	0	0.5	0.5	192	192
2010			10	12	11	10	-13	-26	-28	-10	-26	-10	220	150
2011			-12	-10	10	-16	-7	-19	-20	-38	-33	-28	115	112
Average			16.5	10.7	14.67	18.3	15.5	8.21	-0.6	-1.667	-1.04	7.08	196	193

Table 2: Spatial and seasonal fluctuation of GWT Depth for selected Piezometer sites.

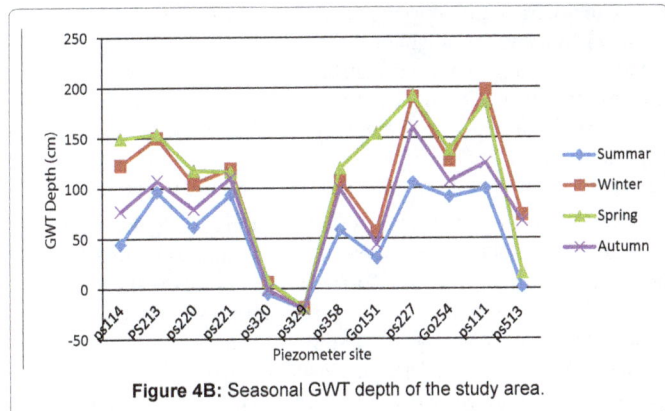

Figure 4B: Seasonal GWT depth of the study area.

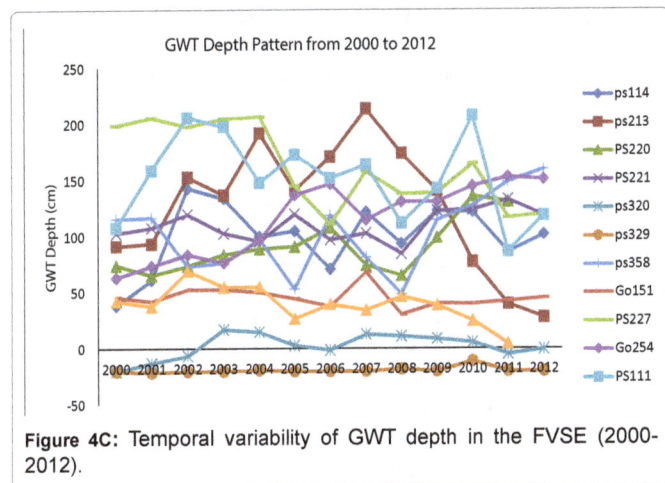

Figure 4C: Temporal variability of GWT depth in the FVSE (2000-2012).

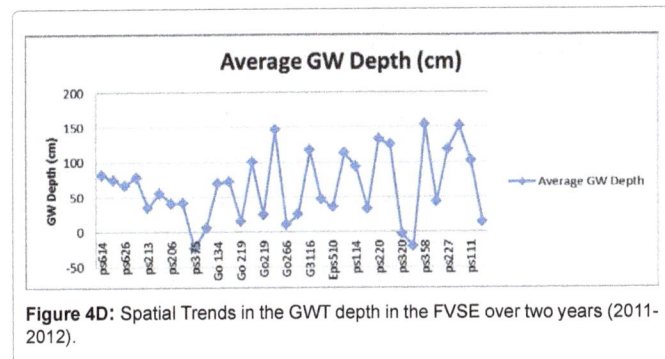

Figure 4D: Spatial Trends in the GWT depth in the FVSE over two years (2011-2012).

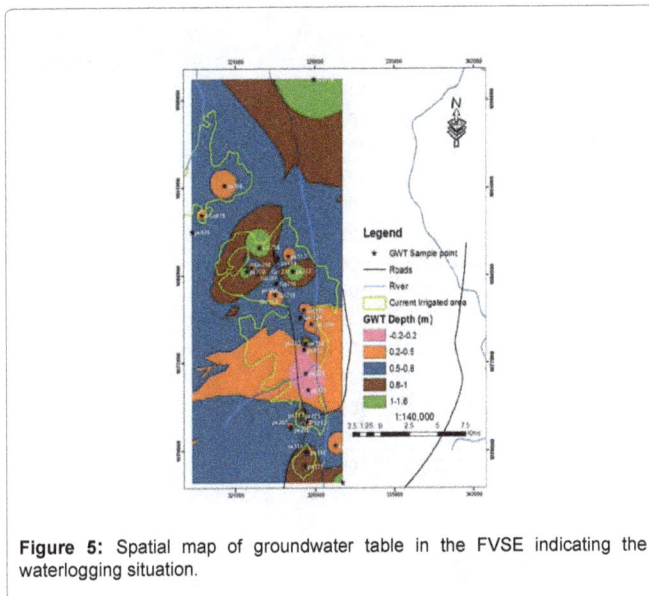

Figure 5: Spatial map of groundwater table in the FVSE indicating the waterlogging situation.

upstream plateaus. From this result, it could be concluded that there was a change of GW flow pattern in this period, from the upstream to the downstream of the plantation; Thereby disturbing the normal drainage system of the area since water is usually drained toward the downstream. This condition is especially challenging to water managers of the Sugar Estate (Table 3).

During the summer (July) season, the GWT depth varied from -0.06 to 0.93 m, with average value of 0.50 m below the ground surface

(Figure 4B). Almost more than half of the plantation has GWT depth <0.50 m and the entire selected cane plantation fields were severely waterlogged during this season. According to Figure 4B, more than 92% of the piezometers have GWT depth <1 m below ground. This implies that the period is characterized by Severe waterlogging conditions in the area (Figure 4C) whereby GWT depth as low as 0.05 m below the ground surface has been recorded in Ps-320 surprisingly Ps-329 and Ps-513 have GWT depth (-0.2 and -0.-06 m), respectively due to the nature of the basement rock. Most irrigation fields have shown an extremely high rise in GWT depth compared to the preceding winter and autumn seasons (Figures 4B and 4C). This might be the implication of the response of the GW depth to rainfall is quick and significant. Spring season, this period is the beginning of the irrigation season, following the main rainy (summer) season. The GWT depth varied from -0.02 to 1.86 m (average ~ 0.99 m) below the ground surface and 0.11 m above the ground surface (Figure 4B). About 62% of the selected cane plantation fields have GWT depth <1.0 m and hence are most critically (severely) waterlogged whereas 39% have the depth (1 m <GWT<2 m) and critically waterlogged.

The GWT depth was slightly reduced compared to the previous summer season. Ps-151, Ps-221, Ps-358, Ps-227, Ps-251, Ps-111 and Ps-231 have relatively shallow GWT depth compared with the other piezometer sites or cane plantation fields. During this period, Ps-320, Ps-329, Ps-151 and Ps-513 had very shallow GWT depths (<2 m) during the previous (summer) season. GWT depth during the summer season (Figure 3), have shown greater reduction compared to the other parts (Figure 4D). This showed that the GWT depths of the area would have been reduced significantly if drainage systems were effective. This ineffectiveness of the drainage system of the area, as discussed earlier, had an adverse effect on the GWT depth condition during the subsequent periods (winter and autumn).

The analysis of the data showed that the GWT depth in the FVSE was increasing during the spring season, which resulted in an

FID	X	Y	Z	Average Yearly GW Depth (cm)
ps614	324770	1085353	1458	82.08
ps604	324630	1083347	1461	74.71
ps626	317112	1087492	1494	66.54
ps115	327163	1069866	1465	78.29
ps213	327239	1072310	1546	34.67
ps205	325865	1071779	1591	55.79
ps206	325878	1071933	1592	41.13
ps356	320045	1091201	1446	41.29
ps375	327007	1078071	1492	-23.42
ps412	327029	1078062	1456	6.31
Go 134	326714	1080603	1463	70.04
Go216	324630	1083347	1468	72.25
Go 219	324517	1082446	1480	15
G0207	323009	1084158	1477	101.13
Go219	324517	1082446	1480	24.63
Go 227	326094	1084349	1418	146.25
Go266	322454	1084656	1476	10.29
G03110	327693	1080101	1459	25.42
G3116	327922	1099667	1439	117.21
EGo819	318010	1088821	1483	47.58
Eps510	329783	1070440	1564	35.75
Eps404	330413	1067462	1597	113.88
ps114	327278	1069859	1534	94.08
PS213	327239	1072310	1546	33.71
ps220	322044	1084333	1486	133.42
ps221	326804	1072774	1558	125.67
ps320	327382	1074827	1508	-2.67
ps329	327183	1076187	1493	-20
ps358	327152	1078545	1462	154.58
Go151	327019	1081228	1454	43.63
ps227	326094	1084349	1418	118.54
Go254	323156	1086264	1437	152.23
ps111	327065	1068758	1458	102.92
ps513	325645	1085546	1445	14.75

Table 3: Average Yearly GWT depth of Piezometer sites of 2011 and 2012 years.

average GWT rise of 1.5 m during this period throughout the selected irrigation fields. But the GWT decreased during the winter season. This was the implication of groundwater recharge. High GWT subsurface soil water in the FVSE has raised gradually near enough the nature of surface level to cause the moisture to move up by capillary action (the hot moisture evaporates this moisture, leaving). The magnitude and extent of waterlogging depends upon the soil structure which facilitates movement of soil water by capillary action. This can be observed from (Figure 4C) that the contribution of groundwater inflow in the study site lead to continuous accumulation of GW in the study area, which the GWT kept moving upward. Long-term monitoring provides evidence of this situation. Figure 4C showed the temporal variability of GWT during 2000-2012.

It shows that GWT depth in 2000 and 2012 were at 0.64 and 0.7 m respectively, which indicates a rise in the table of 0.06 m in a span of 12 years. However, the maximum and minimum rises in the GWT were recorded at Ps-320 and Ps-329 (-0.2 m) and Ps-111 and PS-227 (2.06 m), respectively. A continuously rising GWT caused waterlogging and enhanced flood problems in the studied piezometer sites, particularly during the spring season. These floods were further accented by man-made barriers such as roads, canals obstructing the natural flow of water. Waterlogging and flooding in the agricultural areas caused major damage to crops and soil fertility.

Trends in the groundwater depth in the FVSE of piezometer sites (2010-2012)

In Fincha'a Valley problems due to irrigation are reported with regard to waterlogging. The ground water level within the FVSE showed a rising tendency. This might be due to irrigation (seepage losses out of reservoirs and channels, over watering, etc) but also due to runoff from the upstream area. What has been reported before and confirmed in the field work are problems with a high water table and waterlogged fields. Especially on fields with heavy clayey soils problems with waterlogging have been discovered during the field study and the analysis in a soil.

The result showed that on some plantation fields the GWT was very shallow (GWT <2 m). The same report indicated that in some fields downward percolation of irrigation water below the root zone, especially in soils classified as heavy black cotton soils was so slow that it caused temporary storage within the root zone (in some places up to 10 days after irrigation). These drainage problems have even become one of the major factors in determining the composition of cane variety. In fields located near irrigation canals and drains suppressed cane growth due to seepage and GWT rise could be observed and cane loss due to this problem was economically significant. The measurement of GWT of the FVSE showed the maximum GWT depth values in all years and months.

The results indicated that fields which are located at low topography and on Vertisols were affected by rising GWT depth more than fields near to the distribution canals and nearby drains. Even this short measurement time series of 2010 to 2012 showed significant increasing trends of rising GWT. Sugarcane plantation fields of low topography, Vertisols and flat slope (ps-329, ps-320) were the most affected fields. The major cause for the rise of GWT depth in the area might be seepage from unlined irrigation canals, geological conditions, agricultural runoff, topography and soil type, coupled with poor drainage systems.

The GWT depth within the FVSE showed a rising tendency (Figure 4C). This might be due to irrigation (seepage losses), agricultural runoff and over use of irrigation water. The average GWT depth for selected irrigation fields varied from -0.27 to 1.54 m (Ps-320), (Figure 4D) what had been observed during the field works and confirmed from the results were problems with a high GWT rise and waterlogging condition. Especially on fields with heavy clayey soils (Vertisols) problems with waterlogging have been discovered during the field observation and its trend or extent was confirmed from the results of analyses. However, the lighter soils (Luvisols) light in relation to the heavy soils of the Sugar Estate seem to be of very good quality and not prone to negative impacts of waterlogging under the given conditions. The investigation showed that most of selected irrigation fields had very shallow (GWT <2 m) and in some irrigated fields downward percolation of irrigation water below the root zone, especially in soils classified as heavy black soils was so slow that it caused temporary storage within the root zone in some places after irrigation (Ps-205, Ps-206, Ps-375, Ps-329 and Ps-320).

These drainage problems have even become one of the major factors in determining the composition of cane variety. In fields located down irrigation canals and drains suppressed cane growth due to seepage and ground water table rise could be observed. The measurement of GWT in the FVSE showed the maximum GWT depth values in all years and months. The results indicated that fields which are located at lower topography are affected by rising GWT depth more than fields nearby distribution canals and fields nearby drains. Even this short measurement time series of 2010 to 2012 showed significant increasing trends of rising GWT.

Conclusion

In the present study, waterlogged irrigation fields have been delineated using GIS and remote sensing and other ancillary information. From this study it inferred that GIS and remote sensing data, firsthand knowledge of waterlogging area can be obtained. To confirm the results obtained from image processing, other data such as topography and groundwater level information are very much required. Groundwater maps and satellite derived waterlogged maps may be used to zeroing in to sub areas for further investigation. Waterlogged area was verified in the field through ground truth at the selected irrigation fields/pizometer sites through field observation.

From this field visit and analysis result it was observed that a possible cause of waterlogging could be poor irrigation water management, nature of basement rock and/or geology, nature of soil, slope, poor and neglected drainage design condition of the plantation field. The entire selected sugarcane plantation fields (area) were under critical to severe waterlogging conditions. The study revealed that the GWT rise varied from 2.1 to 57.7% and are under severe and critically waterlogged, respectively. The severely affected areas are where drainage is not sufficiently designed. This study result clearly revealed that GWT depth in the FVSE was extremely shallow, at all seasons, exceeding the critical depth (1.5 m) recommended for sugarcane. It was characterized by a rise during summer and spring, and a gradual decline in winter and autumn due to water uptake by plants, decreased rainfall, increased ET, increased relative humidity and GW discharge to streams and wetlands. Consequently, the direction of GW flow pattern in the cane fields, as suggested earlier, was subject to change in summer and spring based upon the seasonal GWT depth status and topography of the area. In post season the rise of GWT depth during the rainy season was not compensated by the fall of GWT depth during the non-rainy periods.

This was mostly due to the failure of the surface drainage system of the area to drain excess water from the fields. As attempt was made through application of geophysical (resistivity) method identified the root cause for the rise in GWT depth of the study area in addition to topography, soil, poor water management and drainage, the nature of the basement rock and the recharge from direct irrigation and runoff from surrounding upstream escarpments could be the main causes responsible for the rise and fluctuation of GWT depth. Therefore, detailed investigations that include the entire possible causes of GWT rise and adoption of a feasible management strategy to limit a further rise of GWT depth in the area was highly suggested.

Intercropping with other low water crops like sorghum was suggested, as it could reduce percolation rates considerably. Long-term over irrigation had a cumulative effect on the rise of the GWT and could cause water logging and associated problems even if the border drains were effective in stopping the incoming runoff. Thus, efforts on the management of water resources, especially irrigation and drainage, in such areas were extremely important for the sustainability of irrigated agriculture. Moreover, reducing canal water releases into non-cane fields could also reduce net recharge to aquifer and integrated watershade management physiographic/soil/hydrological differences in the irrigation fields and thus is useful to carry out further investigations.

Acknowledgements

The authors are grateful to the Fincha'a Valley Sugar Estate Irrigation and Civil Engineering division for material and technical support. Our special thanks goes to the technical staff of the Research Directorate (Fincha'a) for providing the necessary support during data collection.

References

1. Kahlown MA, Ashraf D, Haq Z (2005) Effect of shallow groundwater table on crop water requirements and crop yields. Journal of Agriculture and Water Management 76: 24-35.

2. Kahlown MA, Azam M (2002) Individual and combined effect of waterlogging and salinity on crop yields in the Indus basin. Journal of Irrigation and Drainage 51: 329-338.

3. Kaul Harshika A, Sopan T Ingle (2011) Severity Classification of Waterlogged Areas in Irrigation Projects of Jalgaon District, Maharashtra. Journal of Applied Technology in Environmental Sanitation 1: 221-232.

4. Asmuth JR, Knotters M (2004) Characterizing groundwater dynamics based on a system identification approach. Journal of Hydrology 5: 34-48.

5. DNRE (Department of Natural Resources and Energy) (2001) Victorian Catchment indicators. Our commitment to reporting on catchment condition. DNRE. Victoria.

6. Tanji KK, Kielen NC (2002) Agricultural drainage water management in arid and semi-arid areas. FAO of the UNS. Irrigation and Drainage paper 61, ISBN 92-5-104839-8. Rome. Italy.

7. Megersa Olumana, Dilsebo H (2010) Characterization of the responses of groundwater monitoring piezometers installed at WSSE. 2nd biannual conference of Ethiopian sugar industry on theme 'sugarcane production & climate'. Adama, Ethiopia.

8. Khan MN, Rastoskuev VV, Sato Y, Shiozawa S (2005) Assessment of hydro saline land degradation by using a simple approach of remote sensing indicators. Journal of Agriculture and Water Management 77: 96-109.

9. Chaudhari VM, Chaudhari Viru R, Neeti N, Bothale RV, Srivastava YK, et al. (2008) Assessment of surface and sub-surface waterlogged areas in irrigation command areas of Bihar state using remote sensing and GIS. Journal of Agricultural Water Management 95: 754-766.

10. Mandal AK, Sharma RC (2001) Mapping of Waterlogged Areas and Salt affected soils in the IGNP Command Area. Journal of the Indian Society of Remote Sensing 29: 239-235.

11. Mothikumar KE, Bhagwat KA (1989) Delineation and Mapping of Salt Affected Lands in Pariej village of Kheda district (Gujarat) by Remote Sensing. Journal of the Indian Society of Remote Sensing 19: 40-57.

12. FAO/UNEP (Food and Agricultural Organization) (1984) Provisional methodology for assessment and mapping of desertification. Rome. p. 84

13. Masoudi M, Patwardhan A, Mand Gore SD (2006) A new methodology for producing of risk maps of soil salinity, Case study: Payab Basin, Iran. Journal of Applied Sciences and Environmental Management 10: 9-13.

14. Megersa Olumana (2004) Evaluation of the infield water application performance of sprinkler irrigation system at Fincha'a Sugar Estate. MSc Thesis Submitted to Arbaminch University, Ethiopia.

15. Mohamedin AAM, Awaad MS, Ahmed AR (2010) The Negative Role of Soil Salinity and Waterlogging on Crop Productivity in the Northeastern Region of the Nile Delta, Egypt. Research Journal of Agriculture and Biological Sciences 6: 378-385.

16. Adaman F, Madra Y (2003) A participatory Framework for Poverty Eradication in I.H. Unver, Rajiv. Gupta K, Kebaroglu A (eds.), Water, Development and Poverty Reduction, Kluwer Academic Publishers, London.

17. Ademe Adenew (2001) Summary of Metrological Data (1979 - 2000) Fincha'a Research Station, Agriculture, Ethiopia. pp. 272.

18. Amdihun A (2007) GIS and remote sensing integrated environmental impact assessment of irrigation project in Fincha'a Valley area, Ethiopia. Journal of Catchment and Lake Research.

19. Kahlown MA, Lqbal M, Skogerboe GV, Rehman SU (1998) Waterlogging, salinity and crop yield relationships. Mona Reclamation Experimental Project. WAPDA. Report (233).

20. Hutchinson MF (1989) A new procedure for gridding elevation and stream line data with automatic removal of spurious pits. Journal of Hydrology 106: 211-232.

7

Prediction of Stream Flow and Sediment Yield of Lolab Watershed Using SWAT Model

Sarvat Gull, Ahangar MA and Ayaz Mohmood Dar*

Department of Civil Engineering, National Institute of Technology, Srinagar, Jammu and Kashmir, India

Abstract

The SWAT model was used to estimate the runoff and sediment yield of Lolab watershed. The model was calibrated, validated, and assessed for evaluation to model ambiguity using Nash–Sutcliffe coefficient (N_{SE}) and coefficient of determination (R^2). Ten highly sensitive parameters were recognized for stream flow simulation of which CN2 (Initial SCS CN II value) factor was the most sensitive one and four highly sensitive parameters were recognized for sediment yield simulation of which SPCON (Linear parameters for sediment re-entrainment) was most sensitive one. The model was calibrated for a time period between 1993 to 2000 and validated from 2001 to 2004 for flow and sediment yield. The predicted and observed stream flow and sediment yields generally matched well. The results of the model calibration and validation showed reliable estimates of monthly stream flow (R^2=0.74 and E_{NS}=0.68) and yearly stream flow (R^2=0.90 and E_{NS}=0.68) during the calibration period and monthly stream-flow (R^2=0.85 and E_{NS}=0.83) and yearly stream-flow (R^2=0.99 and E_{NS}=0.91) during the validation period. For sediment yield, this study shows antremendous model efficiency of monthly sediment yield (R^2=0.80 and E_{NS}=0.79) and yearly sediment yield (R^2=0.86 and E_{NS}=0.78) during the calibration period and monthly sediment yield (R^2=0.88 and E_{NS}=0.86) and yearly sediment yield (R^2=0.83 and E_{NS}=0.58) during the validation period. This study showed that the SWAT model is competent of predicting sediment yields and hence can be used as a tool for water resources planning and management in the study watershed.

Keywords: SWAT model; Model evaluation; Stream flow; Sediment yield

Introduction

A Watershed is an area of land that drains all the streams and rainfall to a common outlet. Essentially a watershed is all the land and water area which contributes runoff to an outlet in the main flow channel [1]. Watersheds are significant because the stream flow and the water quality of a river are affected, human-induced or natural, happening in the land area "above" the river-outflow point. The environment worsening of a watershed is a common occurrence in most parts of the world. Amongst several causes the major one are improper and unwise utilization of watershed resources observed in developing countries [2]. Watershed management implies rational utilization of land, soil, and water resources for optimum and sustained production with minimum hazards to natural resources and environment [3]. Soil erosion is a serious global issue because of its rigorous adverse economic and environmental impacts. Economic impacts on productivity may be due to direct effects on crops/plants both on-site and off-site, and environmental consequences are primarily off-site related to the damage to civil structure, siltation of water ways and reservoirs, and additional costs involved in water treatment. Even though the adverse influences of soil erosion on soil degradation have long been recognized as a key problem for human sustainability, estimation of soil erosion is often complicated due to the complex interplay of many factors such as climate, land cover, soil, topography, lithology, and human activities. This study employs the SWAT (Soil and Water Assessment Tool) model to take advantage of its integration with GIS (Geographic Information System) and locally available data and data from similar areas that can be used to calibrate and validate the model, so that runoff and sediment yield from the watershed can be predicted and most problematic sub-basins can be identified.

Objectives of the Study

1) To analyze the Land use/Land cover of Lolab watershed.

2) To estimate the catchment runoff and sediment yield of Lolab watershed of Pohru catchment using SWAT model.

3) To evaluate the performance of Swat model by comparing its predicted flow and sediment yield with corresponding observed values at the study watershed.

4) To identify the most problematic sub-basins with respect to sediment contribution.

Methods and Materials

Lolab watershed is one of the watersheds of Pohru catchment with an area of about 45,250 hectares. The study area lies between 34° 41' to 34° 24' N Latitude and 74° 09' to 74° 23' E Longitude. The watershed can be divided into three distinct physiographic units i.e., the Mountains, the Karewas and the Flood plains. Elevation in the watershed varies from approximately 1,500 to 3,900 meters above mean sea level.

Model description

Arc SWAT: ArcSWAT is a public domain graphical user interface program. It is designed to link the hydrologic model SWAT (Soil and Water Assessment Tool) and the GIS package ARCINFO. The SWAT model is limited in that it does not explicitly allow for the inclusion of spatial data as model inputs. Data must be processed into a form

***Corresponding author:** Ayaz Mohmood Dar, Department of Civil Engineering, National Institute of Technology, Srinagar, Jammu and Kashmir, India E-mail: ayazmohmood@hotmail.com

that the model can use. Processing these data, even with the use of a GIS is tiresome and time consuming due to the large number of model parameters required for executing SWAT. The development of ArcSWAT aims at an effective use of spatial data to enhance hydrological modeling. The interface performs the following tasks

- To streamline GIS processes tailored toward SWAT modeling needs.

- To automate data communication between Arc/Info and SWAT.

- To provide a user-friendly data entry and editing environment for SWAT.

The SWAT model: The Soil and Water Assessment Tool (SWAT) is a river basin scale, continuous time and spatially distributed physically based model developed to predict the impact of land management practices on water, sediment, and agricultural chemical yields in complex catchments with varying soils, land use and management conditions over long periods of time [4,5]. In this study, the Arc SWAT 2005 version of the SWAT model is applied to predict stream flow and sediment yield. SWAT uses Hydrologic Response Units (HRUs) to describe spatial heterogeneity in terms of land cover, soil type and slope within a catchment. The SWAT model uses two steps for the simulation of hydrology: the land phase and routing phase. The land phase controls the amount of sediment, nutrient and pesticides loading to the main channel in each sub-basin. SWAT offers two methods for estimating surface runoff: The Soil Conservation Service (SCS) Curve Number (CN) procedure and the Green and Ampt infiltration method. Using daily or sub-daily rainfall amounts, SWAT simulates surface runoff volumes and peak runoff rates for each HRU. SCS curve number method is less data intensive than the Green-Ampt method [6]. In this study, the SCS curve number method was used to estimate surface runoff volumes because of the unavailability of sub-daily data for the Green and Ampt method. Sediment yield is estimated using a Modified Universal Soil Loss Equation (MUSLE).

Preparation of model inputs

The basic spatial input datasets used by the model include the Digital Elevation Model (DEM), stream network, land use/cover data, soil data and climatic data. DEM was derived from the NASA 15-meter Shutter Radar Topography Mission (SRTM) dataset. The DEM was used to delineate the catchment and analyze the drainage patterns of the land surface as well as derive slope parameter. The stream network data set was digitized from topographic map of scale (1:50000 and 1:250000) of the study area. The stream network dataset was superimposed onto the DEM to define the location of the stream network. Burning-in stream network operation is most important in situations where the DEM does not provide enough detail to allow the interface to accurately predict the location of the stream network. The land use is one of the most important factors that affect runoff, evapotranspiration, and surface erosion in a watershed. The land use map was obtained from the Department of Agricultural Engineering, SKUAST-K. Agriculture was the dominant land use category in the Lolab watershed followed by the sparse forest.

Since SWAT has pre-defined land use types which identified by four-letter codes, the Land use/Land cover map was reclassified in order to correspond with the parameters in the SWAT database.

The SWAT model requires different soil textural and physical-chemical properties such as soil texture, available water content, hydraulic conductivity, bulk density, and organic carbon content

for different layers of soil. In this particular study, Soil Conservation Department, Kashmir global available soil data (1:50,00,000 scales) in vector format had been obtained from Earth Science Department, Kashmir University. It is converted into a grid format for SWAT model input parameters. A total of three soil texture classes were delineated out of which the fine loamy soil was the major class which accounted for 79.58 percent of the total area of the watershed and was followed by fine silty soil with 16.53 percent of the total area.

Meteorological data is needed by the SWAT model to simulate the hydrological conditions of the basin. The meteorological data required for this study were obtained from Indian Meteorological Department. The meteorological data collected were precipitation, maximum and minimum temperature, wind speed and sunshine hours. The records between 1993 and 2004 were obtained. The problem in the weather data was inconsistency in the data record, in some periods there is a record for precipitation but temperature data are missing, and vice versa. The hydrological data was required for performing sensitivity analysis, calibration, and validation of the model. The hydrological data was collected from Irrigation and Flood Control Department, Kashmir. Sediment data was obtained from Soil and Water Conservation Department, Jammu and Kashmir.

Arc SWAT model setup

The ArcSWAT interface was used for the setup and parameterization of the model. Digital Elevation Model was imported into the SWAT model to automatically delineate the watershed into several hydrologically connected sub-watersheds. After the DEM grid was loaded and the stream networks superimposed, the DEM map grid was processed to remove the non-draining zones. The initial stream network and sub-basin outlets were defined based on drainage area threshold approach. Besides those sub-basin outlets created by the interface, outlets were also manually added at the gauging stations where sensitivity analysis, calibration and validation tasks were later performed. Then watershed delineation activity was finalized by calculating the geomorphic sub-basin parameter. A total of 43 sub basins were created.

The land use and the soil data in a projected Grid file format were loaded into the Arc SWAT interface to determine the area and hydrologic parameters of each land-soil category simulated within each sub-watershed. The DEM data used during the watershed delineation was also used for slope classification. The multiple slope discretization operation was preferred over the single slope discretization as the sub-basins have a wide range of slopes between them. Based on the suggested minimum, maximum, mean and median slope statistics of the watershed, three slope classes (0-10, 10-20, and >20) were applied and slope grids reclassified. Then land use, soil and slope grids were overlaid. In multiple HRU definition, a threshold level was used to eliminate minor land uses, soils, or slope classes in each sub-basin. Land uses, soils or slope classes which cover less than the threshold level was eliminated and the area of the remaining land use, soil, or slope class was reapportioned so that 100% of the land area in the sub-basin was modeled. The threshold levels set is a function of the project goal and amount of detail required. In the SWAT user manual, it is suggested that it is better to use a larger number of sub-basins than larger number of HRUs in a sub-basin; a maximum of 10 HRUs in a sub-basin is recommended. Hence, taking the recommendations into consideration, 5%, 5%, and 5% threshold levels for the land use, soil and slope classes were applied, respectively so as to encompass most of spatial details. A total of 244 HRU's were created. The climatic variables required by SWAT daily precipitation, maximum and minimum temperature, solar radiation, wind speed and relative humidity were prepared in the appropriate dbase format and imported into the model.

SWAT is a complex model with many parameters that makes manual calibration difficult. Hence, sensitivity analysis was performed to limit the number of optimized parameters to obtain a good fit between the simulated and measured data. Sensitivity analysis helps to determine the relative ranking of which parameters most affect the output variance due to input variability [7] which reduces uncertainty and provides parameter estimation guidance for the calibration step of the model. SWAT model has an embedded tool to perform sensitivity analysis and provides recommended ranges of parameter changes. SWAT2005 uses a combination of Latine Hypercube Sampling and One-At-a-Time sensitivity analysis methods (LH-OAT method) [8]. The concept of the Latin-Hypercube Simulation is based on the Monte Carlo Simulation to allow a robust analysis but uses a stratified sampling approach that allows efficient estimation of the output statistics while the One-Factor-At-a-Time is an integration of a local to a global sensitivity method [8]. In local methods, each run has only one parameter changed per simulation which aides in the clarity of a change in outputs related directly to the change in the parameter altered [8].

Model calibration and validation

There are three calibration approaches widely used by the scientific community. These are the manual calibration, automatic calibration, and a combination of the two. Manual calibration is the most widely used approach. However, it is tedious, time consuming, and success of it depends on the experience of the modeler and knowledge of the watershed being modeled [9]. Automatic calibration involves the use of a search algorithm to determine best-fit parameters. It is desirable as it is less subjective and due to extensive search of parameter possibilities can give results better than if done manually.

The manual calibration approach helps to compare the measured and simulated values, and then to use the expert judgment to determine which variable to adjust, how much to adjust them, and ultimately assess when reasonable results have been obtained. The auto- calibration technique is used to obtain an optimal fit of process parameters which is based on a multi-objective calibration and incorporates the Shuffled Complex Evolution Method algorithms [8]. The model was calibrated in order to make the simulation result more realistic for independent calibration period (1993 to 2000). In this study, both the manual and auto-calibration techniques were employed to get the best model parameters.

In order to utilize the calibrated model for estimating the effectiveness of future potential management practices, the model was tested against an independent set of measured data. This testing of a model on an independent set of data set was used. As the model, predictive capability was demonstrated as being reasonable in both the calibration and validation phases, the model can be used for future predictions under different management scenarios. In this study, the model was validated with independent validation period (2001-2004). The simulation flowchart is shown in Figure 1.

Results and Discussion

Sensitivity analysis and parameters calibration

The first simulation using default parameters was not able to correctly reproduce the runoff in the watershed because the actual discharge peaks were underestimated and the base flow was overestimated. Therefore, parameters calibration was need after identifying the most sensitive parameters for runoff shown in Table 1.

The CN2 was increased by 7% of the original value to increase the runoff and decrease infiltration. SOL_AWC was reduced by 20% in all soils to increase the movement of water within the soil profiles. The default value of SOL_K for the soil profiles resulted in an overestimation of the lateral flow in the watershed; therefore, the SOL_K was decreased by 13% to reduce the lateral flow in the soil profiles. SLSUBBSN was also used to control the estimates of lateral flow, and the calibrated values were varied from 25 to 228 m for the individual sub-basins. The simulated runoff become closer to the observed runoff when the ESCO was adjusted to 0.15 and the EPCO was adjusted to 0.38 because these changes allow us to modify the depth distribution of water in the soil layers to meet soil evaporation and plant uptake demands. The default value of ALPHA_BF leaded to large base flow; however, the Lolab watershed has a characteristically rapid response of runoff to rainfall, and the watershed has zero or a low base discharge during dry seasons. Therefore, adjusting ALPHA_BF to 0.70 caused the simulated discharge recession curve to be steeper than when using the default value, which represents the faster drainage behavior of the watershed. GWQMN and GW_REVAP affect the amount of groundwater flow and control the upwelling of groundwater into the unsaturated soil zone. The effects of these parameters on base flow also affect runoff, and low values of GWQMN correspond to high runoff [10]; thus, this parameter was adjusted to 37 to realistically predict daily runoff. In addition, SOL_Z and CH_K2 were determined to be 2,400 mm and 31 mm h-1 to obtain better model accuracy. Table 2 presents an overview of the parameter changes applied during sediment calibration. The parameters were modified based on appropriate ranges as defined in the SWAT model documentation. The values of SPCON and SPEXP were adjusted to increase channel sediment transport capacity and lower the amount of sediment deposition. Two other sensitive parameters affecting sediment transport at the watershed level, CH_EROD and CH_COV, were adjusted to 0.55 and 0.30 respectively (Table 2).

Evaluation of stream flow simulation

The observed and simulated monthly and yearly stream flow for the calibration period from 1993-2000 are shown in Figures 2 and 3. The statistical results for the model performance displayed satisfactory efficiency with $R^2=0.74$ and $N_{SE}=0.68$ for monthly stream flow and $R^2=0.90$ and $N_{SE}=0.68$ for annual flows between the simulated and observed data shown in Table 3. However, the simulated flow was generally lower than the corresponding observed values during periods with concentrated rainfall. The calibrated model was then run from 2001-2004 to validate the model. The observed and simulated monthly and yearly stream flow for the validation period from 2001-2004 are shown in Figures 4 and 5. The statistical analysis results also demonstrated reasonable agreement between the observed and simulated stream flow with monthly R^2 value of 0.85 and N_{SE} value of 0.83 and yearly R^2 value of 0.99 and N_{SE} value of 0.91. Although the statistical evaluation showed the satisfactory stream flow simulation for both calibration and validation periods, SWAT tended to underestimate the runoff during high-flow periods. This could be partly because the present curve number technique is unable to generate accurate runoff prediction for a day that experience several storms. When several storms occur during a single day, the soil moisture level and the corresponding runoff curve number vary from storm to storm [11]. However, SCS-CN methods define a rainfall event as the sum of all rainfall that occurs during one day, and this might lead to underestimation of runoff [12].

Evaluation of sediment load simulation

The results obtained during calibration periods of sediment load simulation are shown in Figures 6 and 7. The majority of sediment

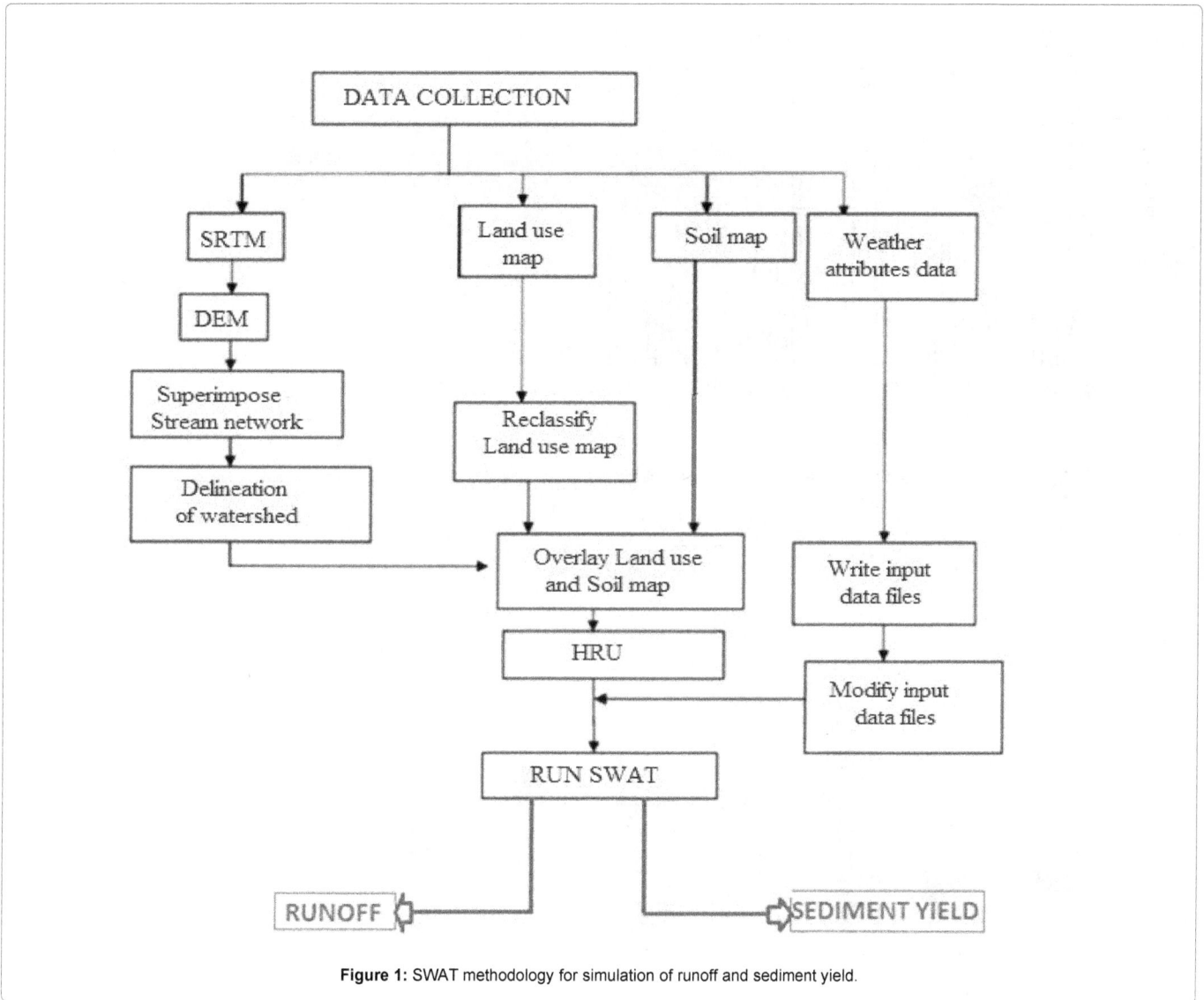

Figure 1: SWAT methodology for simulation of runoff and sediment yield.

The most sensitive parameters	Parameter definition	Range	Ranks of the parameters	Calibrated value
CN2	Initial SCS CN II value	± 25%	1	7%[a]
SOL_AWC	Soil available water capacity (%)	± 25%	2	-20%[a]
SOL_K	Saturated hydraulic conductivity(mm/h)	± 15%	3	-13%[a]
SOL_Z	Depth from soil surface to the bottom of the layer(mm)	0 to 3000	4	2400
ESCO	Soil evaporation compensation factor	0.1 to 1	5	0.15
ALPHA BF	Base flow alpha factor(days)	0 to 1	6	0.7
GWQMN	Threshold depth of water in a shallow aquifer for return flow (mm)	0 to 1000	7	32
SLSUBBSN	Average slope length (m)	10 to 300	8	25-228 (varies by subbasins)
EPCO	Plant evaporation compensation factor	0.01 to 1	9	0.38
CH_K2	Channel effective hydraulic conductivity(mm/h)	0 to 500	10	31

Table 1: Sensitive parameters for stream flow simulation and calibrated values.

The most sensitive parameters	Parameter definition	Range	Ranks of the parameters	Calibrated value
SPCON	Linear parameters for sediment re-entrainment	0.0001-0.01	1	0.008
CH_EROD	Channel erodibility factor	0-1	2	0.55
SPEXP	Exponent parameter for sediment re-entrainment	01-Feb	3	1.45
CH_COV	Channel cover factor	0-1	4	0.3

Table 2: Sensitive parameters for sediment load simulation and calibrated values.

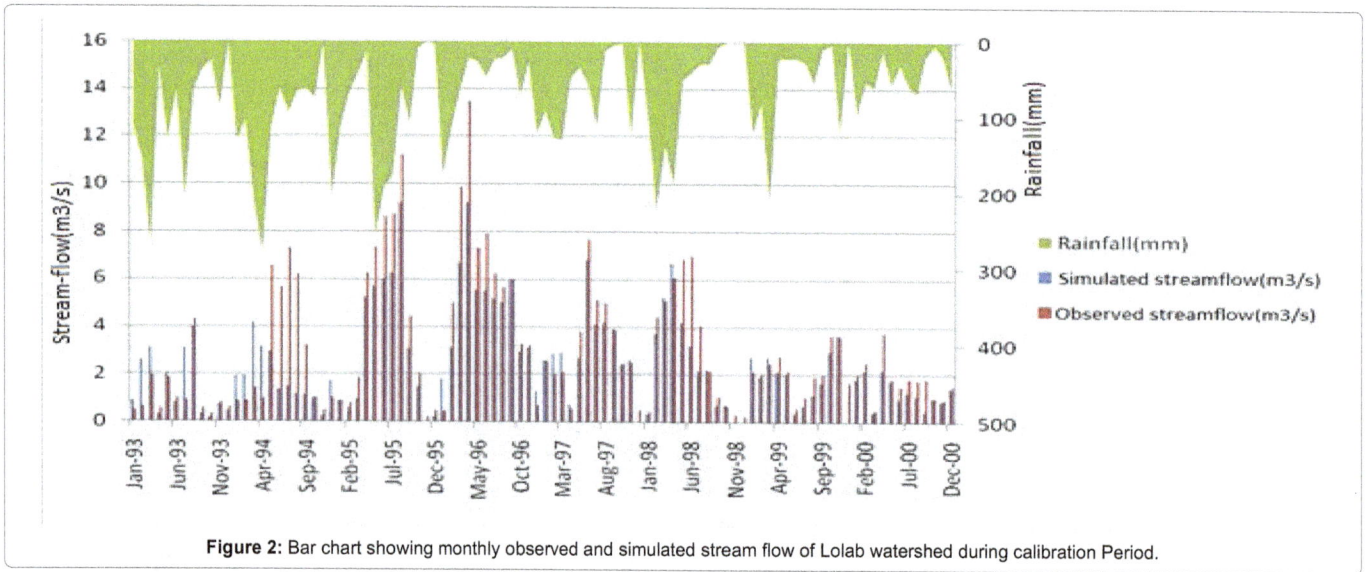

Figure 2: Bar chart showing monthly observed and simulated stream flow of Lolab watershed during calibration Period.

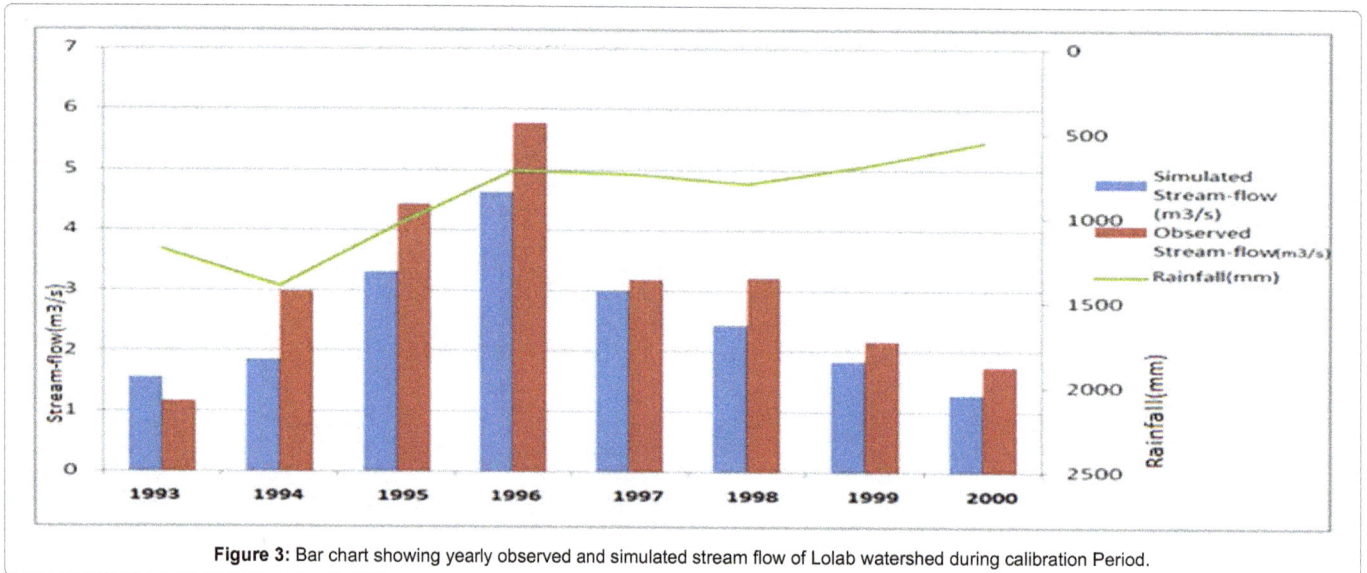

Figure 3: Bar chart showing yearly observed and simulated stream flow of Lolab watershed during calibration Period.

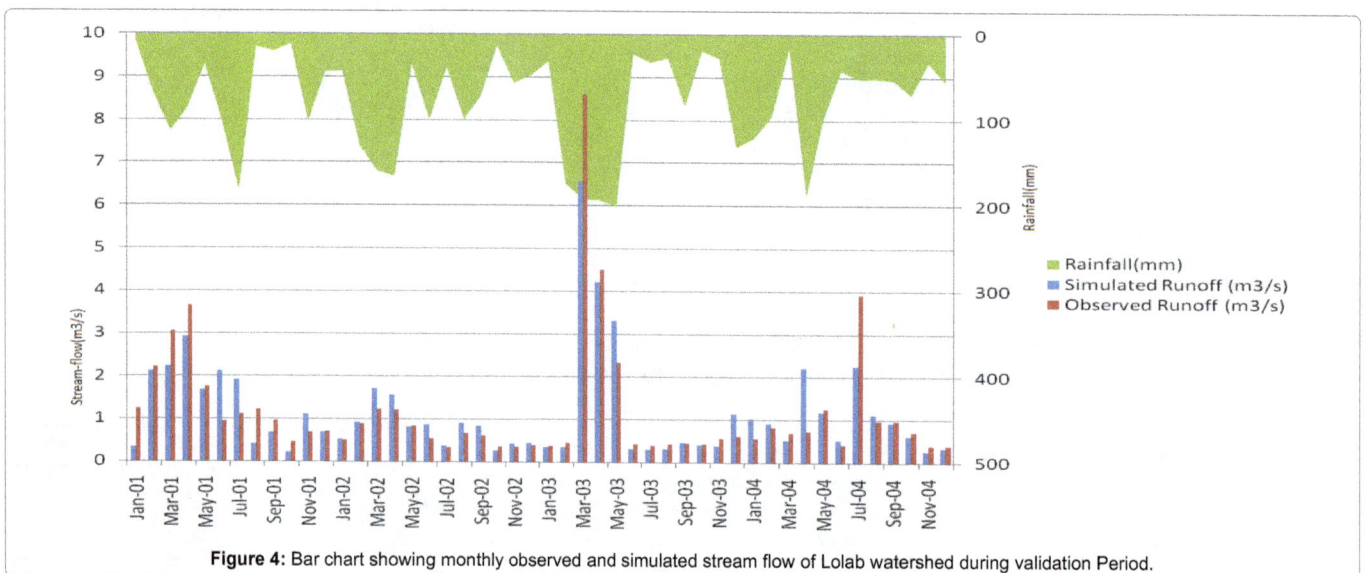

Figure 4: Bar chart showing monthly observed and simulated stream flow of Lolab watershed during validation Period.

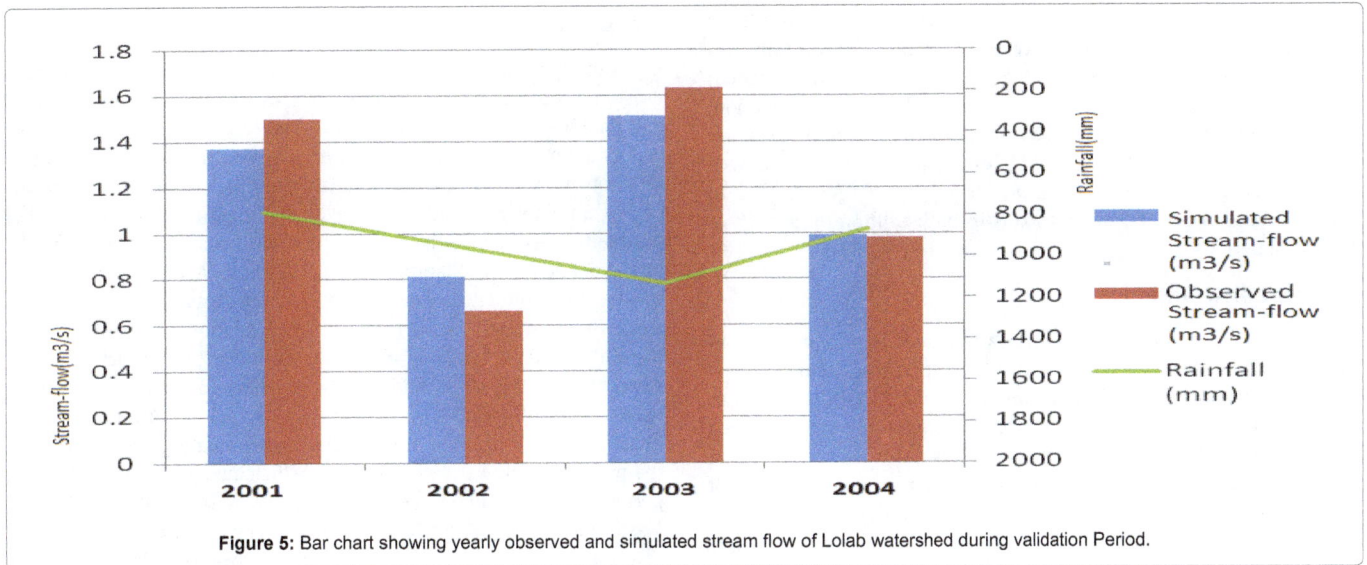

Figure 5: Bar chart showing yearly observed and simulated stream flow of Lolab watershed during validation Period.

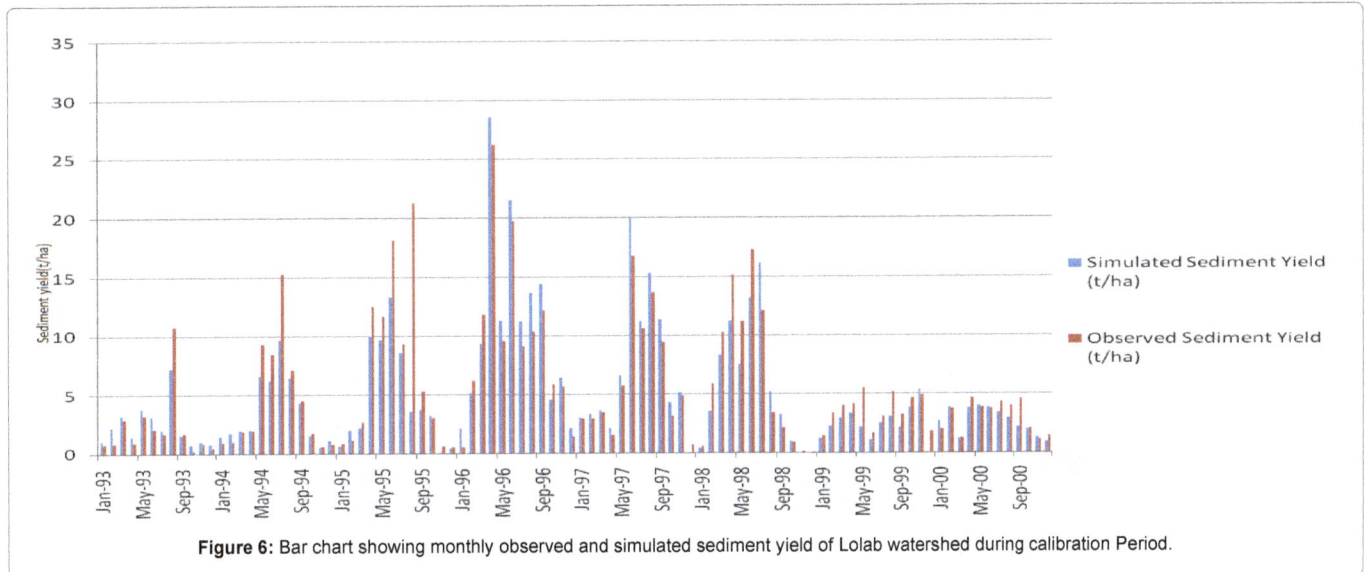

Figure 6: Bar chart showing monthly observed and simulated sediment yield of Lolab watershed during calibration Period.

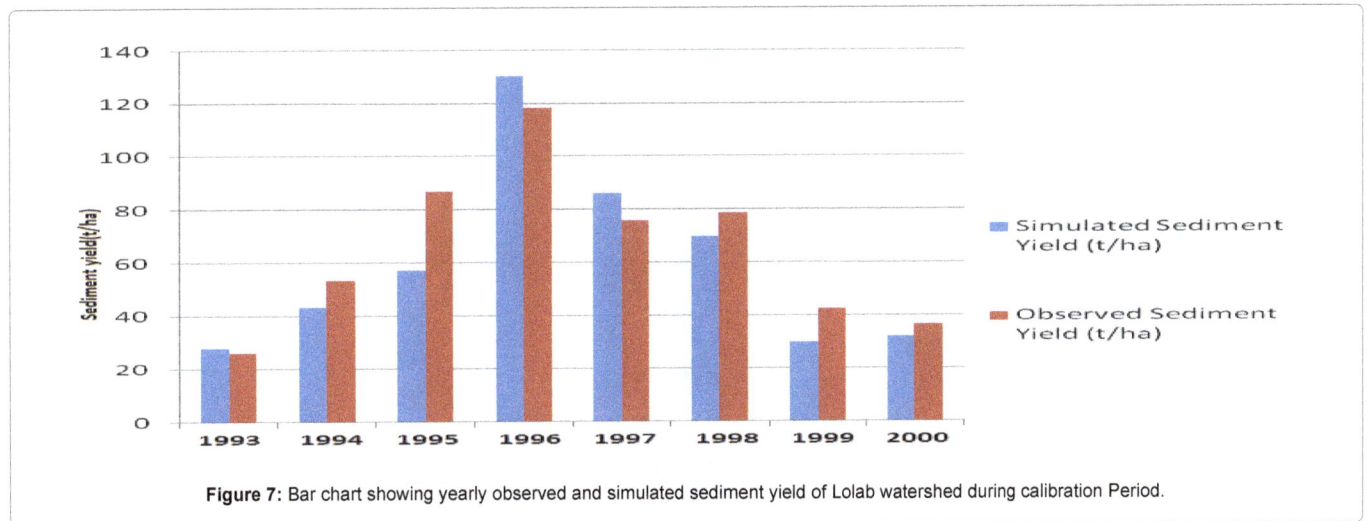

Figure 7: Bar chart showing yearly observed and simulated sediment yield of Lolab watershed during calibration Period.

transport occurs during high-flow events; therefore, capturing these events accurately is vital to model calibration [13]. The monthly and yearly R^2 and the N_{SE} demonstrated that the correlation and agreement between them was acceptable, shown in Table 4. For validation period, the observed and simulated sediment loads during 2001-2004 are shown in Figures 8 and 9. Although the validation period showed a better match than the calibration period, the SWAT tended to underestimate many of sediment loads in both calibration as well as validation periods.

Sediment distribution

The spatial variability of sedimentation rates was identified and based on which the potential area of intervention was identified. The average annual yield of sedimentation for each sub-basin was used to generate sediment source map shown in Figure 10. The study showed that Sub-basins 24, 25, 28 and 38 of Lolab watershed at the existing condition generates maximum annual average sediment yields of 139.54 t/ha, 132.98 t/ha, 138.06 t/ha, 129.23 t/ha respectively. This was attributed due to the topographic slope of these sub-basins. It was a land with more than 42.72% of which has a slope greater than 23 degrees.

Conclusion

Although the problem of soil erosion is recognized from gross erosion estimates and field observations, quantitative information and data are required at micro watershed level to develop alternative watershed management plans and for decision making. In this study, attempts were made to characterize the Lolab watershed in terms of stream flow and sediment yield and identification of potential sediment source areas. In this study the performance of SWAT model was evaluated using Standard calibration and validation statistics. A good agreement between monthly and yearly measured and simulated stream-flow was demonstrated by correlation coefficients Viz. Monthly $R^2=0.74$ Yearly $R^2=0.90$ and Nash-Sutcliffe model efficiency with Monthly $E_{NS}=0.68$ and Yearly $E_{NS}=0.68$ for calibration period while as Monthly $R^2=0.85$, Yearly $R^2=0.99$ and Monthly $E_{NS}=0.83$, Yearly $E_{NS}=0.91$ for validation period. In simulating sediment yield, the monthly and yearly correlation coefficients i.e., R^2 for calibration period were 0.80 and 0.86 respectively while as the monthly and yearly Nash-Sutcliffe model efficiency coefficients i.e., E_{NS} were 0.79 and 0.78 respectively. The values of monthly and yearly correlation coefficients i.e., R^2 for validation period were 0.88 and 0.83 respectively while as

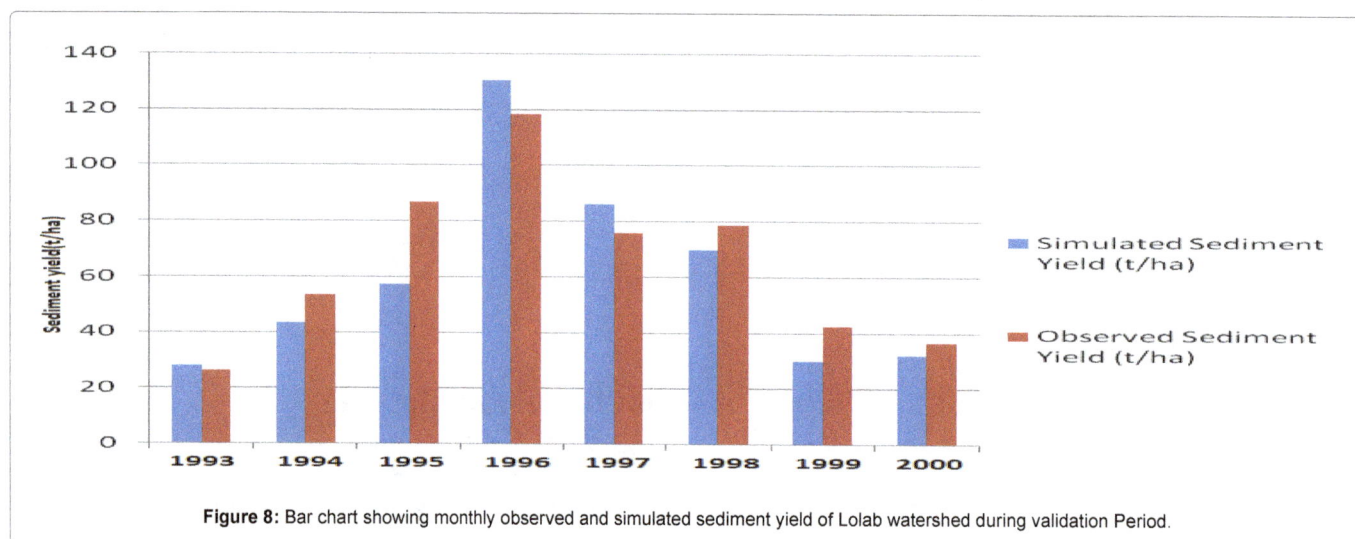

Figure 8: Bar chart showing monthly observed and simulated sediment yield of Lolab watershed during validation Period.

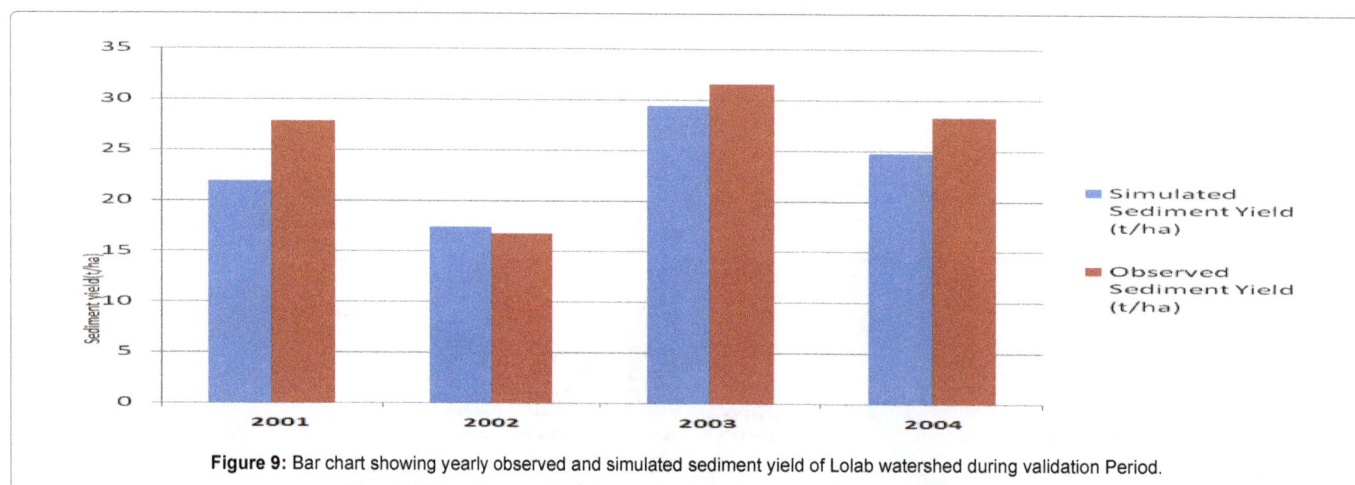

Figure 9: Bar chart showing yearly observed and simulated sediment yield of Lolab watershed during validation Period.

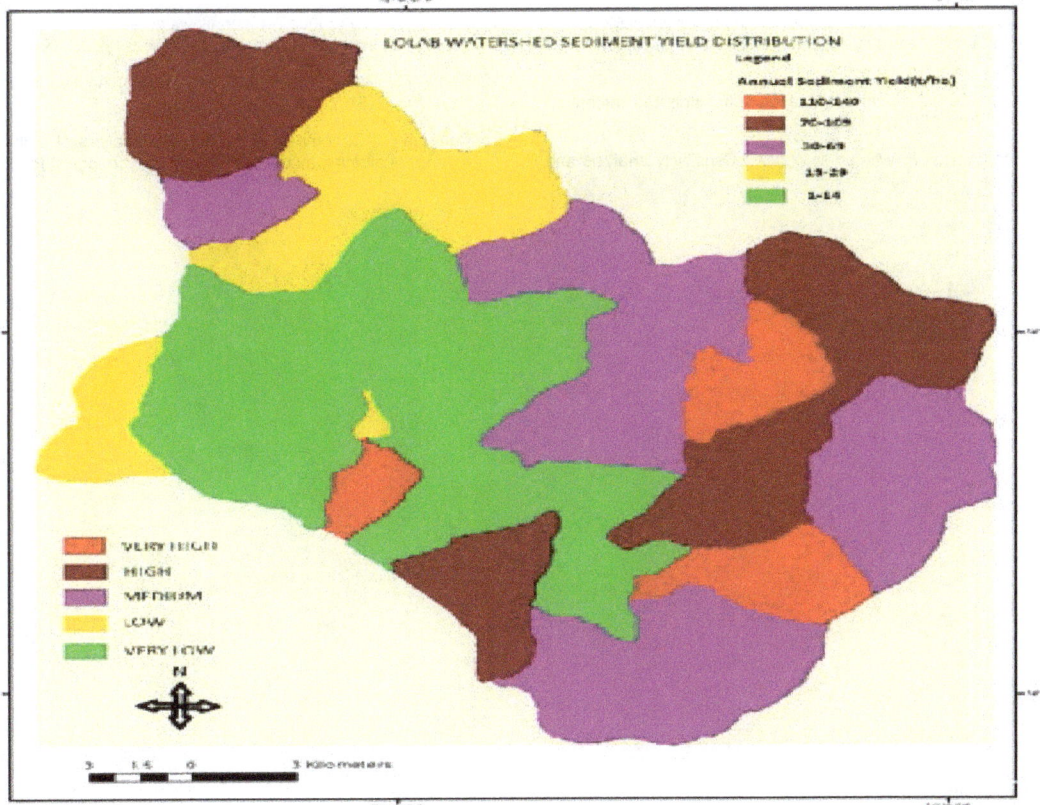

Figure 10: Shows the spatial variability of sedimentation rate.

	Nash-Sutcliffe model efficiency (N$_{SE}$)		Coefficient of Determination (R$_2$)	
	Monthly	Yearly	Monthly	Yearly
CAL	0.68	0.68	0.74	0.9
VAL	0.83	0.91	0.85	0.99

Table 3: Model efficiencies for stream flows during calibration and validation periods.

	Nash-Sutcliffe model efficiency (N$_{SE}$)		Coefficient of Determination (R$_2$)	
	Monthly	Yearly	Monthly	Yearly
CAL	0.79	0.78	0.80	0.86
VAL	0.86	0.58	0.88	0.83

Table 4: Model efficiencies for Sediment yield during calibration and validation periods.

the monthly and yearly Nash-Sutcliffe model efficiency coefficients i.e., E$_{NS}$ were 0.86 and 0.58 respectively. A good performance of the model in the Validation period indicates that the fitted parameters during calibration period can be taken as a representative set of parameters for Lolab watershed and further simulation and evaluation of alternative scenario analysis can be carried out for other periods using the SWAT model. The SWAT model prediction verified that the Lolab watershed is highly erosion potential area contributing high sediment yield exceeding the tolerance limit (soil formation rate) in the study area.

In general, the methodology presented in this paper can be used in other watersheds, and the results are instructive for further use of SWAT in evaluating different management scenarios in hilly-gullied regions of Kashmir. Sub-basin no's 24, 25, 28, 38 of Lolab watershed at the existing condition generates a maximum annual average sediment yield, this can be reduced by using sediment yield intervention strategies such as land slope stabilization, construction bench terraces, changing the land use of steep area and afforestation.

Acknowledgements

We are thankful to the Department Earth Sciences, Kashmir University and Department of Irrigation and Flood Control, Jammu and Kashmir for providing the moral support as well as necessary data. We are also greatly thankful to the National Institute of Technology, Srinagar for providing us the lab facilities and our regards also go to the Meteorological Department, Kashmir as well as SKUAST-K for a moral support.

References

1. Rajora R (1998) Integrated watershed management. Chapter 24. Rawat Publications, Jaipur, India.

2. Food and Agriculture Organization (FAO) (1985) Guidelines for economic appraisal of watershed management projects. In: Gregersen HM, Brooks KN, Dixon JA, Hamilton LS. F'AO Conservation Guide No 16, Rome, Italy.

3. Pulle AA (1988) African Rice (Oryza glaberimma Stued.): Lost crop of the enslaved Africans Discovered in Suriname. An article published by Tinde Van Andel, Mar 5, 2010.

4. Arnold JG, Srinivasan R, Muttiah RS, Williams JR (1998) Large area hydrologic modeling and assessment - Part 1: Model development. Journal of the American Water Resources Association 34: 73-89.

5. Setegn SG, Srinivasan R, Dargahi B (2008) Hydrological Modeling in the Lake Tana Basin, Ethiopia Using SWAT Model. The Open Hydrology Journal 2: 49-62.

6. Fontaine TA, Cruickshank TS, Arnold JG, Hotchkiss RH (2002) Development of a snowfall-snowmelt routine for mountainous terrain for the soil water assessment tool (SWAT). Journal of Hydrology 262: 209-223.

7. Van Griensven A (2005) Sensitivity, auto-calibration, uncertainty and model evaluation in SWAT2005.

8. Van Griensven A, Green CH (2007) Auto-calibration in Hydrological Modeling: Using SWAT2005 in small-scale watersheds. Environmental Modeling & Software 23: 422-434.

9. Eckhardt K, Arnold JG (2001) Automatic calibration of a distributed catchment model. Journal of Hydrology 251: 103-109.

10. Kannan N, White SM, Worrall F, Whelan MJ (2007) Sensitivity analysis and identification of the best evapotranspiration and runoff options for hydrological modelling in SWAT-2000. Journal of Hydrology 332: 456-466.

11. Kim NW, Lee J (2008) Temporally weighted average curve number method for daily runoff simulation. Hydrological Processes 22: 4936-4948.

12. Chow VT, Maidment DR, Mays LW (1988) Applied Hydrology. McGraw Hill, New York, USA, pp: 47-61.

13. Benaman J, Shoemaker CA (2005) An analysis of high-flow sediment event data for evaluating model performance. Hydrological Processes 19: 605-620.

8

Climate Change Impact on Design Storm and Performance of Urban Storm-Water Management System - A Case Study on West Central Mountain Drainage Area in Canada

Sadik Ahmed and Ioannis Tsanis*

Department of Civil Engineering, McMaster University, 1280 Main Street West, Hamilton, Ontario L8S 4L7, Canada

Abstract

A number of future climate projections indicate a likelihood of increased magnitude and frequency of hydrological extremes for many regions around the world. The urban storm-water management infrastructures are designed to mitigate the effect of extreme hydrological events. Changes in extreme rainfall events will have a significant implication on the design of storm-water management infrastructures. This study assessed the potential impact of changed rainfall extreme on drainage systems in the West Central Mountain drainage area located in Southern Ontario, Canada. First, the design storms for the study area were calculated from observed rainfall data and the North American Regional Climate Change Assessment Program (NARCCAP) climate simulations based on SRES A2 Scenario. Frequency analysis was performed on the annual maximum time series data by using the best fitted distribution among twenty seven distributions. The Pearson chi-square test and Kolmogorov-Smirnov were used to test the goodness of fit of each distribution. The results show that L-moment Pareto distribution was selected the most often for data from six RCM+GCM pairs. Overall increase of storm depth in the future is highest when the distributions were identified by the Kolmogorov-Smirnov test. The design storm depths calculated from the observed and climate model simulated data were used as input into an existing PCSWMM model of the study area for flow simulation and hydraulic analysis for the storm-water management system, specifically storm sewer and detention pond. The results show an increase in design storm depths under projected climatic change scenarios that suggest an update of current standard for designing both the minor system and detention pond in the study area. The assessment results of storm water management infrastructures indicate that performance of the detention pond as well as the storm sewer network will deteriorate under future climate condition.

Keywords: Climate change; Storm-water management; Frequency analysis; Detention pond; Storm sewers; Canada

Introduction

The release of greenhouse gases and aerosols due to anthropogenic activities are changing the amount of radiation coming into and leaving the atmosphere. These are, in turn, changing the composition of atmosphere that may influence temperature, precipitation, storms and sea level. Observed increases in global average air and ocean temperatures, melting of polar ice and significant increases in net anthropogenic radiative forcing revealed that our global climate system is undergoing substantial warming [1]. An increased intense of 'dry and hot' extremes for many regions around the world was revealed by a number of studies on different climate model projections [2-6]. It is well known that increasing temperatures tend to increase evaporation which leads to more precipitation; so the changes in global temperature will have a significant effect on increasing magnitude and frequency of extreme precipitation events. These changes in temperature and precipitation will significantly affect frequency and severity of floods. Therefore, the design standards of storm-water management infrastructure, such as storm-water detention pond, and storm sewer have to adapt to the changing hydrologic process under future climate.

The storm-water infrastructures in an urban area are usually designed based on the rainfall depth calculated employing statistical analyses of observed precipitation data. The rainfall depths are calculated from the historic rainfall time series without considering climate change impact i.e., based on the assumption of a stationary climate. But, the climate is now non-stationary [7,8] because of the anthropogenic force. So, the designing of storm-water management infrastructure based on design storm considering the assumption of non-stationary climate will not be able to manage extreme events

in future climate. The importance of developing design standard for addressing the climate change was indicated by many researchers [9-11]. Forsee and Ahmed [12] explored the projected changes in design-storm depths for Pittman watershed in Las Vegas using five NARCCAP data sets, and they showed a significant increase in case of three GCM+RCM pairs. Zhu et al. [13] investigated the potential changes in IDF curve due to climate change impact for six regions in the United States. They found strong regional patterns and increase in the intensity of extreme events under future climate for most of the study sites. Mailhot et al. [14] investigated the climate change impact in IDF curves for Southern Quebec using the Canadian Regional Model projections. The study results show that return period of 2 hour and 6 hour storm events will be approximately halved and return period of 12 hour and 24 hour storm events will decrease by one third. Coulibaly et al. [15] found significant increases in storm depth in 2050s and 2080s in Grand River, Kenora and Rainy River region in Canada by analyzing the storm depth calculated from climate simulations. In most of the studies, frequency analysis was performed on the annual maximum

---

*Corresponding author: Ioannis Tsanis, Department of Civil Engineering, McMaster University, 1280 Main Street West, Hamilton, Ontario L8S 4L7, Canada
E-mail: tsanis@mcmaster.ca

precipitation time series by fitting only one to three distributions for design storm depth calculations. For example, the Log-Pearson Type III for NARCCAP future precipitation time series was used by Moglen and Vidal [11], generalized extreme value was used by some studies [12,14], Extreme value type I (EV I) was used by Zhu et al. [13], Gumbel and generalized extreme value were used by Zhu [16]. This study explored the climate change impact on design storm depth calculated by employing frequency analyses of NARCCAP precipitation data sets. In this study, twenty seven distributions were tested for the observed, NARCCAP current and future dataset, and the best among the fitted distribution was used for frequency analysis to calculate design storm depths. Two statistical tests were used to test the goodness of fit at a 95% confidence level. The source of uncertainty involved in climate change impact studies are resulted from climate model projections, the hydrologic model and data downscaling techniques. The main sources of uncertainty, climate model projections, are derived from three main sources: forcing, model response and internal variability [17]. The climate change impact assessment using climate model data should consider multiple scenarios due to uncertainty in climate model projections. NARCCAP data provide several RCM+GCM pairs, and in this study six pairs of climate projection datasets were used for design storm depth calculation. All the NARCCAP dataset are provided at grid scale. One of the main challenges in climate change impact assessment is bridging the gridded climate change projections with the historic observation at meteorological station. A number of dynamical and statistical downscaling methods are available to downscale climate model gridded data at the target point locations [18-22]. A simple method for transposing gridded climate projections to station scale is the use of delta change factor [13]. In some studies delta change factors have been applied to precipitation time series [22-25], and in other studies it has been applied to design storm depth [12,13]. The delta

change method was applied to transpose design storm depth calculated from gridded NARCCAP data to Hamilton Airport meteorological station.

The design and operation of urban drainage system is associated with local rainfall characteristics, i.e., design storm depth [23]. The design criteria of the urban drainage management infrastructure must be revised with the consideration of possible impact of climate change [10]. Moglen and Vidal [11] examined the changes in detention basin performance under several climate change scenario at a study location north of Washington, DC, and indicated that in most cases, the performance of detention basin would be inadequate under future climate condition. Forsee and Ahmad [12] also revealed the inadequate performance of detention basin under future climate condition in a watershed in Las Vegas Valley, Nevada. There are other studies showing inadequate performance of storm sewer and combined sewer under future climate condition [23,25,26]. This study investigated the performance of storm water management system at a study location in the City of Hamilton, Canada using several different climate projections. The following section details the study location.

Study Area and Data

Study area

The study area (Figure 1), West Central Mountain drainage area, is a part of Red Hill Creek watershed located in the City of Hamilton, Southern Ontario, Canada. The modeling area is about 525 ha. The climate of Hamilton is humid-continental and characterized by changeable weather patterns. However, its climate is moderate compared with most of Canada. The daily average temperature in this area is 7.9 °C based on the data from 1981 to 2010 at Hamilton Airport,

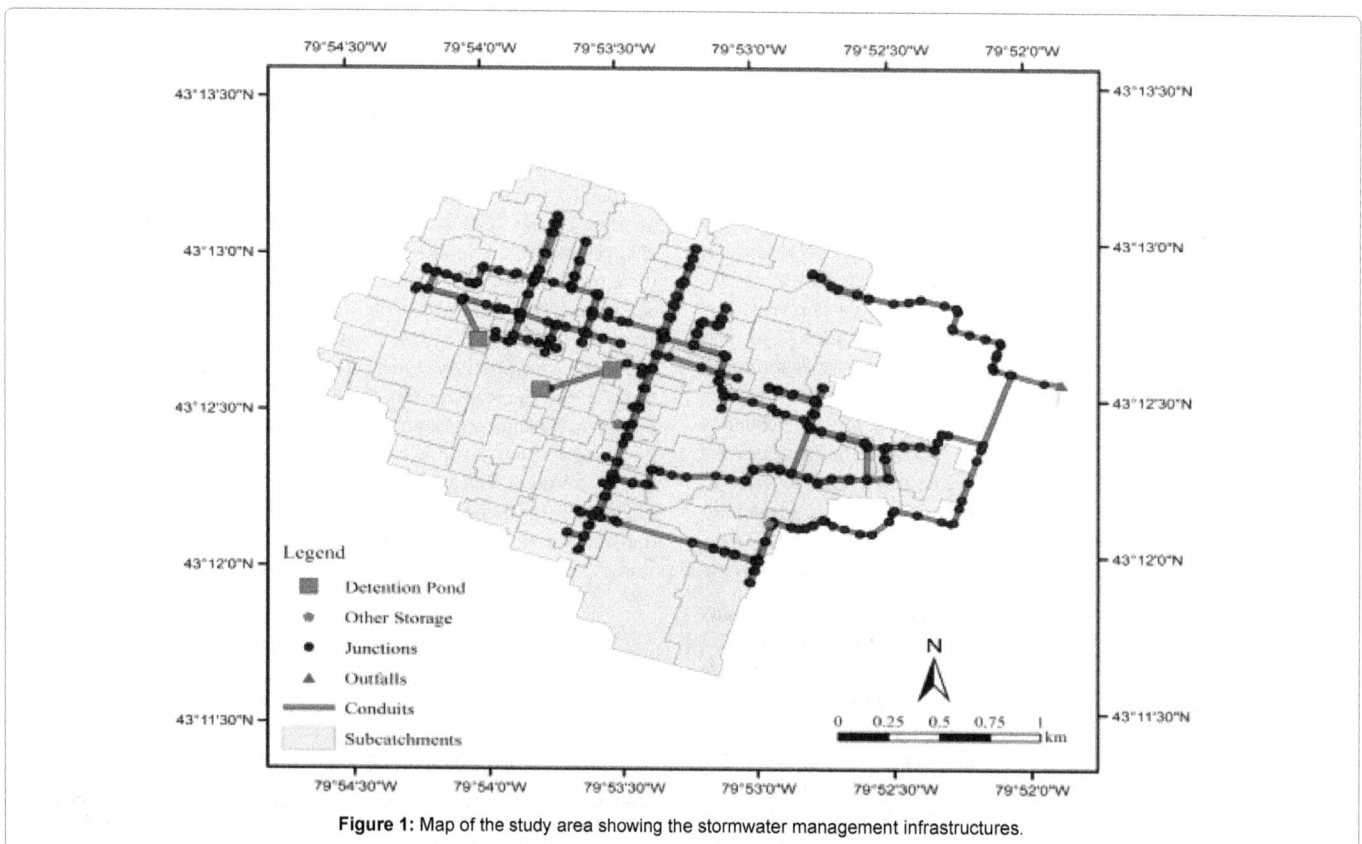

Figure 1: Map of the study area showing the stormwater management infrastructures.

and extreme maximum 37.4 °C and extreme minimum temperature -30 °C were observed on 7 July, 1988 and 16 January, 2004 respectively. The yearly average rainfall and precipitation (rain and snow) are 791.7 mm and 929.8 mm based on data from 1981 to 2010 at Hamilton Airport, and the maximum daily rainfall and precipitation 107 mm were observed on 26 July, 1989. Grillakis et al. [27] analysed observed meteorological data over a twenty year period (1989-2008) from Hamilton Airport to show the interannual trend of precipitation and temperature, and revealed an increase of precipitation 3.5 mm/year and average temperature 0.041°C/year.

Observed meteorological data

The observed hourly rainfall data for 30 years, from 1971 to 2000, were obtained from meteorological station, namely Hamilton Airport meteorological station with latitude and longitude 43 10 25.00 N and 79 56 06.00 W. The hourly rainfall time series of this station was used to calculate the design storm because City of Hamilton uses the design storm calculated from this meteorological station for the study area. This hourly observed precipitation time series was provided by Ontario Climate Center, Environment Canada.

NARCCAP climate data

The climate data sets used in this research were obtained from The North American Regional Climate Change Assessment Program [28-30]. NARCCAP is an international program to produce high resolution climate change simulations covering the conterminous United States and most of Canada. It provides the data sets in order to investigate uncertainties in regional scale projections of future climate and generate climate change scenarios for use in impacts research. The climate data sets are generated by running a set of regional climate models (RCMs) driven by a set of atmosphere-ocean general circulation models (AOGCMs). The AOGCM involves coupling comprehensive three-dimensional atmospheric general circulation models, with ocean general circulation models, with sea-ice models, and with models of land-surface processes. RCM enhance the simulation of atmospheric circulations and climatic variables at fine spatial scales. This study uses the precipitation time series provided by six different RCM+GCM pairs. NARCCAP provides complete data for current and future for these six RCM+GCM pairs, and these six pairs include two pairs of each three RCMs. Table 1 provides the names of the RCMs and GCMs/drivers used in this study.

The spatial resolution of all NARCCAP data sets is 50 km and the temporal resolution of precipitation time series is 3 hour [30]. NARCCAP provides precipitation time series data of time span 33 years for both current (1968-2000) and future (2038-2070) period. First three years of each simulation are spin-up periods [31] and the data of the spin-up period has been discarded. Therefore, the precipitation time series data of time span 30 years for both current (1971-2000) and future (2041-2070) period are actually considered in this study. All the NARCCAP future simulations are driven by a GCM with greenhouse gas and aerosol concentration based on A2 emission scenario described

in the Special Report on Emissions Scenarios (SRES) [32]. The A2 scenario was preferred from an impacts and adaptation point of view. Data are stored in the NetCDF files in 2D arrays. The array dimensions are named "xc" and "yc" within the file. The array dimensions (yc, xc) are found from the grid cell maps for each RCMs. The array dimensions (yc, xc) of nearest point of Hamilton Airport for CRCM, HRM3 and RCM3 are (51,100), (57, 105) and (44, 94) respectively.

Methodology

The method used in this study can be described as a two-step procedure. At first an extensive frequency analysis was performed on the observed, NARCCAP current and future period data sets for design storm calculation. Then, the storm information was transformed into runoff and hydraulic information by employing a fully featured urban drainage system modeling tool.

Design storm

Frequency analysis: A design storm can be represented by a value of rainfall depths or intensity (presented by IDF curves) or by a design hyetograph specifying the time distribution of rainfall during a storm. Design storm depths associated with different duration (3 h, 6 h, 12 h and 24 h) and return period (2 yr, 5 yr, 10 yr, 25 yr, 50 yr and 100 yr) were calculated for historic observations at station scale and climate model simulations at grid-scale. Data of the each time series were aggregated into 3-, 6-, 12- and 24 h duration on an annual basis, and the yearly maximum value for each duration was determined from the aggregated time series to generate time series of annual maximum rainfall depth. Frequency analysis was performed on these annual maximum time series data by using the best fitted distribution among twenty seven distribution as shown in Table 2 as well as Extreme Value type 1 (EV1) which is Gumbel distribution. Environment Canada provides the design storm information in the form of IDF curves and uses Gumbel Extreme Value distribution to fit the annual extremes of rainfall for the study area. Therefore, Extreme Value type 1 (EV1) was used for frequency analyses together with the best fitted distribution. Pearson chi-square test and Kolmogorov-Smirnov were used to test the goodness of fit of each distribution. The best fitted distribution is the distribution that attained the highest percentage of a. The percentage value of 'a' for Chi-square test (equation 1) and Kolmogorov-Smirnov (equation 2) are defined by the following two equations:

$$a_{attained} = 1 - x^2(m = k - r - 1, q) \qquad (1)$$

$$a_{attained} = 1 - x^2(m, q) \qquad (2)$$

where m are the degrees of freedom of chi square test, k is the number of bins used in chi square test, r is numbers of parameters of the distribution and q is the Pearson parameter. Kozanis et al. [33] described the theoretical background of all the tested distributions. The statistical analysis software, Hydrognomon [33], was used to find the best fitted distribution among 27 statistical distributions based on the criteria given in equation 1 and 2 for both observed and climate data.

RCM+GCM Pairs	RCM	GCM/Drivers
CRCM+CCSM	Canadian Regional Climate Model [34]	Community Climate System Model [38]
CRCM+CGCM3	Canadian Regional Climate Model [34]	Third Generation Coupled Global Climate Model [39]
HRM3+GFDL	Hadley Regional Model 3 [35]	Geophysical Fluid Dynamics Laboratory GCM [40]
HRM3+HADCM3	Hadley Regional Model 3 [35]	Hadley Centre Coupled Model, version 3 [41,42]
RCM3+CGCM3	Regional Climate Model version 3 [36,37]	Third Generation Coupled Global Climate Model [39]
RCM3+GFDL	Regional Climate Model version 3 [36,37]	Geophysical Fluid Dynamics Laboratory GCM [40]

Table 1: List of RCM+GCM Data Pairs used in this study.

	CrcmCcsm				CrcmCgcm3				Hrm3Gfdl				Hrm3Hadcm3				Rcm3Cgcm3				Rcm3Gfdl			
Distribution	3	6	12	24	3	6	12	24	3	6	12	24	3	6	12	24	3	6	12	24	3	6	12	24
Normal						√																		
LogNormal											+,√	+,√												
Galton																								
Exponential										*											√x			
Gamma		+										*x	*											
Pearson III										x			√						x		x			
LogPearson III								*																
Gumbel EV 1 Max		*	x		+	*							+		x		*	+				*		
EV2-Max	+	+							+	+							*							+
Gumbel EV 1 Min							+																	*
Weibull							*																	
GEV Max			√																					
GEV Min	x																							
Pareto							√						*x		+									
L-Moments Normal	√					x	x																	
L-Moments Exponential			+										+	*	+				+		+		*	
L-Moments EV1 Max	*	√x				+																		
L-Moments EV2 Max			*						√*x									+				+		
L-Moments EV1 Min																								
L-Moments EV3 Min																								x
L-Moments GEV Max						√					x								*					√
L-Moments GEV Min			√										√	√	√	x	x				√	x		
L-Moments Pareto			x	*,x			√,x							x	x		√	√	√x	+,√			√	
GEV-Max (k spec.)										√														
GEV-Min (k spec.)								+										*					+	
L-Moments GEV-Max (k spec.)				*							*					*			*					
L-Moments GEV-Min (k spec.)																								

Table 2: Best fitted distribution for NARCCAP data for different duration [Case 1 (current x, future √), case 2 (current *, future +)].

Three sets of storm depths were calculated: (1) Case 1: storm depth with best fitted distribution tested by Chi-square test (2) Case 2: storm depth with best fitted distribution tested by Kolmogorov-Smirnov, and (3) storm depth with Extreme Value type 1 (EV1) distribution.

Delta change factor: The climate models (RCMs) provide gridded data; those are areal average and not point estimates [43]. The systematic difference between climate model simulated and observed precipitation is a problem for using RCMs for hydrological purposes [44]. The storm depth values calculated from the NARCCAP datasets are for grid scale. Delta change factor can be applied to discrete totals i.e., design storm depths [12] to transpose projected future change in climate onto point observation. The assumption in this conversion is that areal-to-point relationships of precipitation remain constant in future climates [14]. The delta change factor application procedure (presented by equations 3, 4 and 5) described by Zhu et al. [13] to adjust the historic station scale intensities/depths to produce future station-scale values for the same duration and return period will be used in this study:

$$I_F^{(g)} = I_H^{(s)} \left[1 + \Delta_{F-H}^{(g)}(T,d) \right] \qquad (3)$$

$$\Delta_{F-H}^{(g)}(T,d) = \frac{I_F^{(g)}(T,d) - I_H^{(g)}(T,d)}{I_H^{(g)}(T,d)} \qquad (4)$$

$$I_F^{(s)}(T,d) = I_H^{(s)}(T,d) \frac{I_F^{(g)}(T,d)}{I_H^{(g)}(T,d)} \qquad (5)$$

Where, T and d denote return period and duration respectively, H and F denote historic and future, and s and g denote station and grid respectively.

The point estimates of storm depth for all six RCM+GCM pairs for all three cases are presented in Table 3.

Hydrologic and hydraulic modeling

A large number of hydrological models are used in different countries for different purposes. 'Although hydrological models have been around for quite some time, there is yet to be one exclusive model that can stand apart from the rest and be declared best at modeling in all aspects of the hydrologic system [45]. Considering the urban hydrological and hydraulic modeling capabilities, this study aimed to use PCSWMM 2D Professional, a leading decision support system for US EPA SWMM. PCSWMM also contains a flexible set of hydraulic modeling capabilities used to route runoff and/or external inflows through the drainage system network of natural channels, pipes, storage/treatment units, diversion structures [46]. This study used an existing model, developed using PCSWMM, of the study area. The existing model of the study area was provided by the City of Hamilton. The models that contain proposed detention pond/ storm water management facilities considering the future development are used for minor system/ storm sewer and detention basin performance assessment. The model contains 126 sub-catchments with 172.2 ha impervious area out of 525.06 ha total area. The models used curve number infiltration method and dynamic wave routing method. Three detention pond (pond 1, pond 2 and pond 3) elements were selected for analyses of detention pond performance. The contributing area of pond 1, pond 2 and pond 3 are 44.77 ha (11 sub-catchments, 13.06 ha impervious area), 15.36 ha (8 sub-catchments, 7.7 ha impervious area), 37.63 ha (8 sub-catchments, 13.77 ha impervious area) respectively.

Return Period	Duration (h)	Observed			CrcmCcsm			CrcmCgcm3			Hrm3Gfdl			Hrm3Hadcm3			Rcm3Cgcm3			Rcm3Gfdl		
		Case 1	Case 2	Case 3	Case 1	Case 2	Case 3	Case 1	Case 2	Case 3	Case 1	Case 2	Case 3	Case 1	Case 2	Case 3	Case 1	Case 2	Case 3	Case 1	Case 2	Case 3
2 yr	3	31.1	31	32.5	33.1	34	34.6	35	34.7	37.2	31.9	31.2	33.3	36.8	34.7	38.7	30.8	31.7	35	37.4	38.6	41.2
	6	38.3	37.1	39.5	40	39.3	41.9	43.3	42.4	44.9	41.3	39.7	41.5	37.8	40.2	42.7	42.9	39.1	42.6	42.9	43.1	47.3
	12	43.5	43.8	45.1	44.8	45.8	47.3	52	52.4	52.6	51.1	51	51.2	47.7	47.2	48.4	42.1	46.6	50.7	60.8	52.2	55.7
	24	53.3	50.6	52.4	61.9	58.4	57.9	65.8	57.3	60.9	62.9	59.7	62.4	55	53.2	55.7	62.1	58.4	63.1	62	60	66.5
5 yr	3	42	42.1	46.7	42.9	43.8	48.9	47.2	47.1	55	41.8	42.4	44.7	50	49.2	59.7	44.1	46.5	54.4	60.6	58.3	65.3
	6	52.6	53.7	55.3	53.9	55.9	58.3	60.7	61.8	63.3	52.9	57.4	56.5	55.3	56.7	60.7	58.4	57.6	59.4	65	70.8	72.5
	12	56.9	59.2	61.9	60.1	62.4	64	65.1	69.2	71.4	66.6	68	70.2	64.8	64.5	68.8	61.2	63.4	70	79	80.2	82.7
	24	72	68.7	70.6	80.9	73.7	76.1	85.3	82.5	81.8	86.3	82.3	85.2	78.5	75.4	76.2	83.5	79.8	80.8	93.5	91.8	98.4
10 yr	3	51.1	51.5	56.2	51.7	51.8	58.3	60.1	60.6	67.8	49.9	52	52.4	64.1	65.1	73.9	58	60.4	67.3	79.9	74.9	81.6
	6	63.5	65.9	65.8	65.2	67.9	69	74.1	76	75.8	61.9	69	66.4	71.2	70.1	72.6	68.4	71.6	70.6	85.5	92.2	89.9
	12	68	71	73	72.2	74.4	75	75.9	80.5	83.8	77.7	80.2	82.7	76.7	78.7	82.5	81	80.4	82.7	89.6	101.3	101.3
	24	83.3	82.4	82.7	89.1	85	88	94.2	99.1	95.9	101.1	100	100.4	93.2	91.6	89.9	95.5	95.5	92.5	120	121.1	120.6
25 yr	3	65.6	66.5	68.1	66.3	64.6	69.6	83.6	86.3	83.4	62.3	67.1	61.9	90.6	94.6	91.7	85.3	83.8	83.8	107.8	100.8	102.7
	6	79.2	82	79.1	82.6	83.9	82.5	93.3	94.8	91.5	76	82.1	78.9	95.5	88.2	89.6	81.3	89.9	84.9	121.1	121.1	112.2
	12	85.2	88	87.1	89.4	90.8	88.9	92.8	96.4	99.2	92.5	97.3	98.6	92.8	100.6	99.8	115.4	108.6	98.8	103.5	130	125.3
	24	96.5	102.2	97.9	95.3	103	103	102.8	120.3	113.2	119.1	126.1	119.5	109	113.8	107	107.2	115.9	107.2	161.1	170.7	149.5
50 yr	3	79	80.3	76.9	81	76.5	78.1	107.6	113.1	95.4	73.7	81	68.9	118.4	125.5	104.8	115.3	106.5	95.9	131	124.4	118.1
	6	92.3	94.1	89	98.3	95.5	92.6	109	108.9	103.1	89.3	90.1	88.1	116.7	104.8	100.9	90.6	103.2	95.4	156.6	143.2	129.1
	12	100.7	102.2	97.6	103.4	104.1	99.4	108.2	109.6	110.9	104.5	111.3	110.6	106	120.5	112.8	148.6	134.5	110.9	115.1	152.6	143.4
	24	105.7	118.8	109.2	97.8	119	114.2	108.1	136.3	126.4	132	148.4	133.7	118.7	131.7	119.8	113.9	131.7	118.1	198.1	217.9	171.2
100 yr	3	94.9	99	85.6	99	92.1	86.8	137.5	150.8	107	86.8	100	75.8	154.2	169.3	117.9	157.1	137.4	107.9	156	156.3	133.8
	6	106.7	106.2	98.7	116.8	107.4	102.4	127	123	114.6	105.2	96.5	97.2	140.8	124.7	110.3	99.9	116.3	105.8	202.4	165.3	145.7
	12	118.9	117.9	107.9	118.1	117.9	109.5	126.2	124.5	122.3	117.3	126.7	122.2	120.4	143.7	125.6	190.1	164.9	122.7	128.2	176.1	161.5
	24	114.5	137.1	120.4	98.8	137.6	125.3	112.6	152.7	139.1	144.8	173.4	147.9	126.4	150.8	132.4	119.2	147.7	128.8	241.7	276.1	193.2

Table 3: Design storm depths (in mm) calculated from observed data and NARCCAP future datasets.

Design Storm	Observed	CrcmCcsm	CrcmCgcm3	Hrm3Gfdl	Hrm3Hadcm3	Rcm3Cgcm3	Rcm3Gfdl	Average
24 hr 25 yr for detention pond	102.2	103	120.3	126.1	113.8	115.9	170.7	125
24 hr 5 yr for storm sewer	68.7	73.7	82.5	82.3	75.4	79.8	91.8	80.9

Table 4: Design storm depths (in mm) used for detention pond and storm sewer performance analysis.

The City of Hamilton used 6hour Chicago and 24 hour SCS storm distribution for this study area and found 24 hour SCS distribution to be the governing condition [47]. This study used 24 hour SCS storm distribution for both storm sewer and detention pond performance analysis. The 24 hr -25 yr and 24 hr -5 yr design storm depths (only for case 2, shown in Table 4) were used for detention ponds and storm sewer performance analysis respectively. The last column of the Table 4 provides the average of design storm calculated from six RCM+GCM pairs. A number of hydrologic and hydraulic parameters used by Moglen and Vidal [11] and Berggren et al. [23] as well as other parameter as described in result and discussion section were used for detention ponds and storm sewer performance analysis.

Results and Discussion

Design storm

Design storm depths were calculated for four different duration (3 hr, 6 hr, 12 hr and 24 hr) and six different return periods (2 yr, 5 yr, 10 yr, 25 yr, 50 yr and 100 yr) for observed time series and NARCCAP current and future simulations of six different RCM+GCM pairs. Therefore, a total of 52 (4 observed, 24 NARCCAP current and 24 NARCCAP future) annual maximum time series were used for frequency analysis. The best fitted distribution among twenty seven distributions for NARCCAP current and future datasets are listed in the Table 2. For example, the best fitted distribution for NARCCAP current data in Case 1 was identified by 'x' mark in Table 2 and is GEV Min for CrcmCcsm 3h storm. As two tests were used to test the goodness of fit of each distribution, Table 2 provides

96 selections for 48 NARCCAP datasets. Table 2 shows that L-moment Pareto distribution was selected 14 times (the highest), that is 14.6% of the total selections. Gumbel EV1 Max was selected for 9 times that is 9.4% of the total selection. Therefore, only Gumbel EV1 Max used by different stakeholders for design storm calculation for this study area is not appropriate for climate change impact study. Four distributions namely Galton, L-Momnet EV1 Min, L-Moments EV3 min and L-Moments GEV-Min (k spec.) were not selected as best fitted distribution for any climate data sets. L-moment Pareto distribution was selected 7 times (the highest), for both current and future climate datasets. L-moment Pareto was also selected 12 times (the highest), when Chi-square test was used to test the goodness of fit. Both L-Moment Exponential and Gumbel EV1 Max were selected 7 times (the highest), when Kolmogorov-Smirnov was used to test the goodness of fit. This study identified the best fitted distribution for observed and NARCCAP datasets, and used them for design storm calculation to minimize the uncertainty related to appropriate distribution selections. The design storm depths calculated from observed data and NARCCAP future datasets are presented in Table 3. It is mentionable that the delta change factor was applied on the datasets to get the design storm values for NARCCAP datasets presented in the Table 3. Table 3 shows that there is a significant increase in design storm depths for all six RCM+GCM pairs. Results in the Table 3 also show the overall variability of the design storm depths calculated from the climate data. For example, 3 hr-2 yr storm depths calculated from six RCM+GCM pairs in case 2 are 34, 34.7, 31.2, 34.7, 31.7 and 38.6 mm with mean 34.2 mm and coefficient of variation 7.1%; 3 hr-100 yr storm depths are 92.1, 150.8, 100. 169.3, 137.4 and 156.3 mm

with mean 134.3 mm and coefficient of variation 21.4%. The calculated coefficient of variations also show that the variability increases with the increase of return period. The increase in design storm depths under future climate conditions are also shown in the Figure 2. Figure 2 shows the scatterplot of all design storm depths (in Table 3) calculated from observed data and NARCCAP future datasets. The scatterplots in Figure 2 shows that the data are more dispersed from the 45-degree line for higher values. It revealed that the increase of design storm depth under future climate is higher for higher values. It is notable that the higher values may represent storm depths for higher return period or higher duration. The linear trendlines in Figure 2 also shows overall increase of storm depth is higher for case 2 (when distribution were identified by Kolmogorov-Smirnov test) than other two cases, lowest for case 3 (when frequency analysis was performed using Gumbel EV1 Max). Figures 3, 4 and 5 show the difference between design storm depths calculated from observed data and NARCCAP future datasets for different return period and different duration. Here, the positive values refer to an increase of storm depths in future. Visual inspection of these figures revealed that the difference (increase) of design storm depths increase with the increase of return period overall. For example, design storm depths increased by 15.6%, 20%, 22.8% for 24 hr storm of return period 2 yr, 25 yr and 100 yr respectively for case 1, these increase are 14%,

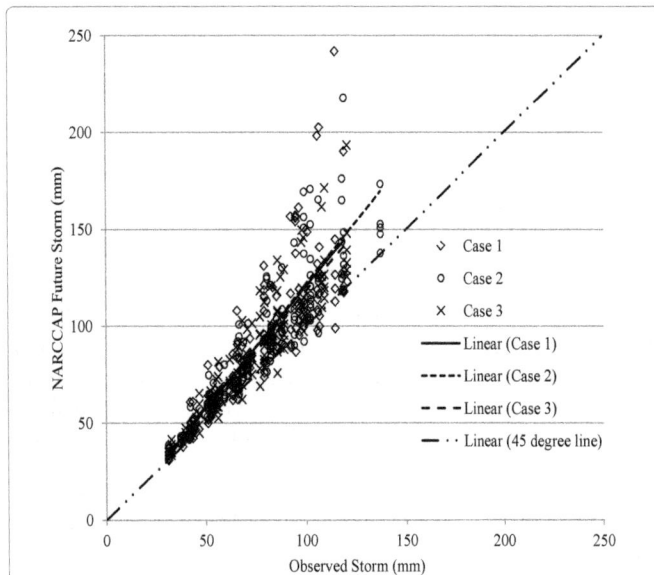

Figure 2: Scatterplot of design storm depths calculated from observed data and NARCCAP future datasets.

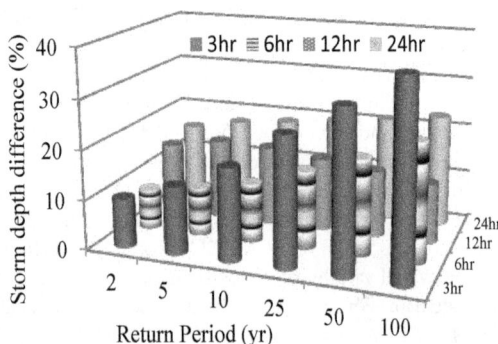

Figure 3: Difference between design storm depths calculated from observed and NARCCAP future datasets for case 1.

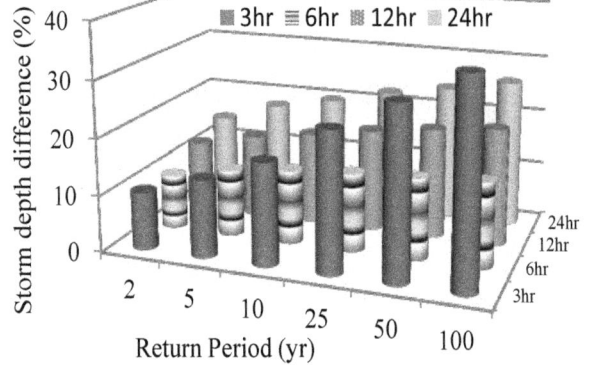

Figure 4: Difference between design storm depths calculated from observed and NARCCAP future datasets for case 2.

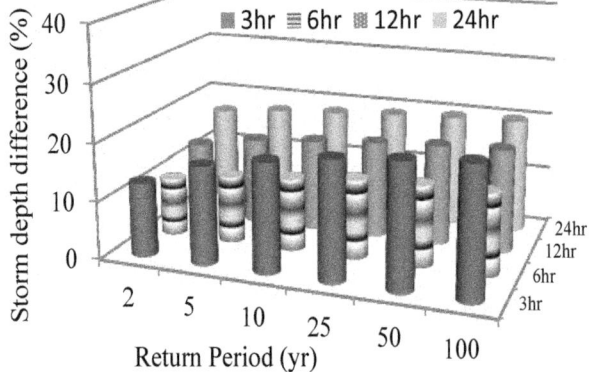

Figure 5: Difference between design storm depths calculated from observed and NARCCAP future datasets for case 3.

22.3% and 26.2% for case 2, and 16.6%, 19.1% and 20% for case 3. The increasing trend in case 3 is not as significant as other cases; the reason might be that the Gumbel EV1 Max is not the best fitted distribution for most of the datasets for case 3. Considering only the 3 hr and 24 hr duration storm, Figures 3 and 4 shows that the increase is higher for shorter duration with higher return period and also higher for longer duration with lower return period. For example, the increase of storm depths is 38.8% and 22% for 3 hr and 24 hr storm of 100 year return period respectively, 9.9% and 15% for 3 hr and 24 hr storm of 2 year return period respectively for case 1; 35.7% and 26.2% for 3 hr and 24 hr storm of 100 year return period respectively, 10.2% and 14% for 3 hr and 24 hr storm of 2 year return period respectively for case 2. Figures 3 and 4 also show that overall increase of storm depths under future condition is higher in case 2 than that in case 1. Considering this issue and sustainable storm water infrastructure design, the design storm depths calculated in case 2 will be used for investigation of detention pond and storm sewer performance study.

Detention pond

The 24 hr 25 yr storm depths listed in the Table 4 were used as input in the PCSWMM model, simulation were performed and the following metrics were collected: Average depth (m), maximum depth (m), maximum total inflow (m³/s), average volume (1000 m³), average percent full (%), max volume (1000 m³), max percent full (%) and max outflow (m³/s). These metrics for three detention ponds are reported

Features	Metric	Observed	CrcmCcsm	CrcmCgcm3	Hrm3Gfdl	Hrm3Hadcm3	Rcm3Cgcm3	Rcm3Gfdl	Average
Pond 1	Avg Depth (m)	0.64	1	1.06	1.08	1.05	1.05	1.2	1.08
	Max Depth (m)	1.45	1.01	1.11	1.14	1.08	1.08	1.2	1.14
	Max Total Inflow (m³/s)	9.28	1.01	1.25	1.33	1.16	1.19	1.97	1.32
	Average Volume (1000 m³)	6.06	1	1.08	1.11	1.05	1.06	1.24	1.1
	Avg Percent Full (%)	31	1	1.1	1.1	1.06	1.06	1.23	1.1
	Max Volume (1000 m³)	15.35	1.01	1.15	1.19	1.09	1.11	1.27	1.18
	Max Percent Full (%)	79	1	1.14	1.19	1.09	1.1	1.27	1.18
	Max Outflow (m³/s)	1.2	1.03	1.74	1.99	1.47	1.56	2.44	1.95
Pond 2	Avg Depth (m)	0.34	1.03	1.12	1.15	1.09	1.09	1.38	1.15
	Max Depth (m)	1.59	1	1.11	1.15	1.07	1.08	1.54	1.14
	Max Total Inflow (m³/s)	4.1	1.01	1.21	1.28	1.13	1.16	1.89	1.27
	Average Volume (1000 m³)	0.88	1.01	1.11	1.14	1.07	1.09	1.42	1.14
	Avg Percent Full (%)	10	1	1.1	1.1	1	1.1	1.4	1.1
	Max Volume (1000 m³)	4.84	1.01	1.15	1.21	1.1	1.11	1.81	1.2
	Max Percent Full (%)	53	1.02	1.15	1.23	1.11	1.13	1.83	1.21
	Max Outflow (m³/s)	1.59	1.03	1.56	1.66	1.37	1.43	2.23	1.66
Pond 3	Avg Depth (m)	0.59	1	1.08	1.12	1.05	1.07	1.27	1.1
	Max Depth (m)	1.62	1	1.14	1.2	1.08	1.1	1.23	1.19
	Max Total Inflow (m³/s)	8.53	1.01	1.25	1.33	1.16	1.19	1.96	1.31
	Average Volume (1000 m³)	4.31	1	1.09	1.12	1.06	1.07	1.3	1.12
	Avg Percent Full (%)	26	1	1.12	1.12	1.08	1.08	1.31	1.12
	Max Volume (1000 m³)	12.7	1.01	1.17	1.25	1.1	1.12	1.29	1.24
	Max Percent Full (%)	77	1.01	1.18	1.26	1.1	1.13	1.3	1.25
	Max Outflow (m³/s)	1.5	1.03	1.87	2.17	1.53	1.64	2.41	2.11

Table 5: Detention Pond Performance Ratios (Future values normalized by observed performance values) for 24 hr 25 yr design storm.

in Table 5. The third column in the Table 5 shows performance values using the design storm calculated from observed data. All other values in the Table 5 are detention pond performance values for NARCCAP future storm normalized by the values in the column 3. Almost all the performance ratios greater than 1 for all six RCM+GCM pairs and average value indicate that the detention ponds will not perform as expected under future climate. The performance ratios of all eight metrics for RCM3+GFDL are highest among the ratios for all six RCM+GCM pairs, that indicates the worst performance of all detention ponds under RCM3+GFDL future scenario. The performance ratios for RCM3+GFDL models varies from 1.2 for average depth to 2.44 for maximum outflow for pond 1, i.e., average depth increase by 20% and maximum outflow increase by 144% under future climate presented by RCM3+GFDL models. The very high increase in the uncontrolled peak discharge indicates the vulnerability of flooding in the downstream of the detention pond. One model, CRCM+CCSM, among the six pairs shows no change for some metrics and insignificant (only 3% for maximum outflow) change for some metrics for all three ponds. Using the future to present performance ratio greater than 1 (i.e., future condition are greater than present conditions), the increases are observed in 93% of all the metrics for all 3 ponds. Results in the Table 5 show that the performance ratios varies from 1.08 for average depth for pond1 to 2.11 for maximum outflow for pond 3 for average design storm, i.e., average depth increase by 8% and maximum outflow increase by 111% under average future climate condition. The performance ratios varies 1.08-1.95, 1.10-1.66 and 1.10-2.11 for average future climate condition for pond 1, pond 2 and pond 3 respectively, the performance ratio varies 1.2-2.44, 1.38-2.23 and 1.27-2.41 for highest increased 24 hr 25 yr design storm by RCM3+GFDL models.

Figure 6 presents the time series plot of inflow, outflow, storage volume and depth for detention pond 1. These time series data were produced by inputting design storm depth from observed data and average (listed in Table 4) of design storm from 6 RCM+GCM pairs. Figure 6 shows that maximum inflow increased from 9.28 m³/s for observed to 12.21 m³/s for NARCCAP average that is an increase of 32%. The outflow from the pond increased from 1.198 m³/s for observed to 2.331 m³/s for NARCCAP average, i.e., the controlled peak flow will be increased by 95% under future average climate condition. Figure 6 shows that the maximum storage volume and maximum depth will increase by 18% and 14% respectively. The maximum values obtained from the simulated time series, the maximum storage volumes are 15347 m³ and 18175 m³, and the maximum depths are 1.45 m and 1.65 m for observed and NARCCAP average respectively.

Storm sewer

The 24 hr - 5 yr storm depths listed in the Table 4 were used for storm sewer performance analysis. These design storm depths with SCS storm distribution was inputted in the PCSWMM model. Then, a number of hydraulic parameters were obtained from the PCSWMM generated status files. The parameters, maximum water level and pipe flow ratio, used by Berggren et al. [23] for measuring hydraulic impact were calculated. Pipe flow ratio is the ratio of the actual maximum flow rate and the flow rate when the pipes were running full in the system.

At the outset, the number of nodes flooded and surcharged observed/baseline scenario and future climate were compared. The number of node flooded and surcharged for 24 hr - 5 yr SCS storm are presented in Table 6. Flooding refers to all water that overflows a node, and surcharge occurs when water rises above the crown of highest conduit. There was only one node flooded under present climate condition. The number of flooded node increased under future climate condition ranging from 4 for CRCM+CCSM models to 72 for RCM3+GFDL models, and 17 for average design storm calculated from 24 hr 5 yr design storm of 6 RCM+GCM pairs. There were 58 nodes surcharged for observed/baseline condition, these numbers increased

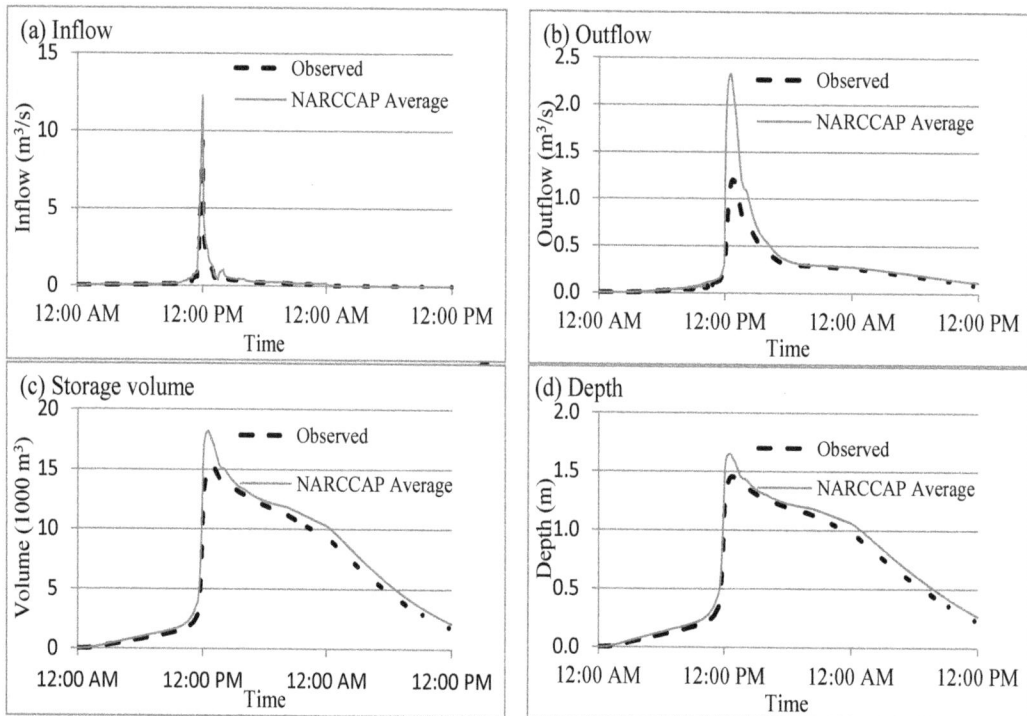

Figure 6: Plots showing time series of inflow, outflow, storage volume and depth from 25 year return period storm for detention pond 1 (observed/baseline values obtained using storm depths calculated from observed data).

Features	Observed	CrcmCcsm	CrcmCgcm3	Hrm3Gfdl	Hrm3Hadcm3	Rcm3Cgcm3	Rcm3Gfdl	Average
Node Flooded	1	4	22	18	15	15	72	17
Node Surcharged	58	92	146	143	98	125	189	131

Table 6: Number of node flooded and surcharged.

Features		CrcmCcsm	CrcmCgcm3	Hrm3Gfdl	Hrm3Hadcm3	Rcm3Cgcm3	Rcm3Gfdl	Average
Max Water Level	mean difference (m)	0.42	1.28	1.27	0.57	0.97	2.6	1.07
	mean difference (%)	26	79	79	36	60	162	66
Pipe Flow ratio	mean difference	0.08	0.2	0.2	0.1	0.16	0.31	0.18
	mean difference (%)	10	25	25	13	20	39	23

Table 7: Difference between the observed/baseline scenario and future climate for maximum water level and pipe flow ratio.

under future climate with the smallest for CRCM+CCSM models which are 92, the largest for RCM3+GFDL models which is 189, and 131 nodes will be surcharged for average future climate condition.

Then, The difference between the observed/baseline scenario and future climate for maximum water level and pipe flow ratio are presented in Table 7. The mean difference between the observed/baseline scenario and future climate for maximum water level at all the nodes varies from 0.42 m for CRCM+CCSM models and 2.62 m for RCM3+GFDL, and the difference between observed and climate average is 1.07, i.e., the maximum water level increase on an average of 26% for CRCM+CCSM models, 162% for RCM3+GFDL models and 66% for average design storms under future climate. Similarly, The mean difference between the observed/baseline scenario and future climate for pipe flow ratios varies from 0.08 m for CRCM+CCSM models and 0.31 m for RCM3+GFDL, and the difference between observed and climate average is 0.18, i.e., the pipe flow ratios increase on an average of 10% for CRCM+CCSM models, 39% for RCM3+GFDL models and 23% for average design storms under future climate.

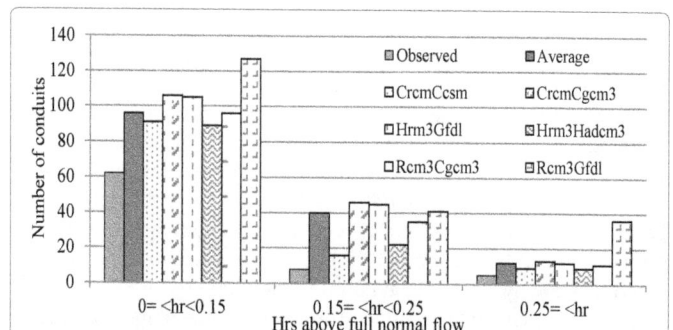

Figure 7: Number of conduits above full normal flow for 5 year return period storm.

Figure 7 presents the number of conduits above full normal flow and Figure 8 presents the number of conduits for capacity limited. These numbers for the observed/baseline period and future period are categorized for three durations: 0=<hr<0.15, 0.15=<hr<0.25 and 0.25=<hr. The numbers are always higher for all categories for all six

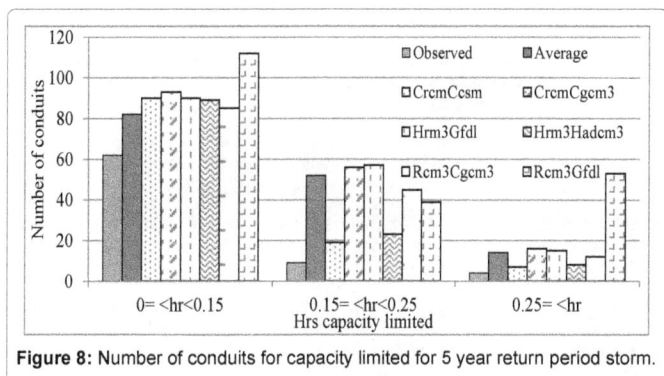

Figure 8: Number of conduits for capacity limited for 5 year return period storm.

RCM+GCM models. The number of conduits above full normal flow and for capacity limited for RCM3+GFDL are the highest among the six RCM+GCM pairs for durations 0=<hr<0.15 and 0.25=<hr.

The higher numbers are observed for CRCM+CGCM and HRM3+GFDL models for second category. The numbers of conduits above full normal flow are 62, 8 and 5 for observed and 96, 40 and 12 for future average climate for three categories, i.e., the numbers increase by 55%, 400% and 140%. The numbers of conduits for capacity limited are 62, 9 and 4 for observed and 82, 52 and 14 for future average climate for three categories, i.e., the numbers increase by 32%, 477% and 250%.

Figure 9 shows the spatial distribution of number of nodes flooded, number of nodes surcharged and pipe flow ratio, and it contributes to the understanding of most vulnerable locations in the study area under future climate condition.

Conclusions

This study explored the potential impact of climate change on the design storm depths and consequent effect on the performance of detention pond and storm sewer network under future climate condition at a study area located in the City of Hamilton, Ontario, Canada.

The best fitted distribution among twenty seven distributions for observed and NARCCAP datasets for design storm calculation were identified in this study. The precipitation time series provided by six different RCM+GCM pairs were used in frequency analysis; two statistical tests were used to test the goodness of fit of each distribution. The delta change factor was used to convert the storm depths calculated from gridded data to station scale values. The results show that there is an overall significant increase of design storm depths for all six RCM+GCM pairs. The visual inspection of scatter plots revealed that the increase of design storm depths under future climate condition is higher for higher values. Visual inspections also revealed that increase of design storm depths also increase with the increase of return period overall. The results also show overall increase of storm depths in future is higher in the case when distributions were identified by Kolmogorov-Smirnov test. The design storm depths calculated using the distribution identified by Kolmogorov-Smirnov test are suggested to use for investigation of stormwater management infrastructure performance study for sustainable infrastructure design.

The 24 hr - 25 yr and 24 hr - 5 yr design storm depths were inputted in the PCSWMM model for analyses of detention pond and storm sewer network performance respectively under future climate condition. The deteriorated performance of three detention ponds were indicated by the performance ratio calculated from eight metrics. The time series plot of inflow, outflow, storage volume and depth also shows increase of the metrics. Results also indicate the worst performance of all detention ponds under RCM3+GFDL future scenario. A number of hydraulic parameters were used to assess the system capacity, and all the parameters show deteriorated performance under future climate condition. Similar to detention pond, the worst performance of the storm sewer network were observed under RCM3+GFDL future scenario. Overall, the urban drainage management infrastructures designed based on current climate condition will not be able to cope with the increased design storm depth under future climate condition. The findings of this study would encourage municipalities and other stakeholders for considering climate change impact in planning and

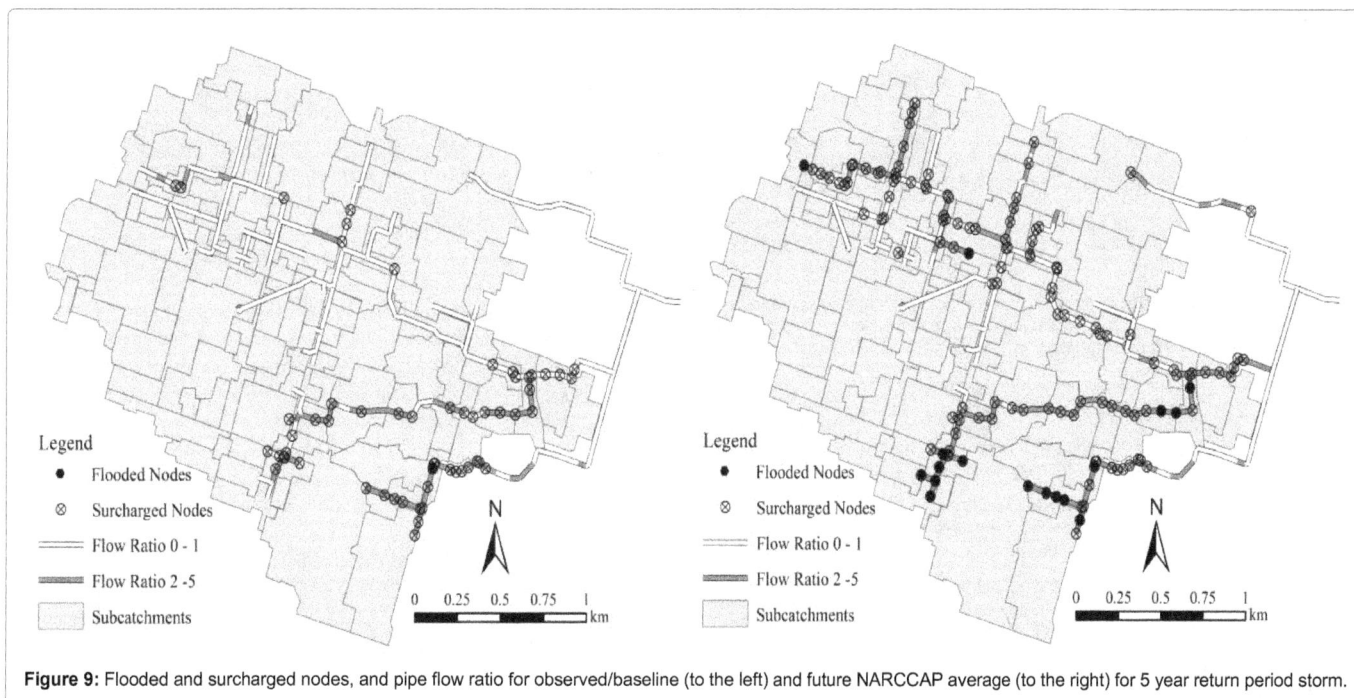

Figure 9: Flooded and surcharged nodes, and pipe flow ratio for observed/baseline (to the left) and future NARCCAP average (to the right) for 5 year return period storm.

designing of drainage management infrastructures to ensure that they will work effectively in future.

Acknowledgement

We wish to thank the North American Regional Climate Change Assessment Program (NARCCAP) for providing the data used in this paper. The authors gratefully acknowledged City of Hamilton, Environment Canada and Computational Hydraulics International (CHI) for providing data and models for the study area. Finally, support from the National Science and Engineering Research Council (NSERC) of Canada in the form of Discovery Grant (RGPIN04808-14) is greatly acknowledged.

References

1. IPCC (2014) Climate Change 2014: Synthesis Report. Contribution of Working Groups I, II and III to the Fifth Assessment Report of the Intergovernmental Panel on Climate Change. In: Pachauri PK, Meyer LA (eds.) IPCC, Geneva, Switzerland, pp: 151-175.

2. Beniston M, Stephenson DB, Christensen OB, Ferro CAT, Frei C, et al. (2007) Future extreme events in European climate: an exploration of regional climate model projections. Climatic Change 81: 71-95.

3. Christensen JH, Christensen OB (2003) Climate modelling: severe summertime flooding in Europe. Nature 421: 805-806.

4. Kundzewicz ZW, Radziejewski M, Pinskwar I (2006) Precipitation extremes in the changing climate of Europe. Climate Research 31: 51-58.

5. Semmler T, Jacob D (2004) Modeling extreme precipitation events-a climate change simulation for Europe. Global and Planetary Change 44: 119-127.

6. Tsanis IK, Koutroulis AG, Daliakopoulos NI, Jacob D (2011) Severe climate-induced water shortage and extremes in Crete: A letter. Climatic Change 106 (4): 667-677.

7. Brown C (2010) The end of reliability. J Water Resour Plann Manage 136: 143-145.

8. Karla A, Ahmad S (2009) Using Oceanic-atmospheric oscillations for long lead time streamflow forecasting. Water Resour Res 45(3): DOI:10.1029/2008WR006855.

9. Guo YP (2006) Updating rainfall IDF relationships to maintain urban drainage design standard. J Hydrol Eng 11 (5): 506-509.

10. Mailhot A, Duchesne S (2010) Design criteria of urban drainage infrastructures under climate change. J Water Resour Plann Manage 136 (2): 201-208.

11. Moglen GE, Vidal GER (2014) Climate change impact and storm water infrastructure in the Mid-Atlantic region: design mismatch coming. J Hydrol Eng 19 (11): DOI:10.1061/(ASCE)HE.1943-5584.0000967.

12. Forsee WJ, Ahmad S (2011) Evaluating urban storm-water infrastructure design in response to projected climate change. J Hydrol Eng 16 (11): 865-873.

13. Zhu J, Stone MC, Forsee W (2012) Analysis of potential impact of climate change on intensity-duration-frequency (IDF) relationships for six regions in the United States. J Water and Climate Change 3 (3): 185-196.

14. Mailhot A, Duchesne S, Caya D, Talbot G (2007) Assessment of future change in intensity-duration-frequency (IDF) curves for Southern Quebec using the Canadian Regional Climate Model (CRCM). J Hydrol 347: 197-210.

15. Coulibaly P, Shi X (2005) Identification of the effect of climate change on future design standards of drainage infrastructure in Ontario. Highway Infrastructure Innovation Funding Program, Ministry of Transportation of Ontario, Canada.

16. Zhu J (2013) Impact of climate change on extreme rainfall across the United States. J Hydrol Eng 18 (10): 1301-1309.

17. Deser C, Phillips A, Bourdette V (2012) Uncertainty in climate change projections: the role of internal variability. Climate Dyn 38: 527-546.

18. Dibike YB, Coulibaly P (2005) Hydrologic impact of climate change in the Saguenay watershed: comparison of downscaling methods and hydrologic models. J Hydrol 307 (1-4): 145-163.

19. Fowler HJ, Blenkinsop S, Tebaldi C (2007) Linking climate change modeling to impacts studies: Recent advances in downscaling techniques for hydrological modeling. Int J Climatol 27 (12): 1547-1578.

20. Kalra A, Ahmad S (2011) Evaluating changes and estimating seasonal precipitation for Colorado River Basin using stochastic nonparametric

21. Praskievicz S, Chang HJ (2009) A review of hydrological modeling of basin-scale climate change and urban development impacts. Progress in Physical Geography 33 (5): 650-671.

22. Prudhomme C, Reynard N, Crooks S (2002) Downscaling of global climate models for flood frequency analysis: Where are we now? Hydrol. Processes 16 (6): 1137-1150.

23. Berggren K, Olofsson M, Viklander M, Svensson G, Gustafsson A (2012) Hydraulic Impacts on Urban Drainage Systems due to Changes in Rainfall Caused by Climatic Change. J Hydrol Eng 17 (1): 92-98.

24. Olsson J, Berggren K, Olofsson M, Viklander M (2009) Applying precipitation model climate scenarios for urban hydrological assessment: A case study in Kalmar City, Sweden. Atmos Res 92 (3): 364-375

25. Semadeni-Davies A, Hernebring C, Svensson G, Gustafsson L (2008) The impacts of climate change and urbanisation on drainage in Helsingborg, Sweden: Combined sewer system. J Hydrol 350: 100-113.

26. Fortier C, Mailhot A (2015) Climate change impact on combined sewer outflows. J Water Resour Plann Manage 141(5), DOI:10.1061/(ASCE)WR.1943-5452.0000468.

27. Grillakis MG, Koutroulis AG, Tsanis IK (2011) Climate change impact on the hydrology of Spencer creek watershed in Southern Ontario, Canada. Journal of Hydrology 409: 1-19.

28. Mearns, L.O., et al. (2007), updated 2012. The North American Regional Climate Change Assessment Program dataset, National Center for Atmospheric Research Earth System Grid data portal, Boulder, CO. Data downloaded 2014-07-07. [DOI:10.5065/D6RN35ST].

29. Mearns LO, Gutowski WJ, Jones R, Leung LY, McGinnis S, et al. (2009) A regional climate change assessment program for North America. EOS, 90: 311-312.

30. NARCCAP (2013) North Americal Regional Climate Change Assessment Program. <http://www.narccap.ucar.edu/> (January 26, 2013).

31. Mailhot A, Beauregard I, Talbot G, Caya D, Biner S (2012) Future changes in intense precipitation over Canada assessed from multi-model NARCCAP ensemble simulations. Int J Climato 32: 1151-1163.

32. Nakicenvoic N, Davidson O, Davis G, Grübler A, Kram T, et al. (2000) Special Report on Emissions Scenarios. A Special Report of Working Group III of the Intergovernmental Panel on Climate Change. Cambridge University Press Cambridge, 599.

33. Kozanis S, Christofides A, Efstratiadis A (2010) Scientific documentation of the hydrogram software version 4, Athens, pp 173.

34. Music B, Caya D (2007) Evaluation of the hydrological cycle over the Mississippi River Basin as simulated by the Canadian regional climate model (CRCM). J Hydrometeor 8: 969-988

35. Jones R, Noguer M, Hassell D, Hudson D, Wilson S, et al. (2004) Generating high resolution climate change scenarios using PRECIS. Met Office Hadley Center. Exter p 40.

36. Elguindi N, Bi X, Giorgi F, Nagarajan B, Pal J, et al. (2007) RegCM Version 3.1 User's Guide, Trieste, Italy. <https://users.ictp.it/RegCNET/regcm.pdf> (July 9, 2015)

37. Giorgi F, Marinucci MR, Bates GT (1993) Development of second generation regional climate model (RegCM2) I: boundary layer and radiative transfer processes. Mon Weather Rev 121: 2794-2813.

38. Collins WD, Bitz CM, Blackmon ML, Bonan GB, Bretherton CS, et al. (2006) The community climate system model version 3 (CCSM3). J Climate 19: 2122-2143.

39. Flato GM (2005) The Third Generation Coupled Global Climate Model (CGCM3). <http://www.ec.gc.ca/ccmac-cccma/default.asp?n=1299529F-1> (August 9, 2015).

40. GFDL GAMDT (2004) The new GFDL global atmospheric and land model AM2-LM2: Evaluation with prescribed SST simulations. J Climate 17: 4641-4673.

41. Gordon C, Cooper C, Senior CA, Banks H, Gregory JM, et al. (2000) The simulation of SST, sea ice extents and ocean heat transports in a version of the Hadley Centre coupled model without flux adjustments. Climate Dynamics 16: 147-168

42. Pope VD, Gallani ML, Rowntree PR, Stratton RA (2000) The impact of new physical parameterizations in the Hadley Centre climate model-HadAM3. Climate Dynamics 16: 123-146.

43. Chen C, Knutson T (2008) On the verification and comparison of extreme rainfall indices from climate models. J Clim 21 (7): 1605-1621.

44. Leander R, Buishand TA (2007) Resampling of regional climate model output for the simulation of extreme river flows. J Hydrol 332: 487- 496.

45. Sharma M (2009) Comparison of downscaled RCM and GCM data for hydrologic impact assessment M.A.Sc. Thesis, Dept. of Civil Engineering, McMaster University, Hamilton, Ontario, Canada.

46. James W, Rossman LA, James WRC (2010) User's guide to SWMM 5, 13th edition. CHI, Guelph, Ontario, Canada.

47. City of Hamilton (2011) West central mountain drainage assessment supplemental capacity analysis and SWM sizing Mewburn and Sheldon neighbourhoods." AMEC Environment and Infrastructure, Burlington, Ontario, Canada.

9

Adaptability Evaluation and Selection of Improved Tef Varieties in Growing Areas of Southern Ethiopia

Yasin Goa Chondie[1]* and Agedew Bekele[2]

[1]*Areka Agricultural Research Center, Areka, Ethiopia*
[2]*Awassa Agricultural Research Center, Awassa, Ethiopia*

Abstract

Eight Tef varieties including local checks were evaluated with the objective of selecting adaptable, best performing varieties and to assess farmers' criteria for Tef variety selection during 2008 and 2009 cropping season at Areka and Hossana stations of Areka Agricultural Research center in the Southern region of Ethiopia. In the study the Tef varieties namely Koye, Gimbichu, Quncho, Dega Tef, Keytena, Amarach and Ajora-1 were collected from the Federal and regional Research center along with local checks, Ethiopia, and Regional Agricultural Research Institute. These materials were put into trial at Areka Agricultural Research center station farms at Areka and Hossana of Wolayta and Hadiya Zones. The trial was laid out in a randomized complete block design with three replications. Each plot measured 3 m × 3 m with 1 m between plots and 1.5 m between blocks. Sowing was done within the last week of July to 1st week of August 2008 and 2009. Data on various characters, such as plant height, panicle length, days to heading, and days to maturity and grain yield. Data was subjected to analysis of variance and there was highly significant difference (p<0.01) among the varieties for grain yield and some of agronomic traits. The results for the trials indicated that there were significant yield differences between the local check and the released varieties at two stations. At Areka, the combined analysis of variance over years indicated that varieties Koye, Amarach and Quncho gave the highest grain yield viz., 988.7, 984.3 and 958.7 kg/ha respectively. Similarly, at Hosanna, varieties Gimbichu, Quncho and koye out yielded other varieties and had yield advantage of 31.9, 25.14 and 15.14% over local variety, respectively. Both combined across locations over year's analysis and farmers' assessments identified two varieties Quncho and Koye as potential varieties for wider production. This result also indicated that farmers were as capable as Researchers in varietal choice. Therefore, based on objectively measured traits (grain yield, days to maturity, plant height, panicle length, days to heading and farmers' preference, Koye and Quncho are recommended for wider cultivation in Areka and Hossana areas of south Ethiopia while varieties (Amarach and Gimbichu) showed specific adaptability for Areka and Hossana areas respectively.

Keywords: Tef; Agricultural research; Grain yield; Maturing variety

Introduction

Tef (*Eragrostis Tef* (Zucc.) Trotter) is an annual grass crop and important cereal harvested for grain in Ethiopia. Ethiopia is not only the origin of Tef but it is also the center of diversity [1]. Tef is adaptable to a wide range of ecological conditions in altitudes ranging from near sea level to 3000 msl and even it can be grown in an environment unfavourable for most cereal, while the best performance occurs between 1100 and 2950 masl in Ethiopia [2]. In the country, cereals, pulses, oil crops, vegetables and root crops are grown annually on the average, 10 million hectares. Of these 7.6 million is allocated for cereals. Tef, the single dominant, occupies 2,404,674 hectares and the production is about 24,377,495 quintals annually [3]. Tef flour is preferred in the production of enjera, a major food staple in Ethiopia. Tef is also grown on a limited basis for livestock forage in other parts of Africa, India, Australia, and South America. In the U.S., small acreages of Tef are grown for grain production and sold to Ethiopian restaurants (Carlson, Idaho) or utilized as a late planted livestock forage (Larson, Minnesota). According to Wondimu et al. [4], Tef is primarily grown to prepare enjera, porridge and some native alcohols drinks. The straw is used for animal feed. In the 2001/2002 cropping season about 133,882.2 ha was covered by Tef. The nutritional value of Tef grain is similar to the traditional cereals. Tef is considered to have an excellent amino acid composition, lysine levels higher than wheat or barley, and slightly less than rice or oats. Tef contains very little gluten. Tef is also higher in several minerals, particularly iron.

In Southern Nations, Nationalities, and Peoples Regional State (SNNPR), the main Tef producing zones in SNNPR are North Omo,

Gurage, Hadiya, Kembata-Tembaro Alaba and kefico Shekicho [3]. It is greatly valued by farmers and consumers. This crop is important crop for human consumption, source of cash and straw for animal feed and plastering compounds for construction purposes. Tef the most preferred crop because its straw quality for livestock feed, best 'enjera' quality, long seed storability, and drought resistance. The importance of Tef is based primarily on consumer preference for enjera (Ethiopian bread). Its agronomic versatility and reliability even under adverse conditions which suit it well to a country of contrasting and unpredictable environments where water logging, drought, pest and disease are all too common and bring repeated famine also makes this crop very important. The regional average yield of Tef is about 7.39 q/ha in 2001/2002 [3] cropping season.

The yields of Tef are low in Ethiopia as well as in southern region due to different production problems including: lack of improved varieties, non-adoption of improved technologies, disease and pests are some of the most serious production constraints in Tef production in Ethiopia.

*Corresponding author: Yasin Goa Chondie, Areka Agricultural Research Center, PO Box 79, Areka, Ethiopia, E-mail: yasingoac76@yahoo.com

Some varieties of Tef were released by the different regional and federal research centers in Ethiopia; however, most of them were not evaluated around areas of southern Ethiopia and farmers were not participated in varietal improvement and testing process. Participation of farmers' in varietal choice has considerable value in technology evaluation and dissemination. Participatory varietal evaluation and selection is being conducted in some crops like common bean [5] and barley [6]. According to Courtois et al. [7] evaluated the effect of participation of farmers by comparing only the rankings of varieties by farmers and researchers at the same locations and reported a strong concordance between farmers and breeders in environments that have been producing contrasting plant phenotypic performance in rice. Two way feedbacks between farmers and researchers is indeed vital component of highly client-oriented breeding programs in locally important and traditionally cultivated crop [8]. Daniel et al. [9] stated that farmers' selection criteria vary with environmental conditions, traits of interest, ease of cultural practice, processing, use and marketability of the product, ceremonial and religious values. Therefore, the objectives of this study were to evaluate and select improved Tef varieties which are adaptable, high yielding and to assess farmers' criteria for variety selection with the participation of farmers in southern Ethiopia.

Materials and Methods

Study area

The experiment was conducted at Areka Agricultural Research Farm of the Hosanna and Areka stations between end of July and August, 2008 and 2009. Hosanna is located at an altitude of 2290 masl, latitude 07° 5' N, longitude 37° 5' E, temperature: 17.02°C, rainfall: 1500-1800 mm, soil type: Profondic Luvisols (Areka Meteorological Station, 2008). Similarly, Areka is located at an altitude of 1830 masl, latitude 07° 4' 24" N, longitude 37° 41' 30" E, temperature: 20.3° C, rainfall: 1200-1700 mm, soil type: Haplic alisol (Areka Meteorological Station, 2008).

Eight Tef varieties namely Koye, Gimbichu, Quncho, Dega Tef, Keytena, Amarach and Ajora-1 varieties were collected from the Federal and regional Research center along with local checks, Ethiopia, and Regional Agricultural Research Institute. These materials were put into trial at Areka Agricultural Research center station farms at Areka and Hossana of Wolayta and Hadiya Zones during Meher season of 2008-2009. The trial was laid out in a randomized complete block design with three replications. Unit plot size was 9 m² (3 m × 3 m) with spacing of 1 m between plots and 1.5 m between blocks. Planting was done by broadcasting at seed rate of 30 kg/ha. Sowing was done within the last week of July to 1st week of August 2008 and 2009. All other recommended agronomic practices were kept normal and uniform to ensure normal plant growth and development. Seed yield of each plot was recorded and then converted into kg/ha. Data on plant height, panicle length, days to heading, days to maturity and grain yield were collected and subject to statistical analysis using SAS statistical software [10]. The farmers used matrix ranking to assess the most suitable varieties for their areas. The characters scored included; plant height, straw yield, thresh ability, days to maturity, seed colour, lodging, shattering, biomass yield and grain yield.

Results and Discussion

The analysis of variance revealed that there were highly significant (p<0.01) difference among varieties for days to maturity, plant height and panicle length, days to heading and grain yield at Areka (Table 1). These results are further supported by Fentie et al. [11] who reported

considerable variation in the days to maturity, plant height and panicle length, days to heading and grain yield of different Tef varieties when planted over years. Koye gave the highest grain yield (988.7 kg/ha) followed by Amarach (984.7 kg/ha) and Quncho (958.7 kg/ha) at Areka station (Table 1). Varieties koye, Amarach and Quncho had yield advantage of 15.9%, 15.4% and 12.4% over the local check respectively (Table 1).

At Hosanna station, the analysis of variance indicated that there were significant (P<0.01) difference among varieties for grain yield. This also agrees with the findings of Ashamo et al. [12] who evaluated 22 Tef genotypes at four locations and reported that significant variations in grain yield of Tef at all test locations. Similarly, in this study there were significant (P<0.01) difference among varieties for days to maturity, plant height and panicle length, days to heading. These results are in contrast with the earlier findings Fentie et al. [11] who noted that the effect of the different varieties used over years didn't show significant difference for plant height and panicle length. Variety Gimbichu gave the highest grain yield (1656.1 kg/ha) followed by Quncho (1571.3 kg/ha). Gimbichu and Quncho had yield advantage of 31.9% and 25.14% over the local check respectively. Gimbichu variety was found to be the earliest in maturity which was (99.5 days) at Hossana (Table 1). Grain yield was generally higher at Hossana than Areka (1656.1 kg/ha) and (988.7) respectively (Tables 1 and 2).

The combined analysis of variance across locations over years among varieties revealed that there was significant difference for 50% days to heading and maturity, plant height, panicle length and grain yield. Varieties by year interaction indicated that there was highly significant (p<0.01) difference for panicle length and days to heading.

Varieties	PH	PL	DH	DM	GY (kg/ha)	%YA/L	Rank
Koye	78.1bc	29.8d	41.5a	89.8a	988.7a	0.159	1
Gimbichu	79.3bc	27.6d	36.8e	83.2d	927.7ab	-	4
Quncho	95.2a	40.6a	38.5cd	87.8c	958.7a	0.124	3
Degatef	85.7ab	37.8ab	39.8bc	89.5ab	790.7c	-	7
Keytena	73c	31.2cd	41.2ab	87.5c	781c	-	8
Amarach	92a	33.3cd	38.2de	87.8c	984.3a	0.154	2
Ajora-1	96.5a	36.2abc	39.5cd	89.8a	795.7c	-	6
Local	79.5bc	30.03cd	39.98bc	88.5bc	853.3c	-	5
Mean	84.9	33.32	39.4	88	983.3	-	-
CV (%)	11.75	15.92	3.57	1.2	7.49	-	-
LSD (5%)	11.76	6.3	1.7	1.24	78.2	-	-

Table 1: Mean grain yield and agronomic data of Tef varieties tested combined over years (2008 and 2009) at Areka.

Varieties	PH	PL	DH	DM	GY (kg/ha)	%YA/L	Rank
Koye	74.5cd	30.2cd	46.5ab	104.8a	1445.7b	15.14	3
Gimbichu	72.6d	26.9d	41.8d	99.5c	1656.1a	31.9	1
Quncho	90.9a	38.7a	46.2ab	104.3a	1571.3ab	25.14	2
Degatef	78.9bcd	35.3ab	47.3a	104.8a	1258.7c	-	5
Keytena	71.1d	31bcd	45.2bc	102.5b	1240.6c	-	7
Amarach	85.3abc	33.2bc	43.2cd	102.8b	1259.3c	-	4
Ajora-1	87.03ab	34.9ab	46.2ab	105a	1045d	-	8
Local	81.4abcd	33.8bc	45.2bc	104.7a	1255.6c	-	6
Mean	80.24	32.99	45.2	103.4	1341.5	-	-
CV (%)	11.75	11.98	3.96	1.1	9.83	-	-
LSD (5%)	11.12	4.7	2.11	1.33	155.6	-	-

Key: GY=Grain yield (kg/ha), PH=plant height (cm), PL=Panicle length (cm), HD=Days to Heading, MD=Days to maturity and YA/L=-% yield advantage over local variety

Table 2: Mean grain yield and agronomic data of Tef varieties tested combined over years (2008 and 2009) at Hossana.

However, significant difference was not observed in plant height, days to maturity and grain yield. Varieties Gimbichu and Quncho gave the highest grain yield (1343.4 kg/ha) and (1318 kg/ha) respectively. Gimbichu gave the highest grain yield in both years and performed consistently over years at Hossana. In the combined analysis across locations over years, all farmers were consistently selected varieties koye and Quncho higher yielding and very white seed color whereas variety Amarach and Gimbichu gave higher yields than local checks are recommended for Areka and Hossnana areas specifically respectively. They also further argued that the high grain yielding potential of Quncho may be due its tallest plant height and bigger stem resisting relatively lodging compared to other improved varieties. Gimbichu and Quncho gave yield advantage of 21.9% and 19.6% over the local check respectively. Gimbichu, keytena and Amarach varieties took (91.3), (95) and (95.3) days to mature respectively (Table 3).

Farmers group around the stations visited and evaluated the research demonstration field twice at stage of maturity and harvesting for varietal choice. Accordingly, farmers set selection criteria of grain yield, maturity period and seed color. Based on their selection criteria, farmers selected Gimbichu for grain yield and for its short maturity period and ease of thresh ability and Quncho for its high yield, very white seed color and tolerance to long rainfall. Therefore, based on quantitatively measured agronomic traits (grain yield, seed color, and lodging, threshability and maturity date) and farmers' visual observation at field, koye and Quncho are recommended for production in Areka and Hossana areas of south Ethiopia and similar agro ecologies. Whereas varieties Amarach and Gimbichu showed specific adaptation for Areka and Hossana areas; respectively are recommended with their full production packages.

Conclusions and Recommendation

The combined analysis of variance revealed that varieties are significant for days to heading, maturity, panicle length, plant height and grain yield. Varieties Gimbichu, koye, Quncho and Amarach had a grain yield advantage of 21.9%, 15.4%, 19.6% and 6.8% over the local check respectively (Tables 3 and 4). Gimbichu was found to be the earliest maturing variety with higher grain yield. Farmers' main selection criteria were grain yield, biomass yield, straw yield, panicle length, lodging tolerance, thresh ability, maturity date and seed color. Based on their selection criteria, farmers selected Quncho for grain yield; biomass yield; straw yield, shattering resistance, tolerance to long rainfall and very white seed color, koye for grain yield and ease of thresh ability and white seed color. Gimbichu for its short maturity period and

Varieties	PH	PL	DH	DM	GY (kg/ha)	%YA/L	Rank
Koye	76.3cd	30d	44a	97.3a	1272.1a	0.154	3rd
Gimbichu	75.95cd	27.3e	39.3e	91.3d	1343.4a	0.219	1st
Quncho	93.1a	39.6a	42.3c	95.6c	1318.2a	0.196	2nd
Degatef	82.4b	36.5b	43.6ab	97.2ab	1068.6c	-	6th
Keytena	72.05d	31.1d	43.2abc	95c	1054.2c	-	7th
Amarach	88.7a	32.3c	40.7d	95.3c	1176.5b	0.068	4th
Ajora-1	91.8a	35.6b	42.8bc	97.4a	964.5d	-	8th
Local	80.4bc	31.9cd	42.5c	96.6b	1101.9bc	-	5th
Mean	82.6	33.2	42.3	95.7	1162.4	-	-
CV (%)	8.8	7.9	2.4	0.93	9.24	-	-
LSD (5%)	6.03	2.2	0.84	0.74	89.29	-	-

Key: GY=Grain yield (kg/ha), PH=plant height (cm), PL=Panicle length (cm), HD=Days to Heading, MD=Days to maturity and YA/L=-% yield advantage over local variety

Table 3: Mean grain yield and agronomic data of Tef varieties tested across Areka and Hosanna combined over years (2008 and 2009).

Tef varieties in 2008 and 2009	Selection Criteria's										
Areka	GY	MD	BY	SY	SC	TS	LG	SH	PH	Total	Over all Rank
Koye	3	2	2	2	2	3	1	2	2	19	3rd
Gimbichu	3	3	2	2	2	3	1	2	2	20	4th
Quncho	3	2	3	2	3	2	2	2	2	21	2nd
Degatef	1	2	2	2	2	2	2	2	2	17	6th
Keytena	1	2	1	2	1	3	2	2	1	15	8th
Amarach	3	2	2	3	2	3	3	3	3	24	1st
Ajora-1	1	2	1	2	2	3	2	2	3	18	5th
Local	1	2	1	2	2	3	2	1	2	16	7th
Hossana											
Koye	3	2	2	3	2	3	1	2	2	20	3rd
Gimbichu	3	3	2	3	2	3	1	2	3	22	2nd
Quncho	3	2	3	3	3	2	3	2	1	23	1st
Degatef	2	2	1	2	2	2	3	2	2	18	5th
Keytena	1	2	2	2	1	3	2	2	2	17	6th
Amarach	3	2	2	2	2	3	2	2	1	19	4th
Ajora-1	2	2	1	2	2	2	2	2	1	16	7th
Local	1	2	2	2	1	3	2	1	1	15	8th

Key: GY=Grain yield, BY=Biomass yield, SY=Straw yield, SC=seed color, TS=Thresh ability, MD=Days to maturity, SH=Shattering tolerance and LG=Lodging tolerance, Preference scale 0-3, 0=Poor, 1=fair, 2=Good, 3=Very good

Table 4: Matrix ranking of tef varieties at Areka and Hossana stations over years (2008 and 2009).

its higher grain yield for Hosanna areas and Amarach for its high yield, simplicity of threshability and very white seed color particularly to Areka areas. Therefore, based on researchers and farmers' preference, it was concluded that varieties Koye and Quncho are recommended for wider cultivation whereas varieties Gimbichu and Amarach are specifically recommended for Hosanna, Areka areas respectively.

References

1. Vavilove NI (1951) The origin, variation, immunity and breeding of cultivated plants (translated from Russian by K. Starr Chester). The Ronald's Press Co., New York, pp: 37-38.

2. Hailu T, Seyfu K (2000) Production and importance of tef in Ethiopia Agriculture. In: Tefera H, Belay G, Sorrels M (eds.), Narrowing the Rift: Tef research and development- Proceedings of the international Tef Genetics and improvement, 16-19 October 2000, Addis Ababa, Ethiopia.

3. CSA (2003) Urban bi-annual employment unemployment survey. 1st Year, Round 1.

4. Wondimu A, Mekbib F (2001) Utilization of tef in the Ethiopian diet. In: Tefera H, Belay G, Sorrells M (eds.), Narrowing the rift: Tef research and development. Proceedings of the International Workshop on Tef Genetics and Improvement, Debrezeit, Ethiopia, pp: 239-244.

5. Kornegay J, Beltran JA, Ashby J (1996) Farmer selections within segregating populations of common bean in Colombia: Crop improvement in difficulty environments. In: Eyzaguirre P, Iwanaga M (eds.), Participatory Plant Breeding, Proceeding of a workshop on participatory plant breeding, 26-29 July 1995, Wageningen, The Netherlands, IPGRI, Rome, Italy, pp: 151-159.

6. Fufa F, Grando S, Kafawin O, Shakhatreh Y, Ceccarelli S (2010) Efficiency of farmers' selection in a participatory barley breeding programme in Jordan. Plant Breeding 129: 156-161.

7. Courtois B, Bartholome B, Chaudhary D, McLaren G, Misra CH, et al. (2001) Comparing farmers and breeders rankings in varietal selection for low-input environments: a case study of rainfed rice in eastern India. Euphytica 122: 537-550.

8. Getachew B, Hailu T, Anteneh G, Kebebew A, Gizaw M (2008) Highly client-oriented breeding with farmer participation in the Ethiopian cereal tef [Eragrostis tef (Zucc.) Trotter]. African J Agric Res 3: 022-028.

9. Danial D, Parlevliet J, Almekinders C, Thiele G (2007) Farmers participation

and breeding for durable disease resistance in the Andean region. Euphytica 153: 385-396.

10. SAS Institute (2002) SAS System for Windows Release 9.2 Inc., Cary, NC, USA.

11. Fentie M, Demelash N, Jemberu T (2012) Participatory on farm performance evaluation of improved Tef (Eragrostis tef L) varieties in East Belessa, north western Ethiopia. International Research Journal of Plant Science 3: 137-140.

12. Ashamo M, Belay G (2012) Genotype x Environment Interaction Analysis of Tef Grown in Southern Ethiopia Using Additive Main Effects and Multiplicative Interaction Model. Journal of Biology Agriculture and Healthcare 2: 66-72.

10

Evaluating Multi-Scale Flow Predictions for the Connecticut River Basin

Muhammet Omer Dis[1], Emmanouil Anagnostou[1],*, Flamig Zac[2], Humberto Vergara[2] and Yang Hong[2]

[1]Civil and Environmental Engineering, University of Connecticut, Storrs, Connecticut, USA

[2]School of Civil Engineering and Environmental Sciences, University of Oklahoma, Norman, Oklahoma, USA

Abstract

This case study evaluates a computationally efficient distributed hydrological model, named Coupled Routing and Excess Storage (CREST), for flood modeling of basins in the Connecticut River Basin (CRB). Simulation of discharges is performed by forcing CREST with a long record (eight years) of high resolution radar-rainfall data and potential evapotranspiration maps derived from the North American Regional Reanalysis. The model performance is evaluated against observed streamflows obtained from United States Geological Survey gauging stations at outlet and interior points of various CRB sub-basins. CREST parameters were calibrated based on a three year record (2005-2007) and validated for the remaining data period (2003-2004 and 2008-2009). The model performance evaluation is based on different metrics, including the Nash-Sutchliffe Coefficient of Efficiency (NSCE), Mean Relative Error (MRE), Root Mean Square Error (RMSE), and Pearson Correlation Coefficient (PCC). The analysis shows that CREST slightly underestimated the peak flows, but exhibited a generally good capability in simulating the stream flow variability for the CRB basins. Specifically, NSCE, MRE, RMSE, and PCC values of hourly flow simulations varied from 0.31 to 0.58, -0.06 to 0.13, 61 to 121 (mm) and 0.60 to 0.83, respectively. At daily time scale the performance metrics exhibited improved values indicating that CREST has sufficient accuracy for long term multi-scale hydrologic simulations.

Keywords: Distributed hydrological model implementation; Calibration and validation; High temporal resolution; CREST; Connecticut river basin

Introduction

Rainfall-runoff modeling has a long history; the first hydrologist that used rainfall-runoff model was an Irish engineer Thomas James Mulvaney (1822-1892) who published his work in 1851. During the last few decades, a number of conceptual and physically-based models have been developed and used for simulation of floods [1-6]. The analysis of hydrological model simulations and their spatiotemporal fluctuations can be used as vital tools to support management activities such as flood risk management, water supply and improving water quality. According to Brakendridge et al. [7], their study shows that an increasing population density has caused a greater risk regarding the natural processes, such as flooding, which makes it crucial to identify such risky areas. Dehotin and Braud [8] mentioned the importance of distributed hydrologic models, indicating that they are valuable mechanisms to spatially and realistically study the prediction of water balance components. Hydrologic modeling can be employed to evaluate flood mitigation alternatives to flood and drought risks. They can be used to evaluate and study the impact of land use on water resources. Additionally, Moriasi et al. [9] states that hydrological modeling is an essential tool for managing water quantity and quality. The spatial structure of distributed hydrological models can inspect and evaluate the heterogeneities of watershed characteristics and its parameters [8,10].

Even though distributed models capture sufficient details and realistic catchment characteristics, over parameterization can be a problem in calibration. Dehotin and Braud [8] stated the growing of concerns about optimum parameterization. They indicated that the major concern is the contrasting aspects of model complexity versus data availability. Rozalis et al. [11] stated that simplicity versus complexity of hydrological models is still a controversial subject for better representation of a catchment. In their study, a simpler model was used with minimum number of parameters. The smaller number

of parameters does not require calibration to decrease uncertainty over ungauged areas. Moreover, Bergstrom [12] claimed an excessive increase in model complexity does not always improve the quality of the results. Brandt [13] supported this statement and further elaborated in his study that going from complex to simpler model does not affect model performance. Thus, using a model with optimal complexity relative to data availability and resolution is the key to improving hydrologic predictability. Also, it is worth mentioning that one of the important parameterizations in the modeling are the initial conditions and associated parameters sets. The initial condition of a model mostly depends on catchment area, catchment topography, antecedent moisture condition, ground water table position and land use. Hence, for distributed hydrological modeling, parameterizations of these initial conditions and its spatial variability are the fundamental factors for runoff simulations, especially for extreme rainfall events [14]. The influence of initial conditions on the model output reduces with increasing the simulation time period [15]. Additionally, a spin-up or run-up period can be used to diminish the sensitivity to initial conditions. In this study a one-year spin up period was used in the model simulation. Therefore, the number of initialization parameters was efficiently reduced.

Two other significant issues for hydrological modeling are related to calibration and validation methodology, specifically the metrics used to analyze and evaluate simulation results [9,10,12,16,17]. In this study, we evaluated the accuracy of a grid-based distributed hydrologic model,

---

*Corresponding author: Emmanouil Anagnostou, Civil and Environmental Engineering, UCONN, USA, E-mail: manos@engr.uconn.edu

which was calibrated and validated in the Connecticut River Basin. The performance of the model related to calibration and verification was evaluated using the Nash-Sutchliffe Coefficients of Efficiency (NSCE), Mean Relative Error (MRE), Root Mean Square Error (RMSE), and the Pearson Correlation Coefficient (PCC), which compared supplemental normalized flow simulations with obtained hourly observations obtained for the basin. For this analysis, a continuous simulation was applied for the hydrological analysis in the Connecticut River Basin using the Coupled Routing and Excess Storage (CREST) Hydrological Model. CREST model was selected for this study because the model currently represents both a national [18] and global flood simulation system [1]. Similar to other distributed models, catchments are represented as grid cells to simulate spatio-temporal variation of water, energy fluxes, and storage.

One of the purposes of this work is to examine the performance of the CREST model with calibrated parameters over interior sub-basin stations. Calibrating a model at a larger scale basin and using these parameters to model un gauged locations is one of the popular ways in situations where no observations are available. According to Bingeman et al. [19], a calibrated hydrological model based on stream flow data should be applicable on other sub-catchments without recalibration of the parameters set. State variables of basin characteristics related to land cover and soil texture for a given set of watersheds can be applied to other watersheds that are hydrologically similar without extensive recalibration [20]. Kouwen et al. [21] study also confirmed the fact that a parameter set can effectively transferred between watersheds to simulate peak flow without further calibration in southern Ontario. Additionally, Xie et al. [22] transferred calibrated parameters from data-rich areas to data-sparse areas and the results showed promising in estimating daily runoff with transferred variables. Hence, CREST was calibrated against stream flow observations at sub-basin outlets and was used to simulate stream flows based on the same parameter set at its interior nested sub-basins without further calibration.

This case study aims to evaluate the multi-basin scale predictability of stream flow in the Connecticut River Basin using the CREST model following calibration procedures suggested in the literature. Potential impacts of these fluctuations are important for water supplies, during the peak seasons of water demand, and mitigating flooding risk associated with peak flows. This study presents an opportunity to enhance and refine the estimation of hydrological processes via the CREST model and understand its calibration, parameterization and validation procedures over a mid-latitude basin. The main objectives of this work are: (1) use available observations to assess the accuracy of simulations, (2) determine the spatial variability of calibration parameters, and (3) understand the applicability of calibrated parameter values applied to interior catchments in order to predict flows at un gauged basin locations. Based on the topography and USGS observation gauges, the Connecticut River Basin was divided into nine sub-basins for all CREST simulations. Knowledge of the fluctuations and the accuracy of the predictions can assist in performing simulations over un gauged locations. For this purpose, three of the sub-basins, which have enough interior locations for un gauged analysis, were further subdivided into sub-watersheds in order to perform an analysis of parent basin calibration parameter runs.

The paper is organized as follows. Section 2 describes the study area, data, and the CREST hydrological model. Section 3 provides information regarding the model calibration, parameterization and validation with analysis of evaluated statistics. Section 4 provides a discussion of the results of the hydrological simulation using the

CREST model for the Connecticut River Basin. The conclusions and future work are discussed in the last section.

Study Region and Data

The Connecticut River Basin (CRB) is the study basin for this work and the CRB is a major river basin in New England. Runoff from the CRB discharges to the Connecticut River. The Connecticut River starts in Quebec Canada and runs through Connecticut (CT), Massachusetts (MA), New Hampshire (NH), and Vermont (VT) and empties into Long Island Sound in Old Lyme Connecticut. The total watershed area contributing to the Connecticut River is approximately 28,500 km². There are approximately 390 towns and cities located within the watershed with a total population of approximately 2.3 million people. The land uses that are within the watershed consist of forest, agriculture, residential and water. Approximately, 79% is covered by forested, 11% by agriculture and the remaining area is covered by residential and water. The CT River flows for about 660 km and provides hydroelectric power, is navigable up to Windsor Locks, used for irrigation, and is used for recreation [23].

Instantaneous records of river discharges obtained from nine United States Geological Survey (USGS) gauging stations within the Connecticut River Basin were analyzed in this study. The location of the gauging stations and their contributing sub-basins are illustrated in Figure 1. The black dots represent the streamflow gauges, which are labeled with USGS station numbers. The locations of USGS monitoring stations were used to determine the number of sub-basins based on the location and topography and their watersheds area. Given a particular gauging station along with the use of DEM data, we are able

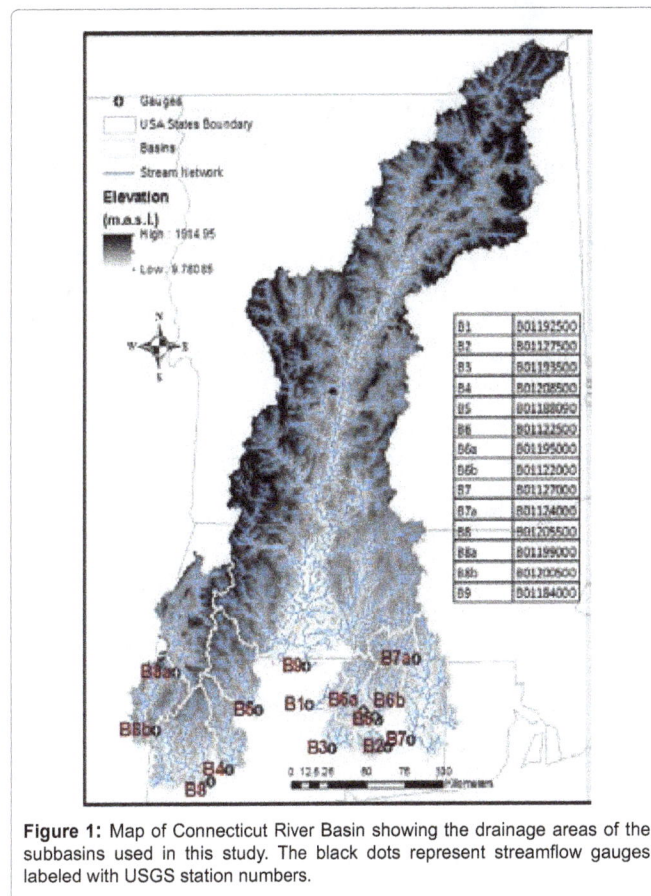

Figure 1: Map of Connecticut River Basin showing the drainage areas of the subbasins used in this study. The black dots represent streamflow gauges labeled with USGS station numbers.

to determine the area contributing runoff at each station. The data from the nine stream gauges were used to calibrate and verify continuous simulations from the CREST model across watersheds with drainage areas ranging from 200 to 25,000 km².

Zanon et al. [24] has argued that the flood response could be reasonably well reproduced when using high resolution rainfall observations. Even though continuous rain gauge data are available for the area, there are several gaps in these data records. As Hirpa et al. [25] reported unequal time period data between observation gauges might affect the certainty of parameter estimations. Additionally, Du et al. [26] have stated that there are many stations in the USA that do not have extensive flow records or have limited flow data. The gauge data measurements are available in 5-min, 15-min, and 30-min intervals, with 15-min intervals being the most common. Thus, the finer time scale gauge measurements were converted into hourly stream flow averages in units of m³/s from January 2002 to December 2009. The input data sets consist of gridded radar-rainfall (mm/h) and Potential Evapotranspiration (PET) (mm/3h) data. The precipitation data was extracted from the WSR-88D Stage IV product obtained from the North American Regional Reanalysis. The Stage IV radar rainfall fields were used to force the model at the hourly time step and 4 km spatial grid resolution. PET data based on the North American Regional Reanalysis (NARR) available at 32 km spatial and 3-hourly temporal resolution [27] were used as forcing variable for the hydrological model.

In addition to the radar rainfall and the PET data, High Resolution Digital Elevation Model (DEM) data was used to delineate the various watersheds and to perform river routing of the stream flow simulations.

Methods of Analysis

Model

A raster-based distributed hydrologic model, Coupled Routing and Excess Storage (CREST) was implemented over Connecticut River Basin. The CREST model is a hybrid modeling strategy, developed by the University of Oklahoma and NASA SERVIR-an acronym meaning

to "to serve" in Spanish- Project Team [1]. The model is spatially distributed rainfall-runoff model dedicated to simulate flow discharges at regional and global scales aimed to represent hydrological processes associated with floods and droughts. Distributed CREST model can be applicable to almost any kind of hydrological problems and provides important advantage over existing models under different land cover and soil type scenarios with user defined spatiotemporal resolutions. The CREST model provides water managers with information about streamflow amounts that can enable better decision-making regarding water resources, floods, and agriculture.

CREST model uses a combination of DEM, PET, and precipitation input data with different user defined spatiotemporal resolutions. The processes modeled include rainfall-runoff generation and capacity for cell-to-cell routing, canopy interception, infiltration, evaporation, recharge baseflow, sub-grid cell variability of soil moisture storage capacity and routing processes at the sub-grid scale. The model controls the maximum storage of the infiltrating water and yield surface runoff generation with connected layers within the soil profile. Cell-to-cell flow routing of direct surface runoff is applied using a kinematic wave assumption. Besides, coupling between the flow simulation and routing component via feedback mechanisms provides realistic applications of the hydrologic variables (i.e. soil moisture). The CREST model has been discussed extensively in previous papers [1,28] and details of the model can be found therein.

CREST model contains several default parameters. These parameters have value ranges initially specified based on land cover and soil type data. The definitions of parameters with their value ranges for CRB are listed in Table 1. The model contains 14 parameters, and three of them are related to initial soil conditions, (the initial value of soil water "iwu", initial value of overland reservoir "iso", and initial value of interflow reservoir "isu") that can be adjusted using warm-up period. The majority of the parameter ranges come from physical considerations. For example, the slope flow speed multiplier (coem) can be regarded as the inverse of manning's roughness, and "river" is similar to "coem" but for river channels [29]. Parameter "under" represents the horizontal velocity of subsurface flow, and hydraulic conductivity is used for this velocity [1,30]. "leako" (leaki) is overland (interflow) reservoir multipliers, whichvaries from 0.01 to 1 [1]. "pwm" is the capacity of the soil to hold water [31]. "pb" is a parameter related to infiltration and described as the exponent of the variable infiltration curve while 'pim' is the percentage of impervious area, which is derived from land cover data [1]. "pke" is a multiplier factor to convert PET to local actual evapotranspiration [29]. "pfc" is the soil saturated hydraulic conductivity which is derived based on soil type data [1]. The initial values of these parameters are adjusted through a calibration procedure utilizing using measured stream flows at gauging stations [6]. This is discussed at the next section.

Model Calibration and Validation

CREST model calibration and validation were carried out over Connecticut River Basin at various selected observation stream gauges (Figure 2). Bergstrom [12] defined model calibration as a process that model parameters are arranged to make model results to meet the measurements. However, if the number of parameters used in the calibration is large, automatic calibration is a better option to reduce labor-intensive. Additionally, automatic approach for model calibration abbreviates the time with the advantage of speed and power of high performance computers as well as the approach eliminates the kinds of subjective human judgments [9,16]. The auto-calibration routine based on Differential Evolution Adaptive Metropolis (DREAM) method

Parameter Name	Description	Suggested range
coem	The slope flow speed multiplier	3.33, 100.0
river	The channel flow speed multiplier	3.33,100.0
under	The horizontal velocity under the ground (Generally hydraulic conductivity used for this velocity)	0.01,51.0
leako	The overland reservoir discharge multiplier	0.01,1.0
leaki	The interflow reservoir discharge multiplier	0.01,1.0
th	The flow accumulation needed for a cell to be marked as a channel cell. If a cells flow accumulation is greater than th then the cells slope flow speed is multiplied by river	0.01,30.0
pwm	The maximum soil water capacity (depth integrated pore space) of the soil layer	0.01,250.0
pb	The exponent of the variable infiltration curve.	0.01,4.0
pim	The impervious area ratio	0.00,100.0
pke	The multiplier to convert between input PET and local actual ET	0.01,1.0
pfc	The soil saturated hydraulic conductivity	0.01,51.0
iwu	The initial value of soil water. (This is a percentage of the pwm)	0.0,100.0
iso	The initial value of overland reservoir	0.01,100.0
isu	The initial value of interflow reservoir	0.01,100.0

Table 1: CREST model parameters with their value ranges.

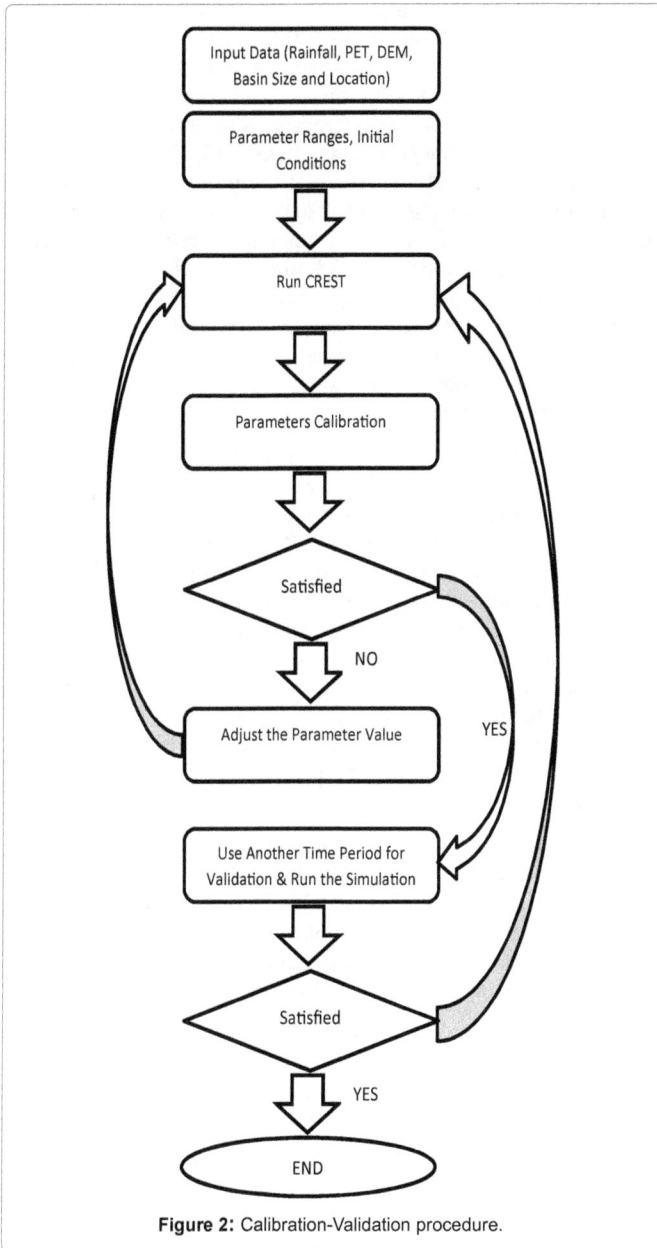

Figure 2: Calibration-Validation procedure.

Parameter Name	01192500	01127500	01193500	01208500	01188090	01122500	01127000	01205500	01184000
coem	74.33	43.52	53.54	50.06	14.55	12.3	18.07	10.39	37.63
river	99.19	80.89	99.06	86.67	79.61	98.53	80.02	90.45	7.16
under	33.69	7.25	38.96	0.9	50.43	4.16	35.3	13.64	17.79
leako	0.02	0.36	0.74	0.4	0.08	0.02	0.1	0.23	0.1
leaki	0.94	0.72	0.84	0.6	0.73	0.53	0.52	0.42	0.23
th	24.76	16.81	15.29	26.82	21.85	13.32	11.62	16.72	13.69
pwm	32.89	3.03	7.59	1.51	2.52	1.91	3.02	7.56	14.3
pb	2.06	0.49	2.45	2.46	0.79	1.31	2.31	1.79	0.38
pim	1.42	0.93	0.99	0.04	0.85	1.08	1.5	1.05	0.69
pke	0.44	0.95	0.99	0.84	0.79	0.17	0.96	0.84	0.72
pfc	44.88	3.48	38.37	50.49	45.68	37.24	35.36	39.72	46.45
iwu	71.33	0.59	91.16	75.92	7.86	91.28	70.48	42.6	89.78
iso	9.17	15.14	2.11	4.77	92.52	40.6	59.05	65.2	80.21
isu	1.45	51.77	12.9	81.45	13.06	48.05	98.21	26.71	85.11

Table 2: Parameters values determined from calibration for each gauged sub basin.

The simulation starts with a warm-up period to reduce the initial condition effect. Specifically, the three parameters related to initial conditions (iwu, iso, and isu) are model states that were adjusted automatically by running CREST with a warm-up period. Marshall and Randhir [23] observed in their study using a 40-year study period over Connecticut River Basin that the maximum snow accumulation occurs in the month of January and decreases exponential through April. To avoid snowmelt effects in parameter calibration the period of January through the middle of March was excluded from the error analysis.

Finally, after parameter calibration, CREST was validated for a four year period. Refsgard [10] defined model validation as a process of illustrating that the calibrated parameter set without further adjustments is capable to reproduce flows in different times other than the calibration period. This process was done again without any interruption during snow accumulation and melting span, but we excluded the snow process period (January 1-March 15) from the analysis.

Evaluation Indexes

Different error metrics have been used in hydrologic model validation exercises of past studies [9,16,32]. Multi criteria error functions aid to explain various aspects of hydrographs. In this study, the model efficiency to predict stream flow at basin outlets was demonstrated qualitatively by plotting time-series of observed and simulated stream flows and determining error statistics based on hydrographs normalized by the corresponding catchment area. The error metrics are listed below:

Nash-Sutcliffe Coefficients of Efficiency (NSCE):

$$NSCE = 1 - \frac{\sum_{i=1}^{n}(Q_{o,i} - Q_{s,i})^2}{\sum_{i=1}^{n}(Q_{o,i} - \bar{Q}_o)^2} \quad \ldots\ldots (1)$$

was devised to calibrate CREST over a number of watershed areas in Connecticut River Basin based on hourly stream flow data for the period 2005 to 2007. The algorithm runs to mitigate sum of squared residuals to estimate posterior parameters using multiple Markov Chain Monte Carlo chains from prior iterations [15].

The numbers of parameters which were subject to calibration were limited to 14 as defined by Wang et al. [1]. Calibrated values are listed in Table 2 along with the acceptable intervals (Table 1) used to constrain the parameters search. Each basin was calibrated and validated separately to capture the spatial variability of the parameters over the region. Even though all parameters are in the acceptable range, the parameter values vary across basins. For example, the slope flow speed multiplier "coem" varied between 10.39 and 74.33 with a mean of 34.93 and 0.61 coefficient of variation. The maximum soil water capacity "pwm" parameter has the maximum coefficient of variation, i.e 1.15.

where $Q_{o,i}$ is measured stream flow of the time i^{th}; $Q_{s,i}$ is simulated stream flow of the time i^{th}; $\overline{Q_o}$ is the average of entire observed streamflow values; and n is the total number of time steps. The value of NSCE varies from $-\infty$ to 1. On the one hand, a value of NSCE ≤ 0 represents that the model does not have capability to use the observed mean as a predictor; on the other hand, NSCE=1 indicates that simulation results are capturing the measurements perfectly.

Mean Relative Error (MRE):

$$MRE = \frac{\sum_{i=1}^{n}(Q_{s,i} - Q_{o,i})}{\sum_{i=1}^{n}Q_{o,i}} \qquad (2)$$

MRE gives an indication of how close predictions are relative to the observations. A value of MRE=0 shows that the simulated total amount of discharges is unbiased to observations.

Root Mean Square Error (RMSE):

$$RMSE(\%) = \frac{100}{\overline{Q_o}} * \sqrt{\frac{\sum_{i=1}^{n}(Q_{o,i} - Q_{s,i})^2}{n}} \ldots \qquad (3)$$

RMSE measures the magnitude of the differences between simulated and observed discharges relative to mean observed discharge value. Therefore, a low RMSE indicates better fit and the value of zero signifies the perfect fit.

Pearson Correlation Coefficient (PCC):

$$PCC = \frac{\sum_{i=1}^{n}(Q_{o,i} - \overline{Q_o})(Q_{s,i} - \overline{Q_s})}{\sqrt{\sum_{i=1}^{n}(Q_{o,i} - \overline{Q_o})^2 * \sum_{i=1}^{n}(Q_{s,i} - \overline{Q_s})^2}} \qquad (4)$$

where $\overline{Q_s}$ is the average of entire simulated streamflow values. PCC represents how well the linear relationship between measurements and predictions. This value ranges from -1 to 1.If PCC=0, then there is no relation between the two variables. The closer either -1 or 1 indicates stronger correlation between them.

Boyle et al. [16] have shown that RMSE is sensitive to the peak flows and can strongly bias recession error characteristics. PCC, on the other hand, reflects the collinearity between simulations and observations and correlation-based measures have excessive sensitivity to peak flows [32]. Moreover, MRE measures the average tendency of simulated against observed data and explains this tendency with overestimations or underestimations [9]. Servat and Dezetter [33] pointed out that even though NSCE error metric shows some weakness with low flows, it is the best objective function to provide extensive information on hydrograph prediction accuracy.

Results

CREST model was applied for the period between 2002 and 2009 to represent the hydrologic response of the Connecticut River Basin and several of its sub-basins. The calibration of the model was carried out over nine sub-basins from 2005 to 2007. Then, CREST stream flow simulations were compared to the measured flows at the outlets of the nine sub-basins and interior locations of three of the catchments over the seven-year time period using 2002 as the model spin-up. Simulations for the various basin sizes are summarized below showing an overall good agreed with observations for different event magnitudes.

In Figure 3, observed and simulated normalized hourly discharges were plotted along with measured rainfall for the period 2004 and 2005 at multiple stations. For brevity, three of the spatially various CRB sub-basins associated with different sizes are illustrated. Results, in the Figure 3, from first and second columns belong to validation and calibration periods, respectively. Three different size basins (small,

middle, and large scale catchments), namely B0128500 (673 km²), B01122500 (1,046 km²), and B01184000 (25,019 km²) catchments, are visualized from top to bottom. As it can be seen from the hydrographs, overall, model performed well over CRB and the variability depicted in the flow simulation is close agreement with observed flows. Even though the timing of the peaks is estimated well, CREST simulations tend to slightly underestimate some of the peak flows. As it can be seen from the hydrographs in April 2005 simulations underestimated observations associated with small amounts of rainfall. However, it would be expected that light rainfall produce smaller runoff and this mismatch in terms of peak discharges can be explained by uncertainties in either rainfall or stream flow observational data. However, although slight underestimation is captured in qualitative stream flow comparisons, the majority of the simulated discharges were attained close to the measurements.

The model outputs from the hydrological simulations are summarized in Table 3A and 3B for calibration and validation periods, respectively. Basin areas, observed and simulated mean discharges, error metrics results (NSCE, MRE, RMSE, and PCC), and percentage of unavailable observations during the analysis are illustrated. During calibration period, as it can be seen from Table 3, NSCE values vary between 0.31 and 0.68 for the nine-basins while MRE results range from -6% to 13%. Additionally, 0.60 is the smallest value with hourly time resolution for PCC metric. The demonstrated NSCE, MRE, and PCC values show that the model simulations have good agreement with measurements in the calibration data period. In terms of RMSE error metric, however, at some stations, it is difficult to appreciate the quality of the results.

In the validation period, on the other hand, Nash is lower (ranging between 0.12 and 0.58) for hourly resolution, while PCC values ranged between 0.42 and 0.77 for the nine-basins. It is interesting enough, during validation, overall MRE exhibited underestimation in the range of 4% to 26%. However RMSE values dropped between 2 and 20% in the basins, which indicate improved random error.

Moriasi et al. [9] have shown that hydrologic model performance is better for longer time steps (i.e. annual versus monthly) and claimed that simulation statistics improve as a function of time resolution. Fernandez et al. [34] supported this statement and reported that NSCE values during calibration in their study in the range of 0.36 and 0.66 for daily and monthly time resolutions, respectively. Hence, daily error metrics were calculated and illustrated in Table 3C.

The study illustrates a better agreement between estimated and measured flows at daily scale and model results become quite realistic. For instance, in B01205500, NSCE values increase from 0.31 to 0.47 during calibration and from 0.16 to 0.30 in validation period for hourly and daily resolutions, respectively. While PCC values improve from 0.60 to 0.70 (calibration period) and 0.46 to 0.59 (validation period) as a function of time resolution in the same catchment, no change is observed in terms of the MRE values, which is expected given that resolution primarily affects the random component of error. Similar improvements are also captured for the other basins (Table 3C). Moreover, RMSE values drop significantly and become less than 100% during both validation and calibration in the basins with the exception of B01193500 (103.91%).

Analysis of the error metrics with their annual fluctuations were, additionally, examined in Figure 4. The figure provides an illustration of simulations versus observation streamflow variability. Results reported in Figure 4A-D show the hourly calculated NSCE, MRE, RMSE, and

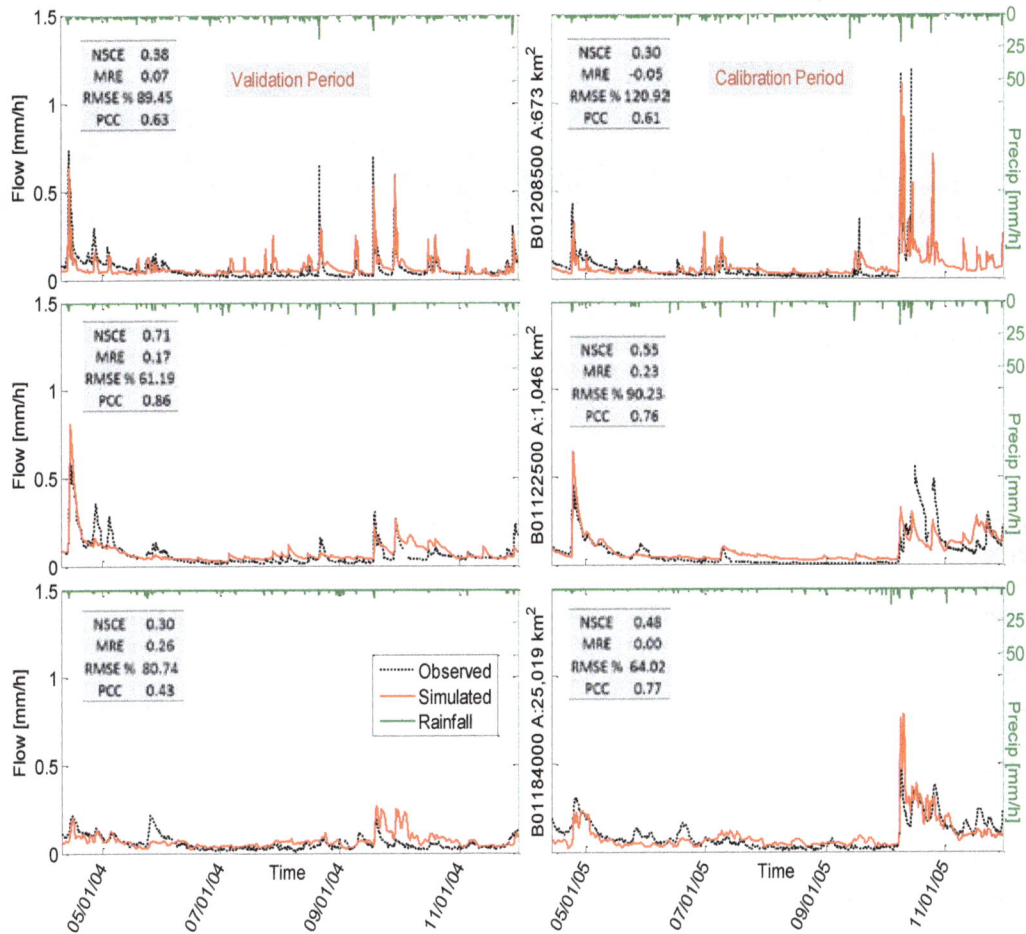

Figure 3: Normalized hydrographs of the model results and observations streamflow along with precipitation; partially calibration and validation period is zoomed in.

PCC statistics, respectively for each year in the 7-year time period. Multiple transparent dots represent the basin statistic and darkness of the dots is increased as a function of watershed size. Analysis of the results acquired for 7-year gives more reasonable amount for smaller basins. The majority of NSCE, MRE, PCC values ranged in acceptable levels during both calibration and validation. As expected, NSCE and PCC values for calibration are better than the validation periods and increase with longer time steps. Moreover, model outputs are quite realistic in between 2004 to 2008, while results diverged in the year 2003 and 2009. Thus, that can yield a bias into the optimization process with low NSCE, RMSE, and PCC.

Three nested subbasins were considered as ungauged interior catchments namely B01122500 (Figure 5), B0112700 (Figure 6), and B01205500 (Figure 7). After model was calibrated for the outlet of the parent basin, the parameters were fixed, and then CREST was used to simulate the interior catchments. Figures 5-7 show the comparison between model outputs and measurements via statistical error functions with annual variability. The catchments were ordered with increasing area. Differences in model responses increase with increasing size of the catchment. Overall, results show that there is good agreement between the predictions and the measurements.

In Figure 5, statistical metrics of NSCE, MRE, RMSE, and PCC

presented for different years and different interior subbasins of parent basin B01122500. The results show that the highest NSCE values are obtained in the larger basin (B01122500) while the small basin (B0119500) exhibits the poorest agreement in terms of the error metric. Basin size dependency is also observed in RMSE metric values with the highest value reported for the smaller basins. MRE results, on the other hand, are within +/- 0.5, and the overall performance of the model in terms of PCC is above 0.5.

To test calibration, validation periods, and resolution effect on the error metrics, Table 4 statistics were calculated and showed the results as a summary. In Table 4, basin statistics are visualized at hourly and daily resolution during calibration and validation periods for ungauged interior subbasins of parent B01122500 basin. As it can be seen from table, number of time steps in the observed data is decreased as a function of catchment area. In addition, it is important to note that discharge point of small basin is at the furthest location from the calibrated downstream outlet. It is expected that the simulation performance is decreased as a function of catchment area. Additionally, slight improvements are observed changing time resolution from hourly to daily.

Table 5 shows calculated basin statistics for interior subbasin of parent B01127000 at hourly time resolution during calibration

Gauge Name	Basin Area	(A)Calibration								
		Q_o[m³/s]		Q_s [m³/s]		NaN Data	NSCE	MRE	RMSE [%]	PCC
	[km²]	μ	σ	μ	σ	[%]				
01192500	190.11	4.38	4.74	4.85	3.89	0.11	0.41	0.11	83.28	0.67
01127500	231.29	5.31	8.55	5.42	7.64	5.29	0.68	0.02	90.59	0.83
01193500	259.00	6.70	10.62	7.11	8.58	4.90	0.42	0.06	121.25	0.66
01208500	673.40	14.03	18.98	13.79	14.20	17.13	0.35	-0.02	109.36	0.61
01188090	979.02	22.32	29.14	25.28	28.85	3.92	0.40	0.13	101.32	0.70
01122500	1,046.36	23.09	26.85	24.75	18.98	1.77	0.64	0.07	70.03	0.81
01127000	1,846.66	41.74	44.58	45.85	34.66	0.10	0.54	0.10	72.34	0.74
01205500	3,998.94	90.83	114.55	89.73	94.74	1.08	0.31	-0.01	104.98	0.60
01184000	25,019.28	690.49	531.64	648.82	452.39	1.22	0.37	-0.06	61.18	0.65

Gauge Name	Basin Area	(B)VALIDATION								
		Q_o[m³/s]		Q_s [m³/s]		NaN Data	NSCE	MRE	RMSE [%]	PCC
	[km²]	μ	σ	μ	σ	[%]				
01192500	190.11	4.20	3.35	4.00	2.40	1.79	0.21	-0.05	71.12	0.50
01127500	231.29	5.84	7.87	5.03	6.54	12.00	0.58	-0.14	87.64	0.77
01193500	259.00	6.63	9.08	6.86	7.10	13.21	0.29	0.04	115.00	0.58
01208500	673.40	17.88	16.33	13.15	9.76	4.24	0.24	-0.26	79.68	0.57
01188090	979.02	24.06	16.83	22.85	16.02	3.64	0.24	-0.05	61.18	0.60
01122500	1,046.36	26.05	23.91	22.31	15.62	5.35	0.44	-0.14	68.54	0.68
01127000	1,846.66	43.90	37.72	43.43	29.58	4.06	0.37	-0.01	68.19	0.63
01205500	3,998.94	97.80	95.21	82.89	57.22	0.02	0.16	-0.15	89.12	0.46
01184000	25,019.28	578.45	455.76	529.10	295.32	3.12	0.12	-0.09	74.09	0.42

Gauge Name	Basin Area	Calibration				(C) Daily Resolution			
						Validation			
	[km²]	NSCE	MRE	RMSE	PCC	NSCE	MRE	RMSE	PCC
01192500	190.11	0.44	0.11	76.75	0.69	0.21	-0.05	64.82	0.53
01127500	231.29	0.72	0.02	79.38	0.85	0.60	-0.14	79.17	0.78
01193500	259.00	0.50	0.06	103.91	0.72	0.35	0.04	97.85	0.62
01208500	673.40	0.47	-0.01	91.83	0.69	0.34	-0.26	69.81	0.67
01188090	979.02	0.44	0.13	93.16	0.73	0.25	-0.05	57.58	0.61
01122500	1,046.36	0.65	0.07	68.51	0.81	0.44	-0.14	67.07	0.69
01127000	1,846.66	0.57	0.10	68.68	0.76	0.41	-0.01	64.84	0.65
01205500	3,998.94	0.47	-0.01	83.77	0.70	0.30	-0.15	65.96	0.59
01184000	25,019.28	0.38	-0.06	59.89	0.65	0.12	-0.09	73.27	0.43

Table 3: Gauged basin statistics determined at hourly time resolution during calibration period in (A) and during validation period in (B). Additionally, their daily statistics are reported in (C).

period in section (A) and during validation period in section (B). Additionally, their daily statistics are reported in part (C). It is observed that, overall NSCE results, in B01124000, are close to each other during calibration and validation periods. MRE value reduces from 0.41 to 0.23 during calibration and validation, respectively. While RMSE drops 20% from calibration to validation period, slight change is obtained in PCC with the value being around 0.7. Slight changes are also obtained as function of the time resolution for all the metrics in calibration an validation periods.

On the other hand, in Figure 6, variability is noted with respect to year for the same parent basin (B01127000) and nested catchment. The same trend is seen in the error metrics results of two basins spanning a 7-yr period with higher accuracy for the large basin (B01127000). This suggests that, overall the model provides a reasonably good description of the fluctuation.

Finally, calibrated largest parent basin (B01205500) is examined with interior nested basins in Figure 7. The smallest basin (01199000) tends to overestimate in terms of MRE results in addition to negative bias for the other two basins. NSCE results are below 0.5 while PCC values are above 0.3 in the 7-yr period. RMSE values improve as function of basin size.

The model efficiency is summarized in Table 6 during validation (section A) and calibration (section B) periods (hourly) as well as daily resolution (section C) for B01199000, B01200500 (considered as internal ungauged basins) and B01205500 (parent basin). MRE results show that the values are in the range of -0.02 and 0.21 for calibration and -0.15 and 0.07 for validation. Correlation is around 0.6 during calibration while it drops to 0.46 for the parent basin and improves to 0.65 for the medium basin (B01200500) in validation period. Specifically for the daily simulations, all RMSE values are below 100%. Examination of these statistics reported in the table shows that observations of the interior basins are well represented with CREST simulations.

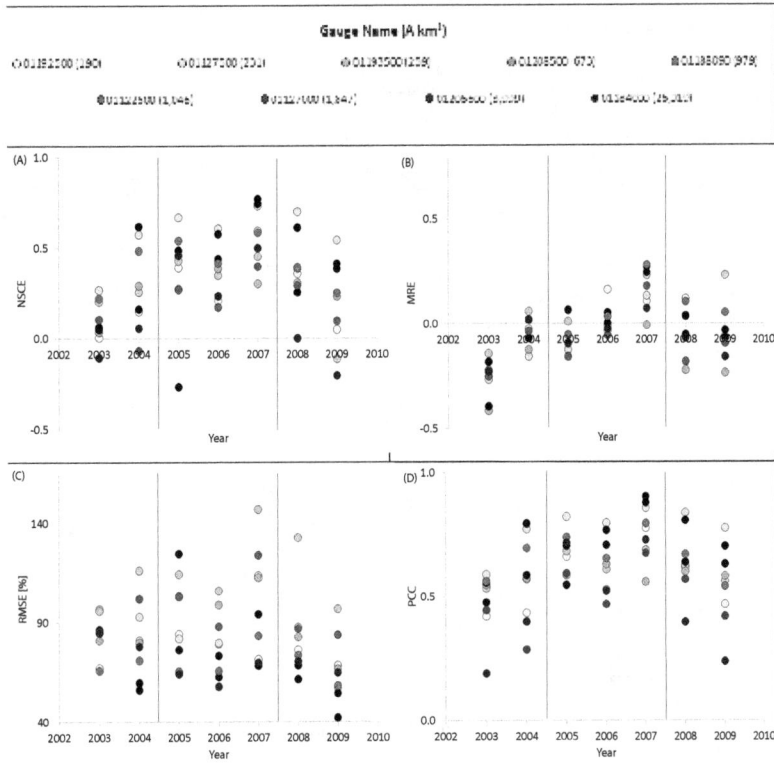

Figure 4: Statistical metrics of (A) NSCE, (B) MRE, (C) RMSE, and (D) PCC presented for different years and basin scales.

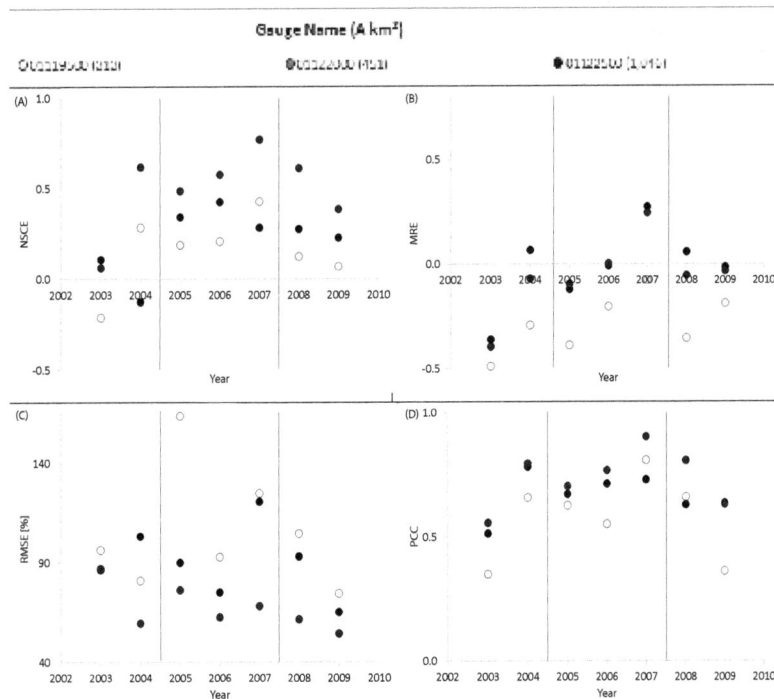

Figure 5: Statistical metrics of (A) NSCE, (B) MRE, (C) RMSE, (D) PCC presented for different years and different interior subbasins of parent basin 01122500.

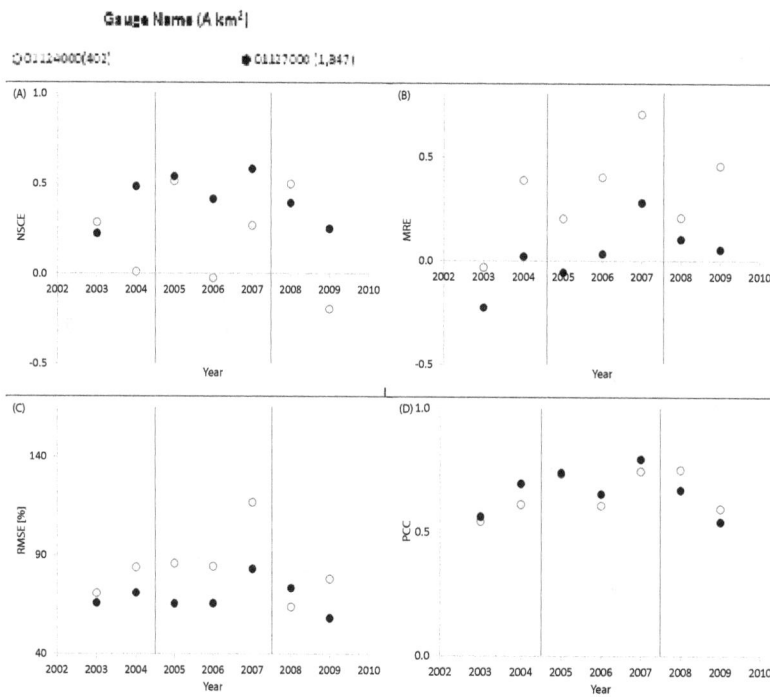

Figure 6: Same as in Figure 5 but for parent basin 01127000.

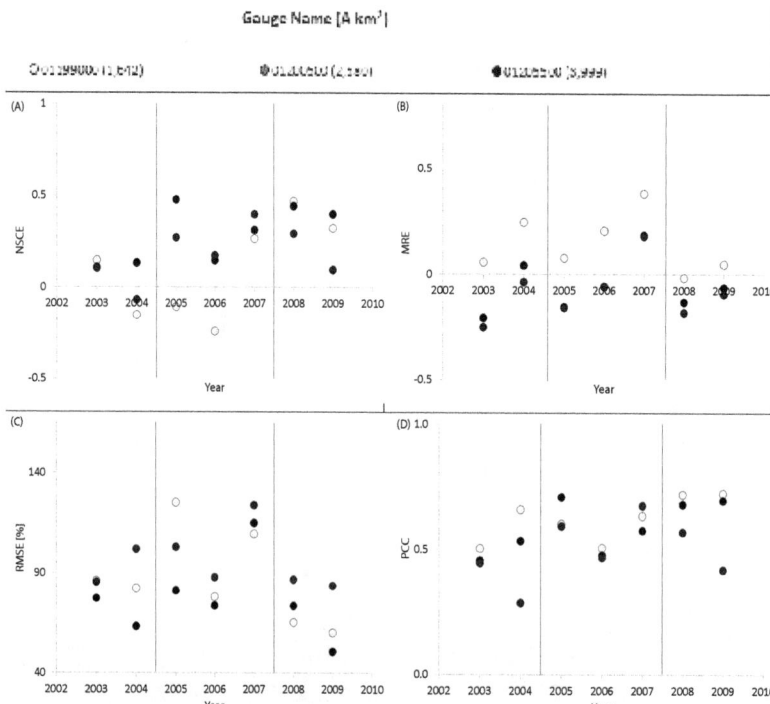

Figure 7: Same as in Figure 5 but for parent basin 01205500.

Gauge Name	Basin Area	(A)Calibration								
		QO[m³/s]		QS [m³/s]		NaN Data	NSCE	MRE	RMSE [%]	PCC
	[km²]	μ	σ	μ	σ	[%]				
01119500	313.39	7.21	12.08	5.60	3.25	10.88	0.26	-0.22	144.00	0.66
01122000	450.66	10.43	12.61	11.18	13.31	2.16	0.39	0.07	94.36	0.71
01122500	1,046.36	23.09	26.85	24.75	18.98	1.77	0.64	0.07	70.03	0.81

Gauge Name	Basin Area	(B)Validation								
		QO[m³/s]		QS [m³/s]		NaN Data	NSCE	MRE	RMSE [%]	PCC
	[km²]	μ	σ	μ	σ	[%]				
01119500	313.39	7.69	7.71	5.03	2.11	16.02	0.09	-0.35	95.40	0.53
01122000	450.66	10.47	10.10	9.80	11.18	2.68	0.15	-0.06	89.14	0.62
01122500	1,046.36	26.05	23.91	22.31	15.62	5.35	0.44	-0.14	68.54	0.68

Gauge Name	Basin area	Calibration				(C) Daily Resolution Validation			
	[km²]	NSCE	MRE	RMSE	PCC	NSCE	MRE	RMSE	PCC
01119500	313.39	0.28	-0.22	135.32	0.67	0.09	-0.35	91.47	0.53
01122000	450.66	0.44	0.07	88.54	0.74	0.18	-0.06	84.68	0.64
01122500	1,046.36	0.65	0.07	68.51	0.81	0.44	-0.14	67.07	0.69

Table 4: Same as in Table 3, but for the ungauged interior subbasins of parent 01122500 basin.

Gauge Name	Basin Area	(A)Calibration								
		QO[m³/s]		QS [m³/s]		NaN Data	NSCE	MRE	RMSE [%]	PCC
	[km²]	μ	σ	μ	σ	[%]				
01124000	401.45	9.32	11.04	13.18	8.97	4.70	0.35	0.41	95.20	0.70
01127000	1,846.66	41.74	44.58	45.85	34.66	0.10	0.54	0.10	72.34	0.74

Gauge Name	Basin area	(B)Validation								
		QO[m³/s]		QS [m³/s]		NaN Data	NSCE	MRE	RMSE [%]	PCC
	[km²]	μ	σ	μ	σ	[%]				
01124000	401.45	9.75	8.72	11.99	7.57	9.25	0.34	0.23	72.72	0.67
01127000	1,846.66	43.90	37.72	43.43	29.58	4.06	0.37	-0.01	68.19	0.63

Gauge Name	Basin Area	Calibration				Validation (C) Daily Resolution			
	[km²]	NSCE	MRE	RMSE	PCC	NSCE	MRE	RMSE	PCC
01124000	401.45	0.38	0.42	92.52	0.71	0.36	0.23	69.86	0.68
01127000	1,846.66	0.57	0.10	68.68	0.76	0.41	-0.01	64.84	0.65

Table 5: Same as in Table 3, but for the ungauged interior subbasin of parent 01127000 basin.

Figures 8 and 9 illustrate the comparison of the simulations in terms of quantiles at the basins outlets and interior ungauged locations, respectively. MRE and RMSE error metrics were calculated at 0.2, 05, 07, 0.9 and 0.95 quantiles and shown with each spatially distributed station with transparency dots. Reduction of the RMSE and increment of the MRE tendency are observed from low to high flows. The poorer performances of the model in terms of MRE and RMSE are obtained at the higher quantiles and considered to be related to inadequate representation of the peak events. Figures confirm the tendency of underestimations at peak discharges and this consistency supports the argument of using frequency analysis with large flow data to improve model results.

Conclusions

This case study has demonstrated the implementation of the CREST model, and its calibration and validation processes at high temporal resolution with spatial distributed gauged observations as well as ungauged/poorly gauged catchments over Connecticut River Basin. The primary objective of this study is calibration and validation processes of distributed hydrological model based on stream flow observations and provide explicitly statistical analysis of the simulations via different objective functions. Another objective of this study is producing flow data for insufficient historical records at gauging stations using calibrated CREST model and predicting flows over un gauged locations of the Connecticut River Basin. In this manner, the basin was partition into 9 sub basins to represent distributed parameters sets via CREST model based on the USGS monitoring stations and their locations, and the model was evaluated with several important statistics with supplemental hydrographs.

Gauge Name	Basin Area	(A)Calibration									
		QO[m³/s]		QS [m³/s]		NaN Data	NSCE	MRE	RMSE [%]	PCC
	[km²]	μ	σ	μ	σ	[%]				
01199000	1,642.05	35.87	39.35	43.37	44.76	6.37	0.00	0.21	109.91	0.58
01200500	2,579.63	57.99	65.09	56.90	42.27	2.36	0.36	-0.02	89.78	0.60
01205500	3,998.94	90.83	114.55	89.73	94.74	1.08	0.31	-0.01	104.98	0.60

Gauge Name	Basin Area	(B)Validation									
		QO[m³/s]		QS [m³/s]		NaN Data	NSCE	MRE	RMSE [%]	PCC
	[km²]	μ	σ	μ	σ	[%]				
01199000	1,642.05	40.75	35.37	43.46	35.54	2.81	0.30	0.07	72.67	0.65
01200500	2,579.63	65.17	54.80	58.52	41.43	8.77	0.32	-0.10	69.13	0.60
01205500	3,998.94	97.80	95.21	82.89	57.22	0.02	0.16	-0.15	89.12	0.46

Gauge Name	Basin Area	Calibration				Validation (C) Daily Resolution			
	[km²]	NSCE	MRE	RMSE	PCC	NSCE	MRE	RMSE	PCC
01199000	1,642.05	0.17	0.21	99.55	0.62	0.36	0.07	68.88	0.67
01200500	2,579.63	0.39	-0.02	86.88	0.63	0.37	-0.10	66.33	0.63
01205500	3,998.94	0.47	-0.01	83.77	0.70	0.30	-0.15	65.96	0.59

Table 6: Same as in Table 3, but for the ungauged interior subbasins of parent 01205500 basin.

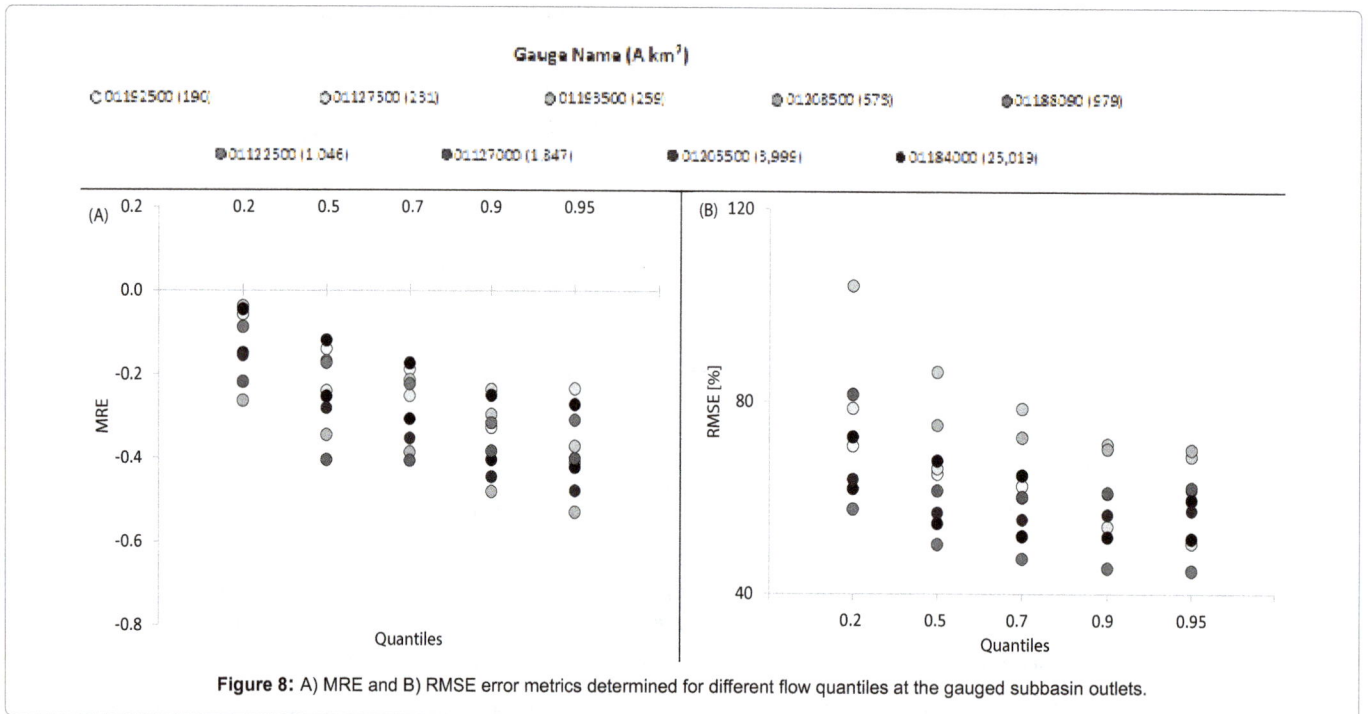

Figure 8: A) MRE and B) RMSE error metrics determined for different flow quantiles at the gauged subbasin outlets.

CREST model was calibrated for three years and validated in four years during the period 2002-2009. Calibration of the model was done over the period 2005 to 2007 and validations were carried out over two distinct periods: 2003 to 2004 and 2008 to 2009.The model was calibrated against 14 parameters within specified limited boundaries while a few of these parameters were adjusted using 2002 as spin up period for plausible initial conditions so that model predicts rainfall-runoff responses closely to the flow measurements and reflects the behavior of watershed system in a various manner. The figures and tables demonstrated the model's accuracy for the simulation of the rainfall-runoff transformation at 9 stations and un gauged interior locations.

In this study, several statistical indicators and supplemental graphical illustrations were applied for evaluation of the CREST model performance. Using multiple statistics helps to cover a different aspect of the hydrographs. Based on the error metrics and graphical comparisons, CREST is considered to provide accurate rainfall-runoff simulations during calibration and validation period. The results also indicate satisfactory in model performance over Connecticut River Basin.

The CREST hydrological model estimates the hourly flow at spatially various catchments well, but with relatively large errors at peak quantiles. Overall, when we take into account statistical metrics and normalized hydrographs of the model results, it can be concluded that

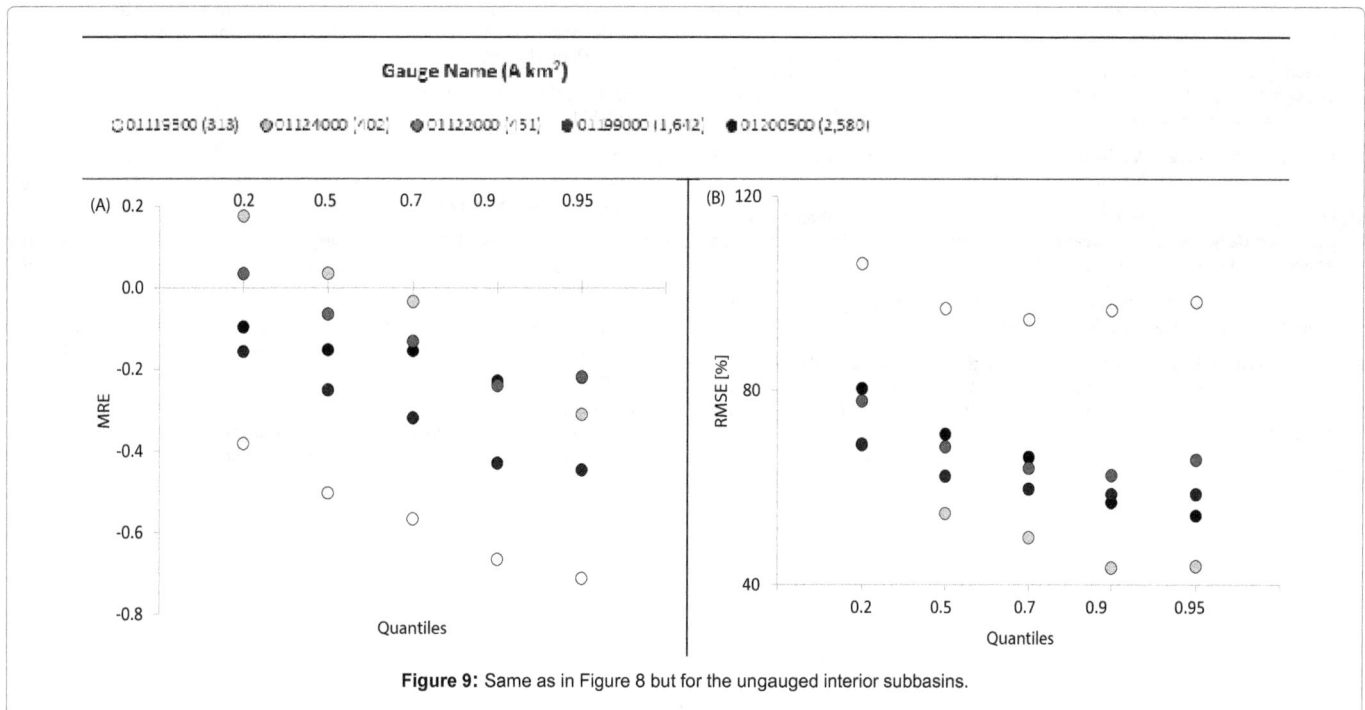

Figure 9: Same as in Figure 8 but for the ungauged interior subbasins.

the CREST is capable of reproducing continuous hourly stream flows in Connecticut River Basin, which allows model use for flood management applications in both gauged and un gauged basins. Results with CREST illustrated that model performance was satisfactory in representing amplitude and timing of the flow peaks, and that the models performs better in predicting the entire hydrograph. The model application and its calibrated parameters can be used for future work such as comparing to satellites-driven flow, flash flood predictions, flood frequency, and future climatic analysis on local and national scale as well.

Acknowledgements

This work was supported by the Northeast Utilities Foundation Endowment in Environmental Engineering. The authors would like to thank the anonymous reviewers for their valuable comments and feedback.

References

1. Wang J (2011) The coupled routing and excess storage (CREST) distributed hydrological model. Hydrological Science Journal 56: 84-98.

2. Odoni N, Lane S (2010) Knowledge-theoretic models in hydrology. Progress in Physical Geography 34: 151-171.

3. Dingman S (2002) Physical hydrology, Upper Saddle River, NJ: Prentice-Hall Inc.

4. Beven KJ (2000) Rainfall-Runoff Modeling.Chichester:John Wiley and sons ltd.

5. Chow V, Maidment D, Mays L (1988) Applied Hydrology. International Editions. Singapore: McGraw-Hill.

6. Crawford N, Linsley R (1966) Digital simulation in hydrology: Stanford Watershed Model IV. Technical Report 39, Palo Alto, CA: Stanford University.

7. Brakenridge G (2003) Flood warnings, flood disaster assessments, and flood hazard reduction: the roles of orbital remote sensing. Pasadena, CA, Jet Propulsion Laboratory, National Aeronautics and Space Administration.

8. Dehotin J, Braud I (2008) Which spatial discretization for distributed hydrological models? Proposition of a methodology and illustration for medium to large-scale catchments.Hydrology and Earth System Sciences 12: 769-796.

9. Moriasi D (2007) Model evaluation guidelines for systematic quantification of accuracy in watershed simulations. American Society of Agricultural and Biological Engineers 50: 885-900.

10. Refsgaard J (1997) Parameterization, calibration and validation of distributed hydrological models.Journal of Hydrology 198: 69-97.

11. Rozalis S, Morin E, Yair Y, Price C (2010) Flash flood prediction using an uncalibrated hydrological model and radar rainfall data in a Mediterranean watershed under changing hydrological conditions.Journal of Hydrology 394: 245-255.

12. Bergstrom S (1991) Principles and confidence in hydrological modeling.Nordic Hydrology 22: 123-136.

13. Brandt M (1990) Simulation of runoff and nitrate transport from mixed basins in Sweden.Nordic Hydrology 21: 13-34.

14. Noto L, Ivanov V, Bras R, Vivoni E (2008) Effects of initialization on response of a fully distributed hydrological model. Journal of Hydrology 352: 107-125.

15. Vrugt J (2009) Accelerating Markov Chain Monte Carlo simulation by differential evolution with self-adaptive randomized subspace sampling.International Journal of Nonlinear Sciences and Numerical Simulation 10: 273-290.

16. 16. Boyle D, Gupta H, Sorooshian S (2000) Toward improved calibration of hydrological models: combining the strengths of manual and automatic methods.Water Resources Research 36: 3663-3674.

17. Beven K (1989) Changing ideas in hydrology-The case of physically-based models.Journal of Hydrology 105: 157-172.

18. Hoedjes J (2014) A Conceptual Flash Flood Early Warning System for Africa, Based on Terrestrial Microwave Links and Flash Flood Guidance.ISPRS International Journal of Geo-Information 3: 584-598.

19. Bingeman A, Kouwen N, Soulis E (2006) Validation of the hydrological processes in a hydrological model Journal of Hydrologic Engineering 11: 451-463.

20. Motovilov Y, Gottschalk L, Engeland K, Rodhe A (1999) Validation of a distributed hydrological model against spatial observations.Agricultural and Forest Meteorology 98–99: 257-277.

21. Kouwen N (1993) Grouped response units for distributed hydrologic modelling. Journal of Water Resources Planning and Management 119: 289-305.

22. Xie Z (2007) Regional Parameter Estimation of the VIC Land Surface Model: Methodology and Application to River Basins in China.Journal of Hydrometeorology 8: 447-468.

23. Marshall E, Randhir T (2008) Effect of climate change on watershed system: a regional analysis Climatic Change 89: 263-280.

24. Zanon F (2010) Hydrological analysis of a flash flood across a climatic and geologic gradient:The September 18, 2007 event in Western Slovenia.Journal of Hydrology 394: 182-197.

25. Hirpa F, Gebremichael M, Over T (2010) River fluctuation analysis: effect of watershed area.Water Resources Research 46: 12529.

26. Du B, Ji X, Harmel R, Hauck L (2009) Evaluation of a watershed model for estimating daily flow using limited flow measurements.Journal of the American Water Resources Association 45: 475-484.

27. Mesinger F (2005) North American Regional Reanalysis.Bulletin of the American Meteorological Society 343-360.

28. Khan S (2011) Satellite remote sensing and hydrologic modeling for flood inundation mapping in Lake Victoria Basin: implications for hydrologic prediction in ungauged basins. IEEE Transactions on Geoscience and Remote Sensing 49: 85-95.

29. USDA-NRCS (1994) State Soil Geographic Database (STATSGO) data users guide, Washington, DC: Data Use Information.National Soil Survey Center Misc. Publ. No. 1492.

30. Nash J (1957) The form of the instantaneous unit hydrograph.IAHS Publ 45: 114-121.

31. Cosby B, Hornberger G, Clapp R, Ginn T (1984) A statistical exploration of the relationships of soil moisture characteristics to the physical properties of soils. Water Resources Research 20: 682-690.

32. Legates D, McCabe G (1999) Evaluating the use of "goodness-of-fit" measures in hydrologic and hydroclimatic model validation.Water Resources Research 35: 233-241.

33. Servat E, Dezetter A (1991) Selection of calibration objective functions in the context of rainfall-runoff modeling in a Sudanese Savannah area. Hydrological Sciences Journal 36: 307-330.

34. Fernandez G, Chescheir G, Skaggs R, Amatya D (2005) Development and testing of watershed-scale models for poorly drained soils.Transactions of the American Society of Agricultural Engineers 48: 639-652.

11

The Need to Enforce Minimum Environmental Flow Requirements in Tanzania to Preserve Estuaries: Case Study of the Mangrove-Fringed Wami River Estuary

Halima Kiwango[1,2]*, Karoli N. Njau[1] and Eric Wolanski[3]

[1]*The Nelson Mandela African Institution of Science and Technology, P.O. Box 447, Arusha, Tanzania*
[2]*Tanzania National Parks, P.O. Box 3134, Arusha, Tanzania*
[3]*Tropwater, James Cook University, Townsville, Qld. 4810, Australia*

Abstract

The importance of restoring and maintaining environmental flows for sustaining the ecosystem integrity of rivers and estuaries has been recognized and given proper attention in policies and legal frameworks in many countries including Tanzania. The Wami River estuary is small but it plays a vital role in processing riverine nutrients, in trapping sediment, in recycling nutrients in the mangroves, and in supporting the ecology of the Saadani National Park and the livelihood of the local communities. The proper functioning of this estuary to a large extent depends on adequate supply of freshwater flows. Our studies reveal that currently the estuary is ecologically healthy but it is threatened by both increasing sedimentation and declining freshwater flow caused by decreasing rainfall - possibly linked with climate change - and by increasing water demand in the watershed for artisanal and large scale agriculture and irrigation schemes. Environmental flow assessment for the Wami River (with exclusion of estuary) has been done and the minimum flows were recommended. However, like in many other rivers in the country, effective implementation of recommended environmental flows remains to be a challenge. In order to maintain a healthy estuarine ecosystem in the future, it is the obligation of the WRBWO now to stick to and enforce the recommendations of its own environmental flow assessment to regulate water usage in the watershed. A similar recommendation also holds for all other rivers and estuaries in Tanzania.

Keywords: Minimum environmental flow requirement; Salinity; Sedimentation; Mangroves; Fisheries; Wildlife; Local economy.

Introduction

Many rivers in the world are suffering from hydrological alterations which result in degradation of aquatic habitats. Changing hydrological regime of the river affect water quality, sediments, nutrients supply and biotic interactions within the river and their associated estuaries. Recognition of the importance of maintaining the natural flow regime for sustaining the ecosystem integrity of rivers and estuaries has led to development of environmental flow (EF) concept and international organizations such as World Conservation Union (IUCN) are emphasizing on the EF as key element of integrated water resources management Dyson et al., [1]. There are various definitions of environmental flow but the Brisbane declaration of 2007 (http://www.watercentre.org/news/declaration) defines EF as "the quantity, timing, and quality of water flows required to sustain freshwater and estuarine ecosystems and the human livelihoods and well-being that depend on these ecosystems". In estuarine environments, EF plays a major role in salinity gradients, mixing patterns and water quality, flushing time (residence time), productivity as well as distribution and abundance of estuarine biota. In Tanzania, EF assessments are supported by policy and legal frameworks such as Tanzania Water Policy of 2002 [2], the Environmental Management Act of 2004 [3] and Water Resources Management Act of 2009 [4]. EF studies have been conducted in some rivers such as Pangani, Wami, Ruvu, Rufiji and Mara [5]. For the Wami River, the study was done in 2007 and 2011 [5,6]. The required minimum flows for both dry and wet years were recommended. However, both studies ignored the freshwater needs of the estuary.

Throughout the world estuaries are known to play an important role in human well-being and the economy, but they are increasingly threatened and, as a response, science-based management and restoration strategies are continuously evolving [7-12].

Furthermore, the functioning of estuarine ecosystems, and their responses to changing abiotic and biotic factors, are system-specific and strongly dependent on the size of the estuary, its latitude, watershed properties and human activities within [12]. Much of the available knowledge is derived from large estuaries where the response to changing environment is slower than that of small estuaries [13].

The Wami estuary falls under the management authority of Saadani National Park (SANAPA) and the whole Wami River under the management authority of the Wami-Ruvu Water Basin Office (WRWBO). The estuary is a lifeline for SANAPA wildlife and people during the dry season where most of the water sources inside the park are dry. It is also the main source of income to Saadani village and adjacent coastal communities through fishery as well as through tourism. Prior to gazetting of the National Park in 2005, the estuarine condition was threatened by increased destruction of mangroves which were heavily exploited for charcoal, fuel wood, and poles, as well as destroyed for creating open inter-tidal areas used for salt production. Local communities have complained about decreasing catch of prawn and fish, local extinction of some fish species, changing water quality particularly increasing water salinity at the estuary as a result of

*Corresponding author: Halima Kiwango, The Nelson Mandela African Institution of Science and Technology, Arusha, Tanzania
E-mail: kiwangoh@nm-aist.ac.tz

decreasing EF, and increasing sedimentation at the mouth of the estuary and in near shore waters where the fisheries are located. However, majority of their claims were based on anecdotal information but there is no sufficient scientific evidence to justify the claims. The Wami River estuary plays a vital role in processing riverine nutrients, in trapping fine sediment, in recycling nutrients in the mangroves, in supporting wildlife and the ecology of the National Park as well as the livelihood of the local communities. Nevertheless, the Wami estuarine ecosystem, its fisheries, the National Park and the local economy are now threatened both by increasing sedimentation and by declining freshwater flow in the Wami River due to decreasing rainfall - possibly linked with climate change - and by increasing water demand in the watershed mainly for artisanal and large scale agriculture and irrigation schemes.

In this paper we show how the estuary is/will be impacted as a result of changing EF and we give emphasis on the importance of effectively implementing the recommended flows by both SANAPA and WRBWO to sustain the estuary. We focus on changing water quality, salinity gradients and mixing patterns, sedimentation, flushing (residence time), nutrient budget and productivity with respect to changing hydrology of the river particularly during wet and dry seasons. We show how the inclusion of estuary to SANAPA in 2005 was important in safeguarding the mangroves which help in trapping sediments and nutrient recycling by crabs as well as in protecting other estuarine wildlife such as hippopotami of which their populations have been observed to increase. We also show that the trapping of sediment by the mangroves help to buffer the sea grass meadows in coastal waters from excessive sediment load, although whether this is enough to preserve the sea grass is unknown and requires further research.

Methods

Study area

Wami River estuary is located in Northern coast of Tanzania between 06°07′213 S, 038°48′965 E and 06°07′155 S, 038°48′886 E [14]. The Wami-Ruvu basin covers an area of 72,930 km² but the Wami River drains about 40,000 km² of that area [15]. The tides are semi-diurnal with a strong diurnal inequality, with spring tides reaching 4 m. The tidal influence extends up to 8 km upstream. The average depth in the estuary is 2.5 m and 3.5 m during dry and wet seasons respectively. It supports extensive mangrove ecosystems and their associated inter-tidal organisms [15,16]. The main fringing vegetation types along the estuary are mangroves, palms and Acacia woodland mixed with grassland. There are eight species of mangroves but the dominant species are *Sonneratioa alba, Avicennia marina, Xylocarpus granatum, Rhizophora mucronata* and *Heritiera litoralis*. Patchy seagrass meadows occur in coastal waters all along the coast (Figure 1).

Hippopotami, crocodiles, and water birds are common along the estuary, while numerous wild animals such as ungulates and colobus monkeys access the upper estuary for drinking freshwater. Though small, the estuary supports one of the important prawn fisheries in Tanzania [16].

Hydrology data

Wami River discharge data from the Mandera hydrometric station (located about 50 km from the estuary) were obtained from WRBWO. The data were processed and used to indicate changes in freshwater flow to the estuary. Local rainfall data were obtained from Tanzania Meteorological Agency while evaporation data were obtained from Nyenzi et al. [17].

Environmental variables

Physical, chemical and biological data were obtained along a transect from the river to offshore at five sites shown in Figure 1 at different times during dry (July-October) and wet seasons (March-June) between 2007 and 2015. Water samples were collected using a Niskin bottle near the surface, at mid-depth and near the bottom at each sampling site. From these samples, water salinity, temperature, dissolved oxygen and pH were measured in situ. Different instruments were used depending on the availability such as the HORIBA model U-10 and BANTE 900P portable multi parameter meters. Salinity was measured using a hand-held refractometer. A secchi disk of 20 cm diameter was used for measurement of water visibility.

Nutrients and total suspended solids (TSS)

Water samples for nutrient and TSS analysis were collected using a Niskin bottle and stored in acid washed 1 L plastic bottles, rinsed with distilled water and re-rinsed with water from the sampling site two to three times. All samples were immediately stored in an iced cool box. In the laboratory, these samples were filtered using BOECO glass-microfibre discs (filters) grade MGC with 0.45 μm pore size and GF/C Whatmann glass-microfiber filters of 4.7 cm diameter. A volume

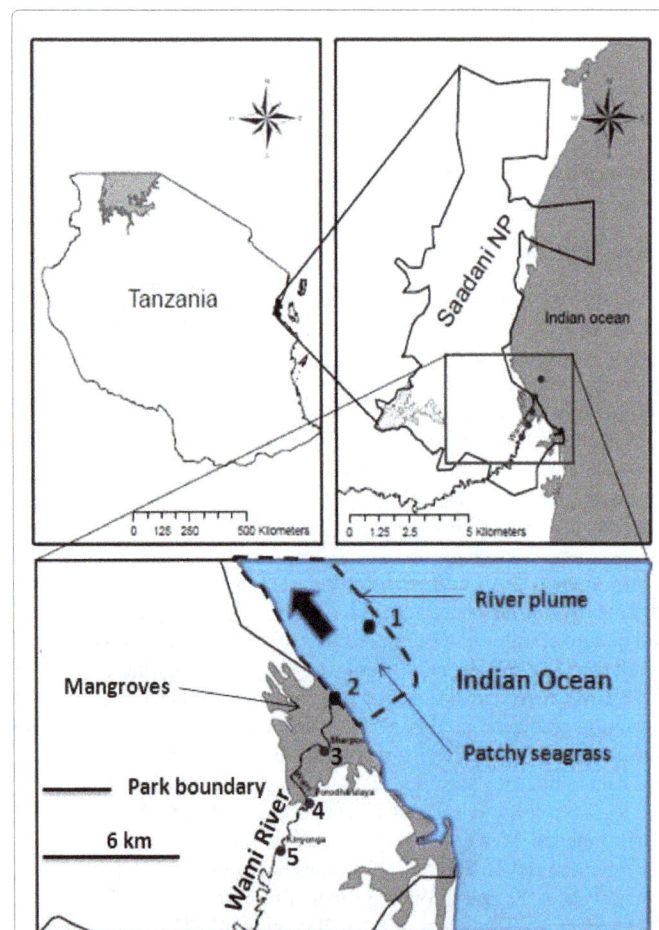

Figure 1: Location of study area and sampling sites 1-5. The arrow indicates the direction of the prevailing net longshore northward current (http://www.gloss-sealevel.org/publications/documents/tanzania_gex2007.pdf). Patchy seagrass meadows occur all along the coast and they are readily apparent in satellite images although they have not been mapped in detail (http://www.seagrasswatch.org/Shop/SG_atlas.pdf). The outline of the river plume (dashed line) is sketched from visual observations.

of 300-400 ml was filtered depending on how turbid the water was. Filtrate was used for nutrient analysis while filter papers containing residue were used for TSS analysis.

Dissolved Inorganic Phosphate (DIP) was determined by using the ammonium molybdate method. The procedures followed were adapted from Murphy and Riley [18]. Dissolved Inorganic Nitrogen (DIN) determination was done following the cadmium reduction method described by Parsons et al., [19]. TSS analysis was performed following the protocol described by APHA [20].

Residence times and the fate of riverine nutrients

The residence time of water in the estuary and the fate of riverine nutrients in the estuary were calculated using the muddy LOICZ model of Xu et al., [21] for the dry season when the estuary was vertically well-mixed, and of Xue et al., [22] during the wet season when the estuary was vertically stratified. The model assumes steady state. The surface area of the estuary was estimated from Google Earth and the volume was calculated based on mean depth for wet season (3.5 m) and dry season (2.5 m). The rainfall and evaporation data together with data obtained on salinity, DIN, DIP and TSS were used in the LOICZ model to calculate the Net Ecosystem Metabolism (NEM; also called (p-r), where p is the production and r is the respiration) and the Nitrogen fixation rate minus the denitrification rate (Nfix-Denit).

Sediment trapping in mangroves

Sediment trapping in mangroves was measured using the method of Golbuu et al., [23]. PVC traps 20 cm long and 2.5" diameter closed at the bottom were deployed on four transects within the mangrove forest. Each transect had 4 traps deployed at a distance of 10 m from the river bank to up to 800 m inside the forest depending on the width of the mangrove forest strip. A hole was dug and a trap was put in up to 15 cm leaving the other 5 cm above the ground level. After a month, all traps were removed, sediments put in a container, dried and weighed.

The role of crabs in recycling mangrove litter

The role of crabs in recycling mangrove litter was measured using the method of Smith et al., [24] and Lindquist et al., [25]. Three plots of 10 × 10 m were established, species identified, counted, height estimated and diameter at breast height (dbh) measured. In each plot 100 new leaves were picked from trees and each leaf was tightly tied with a 1 m string. The leaves were spread evenly throughout the mangrove floor within the plot. This was done at low tide. We returned at low tide again after a tidal cycle and looked for the strings that remained. While a few leaves remained at the surface, most of the leaves either were in crab holes or were absent, having been exported to the estuary by the tides. The mean value of the number of leaves exported to the river for all three plots was calculated.

Movement of hippo in different seasons and their impact on mangroves

Physical observation of hippos' movement during wet and dry season was done to see if they shift from their local territories during changing flows. We also wanted to know if their movements within the mangrove forest impact the vegetation. Therefore we located their tracks along a 2 km stretch along the mangrove-fringed river banks. Whenever tacks were seen, GPS coordinated were taken following the tracks after every 10-20 m intervals for a distance of about 150-200 m. All tacks were then mapped.

Results

Hydrology

Rainfall in coastal Tanzania varies strongly seasonally and inter-annually. Rainfall is bimodal with long rainy (wet) seasons occurring between March and June and the short rainy season occurring between November and December. The long dry season occurs between July and October and the short dry season between January and February. Mean annual rainfall varies between 900-1000 mm [26]. As a result, the Wami River discharge also varied seasonally and inter-annually. From 1954 to 2014, the mean annual discharge varied inter-annually between a maximum value of 241.9 m^3s^{-1} and a minimum discharge of 16.9 m^3s^{-1} and this discharge was not correlated ($r^2=0.04$) with the Southern Oscillation Index (SOI; Figure 2).

Because the residence time of water in the estuary is very short, what controls the flushing of the estuary is not the mean annual flow but the daily mean discharge. Historical data shows that during extremely wet years the discharge exceeded at times 600 m^3s^{-1}, but in dry years that discharge was about 120-150 m^3s^{-1} (Figure 3a). However that discharge was much reduced during the dry season and, in recent years, the minimum discharge varies between 0.2 and 5 m^3s^{-1} during dry and wet years respectively (Figure 3b). Since the data were recorded starting in 1954, there is a decreasing trend in the minimum discharge for both wet and dry years (Figure 3b).

Environmental variables

The Wami Estuary is a warm system throughout the year with temperature range between 27.5°C and 31.9°C (not shown). Dissolved oxygen varies between 6.4 to 11.9 mg l^{-1} during the dry season and 5.4 to 6.7 mg l^{-1} during the wet season (not shown). The estuary was very turbid as the secchi disk readings varied from 0.025 to 0.04 m during

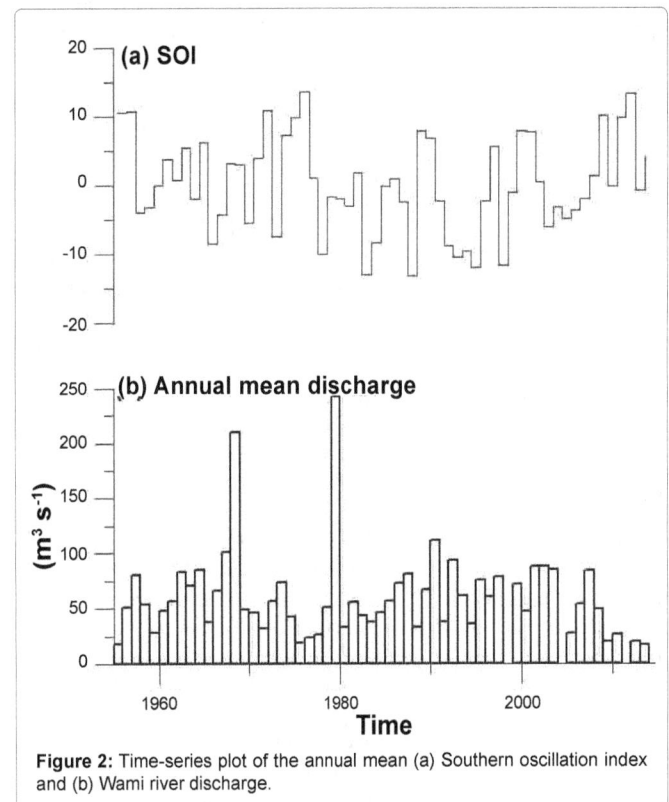

Figure 2: Time-series plot of the annual mean (a) Southern oscillation index and (b) Wami river discharge.

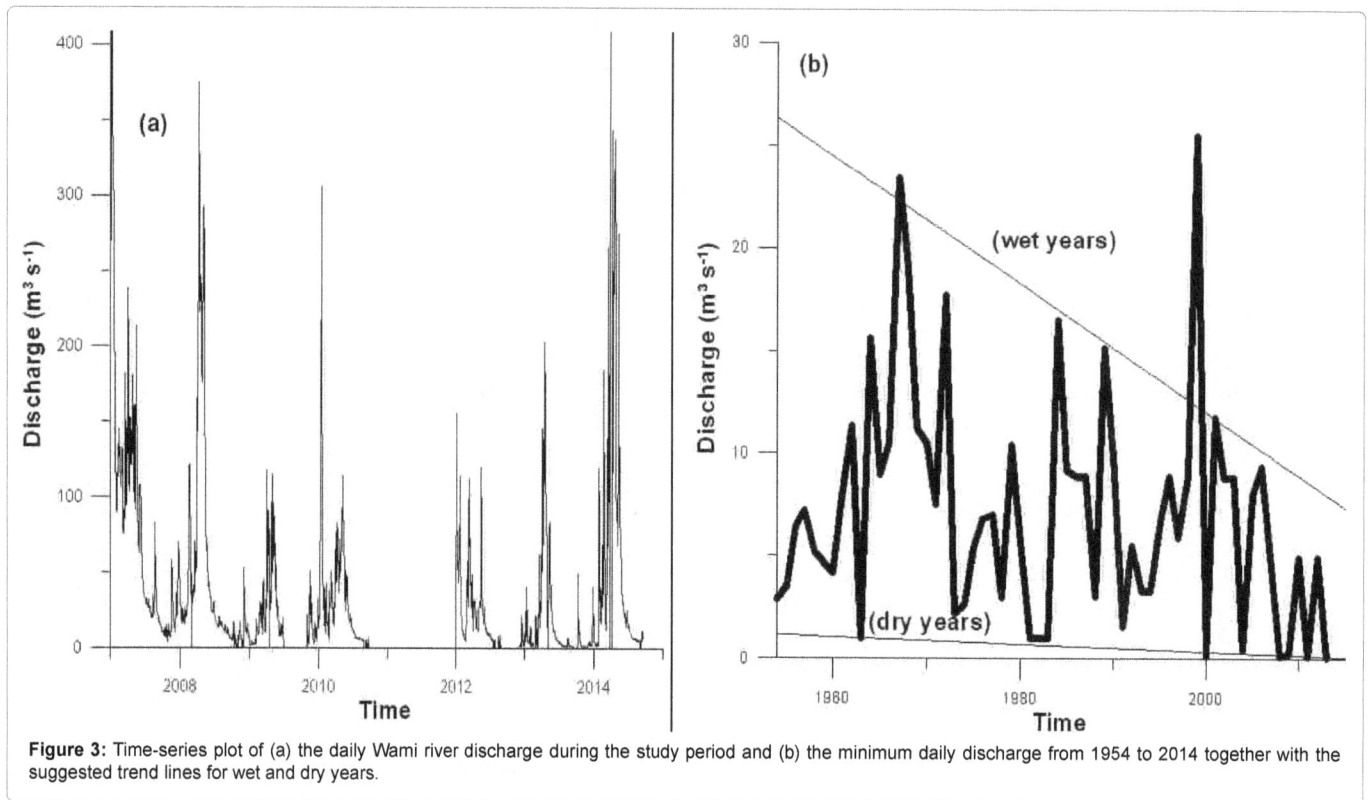

Figure 3: Time-series plot of (a) the daily Wami river discharge during the study period and (b) the minimum daily discharge from 1954 to 2014 together with the suggested trend lines for wet and dry years.

the wet season and 0.2 to 0.7 m during the dry season. By contrast the secchi disk reading at site 1, in coastal waters, was typically about 4 m. The minimum value for pH during both wet and dry seasons was 7.6 and the maximum values varied from 8.1 in the dry season to 8.9 in the wet season (not shown). Higher values of pH were observed in the upper reaches of the estuary and followed a decreasing trend towards the river mouth. This trend coincides with increasing salinity towards the river mouth.

During the dry season, the estuary was vertically well mixed, with salinity of about 30 ppt at the mouth, and salinity reaching the whole estuary (i.e. up to the tidal limit) with salinity up to 7 ppt at the tidal limit (Figure 4a).

In contrast, during the wet season the estuary was highly stratified in salinity with surface salinity of less than 7 ppt and bottom salinity of 35 ppt at high tide at the mouth. This salinity extended only 1-2 km upstream from the mouth and the remaining part of the estuary was freshwater (Figure 4b). At low tide the water was fresh at the mouth and the 1 m thick river plume extended up to 2 km offshore in the Indian Ocean and during our observations it was always deflected northward alongshore by the prevailing net currents sketched in Figure 1.

Nutrients and TSS

In dry season, DIN values shows high variation where low values were obtained in the upper reaches of the estuary (0 µM), increased towards the mid estuary (17.9 µM) and decreased towards the ocean (9.2 µM). On contrary, in wet season the variation is very low compared to dry season with lower values in the upper estuary (0.036 µM) and slightly increased in mid estuary (0.049 µM) and the ocean (0.042 µM). A similar trend was observed for DIP in dry season with low values in the upper estuary (6 µM), higher values in mid estuary (26.7 µM) and decreasing values towards the ocean (16.9 µM). In wet season, the DIP values showed a decreasing trend with higher values in upper estuary

(0.283 µM), and decreasing values in the mid estuary (0.174 µM) and (0.045 µM) towards the ocean. TSS varied between 0-68.56 mg/L in dry season with low values in the upper estuary while in wet season TSS varied between 50-427 mg/L with low values in the ocean and higher values in the mid and upper estuary.

Residence times and the fate of riverine nutrients

The residence time of water, calculated using the LOICZ model, was about 6.9 days during the dry season during dry years and 0.5 days during the wet season during wet years. The fate of riverine nutrients is indicated in the the Net Ecosystem Metabolism (NEM) which was positive (98.3 mmol C/m²/day) in the wet season and negative (-10179.3 mmol C/m²/day) in the dry season. Moreover, the Nitrogen fixation rate minus the denitrification rate (Nfix-Denit) was positive (15.07 mmol DIN/m²/day) in the wet season and negative (-1532.84 mmol DIN/m²/day) in the dry season.

Sediment trapping in mangroves

In the wet season, the Wami River supplied fine sediments at a rate of about 3,763 tons day⁻¹ and about 452 tons day⁻¹ was trapped in mangroves. In the dry season, the riverine fine sediment inflow decreased to about 18 tons day⁻¹, and the fine sediment trapped in mangrove was about 195 tons day⁻¹. Also in recent years sedimentation has been observed in the Wami delta, and in turn this promoted the expansion of mangroves (Figure 5).

The role of crabs in recycling mangrove litter

High percentage of mangrove litter (57%) was consumed and recycled by crabs in their holes in the mangrove soil while 32% of the remaining litter was exported to the estuary and the small percent (11%) remained on the ground to decompose in the mangrove floor to contribute to soil nutrient.

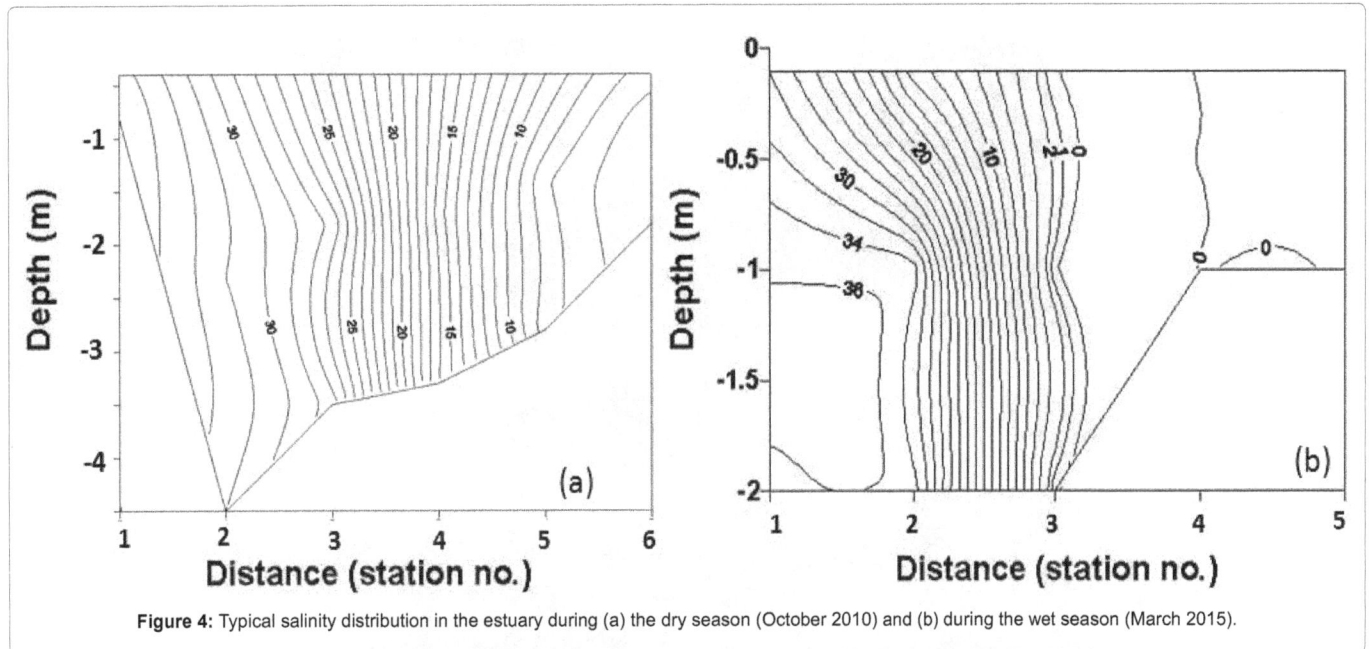

Figure 4: Typical salinity distribution in the estuary during (a) the dry season (October 2010) and (b) during the wet season (March 2015).

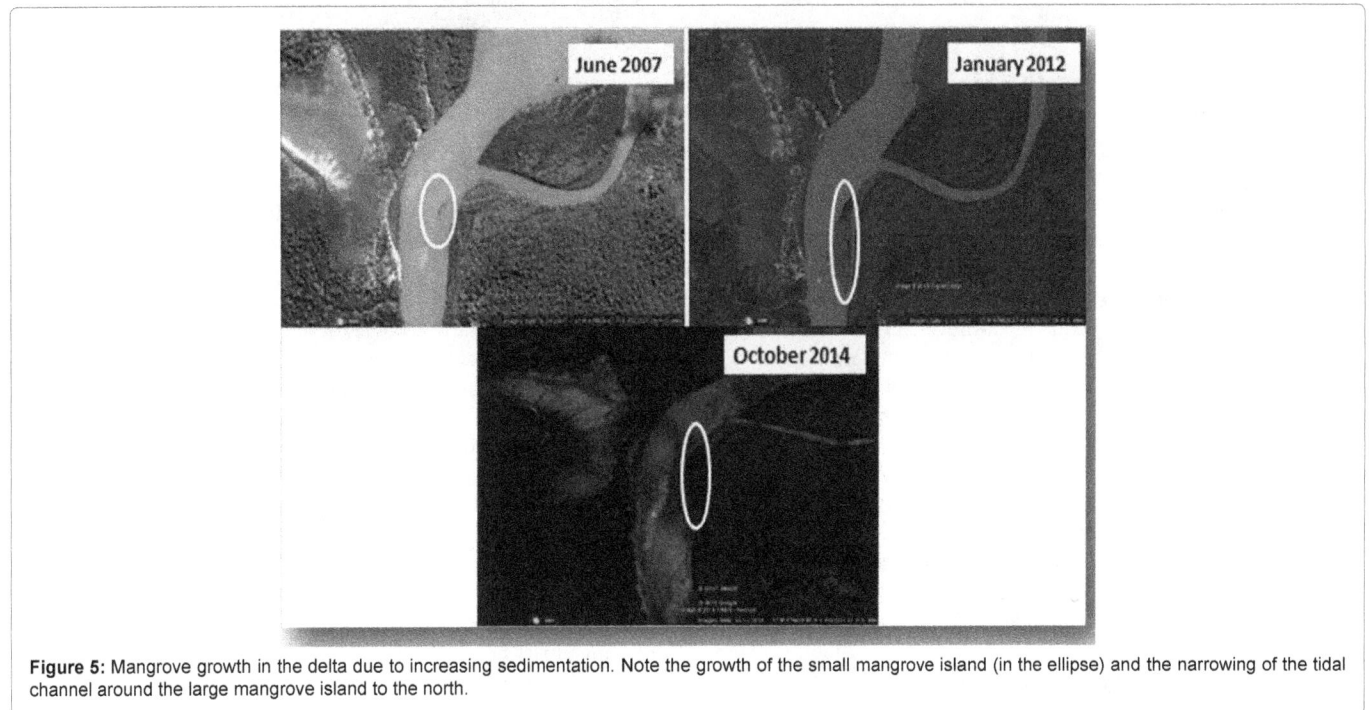

Figure 5: Mangrove growth in the delta due to increasing sedimentation. Note the growth of the small mangrove island (in the ellipse) and the narrowing of the tidal channel around the large mangrove island to the north.

Movement of hippo in different seasons and their impact on mangroves

Hippos are territorial animals only in water but not on land where they normally graze especially during the night. During the dry season hippo were observed to stay in six groups (schools) within the estuary with the first school located in between sampling station 2 and 3. In wet season when the estuary is mainly freshwater dominated, hippo groups were observed to split, form temporary schools and disperse throughout the estuary. Some hippos were located very close to the river mouth.

Hippos and other wildlife (principally elephants and buffaloes) may contribute to bank erosion (Figure 6a) but only at specific points. These are the points where the hippos crossed the mangrove forest on their way to grazing land but they didn't destroy the mangrove vegetation (Figure 6b). Their tracks were fairly straight, indicating that they choose short distances within mangroves and rapidly cross the mangrove forest before they reach the grazing areas; by doing so they create paths which are then used by other animals (Figure 6c).

Discussion

The hydrology of Wami River based on discharge data shows

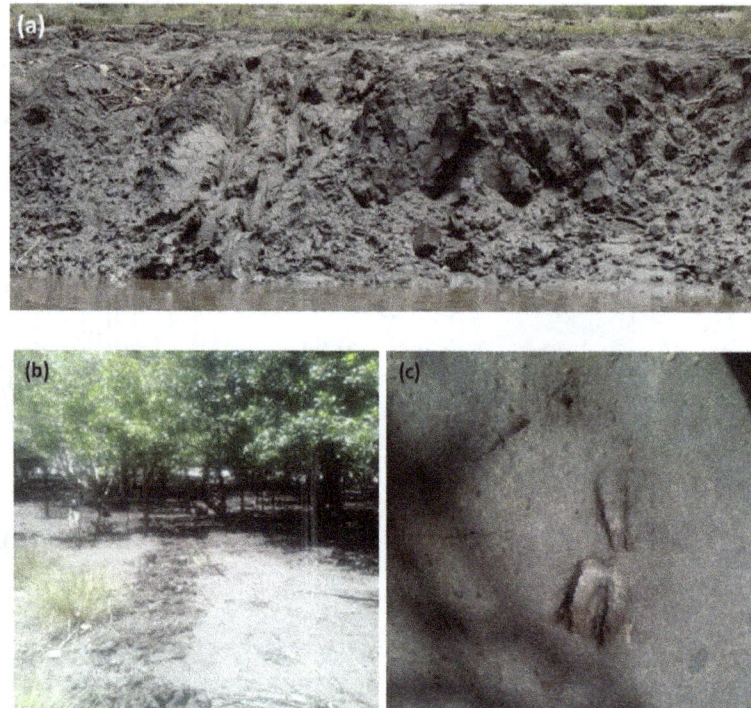

Figure 6: Hippo tracks (a) in the banks of the estuary and (b) in the mangrove forest. (c) Tracks of a small ungulate within a hippo track.

a declining trend, yet there has been increasing demand on water resources within the Wami River watershed for large scale agriculture, irrigation, industrial production and drinking water supply projects [6,26,27]. This declining trend together with increased upstream activities have contributed to the alteration of natural flow and water quality of the Wami River. Four of the six lowest river flows recorded in any year since 1954 occurred in the last five years (Figure 2). Also, there is a decreasing trend of the minimum flows in both wet and dry years (Figure 3). In dry years at present the salinity reaches 7 ppt at the tidal limit and this is a crisis for the wildlife as it cannot drink the water. As the animals in SANAPA could not drink this high salinity water, they either moved in the extremely shallow (depth of season in dry years, or they moved in the upper estuary to drink at low tide during 'normal' years, and other animals migrated out of the park in search of water and some were killed. There are also other indicators of change. Before 2007, the estuary was only moderately turbid with turbidity varying between 75 and 444 NTU during the dry and wet seasons respectively [28]. The turbidity has been increasing even after the EF studies in 2011, where TSS up to 300 mg/L was measured in the closest site to the estuary and in our study the highest value during wet season was about 427 mg/L within the estuary. The main causes of high turbidity in the estuary could be increased sediment load from the watershed from changing land-use, ongoing road constructions in upper regions of the catchment, bank erosion as well as re-suspension of bottom sediments by hippos, but the data are unavailable to quantify the relative importance of these processes. There is no doubt however that the estuary is silting (Figure 5). There were also changes in the pH, and these may be driven by the geochemistry of the Wami River [6,29] since the values of pH increased in the wet season when the system was autotrophic. However, other environmental variables and nutrients are within the acceptable levels as indicated in GLOWS-FIU (2014) [6].

The decreasing trend of minimum flows in the Wami River

implies increasing dry season salinity in the estuary. In turn this will have an impact on the estuarine ecosystem because salinity influences the reproduction, growth, abundance, distribution and diversity of estuarine species [30-35]. While the fluctuations in salinity within the estuary is a common phenomenon [36], different estuarine organisms tolerate differently these fluctuations and most of them can survive within very narrow ranges of salinities [2,37,38]. For example, change in water quality is indicated by changing fish composition and distribution upriver as some of estuarine fish were caught in dry season at Matipwili approximately 20 km from the estuary [6]. The changes in the salinity in the Wami River estuary may have similar impacts to many studied aquatic environments and can be expected to modify the species distribution, the composition and abundance, the mortality of sensitive species, the replacement of freshwater species by salt tolerant species, and the spawning, embryonic development, larvae development and hatching success of some species [32,33,39-41]. Studies on the effect of salinity for specific groups of aquatic biota of Wami estuary need to be done in the future to better understand the impacts of changing salinity patterns.

EF studies conducted in 2007 and 2011 for the Wami River recommended required flows for different selected sites, but excluded the estuary. The recommendations were based on ecological and geomorphological flows in the driest years (Table 1).

During the wet season, the Wami River estuary is flushed in less than a day and the ecosystem appears healthy with no apparent stress to fauna or flora. Such is not the case however in the dry season when the water residence time is typically ~7 days and thus the health of the estuarine ecosystem depends on the daily river discharge – even short periods of river discharge less than 1 m^3s^{-1}, a common occurrence in the dry season (Figure 3), result in excessively high salinity.

Though the estuarine flushing rate was very high in wet season,

Month	Driest year			Maintenance year			Wettest year		
	RAD	AAD	RIP	RAD	AAD	RIP	RAD	AAD	RIP
Oct	3	4.3		13.3	13.3		23	65	
Nov	3	5.9		14	26		23	265.9	
Dec	7.7	15.9		27.3	54.6		59.8	503.9	
Jan	7.7	10.1		32.8	65.7		96.5	412.9	
Feb	7.7	12.3		24.6	49.2		133.3	325.1	
Mar	5.6	5.6		52.4	69.9		170	466.6	
Apr	21.7	102.1	48(T<1 yr)	65	192.9	53(T<1 yr)	170	1240.5	170 (T<1.1 yr)
May	21.7	261.7		65	145.4		170	465.9	
Jun	15.5	42.6		37.5	49.9		91.4	182.8	
Jul	9.2	27.9		20.8	27.7		30.1	60.3	
Aug	3	15.4		14	21.1		23	51.3	
Sep	3	10.4		14	1505		23	61.5	

Table 1: Recommended EF (m^3 s^{-1}) of the Wami River at Mandera. (RAD: Recommended Average Discharge; AAD: Available Average Discharge; RIP: Recommended Instantaneous Peak Discharge) (Source: GLOWS – FIU, 2014).

where EF is high, the estuary was autotrophic as the system produces more than it consumes (NEM=98.32 mmol C/m²/day). In dry season, when EF is reduced the flushing time was much longer than in the wet season but the estuary consumption was higher than production (NEM=-10179.290 mmol C/m²/day). Moreover, Nfix-Denit was negative in dry season (-1532.84 mmol DIN/m²/day) and positive in wet season (15.07 mmol DIN/m²/day) indicating that net denitrification occurred in the dry season and net nitrogen fixation occurred in the wet season. These results indicate that if the EF will continue to decline, there is a high chance of the estuary to become unproductive.

The effect of estuarine water residence time τ_x on NEM has been studied by Swaney et al., (2011) [13] for more than 200 estuaries worldwide using the LOICZ model; a strong negative correlation exists between NEM and τ_x (Figure 7). The results indicate a decreasing net ecosystem metabolism as the residence time increases. However, the Wami River estuary data did not follow that trend line. We suggest that this is because these 200 estuaries plotted in the graphs did not include many small estuaries with very small flushing times, neither do they have mangroves and hippos. Our study suggests that the contribution of mangroves and hippos to nutrient cycling in the estuary is not negligible.

Mangroves in estuaries are known for their contribution to nutrient cycling in the estuary by absorbing some nutrients from water and releasing nutrients through leaf litter as detritus [42]. Detritus derived from mangrove leaf fall provides important source of food for macro-invertebrates such as sesarmid crabs. High concentrations of tannins in mangrove leaves prevents them from being eaten by many estuarine organisms except crabs since tannin interfere with protein digestion. Sesarmid crabs are well known for their ability to transport, retain and consume large quantities mangrove leaves in their burrows [43]. However, in the wet season the effective contribution of mangrove litter to nutrients in the estuary may be smaller than during the dry season because materials are quickly flushed out to the sea; by contrast in the dry season the residence time is large enough for mangrove litter to decompose and contribute to nutrient cycling before they are flushed out. However no definite answer can be given because no studies have been done in Wami mangroves to measure the rate of decomposition of mangrove plant litter and its likely seasonal variation.

Though the mangroves were in theory protected by law since 1994 under the Mangrove Management Project [44], in practice the mangroves of the Wami River estuary continued to degrade due to imbalance between effective law enforcement and increased mangrove harvesting mainly for charcoal and building materials which were exported to Zanzibar. From 1990 to 2005 the mangrove forest cover was reduced by 27% (27.3 ha/y; McNally et al.) [45]. This rate was reduced to 1.8 ha/y in 2005-2010 as the National Park protection laws were progressively implemented by effective law enforcement through regular patrols and arresting of any person who harvest mangrove within the park's boundaries. As a result, at present the mangroves are regenerating quickly and naturally as we observed during our study new mangroves growing in previously clear-felled areas and 2-3 m long branches sprouting from old cuts (Figure 8).

Suspended sediments transported by river flows are an important source of organic and inorganic matter in estuarine ecosystems [46,47]. They tend to settle in streambeds and create microhabitats for aquatic organisms, allow other metals to attach in sediment particles and provide habitats for pathogens [48]. However, the recent increase supply of suspended sediments to the Wami River estuarine systems is large and in most likelihood, by comparison with other estuaries worldwide, results in higher turbidity, which in turn may cause reduction of light penetration and photosynthesis, smothering of benthic organisms, replacement of seagrass by algae, impair predator-prey visibility, alteration of macro-invertebrates and fish spawning habitats [12,49-53]. In turn this may reduce prawn and fish catches by the villagers, but no data are available.

In the wet season, about 12% of the riverine fine sediment inflow was trapped in mangroves. However during the dry season the riverine fine sediment inflow accounted for only about 10% of the sedimentation rate in the mangroves. We suggest that during dry season the sediments originated from the muddy delta as well as bottom resuspension caused by the hippos in the estuary. Thus it appears that during the dry season all the riverine sediment is trapped in the mangroves while during the wet season about 88% of the riverine fine sediment is flushed out of the estuary; from visual observations we suggest that much of that sediment settles in a submerged delta off the river mouth while the remaining fine sediment is transported northward longshore in the river plume and presumably deposits over the patchy seagrass beds in coastal waters (Figure 1). During the dry season 60% of that exported fine sediment returns in the estuary to be trapped in the mangroves. The expansion of mangroves results in increased trapping of fine sediment in the Wami River estuary during a year. However because fine sediment is exported to coastal waters during the wet season, and from visual observations this appears to be increasing due to deforestation in the watershed, it is not known if the increased riverine sediment inflow degrades the

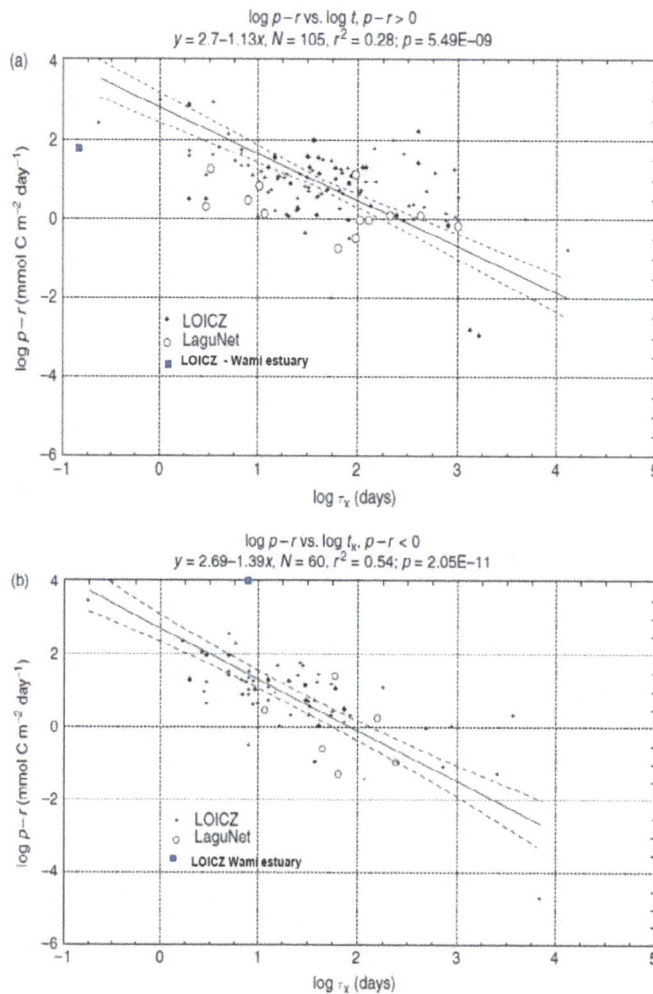

Figure 7: Scatter plot of the Net Ecosystem Metabolism versus the estuarine water residence time τx for (a) autotrophic systems (p-r>0) and (b) heterotrophic systems (p-r<0). (■=Wami River estuary; Swaney et al., [13]).

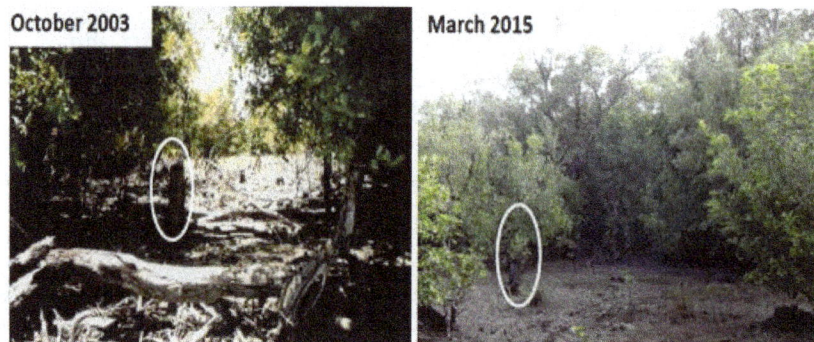

Figure 8: Mangroves recovering in the Wami River estuary since the National Park was gazetted in 2005. Note the tree that has grown by 2015 from the trunk left by loggers in 2003.

seagrass in coastal waters, which supports the local prawn fisheries. Studies are needed on the status of those seagrass meadows.

The hippos, all refugees from outside the National Park, are an important tourism attraction, and thus important to the local economy, but at the same time they are new to the environment and they modify it. Hippos contribute to cycling nutrients from terrestrial land (where they graze) to the estuary (through defecation). While there are no census data, visual observations indicate an increasing population of hippo after protection of the estuary in 2005. The hippos are not destructive of the mangrove forest, but they do create paths which are then used by other animals (Figure 6). They also contribute to bank erosion (Figure 6) and stirring the bottom and this increases the turbidity. Distribution and movement of hippo in the estuary also depends on EF as their movements have been observed to change with changing freshwater flow in different seasons associated with changing salinity patterns within the estuary. However, these movements will only be sustained if the EF recommendations will be effectively implemented. Otherwise under worst case scenarios if there is no enough water, hippos will be forced to change from their territorial behavior and coexist in large groups within the uppermost reaches of the estuary where they may find residual freshwater or they may leave the estuary for other suitable places where they can find enough water. If this situation happens the SANAPA will lose one of its main tourist attractions in the park and consequently the economy of the park will be negatively affected.

On the other hand, the mangroves are more tolerant of salinity fluctuations. They have developed different mechanisms to deal with salinity [54]. However, they require regular flushing with freshwater to balance salinity levels that can be tolerated because excessive salinity is detrimental to mangroves. If the present trends lead to hypersaline conditions in the Wami estuary in the future, then mangrove growth will be impeded and in turn this will prevent fruiting and seed [55-58] and consequently fisheries will be affected.

Changing uses of the land and water resources in the Wami River watershed decrease the freshwater flow and increase the sediment load to the Wami estuary, and this is particularly ecologically important during the dry season to the level that the whole Wami River estuary ecosystem seems to be at a tipping point. This has enormous consequences for the ecology of SANAPA and thus its tourism potential, as well as for coastal fisheries; and in turn this impacts the local communities and their economy. Climate change may exacerbate this crisis because it is predicted to result in increasing frequency of droughts in Tanzania [6,59-61].

Estuary is a very complex ecosystem since its functioning is influenced by both river flow and sea water unlike other upriver sites which are influenced by freshwater flow only. Thus, determination of EF for the Wami estuary is very crucial as the requirement may be different from the river. Despite the fact that, the EF of the estuary has not been done, the recommended EF for upriver sites have not been effectively implemented due to various reasons including insufficient resources and capacity as indicated by Dickens (2011) [5].

The Wami River estuarine ecosystem is now in crisis in the dry season in dry years, and will be increasingly so in the future if effective measures are not taken.

Conclusion and Recommendation

At present, as reviewed by Elisa et al., [62] and Dickens [5] minimum environmental flow requirements are not effectively enforced in Tanzania. Our study shows that the results of this benign neglect are disastrous for the Wami River estuary. This estuary itself falls under the authority of SANAPA, which is enforcing its regulations. As a result the estuary is ecologically healthy but during the dry season it is threatened by increasing salinity due to decreasing freshwater flow in the Wami River. The estuary needs freshwater flow in the Wami River and the management of freshwater flows of the Wami River falls within the WRBWO. WRBWO has carried out EF assessments for the Wami River but it is not effectively enforcing its own recommendation. In order to maintain a healthy estuarine ecosystem it is the obligation of the WRBWO now to stick to and enforce the recommendations of its own environmental flow assessment to regulate water usage in the

watershed. A similar recommendation also holds for all other rivers and estuaries in Tanzania.

Acknowledgements

We would like to extend our sincere gratitude to NM-AIST for proving funds and technical support for this study. Special thanks to SANAPA for facilitating the field work and moral support during the study.

References

1. Dyson M, Bergkamp G, Scanlon J (2008) Flow - The essentials of environmental flows, (2ndedn), IUCN Reprint, Gland, Switzerland.

2. URT (United Republic of Tanzania) (2002) National Water Policy. Ministry of Water and Livestock Development. Dar es Salaam, Tanzania.

3. URT (United republic of Tanzania) (2005) Environmental Management Act, 2004. Government Printers. Dar es Salaam, Tanzania.

4. URT (United Republic of Tanzania) (2009) The Water Resources Management Act, 2009. Government Printers. Dar es Salaam, Tanzania.

5. Dickens C (2011) Critical analysis of environmental flow assessments of selected rivers in Tanzania and Kenya. Nairobi, Kenya: IUCN ESARO office and Scottsville, South Africa, pp. 104.

6. GLOWS–FIU (2014) Climate, Forest Cover and Water Resources Vulnerability, Wami/Ruvu Basin, Tanzania, pp. 87.

7. Costa MJ, Costa JL, Almeida PR, Assis CA (1994) Do eel grassbeds and salt marsh borders act as preferential nurseries and spawning grounds for fish? An example of the Mira estuary in Portugal. Ecological Engineering 3: 187-195.

8. Lindeboom H (2002) The coastal zone: an ecosystem under pressure. In: Field JG, Hempel G, Summerhayes CP (eds) Oceans 2020: Science, Trends, and the Challenge of Sustainability, Island Press, Washington, pp. 49-84.

9. Uncles RJ, Stephens JA, Smith RE (2002) The dependence of estuarine turbidity on tidal intrusion length, tidal range and residence time. Continental Shelf Research 22: 1835-1856.

10. Wolanski E, Boorman LA, Chicharo L, Langlois-Saliou E, Lara R, et al. (2004) Ecohydrology as a new tool for sustainable management of estuaries and coastal waters. Wetlands Ecology and Management 12: 235-276.

11. McKinney RA, McWilliams SR, Charpentier MA (2006) Waterfowl-habitat associations during winter in an urban North Atlantic estuary. Biological Conservation 132: 239-249.

12. Wolanski E, Elliott M (2015) Estuarine Ecohydrology. Elsevier, Amsterdam, in press.

13. Swaney DP, Smith SV, Wulff F (2011) The LOICZ Biogeochemical Modeling Protocol and its Application to Estuarine Ecosystems, Chapter 9.08 (pp. 135-159) In: Treatise on Estuarine and Coastal Science. Wolanski E, McLusky D (eds). Academic Press, Waltham.

14. Mosha EJ, Gallardo G (2013) Distribution and size composition of penaeid shrimps, Penaeus monodon and Penaeus indicus in Saadani estuarine area, Tanzania. Ocean and Coastal Management 82: 51-63.

15. Tobey J (2008) A profile of the Wami River Sub-Basin. A report prepared for the Tanzania Coastal Management Partnership for Sustainable Coastal Communities and Ecosystems in Tanzania.

16. TANAPA (Tanzania National Parks) (2003) Saadani National Park, Management Zone Plan. TANAPA/Department of Planning and Project Development.

17. Nyenzi BS, Kiangi PMR, Rao NNP (1981) Evaporation values in East Africa. Archieves for Meteorology, Geophysics and Bioclimatology 29: 37-55.

18. Murphy J, Riley JP (1962) A modified single solution for the method for the determination of phosphate in natural waters. Analytica Chimica Acta 27: 31-36.

19. Parsons TR, Maita Y, Lalli CM (1984) A manual of chemical and biological methods for seawater analysis. Pergamon Press. Australia.

20. APHA (2005) Standard Methods for the Examination of Water and Wastewater. (21stedn), American Public Health Association, American Water Works Association and Water Environmental Federation, Washington D.C, USA.

21. Xu H, Wolanski E, Chen Z (2013) Suspended particulate matter affects the nutrient budget of turbid estuaries: Modification of the LOICZ model and

application to the Yangtze Estuary. Estuarine, Coastal and Shelf Science 127: 59-62.

22. Xu H, Newton A, Wolanski E, Chen Z (2015) The fate of Phosphorus in the Yangtze Estuary, China, under multi-stressors: hindsight and forecast. Estuarine, Coastal and Shelf Science, in press.

23. Golbuu Y, Victor S, Wolanski E, Richmond RH (2003) Coastal and Shelf Science 57: 941-949.

24. Smith TJ, Boto KG, Frusher SD, Giddins RL (1991) Keystone species and mangrove forest dynamics: The influence of burrowing by crabs on soil nutrient status and forest productivity. Estuarine, Coastal and Shelf Science 33: 419-432.

25. Lindquist ES, Krauss KW, Green PT, O'Dowd DJ, Sherman PM, et al. (2009) Land crabs as key drivers in tropical coastal forest recruitment. Biol Rev Camb Philos Soc 84: 203-223.

26. WRBWO (Wami-Ruvu Basin Water Office) (2008) Wami River Sub-Basin, Tanzania: Initial Environmental Flow Assessment. Final Report. Tanzania Ministry of Water, Morogoro, Tanzania. pp. 144.

27. Madulu NF (2005) Environment, poverty and health linkages in the Wami River basin: A search for sustainable water resource management. Physics and Chemistry of the Earth 30: 950-960.

28. Anderson E, McNally CG, Kalangahe B, Ramadhani H, Mhitu H (2007) The Wami River Estuary: A Rapid Ecological Assessment. Technical Report prepared for the Tanzania Coastal Management Partnership and the Coastal Resources Center, University of Rhode Island.

29. Baumann H, Wallace RB, Tagliaferri T, Gobler CJ (2015) Large Natural pH, CO2 and O2 fluctuations in a temperate tidal salt marsh on diel, seasonal, and interannual time scales. Estuaries and Coasts 38: 220-231.

30. Jiang D, Lawrence AL, Neill WH, Gong H (2000) Effects of temperature and salinity on nitrogenous excretion by Litopenaeus vannamei juveniles. J Exp Mar Bio Ecol 253: 193-209.

31. Bidwell JR, Gorrie JR (2006) The influence of salinity on metal uptake and effects in the midge Chironomus maddeni. Environ Pollut 139: 206-213.

32. Brown AFM, Dortch Q, Dolah FMV, Leighfield TA, Morrison W, et al. (2006) Effect of salinity on the distribution, growth and toxicity of Karenia spp. Harmful Algae 5: 199-212.

33. Kefford BJ, Nugegoda D, Metzeling L, Fields EJ (2006) Validating species sensitivity distributions using salinity tolerance of riverine macroinvertebrates in the Southern Murray-Darling Basin (Victoria, Australia). Canadian Journal of Fisheries and Aquatic Sciences 63: 1865-1877.

34. Pan LQ, Zhang LJ, Liu HY (2007) Effects of salinity and pH on ion-transport enzyme activities, survival and growth of Litopenaeus vannamei postlarvae. Aquaculture 273: 711-720.

35. You C, Jia C, Pan G (2010) Effects of salinity and sediment characteristics on the sorption and desorption of perfluorooctane sulfonate at sediment-water interface. Environmental Pollution 158: 1343-1347.

36. Leroy SAG, Marret F, Gibert E, Chalie F, Reyss JL, et al. (2007) River inflow and salinity changes in the Caspian Sea during the last 5500 years. Quartenary Science Reviews 26: 3359-3383.

37. Dunlop JE, Horrigan N, McGregor G, Kefford BJ, Choy S, et al. (2008) Effect of spatial variation of salinity tolerance of macroinvertebrates in Eastern Australia and implications for ecosystem protection trigger values. Environmental Pollution 151: 621-630.

38. Wolf B, Kiel E, Hagge A, Krieg HJ, Feld CK (2009) Using the salinity preferences of benthic macroinvertebrates to classify running waters in brackish marshes in Germany. Ecological Indicators 9: 837-847.

39. Song J, Fan H, Zhao Y, Jia Y, Du X, et al. (2008) Effect of salinity on germination, seedling emergence, seedling growth and ion accumulation of a euhalophyte Suaeda salsa in intertidal zone and on saline inland. Aquatic Botany 88: 331-337.

40. Muylaert K, Sabbe K, Vyverman W (2009) Changes in phytoplankton diversity and community composition along the salinity gradient of the Schelde estuary (Belgium/ The Netherlands). Estuarine Coastal and Shelf Science 82: 335-340.

41. Zhong Y, Kemp AC, Yu F, Lloyd JM, Huang G, et al. (2010) Diatoms from the Pearl River estuary, China and their suitability as water salinity indicator for coastal environments. Marine Micropaleontology 75: 38-49.

42. Boehm AB, Yamahara KM, Walters SP, Layton BA, Keymer DP, et al. (2011) Dissolved inorganic nitrogen, soluble reactive phosphorous, and microbial pollutant loading from tropical rural watersheds in Hawai'i to the coastal ocean during non-storm conditions. Estuaries and Coasts 34: 925-936.

43. Cannicci S, Burrows D, Fratini S, Smith TJ, Offenberg J, et al. (2008) Faunal impact on vegetation structure and ecosystem function in mangrove forest: A review. Aquatic Botany 89: 186-200.

44. Masalu DC (2009) Report on Environmental Emerging Issues in Tanzania's Coastal and Marine Environments Based on Selected Key Ecosystems.

45. McNally CG, Uchida E, Gold AJ (2011) The effect of a protected area on the tradeoffs between short-run and long-run benefits from mangrove ecosystems. Proc Natl Acad Sci USA 108: 13945-13950.

46. Santschi PH, Hoehener P, Benoit G, Brink MB (1990) Chemical processes at the Sediment-water interface. Marine Chemistry 30: 69-315.

47. Hedges JI, Keil RG (1998) Organic geochemical perspectives on estuarine processes: sorption reactions and consequences. Marine Chemistry 65: 55-65.

48. Labelle RL, Gebra CP, Goyal SM, Melnick JL, Cech I, et al. (1980) Relationship between environmental factors, bacterial indicators and the occurrence of enteric viruses in estuarine sediments. Applied and Environmental Microbiology 39: 588-598.

49. Cloern JE (1987) Turbidity as a control on phytoplankton biomass and productivity in estuaries. Continental Shelf Research 7: 1367-138.

50. Alpine AE, Cloern JE (1988) Phytoplankton growth rates in a light-limited environment, San Francisco Bay. Marine Ecology Progress Series 44: 167-173.

51. Abal EG, Loneragan N, Bowen P, Perry CJ, Perry JW, et al. (1994) Physiological and morphological responses of Zostera capricorni Aschers to light intensity. Journal of Experimental Marine Biology and Ecology 178: 113-129.

52. Wilber DH, Clarke DG (2001) Biological effects of suspended sediments: A review of suspended sediment impacts on fish and shellfish with relation to dredging activities in estuaries. North American Journal of Fisheries Management 21: 855-875.

53. Uncles RJ, Stephens JA (2010) Turbidity and sediment transport in a muddy sub-estuary. Estuarine, Coastal and Shelf Science 87: 213-224.

54. Takemura T, Hanagata N, Sugihara K, Baba S, Karube I, et al. (2000) Physiological and biochemical responses to salt stress in the mangrove, Bruguiera gymnorrhiza. Aquatic Botany 68: 15-28.

55. Ball MC (1998) Mangrove species richness in relation to salinity and water logging: A case study along the Adelaide River flood plain, Northern Australia. Global Ecology and Biogeography Letters 7: 73-82.

56. Ball MC (2002) Interactive effects of salinity and irradiance on growth: Implications for mangrove forest structure along salinity gradients. Trees 16: 126-139.

57. Aziz I, Khan MA (2001) Experimental assessment of salinity tolerance of Ceriops tagal seedlings and saplings from the Indus delta, Pakistan. Aquatic Botany 70: 259-268.

58. Mitra A, Chowdhury R, Sengupta K, Banerjee K (2010) Impacts of salinity on mangroves of Indian Sundarbans. Journal of Coastal Environment 1: 71-82.

59. Boko M, Niang I, Nyong A, Vogel C, Githeko A, et al. (2007) Africa. Climate Change 2007: Impacts, Adaptation and Vulnerability. In: Parry ML, Canziani OF, Palutikof JP, van der Linden PJ, Hanson CE (eds) Contribution of Working Group II to the Foh Assessment Report of the Intergovernmental Panel on Climate Change, Cambridge University Press, Cambridge, UK, pp. 433-467.

60. Tierney JE, Mayes MT, Meyer N, Johnson C, Swarzenski PW, et al. (2010) Late-twentieth-century warming in Lake Tanganyika unprecedented since AD500. Nature Geoscience 3: 422-425.

61. Wolff C, Haug GH, Timmermann A, Damsté JS, Brauer A, et al. (2011) Reduced interannual rainfall variability in East Africa during the last ice age. Science 333: 743-747.

62. Elisa M, Gara JI, Wolanski E (2010) A review of the water crisis in Tanzania's protected areas, with emphasis on the Katuma River-Lake Rukwa ecosystem. Ecohydrology and Hydrobiology 10: 153-166.

12

Identification of Hydraulic Parameters of Wadi El Natrun Pliocene Aquifer Using Artificial Neural Network

Khalaf S*, Ahmed AO, Abdalla MG and El Masry AA

Irrigation and Hydraulics Department, Faculty of Engineering, El-Mansoura University, Egypt

Abstract

Many techniques, approaches and tools were used in this Research to achieve the Methodology. Using artificial neural network to simultaneous hydraulic parameters is one of these techniques. Transmissivity and storativity consider the most important parameter in each aquifer due to the reality of their effect on the aquifer properties. In this research, it is assumed that the transmissivity (T) and the storativity (S), represented by coordinates (X), (Y), hydraulic head (H), and observation times (t). These variables were chosen depending on the literature review. In the present study, the hydraulic head values at each cell (H) and the location of the cells (x, y) are considered as input parameters for finding the unknown parameters. The transmissivity (T) and storativity values (S) at cells are assumed and used in the finite difference method (Forward model) in order to find the value of hydraulic head at that cell. The hydraulic head values were used in the artificial neural networks (Inverse model) to estimate transmissivity (T) and storativity values (S) for Wadi El Natrun Depression. The study is based on coupling of forward model and inverse model. In general, the parameter estimation process consists of identifying a model that would reverse a complex forward relation.

Keywords: Parameter identification; Simultaneous; Inverse model; Neural networks; Groundwater flow; Forward model; Wadi El-Natrun

Introduction

Soft computing techniques analogous to biological nervous system are called as artificial neural networks (ANNs). The reason for wider acceptability of this technique can be attributed to its capability to develop computing tools, which may partially capture amazingly faster and complicated information-processing ability of the brain. Groundwater is an important source of water for drinking, irrigation, and industrial uses. It is also a major source for domestic water requirement. Ground water hydrology examines problems like prediction of groundwater head, distribution of transmissivity, storativity specific yield and estimation of parameters. Involved processes are nonlinear, complex, multivariate with variables having spatial and temporal variability. These are expressed by complex partial differential equations, which are normally solved with considerable approximations using complex numerical models. Precise conceptualization is nearly impossible for groundwater problems as physics of the system cannot be fully understood from the surface. ANNs can be used in groundwater hydrology since they do not require governing equations and their concomitant assumptions.

Strength of ANNs lies in mapping non-linear system data, which are capable of extracting the relation between the input and output of a process without adequate knowledge of the underlying principles. As the computational burden is primarily managed by replacing the numerical model with a surrogate simulator artificial neural network the technique of ANN is quite appropriate for groundwater problems.

Estimation of aquifer parameters demands large time, financial and manpower resources which are always scarce [1]. Therefore, the efforts are on increase to assess these by non-field methods using the technique of ANNs. The determination of aquifer parameters (also termed as inverse problem) has always been a challenge because of its ill-posedness [2-7]. Inversion of the trained feed forward neural network is done to estimate the transmissivity field for synthetic problem [5]. Most of the papers used synthetic or published data to assess parameters in confined aquifer [4-6,8] for the reason of non-availability of sufficient number of patterns. Model sensitivity in terms of number of nodes in hidden layers is carried out by Shigdi et al. [5] using correlation coefficient, Y-intercept, and slope. An approach with a combination of ANNs and type curves based on Papadopoulos and Cooper analytical solution was used by Balkhair [4] whereas ANN and This solution coupling was performed by Lin GF et al. [9]. It was found that Levenberg-Marquardt achieves faster convergence than backpropagation algorithm [7]. To solve the problem of range of aquifer parameters to be used for training [4] carried out training in macro and micro scales. Macro scale is used for very wide and very narrow range whereas micro scale is for middle range of parameters. Lohani et al. [10] presents an efficient and stable artificial neural network (ANN) model for predicting groundwater level in south-east Punjab, India. Lohani et al. [11] used neural network configuration for predicting groundwater level in Amritsar and Gurdaspur districts of Punjab, India is identified. For predicting the model efficiency and accuracy, different types of network architectures and training algorithms are investigated and compared.

The determination of aquifer parameters (also termed as inverse problem) has always been a challenge because of its ill-posedness [2-7,12,13]. ANNs is solving many complex real world-predicting problems. ANNs have been applied to predicting groundwater levels [14,15], precipitation and runoff modeling, and aquifer parameter estimations [3,5,16-19].

*Corresponding author: Dr. Khalaf S, Irrigation and Hydraulics Department, Faculty of Engineering, El-Mansoura University, Egypt
E-mail: samykhalaf2005@yahoo.com

The main objectives of this study may be summarized as follows:

(1) to develop an ANN model for solving the groundwater inverse problem by providing an ANN model to solve complex nonlinear relationships.

(2) to test the performance of the proposed ANN model.

(3) to simultaneously determine the transmissivity and storativity field of an aquifer by using limited observed head values.

Study area (Wadi El Natrun)

The study area lies at El Behera governorate, Western of Nile Delta, Egypt between Longitudes 30° 00' and 30° 33' E and Latitude 30° 20' and 30° 30' N. It parallels to (Cairo-Alex. Desert Highway) Km 90 to Km 110. The study area covers about 2016 Km^2 (Figure 1).

The study area comprises a total area of about 770 km^2. From the hydrological cross section, which pass through Wadi El Natrun area in West-East direction. Pliocene aquifer and from electrical sounding, this aquifer unit is mainly formed of alternating sand and clay and occasionally capped by thin layer of limestone. The Pliocene aquifer is considered as multi-layers' aquifer under confined to semi-confined condition. Most of these waters bearing layers belongs to the Upper and Middle Pliocene. In this study, the information of 111 pumping well and 14 observation well were collected as well as other parameters such as hydraulic conductivity. Table 1 shows the values of hydraulic conductivity, transmissivity, and storativity of Pliocene at Wadi El Natrun area of different authors. According to the analysis of Pliocene aquifer in the area of study, we can notice that, the hydraulic conductivity (K) values ranged between 5 m/day at Wadi El Natrun area to 40 m/day in the east of Wadi El Natrun. The transmissivity (T) values ranged between 500 m^2/day to 1600 m^2/day in different zones of the Depression.

Geologic setting

The study area is occupied sedimentary rocks belonging to the Tertiary and Quaternary Eras. The sedimentary succession comprises several water-bearing formations, which are particularly influenced by structural features and thus affect the groundwater occurrences. However, the surface deposits dominating the area are studied through the geological map which shown in Figure 2. In the study area, Late Tertiary and Quaternary succession were studied by many authors. The subsurface geology of the study area is studied from the well logs as from the Wadi El Natrun deep well. The Pliocene aquifer is local in extent, covers the entire Wadi El Natrun area, and is discontinuously covered by Quaternary deposits of the Pleistocene Aquifer. As a result, the Pliocene aquifer is considered to be partially confined [20]. The Pliocene Aquifer's thickness ranges from 150 to 300 m thick [20].

Hydrologic setting

Hydrogeologic setting in the Wadi El Natrun area is complex, and there is still significant uncertainty as to the precise flow regime and hydraulic connections between the aquifers. Nonetheless, multiple studies have observed both local and regional groundwater flow to be concentrated toward the base of the valley where it discharges to eleven saline lakes [20].

Govering equation

A general form of the governing equation, which describes the three dimensional movement of groundwater flow of constant density through the porous media is [21]:

$$\frac{\partial}{\partial x}\left[Kx\frac{\partial h}{\partial x}\right] + \frac{\partial}{\partial y}\left[Ky\frac{\partial h}{\partial y}\right] + \frac{\partial}{\partial z}\left[Kz\frac{\partial h}{\partial z}\right] - w = Ss\frac{\partial h}{\partial t} \tag{1}$$

Where: Kx, Ky, Kz are values of hydraulic conductivity along the x, y and z coordinate axes (L/t); h: is the potentiometric head (L); w: is the volumetric flux per unit volume and represents sources and/or sinks of water per unit time (t^{-1}); Ss: is the specific storage of the porous material (L^{-1}); and t: is time (t). The first part of this problem was run to get a steady state solution that takes the form:

$$\frac{\partial}{\partial x}\left[Kx\frac{\partial h}{\partial x}\right] + \frac{\partial}{\partial y}\left[Ky\frac{\partial h}{\partial y}\right] + \frac{\partial}{\partial z}\left[Kz\frac{\partial h}{\partial z}\right] - w = 0.0 \tag{2}$$

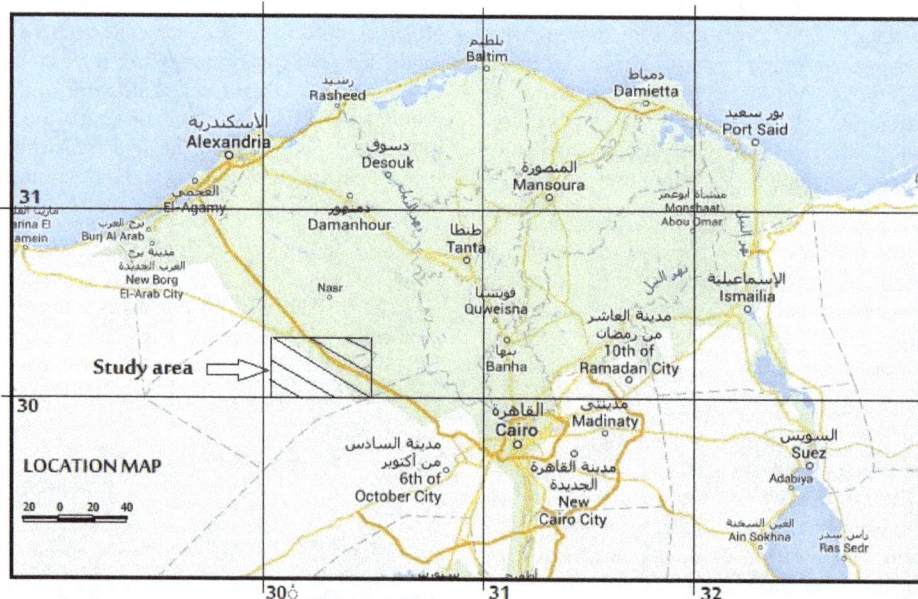

Figure 1: Location map of the study area.

Author	Location of the test zone	Hydraulic parameters		
		K	T	S
Ahmed [27]	North of Wadi El Natrun	-	1043.9	-
Shata et al. [28]	Northeast Wadi El Natrun	11	-	-
Desert research [28]	Northeast Wadi El Natrun	31.9	-	-
General Petroleum [29]	East Wadi El Natrun	26	2600	$3.9*10^{-3}$
Ahmed [27]	2 km from El hamara lake	-	794	-
Saad [30]	East side of Wadi El Natrun	52.9	1291.7	$3.95*10^{-3}$
RIGW [31]	Wadi El Natrun area	9.8	500-327	$1.7*10^{-2}$
Mostafa [32]	East Wadi El Natrun	47	943	$7*10^{-4}$
Saad [30]	East Wadi El Natrun	38.9	695.5	$1.35*10^{-3}$
Ahmed [27]	South Wadi El Natrun	-	1660	-
El-Sheikh [33]	Beni Salama	7.29	838	$8.5*10^{-4}$
Sayed	East Wadi El Natrun (Jacob)	-	3618	0.10466
El Sayed	East Wadi El Natrun (Recovery)	-	2745	-

Table 1: Hydraulic parameters of the Pliocene aquifer.

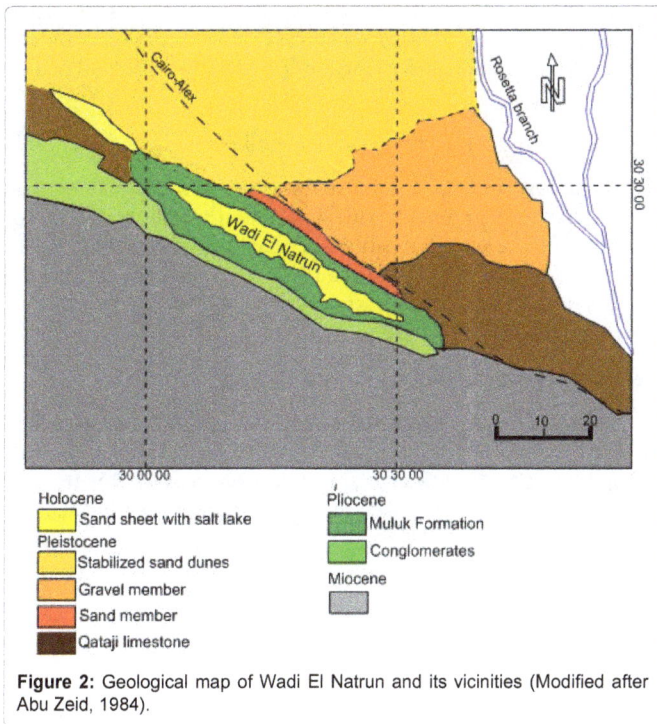

Figure 2: Geological map of Wadi El Natrun and its vicinities (Modified after Abu Zeid, 1984).

From the steady state solution, the hydraulic conductivity for model aquifers can be found. Then the equation is solved for transient case in order to solve for storage coefficient.

Karahan et al. identified transmissivity distribution of a two-dimensional aquifer system under steady-state flow conditions. In that study, randomly selected Cartesian coordinates and error-free hydraulic heads are used as input and corresponding transmissivity values are used as output in the ANN model [18].

Shigidi et al. determined aquifer parameters by using ANN. In order to obtain observed hydraulic heads, they used MODFLOW with a stochastically generated transmissivity distribution [5]. Their network architecture uses the transmissivity values as input and the hydraulic heads as output. The error residuals between computed and observed hydraulic heads are reduced by adjusting the transmissivity field, and

this process is cycled until the convergence is met. On the other hand, Garcia et al. [17] solved the same problem in their previous study- i.e., Shigidi et al., by adding a noise term to the observed hydraulic heads and they obtained more precise results than their previous study [5].

Choose the appropriate architecture of network among the available networks based on the type of the data and the problem. After many trials, Multilayer Perceptron network (MLP) has been chosen because of its high capabilities to generalize well in problems plagued with significant heterogeneity and nonlinearity. The Multilayer Perception (MLP) trained with the back-propagation algorithm is perhaps the most popular network for hydrologic modeling [22]. The majority of the ANN applications in water resources engineering involve the employment of a conventional Feed Forward Back Propagation Method (FFBP).

The important feature of this network is its ability to self-adapt the weights of neurons in intermediate layers to learn the relationship between a set of patterns given as examples and their corresponding outputs. So that after having been trained it can apply the same relationship to new input vectors and produce appropriate outputs from inputs that the system has never seen before, a feature known as the generalizability of an ANN [1]. The artificial neural network with a multilayer back propagation network has been used successfully in several studies [5,14,23-26].

Conceptual model of Wadi el Natrun

On the light of the hydrogeologic properties of the Pliocene aquifer in Wadi El Natrun Depression (chapter five), a pictorial representation (conceptual model) of the water flow system is constructed to this aquifer. The constructed conceptual model depends on the following facts:

1) The aquifer of Wadi El Natrun consists of five Geoelectric layers appear as follows:

- The first Geoelectric layer consist of dry sand, gravel and rock fragments, called the subsurface. The thickness of this layer is ranged between (0.9 m) and (3.8 m).

- The second Geoelectric layer is formed from shaly, sand and presence of shale interbedded with sand and differential amounts of water. The thickness of this layer is ranged between (2.4 m) and (22 m).

- The third geoelectric layer is made up of clayey, sand and gravel and underground water included in this layer, the thickness is ranged between (61.6 m) and (69 m).

- The fourth geoelectric layer is composed of clayey, sand and gravel, and considered the continuation of the third geoelectric layer. It includes the same lithology. These values give indication about the presence of underground water in this layer, as shown as shown the thickness of this layer is ranged between (86.7 m) and (92.4 m).

-The fifth geoelectric layer is barrier separating it and from the fourth geoelectric layer.

At the end, this section includes two water bearing layers, the third and the fourth geoelectric layers, that are saturated with water. These layers are considered the main aquifer of the study area.

2) The groundwater flow generally from NE to SW direction toward Wadi El Natrun Depression. The recharge from irrigation and the evapotranspiration will be neglected. The main discharge source is the present flowing productive well (48 wells).

3) The Pliocene aquifer of Wadi El Natrun occurs under the confining conditions.

4) The base of the aquifer is not detected.

Model application

The model domain was selected to cover 770 km² (55 × 14 km). The model domain was discretized using 144 row and 224 columns square cells. This discrimination produces 32,250 cells in the model layer. The width of the cells along rows (in x-direction) is equal to the length along columns (in y-direction) 250 m as shown in Figure 3.

Estimating of hydraulic head

After finish the calibration of the MODFLOW model, and reached a good match between observed and calibrated head, as shown previous, the model will be ready to estimate the hydraulic distribution for all cells (Figure 4) which will use in the artificial neural network model. The values of hydraulic head in each cell as well as the coordinates of these cells can be exported as a .txt. File which will used later in the next chapter to predict the transmissivity and storativity using ANNs.

Construction of ANN model

Construction ANN model utilization FFBP Neural Networks (MLP) which built in MATLAB program version 7. Different ANN structure had been investigated to find optimum ANN model. The optimum neuron number in each hidden layer was also investigated.

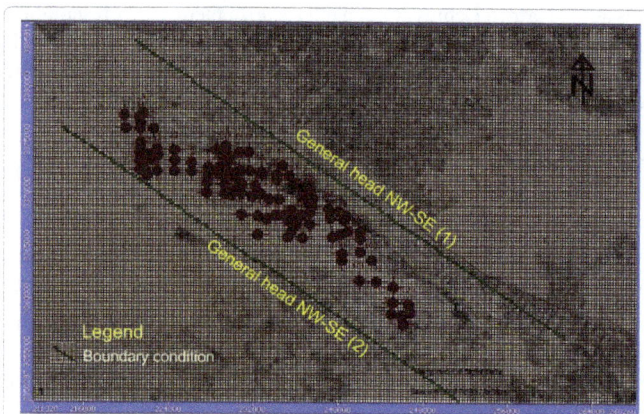

Figure 3: Flow model boundaries, grid, and pumping wells.

Figure 4: Head contour map of the model area.

In the ANN model architecture, the input layer includes 4 input (hydraulic head, Cartesian coordinates, and the time of observation), and the output layer include 2 outputs (transmissivity and storativity). In order to find the best number of hidden layers of ANN model many networks with different structure have been developed. These models had two hidden layers. The number of neuron in each hidden layer is also very important in the structure of ANN. The low number of neurons in each layer decreases the learning performance. In contrast, a high number of neurons increase the training time. Therefore, the optimum number of neurons in each layer is essential to build the ANN structure. In the present research, a different number of neurons (10-17) have been tested to find the best ANN structure. By comparing performance of developed ANN models the optimum number of neurons in each hidden layer for different structure (3 × 6 × 2), (3 × 10 × 2), (3 × 6 × 4 × 2), (3 × 8 × 4 × 2), (3 × 10 × 4 × 2), (3 × 12 × 6 × 2), (3 × 14 × 6 × 2), (3 × 14 × 8 × 2), (3 × 16 × 8 × 2) and (3 × 17 × 8 × 2). They had the same input and output layer, so they just differ with the hidden layers. The best results obtained from the network with two hidden layers with 17 and 8 hidden neurons, respectively.

For present study, FFBP (3 × 17 × 8 × 2) architecture is selected in the model. The logsig transfer function is applied in the first hidden layer and the purelin transfer function is applied in the second hidden layer (Figure 5).

Construction data matrix of ANN model

The matrixes of ANNs are different as compare with other matrixes, because that every parameter value should be arranged at one a row of this matrix. There are three matrixes need to be constructed in older to use in artificial neural networks model. Two matrixes for training the ANNs model, while the third matrix used as input parameters that we want to get its output (Table 2).

Network training, validation, and testing

Generally, the default setting of artificial neural network divided date into three divisions, which are training, validation and testing with a proportion of 60%, 20% and 20% of data respectively. The training set the large proportion of data to learn pattern present in the data.

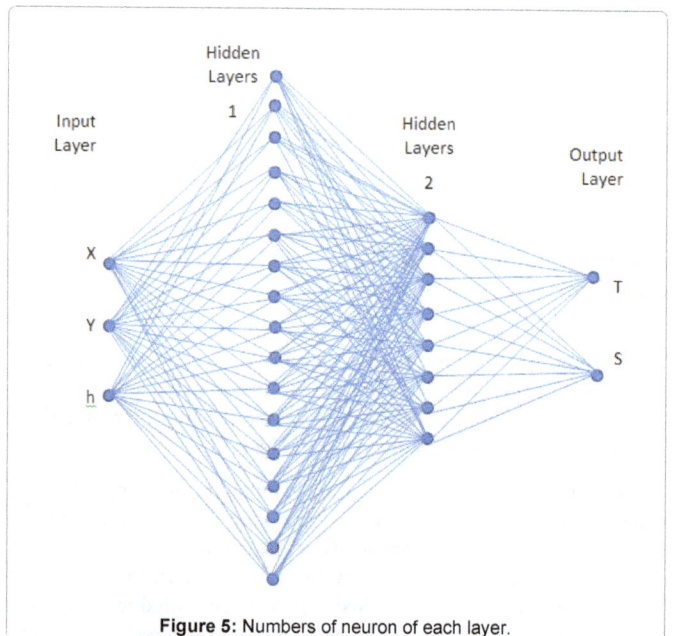

Figure 5: Numbers of neuron of each layer.

N	X	Y	h	T	S
1	219921.6	3378928	-10	1100	0.000155
2	219735.4	3378244	-11	731	0.0000554
3	221375.6	3373596	-18.25	955	0.0000152
4	224584.5	3374936	-12.5	1366	0.00288
5	227479.5	3375523	-8.5	613	0.00265
6	223747	3371761	-18	840	0.00135
7	228461.4	3370956	-16.25	827	0.00183
8	237002.8	3371180	-6.5	1435	0.00177
9	243511.9	3365666	-7.25	650	0.017
10	238908.6	3359950	-17.75	719	0.0075
11	248609.4	3360650	-8	1460	0.00181
12	250137.2	3359494	-6.75	870	0.0017
13	251607.1	3358735	-6.85	718	0.0012
14	252927.7	3357222	-7.25	595	0.0095
15	243511.9	3365666	-8.1	920	0.10466
16	238908.6	3359950	-13	1235	0.0166
17	248609.4	3360650	-7.5	630	0.0845
18	243887.9	3361092	-16	575	0.0124
19	247490.7	3356758	-16.5	535	0.0099
20	249098.6	3356194	-13.5	719	0.0075

Table 2: Data for training ANN.

The conditions to stop training processes were set before the network is trained. Training was controlled by some of conditions as: the maximum number of iterations, target performance which specifies the tolerance between the neural network prediction and actual output, the maximum run time and the minimum allowed gradient and. The overall training of the ANN will involve the following processes; the input values of the first layer are weighted and passed on to the hidden layer; the neurons in the hidden layer will produce outputs by applying an activation function to the sum of the weighted input values; the resulting outputs are then weighted by the connections between the hidden and output layer.

The network technique enable the user to make good training by shows the result with network calculated by submitted a figure shows the training fitting. Figure 6 shows good fitting which mean that the network gives a result data closed to the data that we had insert as output. The training coefficient R represent the fitting of observed invers calculated which is 0.98296, and this number is very closed to 1, so that mean we had good match.

Validation is the simple check of the model which used 20% of data to ensure that the training is able to give accepted result and good match between observed and calculated data model is controlled using the same synaptic weights in the validation part. For validation R=0.898 consider well because that, there is a good fitting between both observed and validate data (Figure 7).

Testing is an important process, which consider as the final check of the network to check the ability of this network to predict and gives accurate. Then we can decide that the model is good and show good match between observed and calculated data. Figure 8 is a simple chart shows the coefficient (R) and the result of the fitting between validate data and the calculated data (Table 3).

Performance: The performance of the ANN model can be quantified by statistical measure addressing the magnitude of the variable. The model can be validated in terms of root mean square error (RMSE), correlation coefficient (R) and scatter index (SI) as follows

$$RMSE = \sqrt{\frac{\sum_{i=1}^{N}(y_i - x_i)^2}{N}} \tag{3}$$

$$R = \frac{\sum_{i=1}^{N}(x_i - x^-) - (y_i - y^-)}{\sqrt{\sum_{i=1}^{N}(x_i - x^-)^2 \sum_{i=1}^{N}(y_i - y^-)^2}} \tag{4}$$

$$SI = \frac{RMSE}{x^-} \tag{5}$$

Where x_i is the observed values at the ith time step, y_i is the simulated values, N is the number of time increment and x^- and y^- are the mean values of observation and simulation, respectively.

The desired results are generated in the output layer. The network achieves the desired learning by adjusting its interconnected weights continuously until there is a close match between the outputs from the neurons and the output from the training data. The difference between the predicted outputs and the original outputs is referred to as error. Figure 9 shows the fitting and the coefficient for all training, validation, and testing.

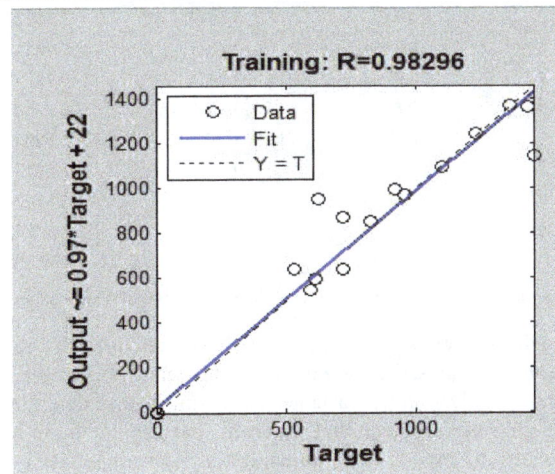

Figure 6: Training chart.

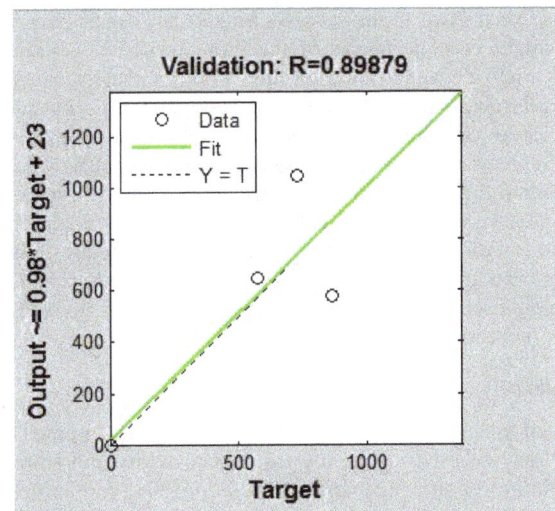

Figure 7: Validation chart.

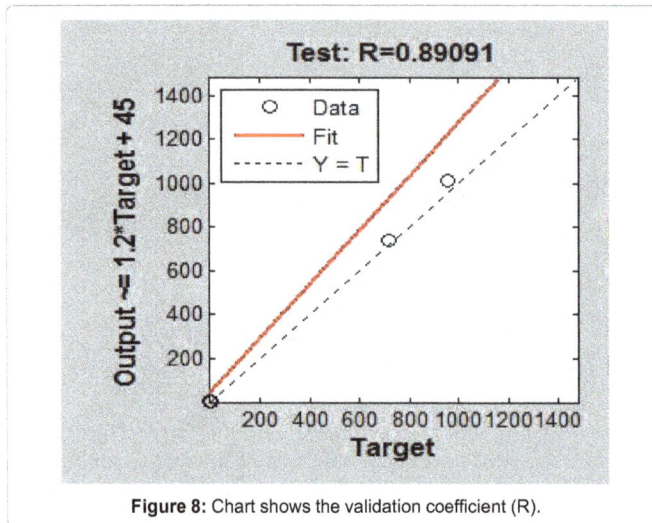

Figure 8: Chart shows the validation coefficient (R).

Models No.	No. of coefficient		
	Training	Validation	Testing
1	0.6361	0.64	0.5443
2	0.7135	0.5443	0.6577
3	0.7974	0.6577	0.7411
4	0.8773	0.7411	0.7974
5	0.8812	0.7562	0.7934
6	0.91545	0.837	0.8253
7	0.9154	0.8423	0.8517
8	0.9626	0.8674	0.8742
9	0.9795	0.8734	0.8734
10	0.9995	0.87824	0.89091

Table 3: Coefficient of training, validation and testing of ANN models.

Sensitivity analysis: Input variables are not, in general, independent- that is, there are interdependencies between variables. Sensitivity analysis rates variables according to the deterioration in modeling performance that occurs if that variable is no longer available to the model. In so doing, it assigns a single rating value to each variable. However, the interdependence between variables means that no scheme of single ratings per variable can ever reflect the subtlety of the true situation. Consider, for example, the case where two input variables encode the same information (they might even be copies of the same variable). A particular model might depend wholly on one, wholly on the other, or on some arbitrary combination of them. Then sensitivity analysis produces an arbitrary relative sensitivity to them. Moreover, if either is eliminated the model may compensate adequately because the other still provides the key information. It may therefore rate the variables as of low sensitivity, even though they might encode key information. Similarly, a variable that encodes relatively unimportant information, but is the only variable to do so, may have higher sensitivity than any number of variables that mutually encode more important information.

Discussion

From the previous section, it is shown that the proposed ANN model may be used to identify the transmissivity and storativity distributions of an aquifer in an inverse modeling framework. For real-world problems, there are some important issues that require further analysis. Note that the same input structure, network

architecture, and solution parameters are used in this section. The contours of predicted vs. actual transmissivities and storativities for noise data conditions can be seen in Figures 10 and 11, respectively. It is clearly seen from Figures 10 and 11 that the increase in the standard deviations may not significantly change the prediction performance of the proposed ANN model. This situation states that the proposed ANN model may have generalization ability. Therefore, it may be used in the real-world parameter estimation problems.

Results of Ann Model

After finish, the training by obtained good match between observed and predicted data with good validation and testing fitting then I saved the network in order to use it in prediction. The saving part of network mean saving of all weights that the network reached for obtaining good results. Then we use the same network to predict the distribution of both transmissivity and storativity. For that, we imported the new matrix of input, which include the distribution of head, which represented by the head (h) and the coordinates (x and y) which obtained from the MODFLOW model. By run the model to simulate the new data. The result appears directly in anew matrix include predicted transmissivity and storativity. Figures 12 and 13 shows the distribution of transmissivity and storativity respectively.

As shown before the actual values of transmissivity ranged between

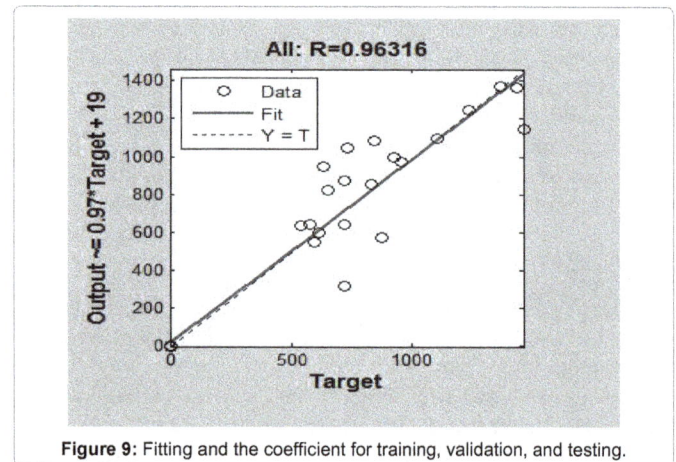

Figure 9: Fitting and the coefficient for training, validation, and testing.

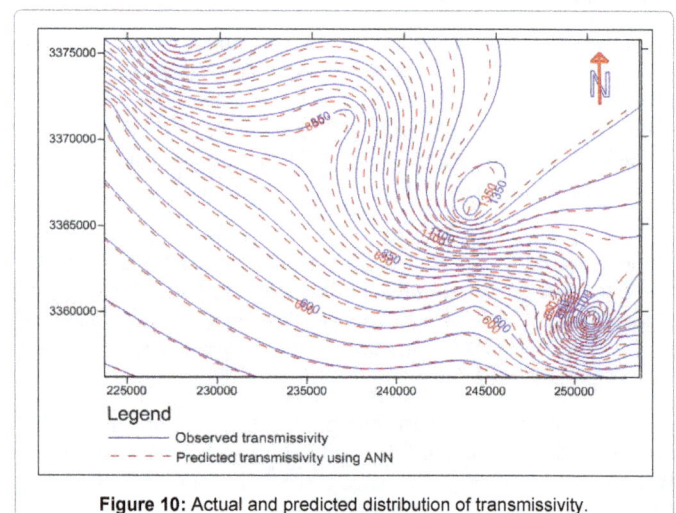

Figure 10: Actual and predicted distribution of transmissivity.

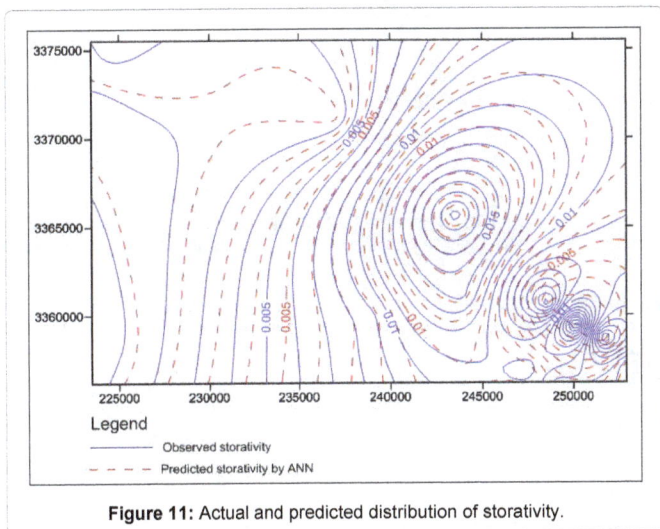

Figure 11: Actual and predicted distribution of storativity.

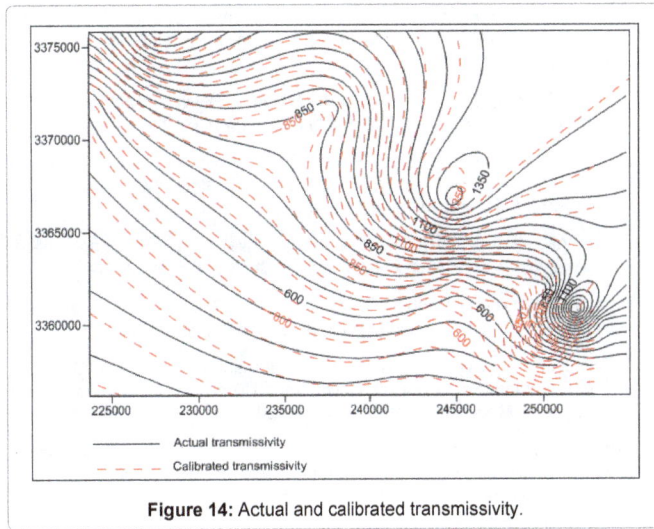

Figure 14: Actual and calibrated transmissivity.

Conclusion

An ANN solution technique is proposed to simultaneously identify the transmissivity and storativity distributions of an aquifer system. A hypothetical aquifer is modeled on MODFLOW for transient flow conditions, and its results are used as observed values to train and validate the developed ANN model. The proposed ANN model requires Cartesian coordinates, sampled piezometric heads, and sampling times as input and associated transmissivities and storativities as output. The number of hidden layer neurons is determined using try and error. The synaptic weights of the ANN model are defaults. To prevent potential over-training problems, the available data set is divided into three parts (training, validation, and testing) and overtraining in the model is controlled using the same synaptic weights in the validation part. After obtaining the synaptic weights, transmissivity and storativity distributions are predicted head values for different simulation times. Results showed that predicted transmissivity and storativity distributions are very close to actual values. In addition, the performance of the ANN model is tested for noise data conditions in the observed head values. For this purpose, the available observed piezometric heads are corrupted with Gaussian noise of zero mean and different standard deviations. Results showed that there is no significant change in the prediction performance of the proposed ANN model and this model may be applied to real aquifer parameter estimation problems.

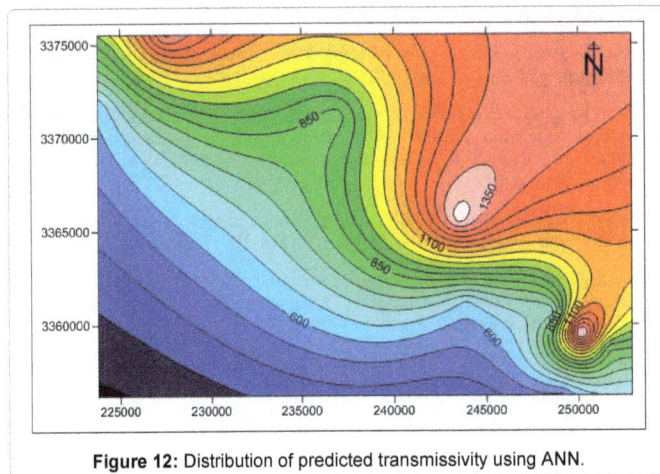

Figure 12: Distribution of predicted transmissivity using ANN.

References

1. Mehrotra P, Quaicoe JE, Venkatesan R (1997) Induction motor speed estimation using artificial neural networks. Canadian Conference on Electrical and Computer Engineering

2. Aziz ARA, Wong KV (1992) A neural network approach to the determination of aquifer parameters. Ground Water 30: 164-166.

3. Zio E (1997) Approaching the inverse problem of parameter estimation in groundwater models by means of artificial neural network. Prog Nuclear Energy 31: 303-315.

4. Balkhair KS (2002) Aquifer Parameters Determination for Large Diameter Wells Using Neural Network Approach. J Hydrology 265: 118-128.

5. Shigdi A, Gracia LA (2003) Parameter estimation in ground-water hydrology using artificial neural networks. J Comput Civil Eng 17: 281-289.

6. Karahan H, Ayvaz MT (2004) Forecasting Parameters Using Artificial Neural Networks. J Porous Media 5: 43-49.

7. Li S, Liu Y (2005) Parameter Identification procedure in Groundwater Hydrology with Artificial Neural Networks ICIC 2005, Part II, LNCS 3645, 276-285.

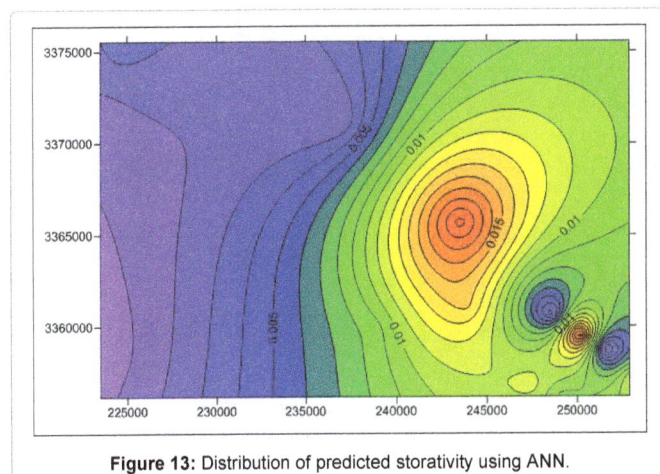

Figure 13: Distribution of predicted storativity using ANN.

500 and 1600 m²/day (Figure 10) and the actual values of storativity randed between 0.001 to 0.05. The values of predicted transmissivity are in the same range of the actual, also the distribution of the predicted transmisivity is closed to the actual one (Figure 14). Also, the values of predicted storativity are in the same range of the actual, also the distribution of the predicted storativity is closed to the actual one (Figure 11).

8. Morshed J, Kaluarachchi JJ (1998) Parameter estimation using artificial neural Network and genetic algorithm for free-product migration and recovery. Water Resour Res 34: 1101-1113.

9. Lin GF, Chen GR (2006) An improved Neural Network approach to the determination of aquifer parameters. J of Hydrology 316: 281-289.

10. Lohani AK, Krishan G (2015) Application of artificial neural network for groundwater level simulation in Amritsar and Gurdaspur districts of Punjab, India. J Earth Sci Climate Change 6: 274.

11. Lohani AK, Krishan G (2015) Groundwater level simulation using artificial neural network in southeast, Punjab, India. J Geology Geophy 4: 206.

12. Harrouni K, Ouazar D, Wrobel LC, Cheng AHD (1997) Groundwater parameter estimation by optimization and DRBEM. Eng Anal Boun Elem 19: 97-103.

13. Maier HR, Dandy GC (1998) The Effect of Internal Parameters and Geometry on the Performance of Back-Propagation Neural Networks: An Empirical Study. Environmental Modeling & Software 13: 193-209.

14. Coppola E, Szidarovszky F, Poulton M, Charles E (2003) Artificial neural network approach for predicting transient water levels in a multilayered groundwater system under variable state, pumping, and climate conditions. J Hydrol Eng 8: 348-360.

15. Daliakopoulos IN, Coulibaly P, Tsanis IK (2005) Groundwater level forecasting using artificial neural networks. J Hydrol 309: 229-240.

16. Lingireddy S (1998) Aquifer Parameter Estimation using genetic Algorithms and Neural networks. Civil Engrg Env Syst 15: 125-144.

17. Garcia LA, Shigidi A (2006) Using neural networks for parameter estimation in ground water. J Hydrol 318: 215-231.

18. Karahan H, Ayvaz MT (2006) Forecasting aquifer parameters using artificial neural networks. J Porous Media 9: 429-444.

19. Maier HR, Dandy GC (2000) Neural Networks for the Prediction and Forecasting of Water Resources Variables: A Review of Modeling Issues and Applications. Environmental Modeling & Software 15: 101-124.

20. Masoud AA, Atwia MG (2011) Spatio-temporal Characterization of the Pliocene Aquifer Conditions in Wadi El-Natrun Area, Egypt. Environmental Earth Sciences 62: 1361-1374.

21. Freeze RA, Cherry JA (1979) Groundwater: Englewood Cliffs, NJ, Prentice-Hall, p: 604.

22. ASCE Task Committee: on Application of Artificial Neural Networks in Hydrology I: Preliminary Concepts. J of Hydrologic Eng 5: 115-123.

23. Coppola EA, Rana AJ, Poulton MM, Szidarovszky F, Uhl VW (2005) A Neural Network Model for Predicting Aquifer Water Level Elevations. Ground Water 43: 231-241.

24. Nayak PC, Rao YRS, Sudheer KP (2006) Groundwater Level Forecasting in a shallow Aquifer Using Artificial Neural Network. Water Resour Manage 20: 77-90.

25. Karahan H, Ayvaz MT (2005) Groundwater parameter estimation by optimization and dual reciprocity finite differences method. J Porous Media 8: 211-223.

26. Karahan H, Ayvaz MT (2008) Simultaneous parameter identification of a heterogeneous aquifer system using artificial neural networks. Hydrogel J 16: 817-827.

27. Ahmed SA (1999) Hydrogeological and isotope assessment of groundwater in Wadi El-Natrun and Sadat City, Ain Shams University, Cairo, Egypt, p: 237.

28. Shata AA, El-Fayoumy IF (1970) Geomorphology, geology, and, the hydrogeology soil of Wadi El-Natrun- Maryut Agriculture Project Sympos. Hydrology of Nile Delta, UNISISCO 2: 385-396.

29. General Petroleum Company (GPC) (1977) Progress Report on geology and Stratigraphy of western Desert, Cairo, Egypt, p: 88.

30. Saad KF (1962) Groundwater hydrology, in preliminary report on the geology, hydrogeology, and groundwater hydrology of Wadi El-Natrun and adjacent area, Part 3, Cairo, U.A.R., Desert Institute, The General Desert Development Organization, p: 61.

31. RIGW/IWACO (1990) Hydrogeological inventory and development plan of groundwater Western Nile Delta region, Report, TN 77.01300-90-02.

32. Mostafa NE (1993) Hydrogeological and Hydro studies on Wadi El-Farigh Area, Western Desert, Egypt. M.Sc. Thesis, Faculty of Science, El-Minufiya University Shibin El-Kom, Egypt, p: 132.

33. El-Sheikh (2000) Hydrogeology of the area north and west of Wadi El-Natrun. M. Sc. Thesis Faculty of Science, Minufiya Univ, Egypt, p: 151.

13

Uncertainties in Techniques used to Determine Areas under Irrigation in the Upper Orange River Basin

Mahasa Pululu S[1]*, Palamuleni Lobina G[2] and Ruhiiga Tabukeli M[2]

[1]*Department of Geography, Faculty of Natural and Agricultural Sciences, Qwaqwa Campus, University of the Free State, Phuthaditjhaba, South Africa*
[2]*Department of Geography and Environmental Sciences, School of Environmental and Health Sciences, Mafikeng Campus, North West University, Mmabatho, South Africa*

Abstract

The paper addresses uncertainties that emanate as a result of methods used to determine irrigation areas in the Upper Orange River catchment area. The largest water user is the irrigation sector. What is not known for all schemes are the return flows but an average estimation of 13% is done for the main irrigation areas. Though several previous studies have addressed water conservation and demand management in the in the Orange-Senqu River catchment area; some pitfalls/caveats remain identified by these studies pertaining to the practical implementation of results. It was the necessary to look into several methods used since the results produced, in some instances differed so much.

An establishment of a standard methodology for the collection of data on irrigation water applied to crops, water use by crops and crop yields is a necessity. Establishment of an inventory GIS Database for irrigation inventory could prove useful if it could enhance the collation and collection of detailed and reliable data about irrigation water use by crops and crop yields. It could lead to documenting best management practices for irrigation in the catchment area. Another path could be to assess and consider various instruments that could be used for water conservation and demand management and further made improvements on water conservation and water demand management (WC/WDM) in the sector.

Keywords: Satellite imagery; Climate change; Crop water requirements; Geographic information systems (GIS); Remote sensing (RS)

Introduction

This chapter provides the characteristics of the study area and describes the methodology employed to determine areas under irrigation in the study area. It concludes by showing inherent uncertainties in each method used to determine the size and allocation of irrigation water for that particular area.

Characteristics of the Upper Orange-Senqu River Basin

Geographical overview: The Upper Orange WMA covers 103 671 km² and is part of the Orange River watercourse. Lesotho has been included in the study area and covers 30 492 km². The total area is 134 163 km² as shown in Figure 1. This area lies between Latitudes (28⁰ 0' 0" and 32⁰ 0' 0" S) and Longitudes (24⁰ 0' 0" and 30⁰ 0' 0" E). The Orange River, (called the Senqu River in Lesotho), originates in Lesotho *Maluti Mountains,* close to the Lesotho's highest peak, *Thabana Ntlenyana* at 3.482 m above sea level. If there were no developments of any nature in the river basin, the average natural run-off would be more than 12000 million m³/a, representing the average river flow that would be evidenced. It now happens that less than half of the natural run-off reaches the river mouth at Alexander Bay due to high levels of developments in the basin [1].

The Upper Orange WMA falls within the Orange River basin which is the largest river basin in South Africa. The Modder-Riet River catchment which is geographically part of the Lower Vaal catchment is included in the Upper Orange WMA because the water resources of the Modder-Riet are augmented by numerous transfers from the Orange River and its catchments. The rest of the Vaal River catchment comprises the most important tributary of the Orange River but is treated as separate WMA's. The WMA extends from the headwaters of the Caledon and Orange Rivers in the east to the confluence of the Orange and the Vaal Rivers, downstream of Vanderkloof Dam in the

west. The water resources and requirements of Lesotho are determined as they affect the water resources management of the Upper Orange WMA. The Upper Orange WMA is upstream of the Lower Orange WMA. The Lower Orange WMA delivers water to Namibia.

The Caledon River forms the north-western boundary of Lesotho with South Africa and is a major tributary of the Orange River. The Caledon River joins the Orange River a short distance upstream of Gariep Dam. Upstream of this confluence the Orange River is joined by the Kraai River at Aliwal North. The main tributaries joining the Orange River downstream of Gariep Dam are the Stormberg and Seacow Rivers. These two tributaries are small in terms of water resources.

Greater Bloemfontein in the Modder River catchment is the largest urban area in this WMA. Kimberley is situated on the northern WMA boundary with the Lower Vaal WMA and is considered to be part of the Lower Vaal WMA. Other significant urban areas are Thaba Nchu, Botshabelo, Wepener, Dewetsdorp, Reddersburg, Edenburg, Jagersfontein, Trompsburg, Brandfort, Dealesville, Petrusburg, Jacobsdal, Koffiefontein, Oppermans, Fauresmith, Trompsburg, Springfontein, Bethulie, Smithfield, Rouxville, Zastron, Vanstadensrus, Barkly East, Ficksburg, Colesberg, Aliwal North and Phillipolis.

**Corresponding author: Pululu Sexton Mahasa, Department of Geography, Faculty of Natural and Agricultural Sciences, Qwaqwa Campus, University of the Free State, Private Bag X13, Phuthaditjhaba 9866, South Africa
E-mail: mahasapululusexton@gmail.com*

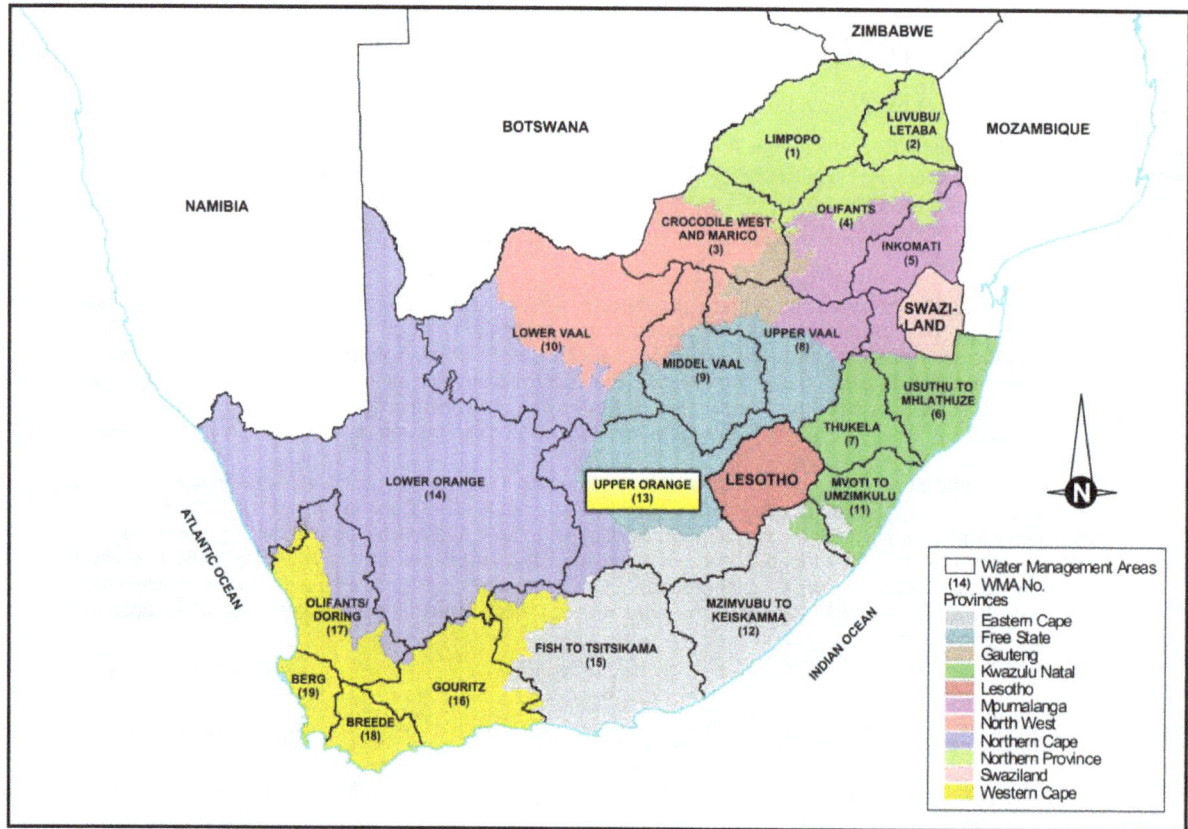

Figure 1: Location of upper orange river.

In Lesotho, the only significant urban area is the capital city, Maseru. Other significant urban areas in the area include Leribe, Mohale'shoek and Mafeteng.

Topography: The Orange River length is reported in the literature to be between 1900 to 2300 km. However, the river length was calculated to be 2415 km from a detailed measurement in Google Earth and Garmin MapSource. This would place the Orange River as the 44th longest river in the world. The mountainous topography in Lesotho results in a sharp river slope with an average of about 3.9 m/km for the first 400 km, but from the South African boarder to Upington bridge the flat topography results in a moderate low slope of about 0.55 m/km, followed by the lower reach (Upington to river mouth) where the river drops on average with 1.04 m/km, including the Augrabies falls [2].

Geology: The geology of the study area is of volcanic origin in the upper reaches in the Lesotho Highlands characterised by young rock types of two series of the Karoo system. Extrusive igneous rocks of the Stormberg series (part of Karoo system) cover the eastern high-lying areas. The north-western part of this WMA is described as compact, dominantly argillaceous strata with small pockets of compact tillite and compact sedimentary and extrusive rocks near the WMA boundary. Compact arenaceous and argillaceous strata (fine sedimentary rocks of the Karoo system) underlie the remainder of the Upper Orange WMA.

The upper layer namely the Lesotho formation comprises of 1,500 m thick basalt lavas underlain by Clarens sandstone formation, Molteno beds and the upper Beaufort beds. Gradients are steep. The middle part of the area is dominated by the consolidated sedimentary rocks of the Karoo succession. The lower part of the area is under Kalahari sand

cover [2]. Moving westwards, the Orange River exposes some of the oldest known rocks as it traverses many geological units in its valley (Orange-Senqu River Commission [3]. Of these geological formations, only the Kalahari sands are water-bearing in primary openings. Groundwater is available mainly in larger dissolution openings and fractures[4].

It is further mentioned that hydrogeological information for the South African part of the Orange River Basin can be obtained from Vegter.

Climate: Considerably the climatic conditions vary from east to west across the Upper Orange WMA. Excluding Lesotho, the mean annual temperature ranges between 18°C in the west to 12°C in the east and averages about 15°C for this area as a whole. In Lesotho, the mean annual temperature ranges from 14°C to below 8°C in the more mountainous parts towards the east with an average of about 11°C for the country as a whole. Maximum temperatures are experienced in January and minimum temperatures usually occur in July. Rainfall is almost all seasonal and most rain occurs in the summer period (October to April). December to March are the peak rainfall months. Rainfall occurs generally as convective thunderstorms and is sometimes accompanied by hail. The mean annual rainfall decreases fairly uniformly westwards over the Upper Orange WMA from the eastern escarpment regions across the central plateau area. Rainfall amounts are highest in Lesotho at approximately 2000 mm per year and decrease to about 200 mm in the west of the study area with a high degree of variability as well (Figure 2). The mean annual rainfall for the area is about 400 mm per year. This makes about 50% of the area referred to as hyper-arid to semi-arid, with an increase in aridity westwards. Due

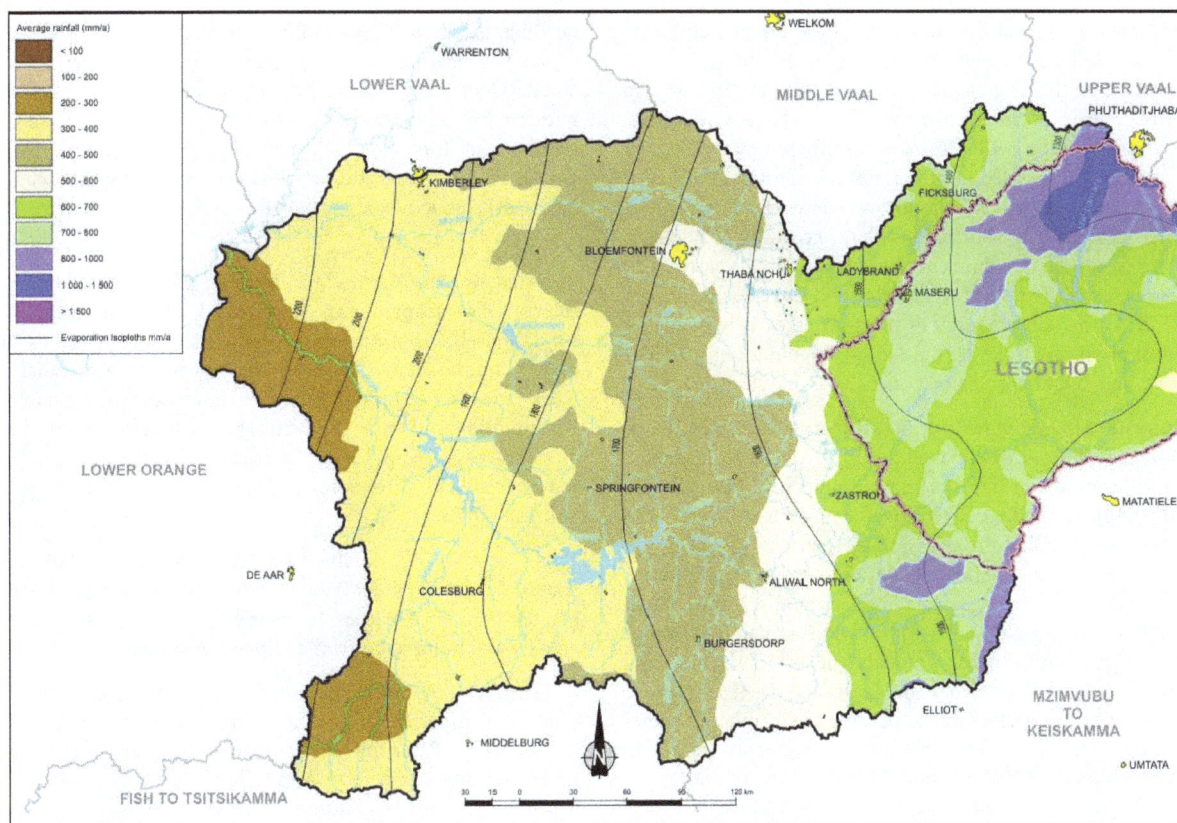

Figure 2: Distribution of annual rainfall of Upper Orange River.

to climatic variations, recorded extremes in runoff have been between 26000 million m³/a and as low as 1100 m³/a [2].

Equally variable is potential evaporation. In Lesotho, a low value of 1200 mm per year has been recorded to a high value of 3 500 mm per year at the river mouth. The calculated evaporation losses from the Orange River ranged from 575 million m³/a at an annual low flow release rate of 50 m³/s to 989 million m³/a at an annual release rate of 400 m³/s [2].

Vegetation: At the highest altitudes in Lesotho, there is Alpine vegetation that comprises of climax heather communities of the pure grassveld veld type (which requires moderate to high rainfall) composed mainly of low woody species interspersed with alpine grasses. High-lying areas consist of grassland habitat while westwards to the False Upper Karoo in the remaining lower altitude; the area has the mixed sour Grassveld. The middle and lower Orange River basin are characterised by a series of karooid vegetation types [3]. According to Mucina and Rutherford [4], moving north to the Modder-Riet catchment, there is still some pure grassveld in the east but most of this catchment is false karoo and false bushveld. In the south and south-western parts of this WMA one finds mainly false karoo together with karoo and karroid and false bushveld. There is also a bit of false karoo along its western edge where it is drier.

Soils: In Lesotho, the main Mountain Black Clays soils are dominant. At high altitudes, these are very shallow and erode easily under marginal overgrazing and cultivation practices. On the summit during summer, soils are often waterlogged and they usually freeze in winter, increasing their susceptibility to erosion [3]. Soil depths are generally moderate to deep over the upper Orange catchment. There are six main soil/texture/relief types that predominate and the section that follows indicates their distribution across the catchment:

- Sandy Loam: In the upper Caledon valley and to the south of the Orange in the western part of the catchment of moderate to deep depth and undulating relief.

- Clay Soil: Confined to the mountainous areas in the eastern portion of the Lesotho of moderate to deep depth and steep relief.

- Sandy Soil: Confined to areas around Bloemfontein and Petrusburg of moderate to deep relief and flat relief.

- Clay Loam (flat relief): Confined to the north-western part of the catchment of moderate to deep depth.

- Clay Loam (Steep relief): Confined to the south-eastern part of the WMA of moderate to deep depth.

- Clay Loam (undulating relief): The predominant soil type in the remainder of the catchment of moderate to deep depth.

Sands or weakly developed soils cover most of the remainder of the Orange River catchment area. With the exception of mainly the Kalahari component, in terms of soil erosion most of the basin is considered to be of medium to high risk [3].

Water Resources: The surface water resources, which naturally occur in the WMA (together with inflows from Lesotho), are already well developed, and with a high degree of utilization. Estimated natural water resources of the Orange River basin are about 12000 million m³/annum (Mm³/a), although currently less than 50% of the available water is abstracted by several developments in the Orange and Vaal catchment areas. Approximately 4000 Mm³/a of the natural runoff originates in the Lesotho Highlands, and approximately 800 Mm³/a originates from another basin downstream of the Orange-Vaal Rivers' confluence. The remaining 6500 Mm³/a is contributed by other areas in the basin to the Vaal, Caledon, Kraai and Middle Orange Rivers. The Vaal River is a major and very important tributary of the Orange River that provides Gauteng with all its water. According to Figure 3, extensive water resource developments have taken place upstream of this confluence, including several large dams and inter-basin transfer schemes [2]. A further classification of the study area is with regard to hydrological zones and these are the Senqu, Caledon, Riet/ Modder and Upper Orange River hydrological zones [5].

Development of the Orange River: In Southern Africa, the Orange River catchment is the most developed of all the rivers, with at least more than thirty-one major dams having a storage capacity of more than 12×10^6 m³ [6]. The Orange River (together with its main tributary the Vaal River) is controlled through storage reservoirs in the upper WMA and in Lesotho, with limited regulation capacity in the Lower Orange WMA. The main storage dams are Gariep and Vanderkloof. The construction of the Gariep and Vanderkloof Dams in the Orange River made a great contribution towards the establishment and maintenance of irrigated crops throughout large sections of the Orange River, however, with a negative impact on the environment. Large-scale infrastructural development (dams, etc.) and water abstraction in the catchment result in only half of the 11500 million m³ annual runoff reaching the Orange River estuary further down of the study area in

the west. Until today most of the Orange's water is used for irrigation farming, i.e. about 2160 Mm³/a to irrigate approximately 180000 ha. Water is also used for the generation of hydropower at Gariep and Vanderkloof dams. As the power is essentially generated with water released for other purposes, this is not regarded as an additional requirement for water [2]. However, the unnatural regulation of flows in the river has numerous effects on the physical, chemical and biological characteristics of the Orange River.

Along the Senqu River in Lesotho, mixed farming charecterised by extensive sheep and cattle farming and wheat ploughing in the river's valley is practiced. About 70% of home gardens of these rural households produce rain-fed vegetables and household and/ or community domestic water is supplied for occasional irrigation when rains are erratic. Almost all home-grown vegetables are for consumption and small quantities are sold at village markets. In the Upper Orange-Senqu basin, the Lesotho Highlands Water Project and the associated huge investment it has impact significantly on Lesotho's political economy [7].

In South Africa in the Upper Orange River WMA the level of economic development along with population numbers are much greater than in Lesotho. Livestock farming is the main economic activity and rain-fed cultivation covers extensive areas [7].

Throughout the catchment a wide variety of crops are grown under irrigation because of the extreme range of climatic conditions of the Orange River that vary from cool temperate and alpine regions through progressively more arid terrain and ultimately through hyper-arid desert. Mostly crop production is rain-fed and is interspersed with

Figure 3: Orange River catchment base map and main hydrological zones.

cropping under irrigation in the more temperate north-eastern sections of the basin. Moving westwards, irrigation supports crop production to the point where rainfall is very unreliable or very low and crops are produced only under irrigation. Predominantly in the more temperate north-western sectors there is mixed cropping with field crops and fodder crops. The main field crops are maize, wheat, dry-bean, potato, soybean, groundnut and cotton while the main fodder crops are lucerne, pastures and maize-silage. Limited areas of orchard crops such as peaches, apples and cherries are grown in the high altitude areas with adequate winter chill. Permanent orchard and vine crops like wine grapes, table grapes, raisin grapes, citrus and dates predominate in the dryer western areas also common to this area is lucerne [6].

Formerly in the Upper Orange, mining activities were a dominant sector but have declined in recent years such that small diamond operations and salt works now remain [7].

Methodology

The aim of this section is to discuss the methodology used in this study to challenge uncertainties in techniques used to determine areas under irrigation in the Upper Orange River Basin. The method made use of Remote Sensing (RS), namely LandSat imagery (Landsat 7-ETM+):

Imagery:

- Spectral Bands: Landsat TM bands (i.e., all seven),

- Datum/ Projection: Self organizing Map (SOM) / World Geodetic System 1984 (WGS84),

- Coverage Date: Scene dependent (nominally 2011 +/- 3 years),

- Coverage: Single Landsat WRS Path/Row,

- Pixel Size: Combination of 30 and 28.5 metres,

- Orientation: Path oriented,

- Interpolation Technique: Cubic Convolution.

Allocation schedules from Department of Water Affairs (DWA) and Orange-Senqu River Commission (ORASECOM), questionnaires and interviews and finally where possible after permission was granted field observations were also used. In order to retrieve key policy information on both external and internal issues from the two countries, four questionnaires were allocated. Each distributor was allocated one questionnaire in this tier level. In total seven water users in the study area and immediate surrounds were considered.

This research used questionnaires, interviews with specialists and key role-players and observations where possible. The questionnaire dealt with general demographic data of respondents intended to ease the mood between the researcher and the respondent as it has no bearing on the study. It further addressed policy, water allocation, financial, technical, human and material resource capacity information of the selected water institutions. In total, there were sixty questions in the questionnaire. Most questions were close-ended and respondents were guided by given options already set out. Open-ended questions allowed respondents freedom to express their views. The water users chosen for interviews were chosen on the basis of a particular status. The inclusion criteria were based on data availability, role in the local economy, etc. Choosing interviewees was on the basis of their working area and expertise in the areas of water management in their respective institutions. Only one interview was chosen per institution. It was fore-thought that the key respondents could confirm or correlate existing information in allocation schedules.

Sampling procedure

The sampling network and strategy was designed to cover wide range of determinant factors (i.e., water allocation needs, number of licenses issued, etc.) at the key locations, which reasonably represented the whole study area. It involved retrieving information from the national, provincial, municipal and various users' inventories in the area.

Data collection

Primary data was collected through field surveys. Allocation schedules indicating water licensing from DWA and ORASECOM were used. The use also included the 2009 -10 field crop boundaries by the Department of Agriculture, Forestry and Fisheries (DAFF) and Water Authorization and Registration Management System (WARMS) database for water use registration conducted in January 2010.

Results from other studies

Of all the Satellite Imagery available, ranging from USGS Landsat imagery (through the GLOVIS viewer), SPOT Image data, Google Earth imagery to ESAD MrSID, only Landsat 7- ETM+ for 2013 up to 2014 was considered for use. Appearances and bands used were: Spectral Bands: 3- Landsat TM bands (i.e., Band 2 (visible green light) is indicated as blue, Band 4 (near-infrared light) is indicated as green and Band 7 (mid-infrared light) is indicated as red). The main purpose was to determine irrigated areas according the current version of the field crop boundary mapping by DAFF [6]. It is further mentioned that much of the USGS archive for this dataset was downloaded to achieve this objective. However, downloading every scene available for processing was impracticable with the resources available.

Selection of Satellite Imagery

Landsat images obtained though they were freely available but the files were very large (between 200 and 300 Mb). Since the images were many then downloading this volume of data (i.e., 550 Gigabytes of raw and processed imagery) was eventually problematic. Other data management requirements included compressing and uncompressing from one format to the other. For example, each image had to be compressed (in GZ format) followed by uncompressing to produce a single .TAR file. A second stage of uncompressing of the .TAR file produced 9 .TIF files. Each TIF file produced a separate band width that consisted of the image. Each file had the image location_date_band number format. In this analysis the six bands used were band 7 (B70), band 5 (B50), band 4 (B40), band 3 (B30), band 2 (B20) and band 1 (_B10). The other bands were inappropriate to be used for the classification of vigorously growing vegetation [6].

The decision was to use LandSat 7 with 'slcoff' to determine irrigated areas. The part of the sensor system controlling the satellite movement on the scanning process was not functional for some time during 2003, and that resulted with images that had strips up to 30% image information missing. For most tiles, a time series of scenes was possible to obtain and missing values for selected were filled because the exercise was furnished with the field crop boundaries by DAFF from 2006 to 2008. In that project, approximately a 30% overlap existed. For each satellite image (or scene), the coverage was called a 'TILE'.

At the same "tile" location for 2009 to 2011, satellite imagery was made available at different times providing up to 20 scenes for each individual tile were made available, though not all of them could be utilised. That made it possible irrigated areas to be classified for the different times of the year. The exercise involved utilising at least 4 'scenes' at each "tile" location so that identification of irrigated crops growing during the different seasons could be done. This was, however, impossible due to unavailability of specific images from the USGS (images covered in cloud and images containing missing data) [6].

Image Processing

ORASECOM [6] mentioned that in order to process satellite imagery, GIS software and the IDRISI image processing was used. Only two paths (Paths 170 to 172) will be discussed and considered in the study area. These represented areas that practised irrigated rainfed and mixed agriculture, and where crop production was supported by irrigation was done due to low annual rainfall (i.e., 400 mm or less).

After the evaluation of several classification methods, only two of the clustering algorithms, namely the Kmeans and Cluster methods could be used. Clustering is ubiquitous in science and engineering, with diverse and numerous application domains, ranging from medicine and bioinformatics to the social sciences. It is mentioned that a "satisfactory" classification is provided by the Kmeans, but is very slow in comparison to the Cluster technique which is very fast. It should also be mentioned that the main aim was to basically determine only two classes- under irrigation or not [6].

Kmeans

Kmeans is one of the simplest unsupervised learning algorithms that solve the well-known clustering problem. The classification a given data set through a certain number of clusters (assume k clusters) fixed a priori is achieved in a simple and easy way. The determination of k centroids, one for each cluster is central to this procedure. These centroids are cleverly placed because of differences in location that lead to differences in the result. So, the best option is that placement should be as far away from each other as possible. For any adjacent points, a relational association for a given data set and the nearest centroid is also placed under consideration. When no point is pending, the first step is completed and an early groupage is done. At this point the re-calculation of k new centroids as barycentres of the clusters that resulted from the previous step is a necessity. Having acquired these k new centroids, a new binding should be determined between the same data set points and the nearest new centroid, thus producing a loop. As a result of this loop observations may indicate a change in location of the k centroids step-wise until no further changes are achieved. In other words centroids are now fixed.

Finally, this algorithm minimises an objective function (i.e., a squared error function). The objective function:

$$W(C) = \frac{1}{2} \sum_{k=1}^{K} \sum_{C(i)-k} \sum_{C(j)-k} \left\| x_i - x_j \right\|^2 = \sum_{k=1}^{K} N_k \sum_{C(i)-k} \left\| x_i - m_k \right\|^2$$

where

m_k is the mean vector of the kth cluster

N_k is the number of observations in k^{th} cluster

The algorithm comprises of the following steps:

1.   Place K points into the space represented by the objects that are being clustered. These points represent initial group centroids.

2.   Assign each object to the group that has the closest centroid.

3.   When all objects have been assigned, recalculate the positions of the K centroids.

4.   Repeat Steps 2 and 3 until the centroids no longer move. This produces a separation of the objects into groups from which the metric to be minimized can be calculated.

The Kmeans Method is classified as either a "fine" or a "broad" type. Using the classification that is "fine" created a larger number of classes (about 40) whereas the "broad" classification provided classes ranging from 10 to 16. Smaller number of classes gave way to problems, due to combinations between larger numbers of pixels with similar signatures in the software, such that both non-irrigated and irrigated areas of healthy growing vegetation were perceived as one class. Based on this, the cluster technique was considered for use also.

Cluster

Clustering can be regarded as the most important unsupervised learning problem; so, as every other problem of this kind, it deals with finding a structure in a collection of unlabelled data. Clustering could be defined as the organisational process of grouping objects into categories whose members have similarity in some way. A cluster is therefore a collection of objects which are "similar" between them and are "dissimilar" to the objects belonging to other clusters.

There are two types of clustering: conceptual clustering and distance-based clustering.

The Goals of Clustering

The determination of the intrinsic grouping in an unlabelled data-set is the central goal of clustering. But what constitutes a good clustering is very difficult to decide? Actually there is no absolute "best" criterion which would be independent of the final aim of the clustering. Consequently, it is the user who should define and decide on this criterion, such that the result of the clustering will suit individual applications and needs.

For instance, for homogeneous categories (data reduction) there could be an interest in obtaining members so as to find "natural clusters" and describe their unknown properties ("natural" data types), in obtaining useful and suitable categories ("useful" data classes) or in obtaining unusual data objects (outlier detection).

Requirements

A clustering algorithm should:

• account for scalability;

• discover clusters with arbitrary shape;

• deal with different types of attributes;

• be able to deal with outliers and related noise;

• have minimal required conditions for domain knowledge in obtaining input parameters;

• exhibit insensitivity to the order of input records;

• show usability and interpretability.

• reflect high dimensionality;

Problems

There are numerous problems associated with clustering. Among them: current clustering fail to satisfy all the required conditions sufficiently (and concurrently); handling large number of data sets and large number of dimensions is often problematic due to time complexity; the effectiveness of the technique is governed by the definition of "distance" (for distance-based clustering); if a distance measure is non-existent it has to be defined and that is not easy, especially in multi-dimensional spaces; the result of the clustering algorithm (that in many cases can be arbitrary itself) can be interpreted in different ways.

Classification of Clustering Algorithms

Clustering algorithms may be categorised as below:

- Exclusive Clustering
- Hierarchical Clustering
- Overlapping Clustering
- Probabilistic Clustering

In the first case data are categorised exclusively, so that if a certain datum was a member of a definite cluster then it belonged there only. On the contrary the second type, the overlapping clustering, utilises fuzzy sets to cluster data, so that each point may be a member of two or more clusters with different degrees of membership. In this case, data will be associated to an appropriate membership value. Instead, a hierarchical clustering algorithm is on the basis of the union of the two nearest clusters. The initial condition is every datum is set as a cluster. The final clusters wanted may be reached after a few iterations. Finally, a completely probabilistic approach is used in the last kind of clustering.

For the study area, the cluster technique created around 60 to 100 classes and clearly separated pixels from irrigated crops and pixels that represented vigorously growing natural vegetation or rain-fed crops. This technique therefore enabled the user to select classes to a certain degree of accuracy that defined irrigated crops only (ORASECOM)[6]. While most practitioners instead continue to use a variety of heuristics that have no known performance guarantees.

Selection of the Best Method

ORASECOM [6] Mentioned that the three techniques (clusters fine and broad, kmeans) in comparison to the Standard False Colour Composite (FCC) may first select the best method. Secondly, identification of the classes showing irrigated areas would be done. This categorised the classified images with the FCC image so as to enable the user to "zoom in" on the same field on all the images and do the comparison. In IDRISI when using the MAP COMPOSER facility to overlay the field crop boundary vector layer, this exercise could be achieved easily.

Identification and selection of the correct class

By zooming in to different locations across the image the user was able to select the classes that best represent irrigation. As mentioned in the previous section, some classes from the clusters classification were chosen to best represent irrigated crops in that tile [6].

These selected classes were then retained, and all other classes ignored, to create a new image that illustrated vigorously growing/ irrigated crops only. Classes would now be shown in the same colour: red. The FCC is included for comparison purposes. This new image, showing vigorously growing/irrigated crops, would be placed in all the tile folders, where image under process was undertaken and represented the last step using the IDRISI software. It should be recalled that a large

number of files were placed in some of the tile folders. This was due to methodology investigated before the current technique was adopted [6].

Finding the most effective classification method was preceded by a great deal of experimentation. These included a method initially explored where the image was firstly divided into four quadrants, then masked with the field crop boundaries, before interpolating to fill in missing data.

ArcGIS

The rasterised file indicating pixels categorised as irrigated/ vigorously growing crops was imported to the ArcMAP module in ArcGIS. An attribute table was made where all classified irrigated areas was labelled "1" and all non-irrigated areas was assigned a "0" and overplayed with the field crop boundary in a process that determined the area of irrigated/vigorously growing crops in each field crop boundary. ORASECOM [6] indicated that this process involved the following steps:

- Convert the field crop boundary from a vector shapefile to a rasterised file. This speeds up the combining of the field crop boundaries and classified rasterised images.

- The Tabulate Area function in the Zonal option under Spatial Analyst Tools in ArcToolbox was used to create an attribute table that contains the area of vigorously growing/irrigated crops in each field crop boundary.

- This attribute table was exported as a DBASE ('.dbf') table.

Classification Process

The classified image clearly showed how the classification process often resulted in varying proportions of each field being classified as vigorously growing. The centre pivot would show approximately 70% classified as vigorously growing, while the centre pivot top centre indicated slightly less than 10% vigorously growing. The '.dbf' file that defines each field with a unique ID also lists the actual area of vigorously growing vegetation (value_1) and non-vigorously growing vegetation (value_0) in each field.

Strict classification versus less strict classification

It is mentioned in ORASECOM [6] that during the class selection process, classification was based on a variable percentage of cover to say if a particular field was irrigated or not. To the east of the Vaal basin, the crops were often grown during the rainy summer months, so only complementary irrigation and sometimes no irrigation was required. The exercise to separate the rain fed crops from those that are irrigated had to use a very strict classification. A low percentage of 30% was used if the classification of a particular image was strict (i.e., only the most vigorously growing vegetation identified). This meant that a field was identified as being under irrigation if 30% or more of the field was classified as irrigated. If the classification was less strict, then a higher cut off percentage of (say 70%) was utilised.

Selection of images and variability within images

Summer images which were inclusive of immense areas of rain-fed crops were rejected. An iterative process was in under consideration on the effect of including/excluding summer images, where the user was able to rapidly assess the results and necessitate a database type analysis. In addition, it was necessary to identify and anticipate that fields near rivers and/or within formal irrigation boards are more likely to be irrigated. Across a single image, different criteria were therefore applied. Again, based on the best achievement it was better to use a buffer system embedded in an ACCESS database that undertook the classification of the fields within these buffers more reliably than those outside the buffer [6].

Interactive classification tool

According to ORASECOM [6], it became apparent that in order for above-mentioned requirements to be achieved, a tool was necessary to enable all that could be done. An Access database with an interactive tool, which allowed a user to determine the effects of remote sensing classification thresholds to distinguish between vigorously growing vegetation and irrigated crops, and location-based screening rules which acknowledge:

- Run-of-river abstractions along major tributaries; and known groundwater abstraction areas;
- Irrigation schemes.

It is further mentioned that the tool could simplify future updating of estimates of irrigation areas as it encapsulates the methodology described above, validates other data sources (such as the WARMS), and to make use of local knowledge about sources of irrigation water.

The steps used in this tool as mentioned in ORASECOM [6] are as follows:

- For each image, importing the '.dbf' file;

- Defining the cut-off percentage. The database calculated the percentage coverage of vigorously growing vegetation for each field, in comparison to the cut-off percentage of that image scene to determine if a field was irrigated or not;

- Considering as "irrigated", all fields under centre pivots as per DAFF field crop boundary data;

- Incorporating the buffer areas that had different decision criteria placed on them. Retaining all fields including those under centre pivots as irrigated when determined by remote sensing methods and defining all other fields not within the buffers as "not irrigated"; and

- Comparing the total area under irrigation per irrigation zone with other sources of information such as the WARMS database and the reports on the Orange River (the *Orange River Development Project, Evaluation of Irrigation water use*) and Vaal River (the *Vaal River Basin Study, Evaluation of Irrigation*).

It was of interest to note that the area of centre pivots differed drastically from the data sources (in Figure 4 for example 4 V6, 45.665 km² is irrigated by centre pivots compared to 12.513 km² of irrigated area in WARMS).

Challenges and Recommendations

➢ Regulation of the water of the Orange-Senqu system is done

ZONEID	COUNTRY	OSZONE	Description	Scheduled Area	WARMS Area	Centre Pivot Ar	OV Study An	RS Area	Screened Area
11	Lesotho	O1	Caledon U/S Welbedacht Dam				149	106	174
12	RSA	O15	Orange-Vaal confluence to Boegoeberg Dam	6 853	16 067	12 592	6 852	11 422	15 576
13	RSA	O16	Boegoeberg Dam to Upington	8 623	1 159	669	8 578	9 807	9 807
14	RSA	O17	Upington to Neusberg	13 163	13 723		13 163	11 536	11 536
15	RSA	O18	Neusberg to Namibia border	9 731	651		9 731	9 217	9 217
16	RSA	O19	Namibia border to Onseepkans	1 045	1 535		1 045	913	913
40	Namibia	O19	Namibia border to Onseepkans					273	273
36	RSA	O2	U/S Gariep, D/S Welbedacht Dam	4 775	3 280	2 239	3 482	1 359	2 979
18	RSA	O20	Onseepkans to Vioolsdrift	835	1 938		351	2 481	1 887
45	Namibia	O20	Onseepkans to Vioolsdrift					317	317
42	Namibia	O21	Vioolsdrift to Orange-Fish confluence	442			210	2 240	2 240
43	Namibia	O21	Vioolsdrift to Orange-Fish confluence	442	316		210	324	324
41	Namibia	O22	Orange-Fish confluence to river mouth					181	181
19	RSA	O22	Orange-Fish confluence to river mouth	761	531	198	750	198	198
20	RSA	O3	U/S Aliwal North, D/S Oranjedraal	1 575	553	186	1 550	134	263
17	RSA	O4	U/S Gariep, D/S Aliwal North	2 560	2 970	911	1 940	854	1 465
21	RSA	O5	Kraai U/S Aliwal North	-999	970	450	-999	498	699
22	RSA	O6	U/S Van der Kloof, D/S Gariep	2 316	2 418	1 997	2 316	1 349	2 345
23	RSA	O7	Canals ex Van der Kloof Dam	17 378	322	4 420	17 378	3 454	4 474
24	RSA	O9	Van der Kloof Dam to Douglas	14 174	19 372	22 004	14 173	17 616	23 011
57	RSA	OLU	Lower Orange D/S Onseepkans		48			2 345	293
44	Namibia	OLU	Lower Orange D/S Onseepkans					392	6
46	RSA	OML	Middle Orange - D/S Boegoeberg		16 641	9		218	218
54	RSA	OMU	Middle Orange - U/S Boegoeberg		1 379			239	239
52	RSA	OUL	Orange tributaries - Van der Kloof to Douglas		311	889		891	889
999	Outside	OUTOFBASIN	Out of Basin					32	0
53	RSA	OUU	Orange tributaries U/s Van der Kloof		11 778	1 080		14 484	3 966
6	RSA	SER	Senqu (RSA)		509	24		993	52
25	RSA	SH	Sak River		16 451			16 071	11 756
26	RSA	V1	U/S Grootdraai Dam to Vaal Dam	7 279	20 674	14 899	7 279	7 667	15 202
27	RSA	V10	Vaal and Riet U/S of confluence with Orange	20 869	10 755	25 491	20 896	7 270	29 595
28	RSA	V2	Wilge, Liebenbergsvlei Rivers	5 517	24 006	15 262	5 517	9 229	16 232
29	RSA	V3	Vaal Dam to Barrage	9 008	14 150	8 640	9 008	8 679	15 671
30	RSA	V4	Barrage to Bloemhof Dam	16 586	43 120	28 585	15 646	21 818	33 981
31	RSA	V5	Sand and Vet Rivers	12 196	15 463	21 301	12 196	19 289	23 199
32	RSA	V6	Bloemhof Dam to Schmidtsdrift	50 277	12 513	45 665	47 392	17 540	69 513
33	RSA	V7	Harts River	1 783	16 526	18 120	1 763	6 073	20 406
34	RSA	V8	Riet River	4 234	6 793	4 714	8 391	1 737	5 303
35	RSA	V9	Modder River	3 564	22 824	22 527	6 371	9 770	27 470

Record: 41 of 50

Figure 4: Comparisons of Irrigated Areas per Irrigation Zone.

by more than thirty-one major dams and is also a highly complex and integrated water resource system with numerous large inter and intra-basin transfers [6].

> This makes the basin to be notably much interconnected" [8].

> An incomplete inventory of Water Users in the Upper Orange River Basin arises as result of the vast nature of the basin.

> Most irrigated land is privately owned and as such access is highly problematic.

> The sensitive nature of the research makes accessibility even restricted because results could largely influence the allocation schedules of irrigation water.

> Until so far there is no established standard methodology for the collection of data on irrigation water applied to crops, water use by crops and crop yields.

> There is also a diverse array of instruments used to support water conservation/water demand management.

> Divulging correct ground information could be rewarded by incentives.

Figure 5 summarises findings of different studies undertaken to determine irrigation demand for the study area. All these studies used different methods and came up with different results in many cases. The studies were conducted by WRP Consulting Engineers (WRP), Development of Reconciliation Strategies for Large Bulk Water Supply Systems: Orange River-Irrigation Demands and Water Conservation / Water Demand Management (Task 8) (ORECON), Orange-Senqu River Commission (ORASECOM) and Orange River Re-planning Study (ORRS). It is worth-noting how results from different vary in this region. The Government of the Kingdom of Lesotho's publication of June 2012: "First Annual State of Water Resources Report (April 01, 2010 - March 31, 2011)" was consulted to provide the irrigation assumed for Lesotho [9].

The words 'equipped for irrigation' and 'believed to be irrigated' were used because, when using remote sensing techniques, it was not all that easy to be absolutely sure that an area is 'equipped for irrigation', and/or it is being irrigated, and, as proposed by consultants, an approach based on evidence was considered; with the possibility of

combination from the from the multi-temporal imagery evidence with other evidence, such as: the issuance of WARMS registrations (these are point locations); the shape of the field crop boundary; mean annual rainfall; proximity to water sources (rivers, farm dams); land slope; and expert knowledge of the agricultural areas under irrigation as per team members. The satellite imagery underwent automatic classification first because, without extensive ground truth, there would be difficulty in distinguishing between land parcels as being non-irrigated, partially irrigated and irrigated, irrespective of using multi-temporal imagery. In this case additional evidence was considered in order to assist with separating out confusing classifications [6].

Testing of Crop Classification Techniques

When ground truth data on crops at a high resolution is available, it may lead to more accurate, or even useful, crop classification. This information was to large extent not available in the catchment area but progress was made using other classification techniques in selected parts of the catchment area [6].

The registered area is 13,620 ha in tile 174-82; the field crop boundary by DAFF identified 21,293 ha; based on RS selection method 6,453 ha was identified. For July 2009 the RS using filled 'slcoff' gave 4,122 ha. (i.e., disregarding 400 ha which was doubted irrigated).

According to ORASECOM [6], identification of these areas was very inconsistent so was the definition of what was considered "apparently set up for irrigation" or "irrigated". Speculations showed that while registration of a large area was done, the rainfall-runoff supplying any of the irrigated areas may be unreliable, giving rise to differences in the extent and location of actual irrigation from year to year.

Ground truthing

The use of remote sensing facilitates improvement on estimating the extent of irrigation areas in the catchment area but ground truthing may not be underestimated. It provides estimates of irrigated areas accurately even for rainfed agriculture. Ground truthing is even more useful for crop classification purposes. An emphasis in the report on the *Promotion of Water Conservation and Water Demand Management in the Irrigation Sector* is placed in collecting and collating of information on irrigated areas including crop types and that this should be done at the water user association level in the South African context, and

REACH NO.	DESCRIPTION OF REACH	WATER ALLOC-ATION (m³/ha/a)	FIELD IRRIGATION REQUIREMENT (mill. M³/a)							IRRIGATED AREAS (ha)							
			WRP (2012)	ORECON Proposed	ORRS		ORASECOM			DWA UPINGTON	WRP (2012)	ORECON Proposed	ORRS		ORASECOM			DWA UPINGTON
					Sched.	Act	WARMS (FROM ORSECOM)	ORRS Sch ha (FROM ORSECOM)	REMOTE SENSING				Scheduled	Act	WARMS (FROM ORSECOM)	ORRS Sch ha (FROM ORSECOM)	REMOTEEN SING	
1	Caledon River: U/S Welbedacht Dam	7 620	71.4	40.3	11.0	28.1	65.1	11.0	25.6		9 364.8	9 930	1 440.0	3 692.0	8 541.2	1 440.0	3 358.6	
2	Caledon River: Welbedacht Dam to Gariep Dam	8 000	18.2	36.5	38.2	27.9	49.1	26.2	25.6		2 275.0	5 835	4 775.1	3 482.0	6 142.6	3 280.4	3 199.3	
3	U/S Aliwal North D/S Oranjedraai	8 000	12.6	6.6	1.9	1.9	8.5	12.6	2.5		1 575.0	877	1 574.6	1 550.0	1 061.6	1 575.0	314.4	
4	Aliwal N to Gariep Dam	8 000	43.3	52.5	20.5	15.5	23.8	20.5	11.7		2 562.5	8 229	2 559.7	1 940.0	2 970.4	2 560.0	1 465.5	
5	Kraai U/S Aliwal N	7 620	11.6	28.0	0.0	0.0	7.4	7.6	5.3		1 524.9	6 341	0.0	0.0	969.5	999.0	699.1	
6	Gariep dam to Vanderkloof dam	11 000	28.4	27.7	25.5	25.5	26.6	26.0	25.8		2 580.9	3 121	2 316.0	2 316.0	2 418.3	2 361.0	2 345.2	
7	Canals ex Vanderkloof dam	11 000	201.7	195.1	191.0	191.0	3.5	191.2	49.2		20 914.9	17 678	17 377.8	17 378.0	322.1	17 378.0	4 474.4	
8	Scholzburg and Lower Riet IBs	9 140	49.4	50.2	41.8	41.8	0.0	0.0	0.0		5 408.1	4 564	4 574.2	4 575.0	0.0	0.0	0.0	
9	Vanderkloof Dam to Orange-Vaal conf:	10000 to11000	170.0	187.4	149.8	149.8	334.0	155.9	301.7		15 845.8	17 465	14 174.1	14 173.0	31 461.5	14 174.0	27 865.3	
10	Krugerdrift dam to Tweerivier gauge - Modder River	8 640	29.4	52.5*	29.1	28.9	197.2	30.8	237.3		3 402.8	7 004	3 364.4	3 340.0	22 823.5	3 564.0	27 470.1	
11	Tierpoort Dam to Kalkfontein Dam: Tierpoort IB	9 000	6.4	8.1*	6.4	6.0	0.0	0.0	0.0		711.1	1 018	708.0	665.0	0.0	0.0	0.0	
12	Kalkfontein Dam to Riet River Settlement: Kalkfontein WUA (canal)	11 000	33.5	56.7*	33.5	33.5	74.7	46.6	58.3		3 510.0	6 187	3 046.3	3 046.0	6 792.6	4 234.0	5 303.2	
14	Douglas weir to Orange-Vaal Conf. (Orange water)	9 140	104.3	104.3	74.2	66.6	98.3	190.7	270.5		11 410.3	11 410	8 113.0	7 285.0	10 755.0	20 869.0	29 594.7	

Figure 5: Summary of Findings of Different Studies Undertaken to Determine Irrigation Demand for the Study Area.

by similar status organisations in the other states. These organisations require this information and usually have a close relationship with farmers in their respective settings [6]. Funding for GIS development and training purposes for these organisations may lead to improved estimation of areas under irrigation, cropping patterns and methods of irrigation used.

Updates of the database of irrigated areas

The field crop boundaries in South Africa could be supplemented with long-term continuous mapping of areas under irrigation in Botswana, Lesotho and Namibia at the catchment level on an annual basis (ORASECOM, 2011). It further indicated that at present, isolated areas in the basin are under crop type mapping, and an expansion is being planned and could be underway soon for the whole catchment. Also under consideration is to have ground truth data from a sample of locations within each satellite image tile. The project may also need to include non-irrigated field crop boundaries. A continuous time series of satellite images is considered a vital requirement to this exercise as well. If all these could be achieved then a large improvement on confidence and estimates of irrigation water use could be determined at a relatively low cost [6].

Water use monitoring system

The potential improvement of irrigation water schedules and water use efficiency can be enhanced by developing a monitoring system for crop water use based on ground based meteorological observations and on near real-time satellite. This coupling could lead continuous mapping of crop cover, a further requirement for mapping crop types. In following on this, the data on climate parameters derived for the statistical downscaling purposes may be utilised for calibrating the development and approach of an operational system. This may serve as platform for an agricultural extension system to farmers under use in different states in the basin or governments' advisory contractors [6].

The Irrigation Scenario Tool produced according to ORASECOM [6] can undergo improvement to include:

- Industrial and domestic demands;

- Introduction of risk based scenario generation;

- The introduction of optimisation based on economic returns; and

- Coupling the Irrigation Scenario Tool (for providing estimates on irrigation water required) to the Water Resources Yield Model.

References

1. DWA-Department of Water Affairs, South Africa (2012) Reconciliation Strategy Report for the Large Bulk Water Supply Systems of the Greater Bloemfontein Area. Prepared by Aurecon in association with GHT Consulting Scientists and ILISO Consulting as part of the Water Reconciliation Strategy Study for the Large Bulk Water Supply Systems: Greater Bloemfontein Area.

2. DWAF-Department of Water Affairs and Forestry, South Africa (2009) Directorate Water Resource Planning Systems: Water Quality Planning. Orange River: Assessment of water quality data requirements for planning purposes. Water Quality Monitoring and Status Quo Assessment.

3. ORASECOM-h Orange-Senqu River Commission (2008) Preliminary Transboundary Diagnostic Analysis.

4. Mucina L, Rutherford MC (2011) The Vegetation of South Africa, Lesotho and Swaziland. Strelitzia 19. South African National Biodiversity Institute, Pretoria.

5. DWA-Department of Water Affairs, South Africa (2012) Development of Reconciliation Strategies for Large Bulk Water Supply Systems Orange River: Surface Water Hydrology and System Analysis Report. WRP Consulting Engineers Aurecon, Golder Associates Africa, and Zitholele Consulting.

6. ORASECOM-Orange-Senqu River Commission. 2011. The Promotion of WC WDM in the Irrigation Sector Report 011/2011.

7. Huggins G, Rydgren B, Lappeman G (2010) The Assessment of Goods and Services in the Orange River Basin. Produced for WRP as part of Support to Phase II ORASECOM Basin Wide Integrated Water Resources Management Plan. Report-WP-WP5-010-2010-fd.

8. DEA-Department of Environmental Affairs, South Africa (2013) Long- Term Adaptation Scenarios Flagship Research Programme (LTAS) for South Africa. Climate Change Implications for Water Sector in South Africa. Pretoria. South Africa.

9. DWA-Department of Water Affairs, South Africa. 2013. Development of Reconciliation Strategies for Large Bulk Water Supply Systems: Irrigation Demands and Water Conservation/ Water Demand Management. Pretoria. WRP Consulting Engineers (Pty) Ltd., Aurecon, Golder Associates Africa, and Zitholele Consulting.

14

Estimating Mean Long-term Hydrologic Budget Components for Watersheds and Counties: An Application to the Commonwealth of Virginia, USA

Ward E Sanford[1]*, David L Nelms[2], Jason P Pope[2] and David L Selnick[3]

[1]Mail Stop 431, U.S. Geological Survey, Reston, Virginia, 20171, USA
[2]U.S. Geological Survey, Richmond, Virginia, USA
[3]U.S. Geological Survey, Reston, Virginia, USA

Abstract

Mean long-term hydrologic budget components, such as recharge and base flow, are often difficult to estimate because they can vary substantially in space and time. Mean long-term fluxes were calculated in this study for precipitation, surface runoff, infiltration, total evapotranspiration (ET), riparian ET, recharge, base flow (or groundwater discharge) and net total outflow using long-term estimates of mean ET and precipitation and the assumption that the relative change in storage over that 30-year period is small compared to the total ET or precipitation. Fluxes of these components were first estimated on a number of real-time-gaged watersheds across Virginia. Specific conductance was used to distinguish and separate surface runoff from base flow. Specific-conductance (SC) data were collected every 15 minutes at 75 real-time gages for approximately 18 months between March 2007 and August 2008. Precipitation was estimated for 1971-2000 using PRISM climate data. Precipitation and temperature from the PRISM data were used to develop a regression-based relation to estimate total ET. The proportion of watershed precipitation that becomes surface runoff was related to physiographic province and rock type in a runoff regression equation. A new approach to estimate riparian ET using seasonal SC data gave results consistent with those from other methods. Component flux estimates from the watersheds were transferred to flux estimates for counties and independent cities using the ET and runoff regression equations. Only 48 of the 75 watersheds yielded sufficient data, and data from these 48 were used in the final runoff regression equation. Final results for the study are presented as component flux estimates for all counties and independent cities in Virginia. The method has the potential to be applied in many other states in the U.S. or in other regions or countries of the world where climate and stream flow data are plentiful.

Keywords: Hydrologic budget; Evapotranspiration; Runoff; Recharge; Hydrograph separation

Introduction

Water-resource managers must allocate both groundwater and surface-water resources to multiple users based on estimates of short-term and long-term water availability. In response to recurring droughts and water shortages, many places often attempt to develop comprehensive water-supply plans. In 2005 in Virginia (USA), localities (counties and independent cities) were required to develop either local or regional water-supply plans in response to the Virginia Local and Regional Water Supply Planning Regulation (9 VAC 25-780). Although recent studies within the state [1-5] focused on the resources of the Virginia Coastal Plain, reliable information is frequently lacking on water availability west of the coastal plain (Figure 1), especially pertaining to long-term fluxes such as recharge to groundwater aquifers.

Flux estimates of components of the hydrologic cycle can be made by creating a water budget in which the various components must balance. Such a water balance approach is reasonably accurate when all of the terms in the budget can be calculated or estimated. This approach is appropriate for the scale of an entire state, such as Virginia, because most other methods used to estimate recharge (such as the use of environmental tracers or water levels) are highly dependent on local measurements in both space and time [5,6]. New datasets, including national climate data sets with a resolution of less than one mile, and cost-effective specific-conductance data for base-flow separation, are now available in the United States to assess water availability at a regional level, such as for the Commonwealth of Virginia. Such assessments would be valuable for water resource managers at the state, county, and local planning levels and the method is applicable to other regions as well.

Water budgets are quantified routinely for watersheds, but to quantify budgets for county units for which managers off make decisions requires results from watersheds to be transferrable to counties through some type of regression. Along these lines, the purpose of this study was to demonstrate such a method by quantifying components of the hydrologic budget on a large number of watersheds across the entire Commonwealth of Virginia, and using the results to estimate hydrologic budget components for all of Virginia's counties and independent cities. These components include precipitation, surface runoff, infiltration, total evapotranspiration (ET), riparian ET, groundwater recharge, and base flow or groundwater discharge, and are calculated using long-term average values (1971-2000) from mean precipitation data, and base-flow separation data from 2007-2008.

*Corresponding author: Ward E. Sanford, Mail Stop 431, U.S. Geological Survey, Reston, Virginia, USA, 20171, E-mail: wsanford@usgs.gov

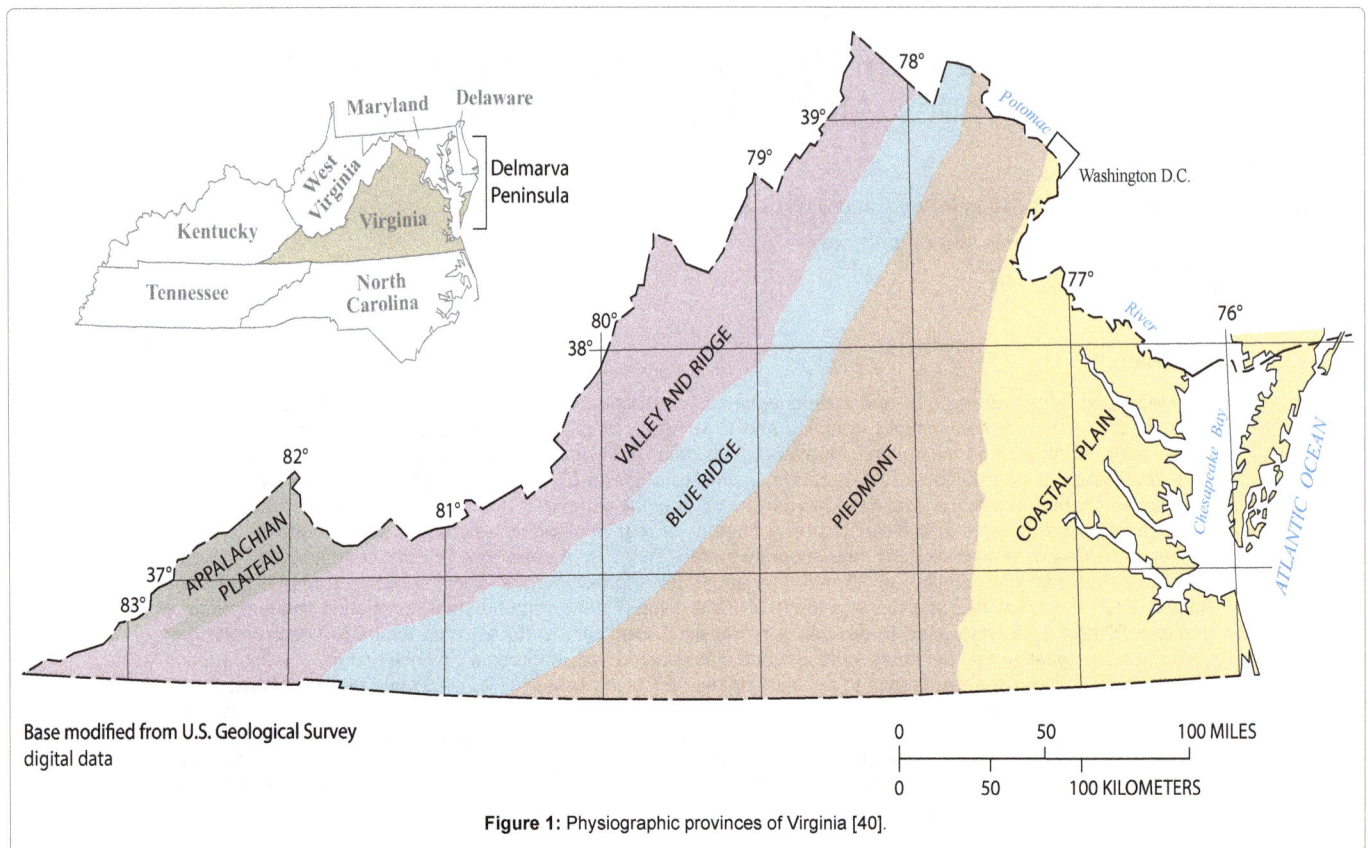

Figure 1: Physiographic provinces of Virginia [40].

The latter were adjusted to long-term conditions based on historical stream flow data. Within watersheds or counties values are expected to deviate, both temporally and locally, from the calculated mean values. A few watersheds with historical specific conductance data from the neighboring states of Maryland and Delaware were included in the analysis to improve estimates of surface runoff and base flow for the Coastal Plain Province. Detailed data associated with the study have been included in an earlier USGS report [7].

Location and setting of study area

The Commonwealth of Virginia is located in the east-central United States, bounded by the Potomac River and Maryland on the northeast, West Virginia on the north and west, Kentucky and Tennessee on the southwest, North Carolina on the south, and the Chesapeake Bay and Atlantic Ocean on the east (Figure 1). Virginia is positioned across five different physiographic provinces: the Coastal Plain Province in the far east, the Piedmont Province in the east, the Blue Ridge Province in the west, the Valley and Ridge Province in the far west, and the Appalachian Plateau in the extreme southwest. Politically, the commonwealth is divided into 95 counties and an additional 39 independent cities (Figure 2). Land surface elevations rise from sea level at the eastern coastline upward through the low-lying plains of the Coastal Plain Province and the rolling hills of the Piedmont Province, to the long, linear ridges of the mountains of the Blue Ridge and Valley and Ridge Provinces. The mountains of the Blue Ridge, Valley and Ridge Provinces, and Appalachian Plateau in Virginia frequently reach up to 600 to 900 meters (m) above sea level, with local relief frequently exceeding 300 m.

The climate of Virginia is diverse and varies from the warm, temperate, eastern coastal areas that have temperatures moderated by

the Atlantic Ocean, to the cooler continental climate of the mountainous provinces in the north and west. Mean annual temperatures range from 15 degrees Celsius (°C) in Virginia Beach in the southeast to 9°C in Highland County in the west. Rainfall patterns vary across Virginia and are affected by topography in the north and west, and by the presence of tropical moisture systems in the south and east. Annual precipitation is lowest in the northern valleys, where average values are less than 100 centimeters per year (cm/yr) at many locations, and highest along the southwestern ridges where average values can exceed 125 cm/yr. Temperature and rainfall are adequate to support a substantial agriculture industry, with crop and pasture lands evenly scattered between forests of mixed deciduous and evergreen trees across most of Virginia. In the mountainous western provinces, though, agriculture is restricted mostly to the valleys, with forests covering most of the ridges. The largest urban and suburban areas have developed around Fairfax County in the north, the Tidewater area of Norfolk and Hampton Roads in the southeast, the capital city of Richmond in the southeastern central region, and Roanoke in the west.

Previous investigations

Regional studies of water-resource characteristics of the Commonwealth of Virginia have previously been delineated by physiographic province. The water resources of the coal-mining areas in the Appalachian Plateau of Virginia have been studied in terms of hydrology [8], effects of mining [9], water quality [10,11], geochemistry [12], and hydraulic characteristics [13]. The water-resource characteristics of the Valley and Ridge, Blue Ridge, and Piedmont Provinces have been studied as part of the USGS Regional Aquifer System Analysis (RASA) program. These studies in the western provinces included that of the hydrogeology [14], groundwater quantity [15], and shallow hydrologic characteristics through stream

EXPLANATION

County-name index

1. Accomack	27. Dinwiddie	51. Lancaster	75. Rappahannock	89. Tazewell
2. Albemarle	28. Essex	52. Lee	76. Richmond	90. Warren
3. Allegheny	29. Fairfax	53. Loudoun	77. Roanoke	91. Washington
4. Amelia	30. Fauquier	54. Louisa	78. Rockbridge	92. Westmoreland
5. Amherst	31. Floyd	55. Lunenberg	79. Rockingham	93. Wise
6. Appomatox	32. Fluvanna	56. Madison	80. Russell	94. Wythe
7. Arlington	33. Franklin	57. Mathews	81. Scott	95. York
8. Augusta	34. Frederick	58. Mecklenberg	82. Shenandoah	
9. Bath	35. Giles	59. Middlesex	83. Smyth	
10. Bedford	36. Gloucester	60. Montgomery	84. Southampton	
11. Bland	37. Goochland	61. Nelson	85. Spotsylvania	
12. Botetourt	38. Grayson	62. New Kent	86. Stafford	
13. Brunswick	39. Greene	63. Northampton	87. Surry	
14. Buchanan	40. Greensville	64. Northumberland	88. Sussex	
15. Buckingham	41. Halifax	65. Nottoway		
16. Campbell	42. Hanover	66. Orange		
17. Caroline	43. Henrico	67. Page		
18. Carroll	44. Henry	68. Patrick		
19. Charles City	45. Highland	69. Pittsylvania		
20. Charlotte	46. Isle of Wight	70. Powhatan		
21. Chesterfield	47. James City	71. Prince Edward		
22. Clarke	48. King & Queen	72. Prince George		
23. Craig	49. King George	73. Prince William		
24. Culpeper	50. King William	74. Pulaksi		
25. Cumberland				
26. Dickenson				

Independent city-name index

96. Alexandria	105. Emporia	120. Norfolk
97. Bedford	106. Fairfax	121. Norton
98. Bristol	107. Falls Church	122. Petersburg
99. Buena Vista	108. Franklin	123. Poquoson
100. Charlottesville	109. Fredericksburg	124. Portsmouth
101. Chesapeake	110. Galax	125. Radford
102. Colonial Heights	111. Hampton	126. Richmond
103. Covington	112. Harrisonburg	127. Roanoke
104. Danville	113. Hopewell	128. Salem
	114. Lexington	129. Staunton
	115. Lynchburg	130. Suffolk
	116. Manassas	131. Virginia Beach
	117. Manassas Park	132. Waynesboro
	118. Martinsville	133. Williamsburg
	119. Newport News	134. Winchester

Base modified from Virginia Department of Conservation and Recreation, Universal Transverse Mercator projection Zone 17N, North American Datum of 1983

Figure 2: Names and locations of counties and independent cities in the Commonwealth of Virginia.

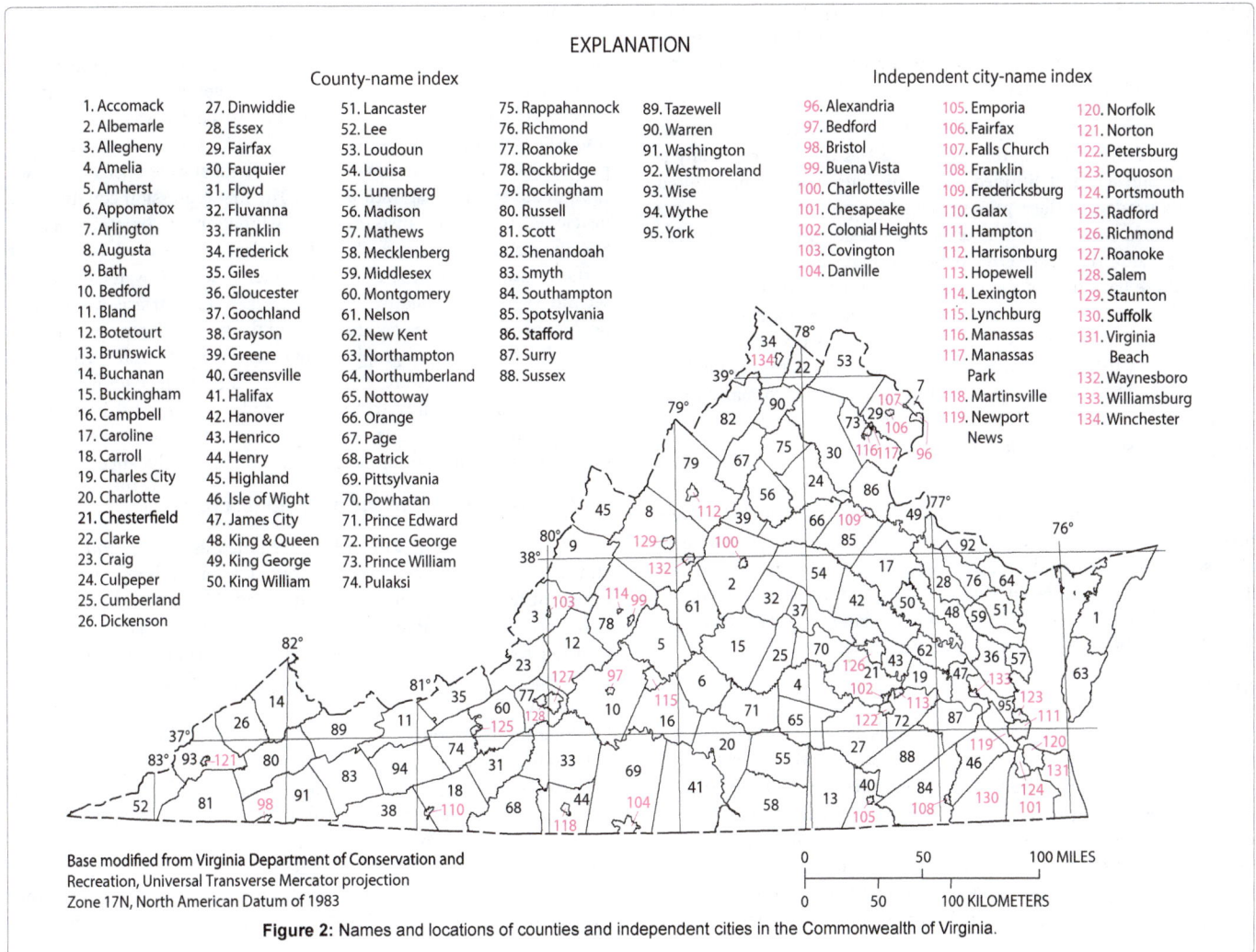

flow recession analysis [16]. In addition, base-flow [17] and low-flow [18] characteristics have been determined for these provinces. In the Coastal Plain of Virginia, descriptions of the hydrogeologic framework, groundwater quality, and groundwater discharge have been published elsewhere [1,4,19]. A similar regression approach was developed for the State of Minnesota [20], although the base-flow evaluation there was done using a physical-hydrograph separation technique. Numerous techniques have been documented in the literature for estimating recharge [5,6,21] but most of these approaches are site- and time-specific field-based methods whose results are difficult to scale-up to long-term mean values for watersheds or counties.

Geologic setting

The geology of Virginia is diverse with rocks and sediments that range in age from the early Proterozoic (>1 billion years old) to Holocene (<10,000 years old). The Coastal Plain is composed of unconsolidated sediments that pinch out at its western edge, but are up to several thousand feet thick at the Atlantic coastline. These sediments were deposited after being eroded from the Appalachian Mountains following the opening of the Atlantic Ocean during the Triassic and Jurassic Periods. The sediments vary in size from clay to gravel and were deposited in fluvial and marine environments as sea levels rose and fell. The hydrologic cycle on the Coastal Plain is impacted by the average grain size of surficial sediment, which can be classified as fine (silt and clay), medium (silt and sand), or coarse (sand and gravel). Average grain size is dependent on the stratigraphic unit exposed locally at the land surface [22].

The Piedmont Province is underlain by polydeformed rocks believed to be of late Proterozoic age that were metamorphosed during the Paleozoic Era. Rock types vary, but the dominant varieties are gneiss, schist, granite, and slate (in the far south central region). During the Mesozoic Era, a number of rift basins opened up in the Atlantic Ocean, parallel to the Mid-Atlantic Ridge; they filled with siliciclastic and carbonate sediments which later were lithified. A few of these Mesozoic Rift Basins are present in the Piedmont Province, the largest being the Culpeper Basin in Culpeper, Fauquier, Prince William, Fairfax, and Loudoun Counties.

The rocks of the Blue Ridge Province are the oldest in Virginia, and most formed during the Proterozoic Era (1.4-0.6 billion years ago). The rocks are predominantly basement granites and gneisses that have been exposed on the land surface by uplift and erosion. The province can be separated into two sections based on the origins of the topography [23]. The section north of the Roanoke River is characterized by a narrow range of high mountains underlain by Precambrian to Cambrian quartzite, phyllite, metabasalt, and granodiorite that form the northwest limb of an anticlinorium [17]. The section south of the Roanoke River is much broader, with steep ridges separated by parallel valleys, high ridges, highlands, plateau, and escarpment. Precambrian gneiss, schist, amphibolite, volcanic and metasedimentary rocks, Cambrian quartzite, and faulted carbonate rocks and shale underlie this section of the Blue Ridge [23].

The Valley and Ridge Province is underlain by layered sedimentary rocks of the Paleozoic Era. The rocks were laid down horizontally as sediments, buried, and lithified, but were later folded and faulted, and finally eroded to their present state of exposure. The rocks vary in composition between carbonate and siliciclastic. Many of the oldest (from the Cambrian and Ordovician Periods) are carbonates and some of these have been dolomitized. The carbonate rocks tend to lie in the valleys of the province, whereas the more resistant sandstones are present along the ridges. Shale's and siltstones occur both in the valleys and on the ridge slopes. Many of the carbonate regions have been karstified by percolating groundwater giving rise to many caves, springs, and sinkholes. The middle and late Paleozoic Era (Devonian though Mississippian) rocks in the province are almost entirely siliciclastic.

The Appalachian Plateau Province is characterized by a well-dissected, mountainous landscape with dendritic drainage formed on almost flat-lying to gently folded Paleozoic sedimentary rocks [24]. The rocks are predominantly siliciclastic in composition, with rock of Pennsylvanian age the most abundant at the land surface. Coal occurs in beds throughout the Pennsylvanian-aged rock.

Methods

The approach taken in this study was based on the principle of mass conservation, both of water and solute, within a watershed. Mass conservation equations were developed for components of the hydrologic budget, including precipitation, surface runoff, and evapotranspiration (ET), infiltration, recharge, riparian ET, and base flow. The use of long-term (30-year) mean averages for precipitation and ET allowed change in storage to be neglected. The components were estimated from (1) external data sources, (2) data collected

from watersheds across Virginia, or (3) solving the mass balance equations when all other components were estimated (Table 1). Data were analyzed from 108 gaged watersheds across the region (Table 2 and Figure 3), and two multiple-parameter regression equations were developed that allowed the results to be transferred from the watersheds to the entire Commonwealth of Virginia. Long-term mean precipitation and stream flow data for individual watersheds were used to estimate evapotranspiration rates. The first regression equation was developed for evapotranspiration as a function of climatic variables. Specific conductance and chloride analyses were used to estimate surface runoff and base-flow components for 48 watersheds. The second regression equation was developed for surface runoff as a percent of precipitation, as a function of the two landscape parameters, bedrock type and physiographic province. Finally, all of the hydrologic budget components were estimated for the entire Commonwealth of Virginia on a locality (county and independent city) basis, using existing precipitation data, the regression equations developed for evapotranspiration and surface runoff, and the mass balance equations.

Budget components of the hydrologic cycle

Individual watersheds can be envisioned as having both a water and solute budget. Each of these budgets has different terms that represent flow into or out of the watershed (Figure 4). In general, the difference between these inflow and outflow terms leads to a change in water stored within the watershed. On a monthly or annual time scale these changes in storage can be significant fractions of the inflow or outflow. The annual change is storage, however, will never exceed the total inflow or outflow for one year. Thus if the water balance is applied to a period of three decades using long term mean inflows and outflows, the change in storage should not exceed 1/30th of the total inflow or outflow for that time period. So for long time periods the change in storage term becomes relatively small and can be neglected and a steady-state condition assumed. This steady-state water-balance approach for long-time periods has been recognized as valid in other hydrologic studies [25,26]. Based on the principle of conservation and

Budget Component	Estimates For Watersheds	Estimates For Localities
Precipitation	1. PRISM climate data (1971-2000)	10. PRISM climate data (1971-2000)
Total Streamflow	2. USGS NWIS Database (1971-2000)	Not applicable as locality and watershed boundaries do not coincide, but represented rather as Net Total Outflow (see below)
Evapotranspiration (total)	3. Precipitation minus streamflow (EQ 3). Regression equation developed for application to localities.	11. Estimated from a regression equation (EQ 15) relating total evaporation estimates from watersheds to climatic' characteristics, with an additional adjustment for percent impermeable surface
Base Flow	4. Estimated from chemical hydrograph using equation 11, assuming 2 different values of runoff concentration. Values were then adjusted for 1971-2000 conditions via a regression equation relating monthly base flow to streamflow.	Not applicable as locality and watershed boundaries do not coincide, but represented rather as Net Groundwater Discharge (see below)
Surface Runoff	5. Streamflow minus base flow (EQ 7)	12. Estimated from a regression equation (Table 3) relating surface runoff as a percentage of precipitation from watersheds to rock type and Physiography, with an additional adjustment for percent impermeable surface.
Evapotranspiration (riparian)	6. Estimated from chemical hydrograph using (EQ 14)	13. Estimated from (EQ 17) relating riparian ET to the estimated fraction of marsh area (FM), which was estimated from (EQ 16) relating FM to the air temperature and topographic slope.
Evapotranspiration (vadose)	7. ET(total) minus ET(riparian) (EQ 3)	14. ET(total) minus ET(riparian) (EQ 3)
Infiltration	8. Precipitation minus surface runoff (assumes negligible ET from precipitation ponded on surface)	15. Precipitation minus surface runoff (assumes negligible ET from precipitation ponded on surface)
Recharge	9. Infiltration minus ET(vadose) (EQ 5)	16. Infiltration minus ET(vadose) (EQ 5)
Net Total Outflow	Not calculated. Equivalent to total streamflow (see above)	17. Precipitation minus ET(total) (EQ 3)
Net Groundwater Discharge	Not calculated. Equivalent to base flow (see above)	18. Net Total Outflow minus Surface Runoff (EQ 7)

Table 1: Methods used in this study for estimating individual components of the hydrologic budgets and numbered according to the order in which they were calculated.

Table 2. Real-time watersheds included in this study. See figure 3 for map locations.
(ET=evapotranspiration, SC=specific conductance, CP=Coastal Plain, VR=Valley and Ridge, BR=Blue Ridge, PM=Piedmont, unk=unknown)

Map number	USGS Gage Number	Stream gage and watershed name and location	Physiographic Province	Area in square kilometers	Flow used to estimate ET	SC probe installed for this study	Samples collected for chloride analysis	SC data used for base flow estimate
1	01487000	Nanticoke River near Bridgeville, DE	CP	194			X	
2	01613900	Hogue Creek near Hayfield, VA	VR	41	X		X	X
3	01614830	Opequon Creek near Stephens City, VA	VR	39			X	
4	01615000	Opequon Creek near Berryville, VA	VR	151	X		X	X
5	01616075	Fay Spring near Winchester, VA	VR	unk		X	X	
6	01616100	Dry Marsh Run near Berryville, VA	VR	28			X	
7	01616500	Opequon Creek at Martinsburg, WV	VR	707	X	X	X	
8	01622000	North River near Burketown, VA	VR	974	X	X	X	X
9	01625000	Middle River near Grottoes, VA	VR	966	X	X	X	X
10	01626000	South River near Waynesboro, VA	BR	329	X	X	X	X
11	01627500	South River at Harriston, VA	BR	549	X	X	X	X
12	01629500	S F Shenandoah River near Luray, VA	VR	3566		X	X	X
13	01630700	Gooney Run near Glen Echo, VA	BR	54			X	
14	01631000	S F Shenandoah River at Front Royal, VA	VR	4232	X		X	X
15	01632000	N F Shenandoah River at Cootes Store, VA	VR	544	X	X	X	X
16	01632082	Linville Creek at Broadway, VA	VR	119			X	X
17	01632900	Smith Creek near New Market, VA	VR	242	X		X	X
18	01633000	N F Shenandoah River at Mount Jackson, VA	VR	1316	X	X	X	X
19	01634000	N F Shenandoah River near Strasburg, VA	VR	1994	X		X	X
20	01634500	Cedar Creek near Winchester, VA	VR	264	X		X	X
21	01635090	Cedar Creek above Hwy 11 near Middletown, VA	VR	396		X	X	X
22	01635500	Passage Creek near Buckton, VA	VR	224	X		X	X
23	01636242	Crooked Run below Hwy 30 at Riverton, VA	VR	122			X	
24	0163626650	Manassas Run at Rt 645 near Front Royal, VA	BR	28			X	
25	01636316	Spout Run at RT 621 near Millwood, VA	VR	54			X	X
26	01643700	Goose Creek near Middleburg, VA	BR	316	X	X	X	X
27	01644280	Broad Run near Leesburg, VA	PM	197		X	X	
28	01646000	Difficult Run near Great Falls, VA	PM	150	X	X	X	X
29	01649500	NE Branch Anacostia River at Riverdale, MD	CP	189				X
30	01651000	NW Branch Anacostia River near Hyattsville, MD	PM	127				X
31	01656000	Cedar Run near Catlett, VA	PM	242	X	X	X	X
32	01658000	Mattawoman Creek near Pomonkey, MD	CP	142				X
33	01660400	Aquia Creek near Garrisonville, VA	PM	91	X	X	X	X
34	01663500	Hazel River at Rixeyville, VA	BR	743		X	X	
35	01664000	Rappahannock River at Remington, VA	BR	1603	X			
36	01665500	Rapidan River near Ruckersville, VA	BR	298	X	X	X	X

Table 2 (continued). Watersheds included in this study. See figure 3 for locations.
(ET=evapotranspiration, SC=specific conductance, CP=Coastal Plain, VR=Valley and Ridge, BR=Blue Ridge, PM=Piedmont)

Map number	USGS Gage Number	Stream gage and watershed name and location	Physiographic Province	Area in square kilometers	Used to estimate ET	SC probe installed for this study	Samples collected for chloride analysis	SC data used for base flow estimate
37	01666500	Robinson River near Locust Dale, VA	BR	464	X	X	X	X
38	01667500	Rapidan River near Culpeper, VA	BR	1212	X	X	X	X
39	01669000	Piscataway Creek near Tappahannock, VA	CP	73		X	X	
40	01669520	Dragon Swamp at Mascot, VA	CP	280		X	X	X
41	01671020	North Anna River at Hart Corner near Doswell, VA	PM	1199		X	X	
42	01671100	Little River near Doswell, VA	PM	277		X	X	
43	01672500	South Anna River near Ashland, VA	PM	1023	X	X	X	X

44	01673000	Pamunkey River near Hanover, VA	PM	2792	X			
45	01673638	Cohoke Mill Creek near Lestor Manor, VA	CP	23		X	X	
46	01674000	Mattaponi River near Bowling Green, VA	PM	666		X	X	
47	01674500	Mattaponi River near Beulahville, VA	CP	1559	X			
48	02011400	Jackson River near Bacova, VA	VR	409		X	X	X
49	02011500	Back Creek near Mountain Grove, VA	VR	347		X	X	X
50	02013000	Dunlap Creek near Covington, VA	VR	420	X	X	X	X
51	02013100	Jackson River BL Dunlap Creek at Covington, VA	VR	1590		X	X	
52	02014000	Potts Creek near Covington, VA	VR	396	X	X	X	X
53	02015700	Bullpasture River at Williamsville, VA	VR	285		X	X	X
54	02016000	Cowpasture River near Clifton Forge, VA	VR	1194	X	X	X	
55	02016500	James River at Lick Run, VA	VR	3556		X	X	X
56	02017500	Johns Creek at New Castle, VA	VR	272	X	X	X	X
57	02018000	Craig Creek at Parr, VA	VR	852	X	X	X	X
58	02020500	Calfpasture River above Mill Creek at Goshen, VA	VR	365	X	X	X	X
59	02021500	Maury River at Rockbridge Baths, VA	VR	852	X	X	X	X
60	02024000	Maury River near Buena Vista, VA	VR	1676	X	X	X	X
61	02025500	James River at Holcomb Rock, VA	VR	8440		X	X	
62	02026000	James River at Bent Creek, VA	VR	9538		X	X	
63	02030000	Hardware River BL Briery Run near Scottsville, VA	BR	300		X	X	
64	02032640	N F Rivanna River near Earlysville, VA	BR	280		X	X	X
65	02039500	Appomattox River at Farmville, VA	PM	785		X	X	
66	02040000	Appomattox River at Mattoax, VA	PM	1878	X	X	X	X
67	02041000	Deep Creek near Mannboro, VA	PM	409	X	X	X	X
68	02042500	Chickahominy River near Providence Forge, VA	CP	650	X			
69	02044500	Nottoway River near Rawlings, VA	PM	821	X	X	X	X
70	02045500	Nottoway River near Stony Creek, VA	PM	1499		X	X	
71	02046000	Stony Creek near Dinwiddie, VA	PM	290		X	X	
72	02047500	Blackwater River near Dendron, VA	CP	751	X	X	X	X

Table 2 (continued): Watersheds included in this study. See figure 3 for locations.

(ET=evapotranspiration, SC=specific conductance, CP=Coastal Plain, VR=Valley and Ridge, BR=Blue Ridge, PM=Piedmont)

Map number	USGS Gage Number	Stream gage and watershed name and location	Physiographic Province	Area in square kilometers	Used to estimate ET	SC probe installed for this study	Samples collected for chloride analysis	SC data used for base flow estimate
73	02049500	Blackwater River near Franklin, VA	CP	1588	X			
74	02051500	Meherrin River near Lawrenceville, VA	PM	1430	X	X	X	X
75	02052000	Meherrin River at Emporia, VA	PM	1927	X			
76	02053800	S F Roanoke River near Shawsville, VA	BR	282	X	X	X	
77	02054500	Roanoke River at Lafayette, VA	VR	658	X			
78	02055000	Roanoke River at Roanoke, VA	VR	994	X			
79	02056000	Roanoke River at Niagara, VA	VR	1318	X	X	X	
80	02056900	Blackwater River near Rocky Mount, VA	BR	298		X	X	
81	02059485	Goose Creek at Rt 747 near Bunker Hill, VA	BR	324			X	
82	02059500	Goose Creek near Huddleston, VA	BR	487	X	X		X
83	02061000	Big Otter River near Bedford, VA	BR	295			X	
84	02061500	Big Otter River near Evington, VA	BR	816	X	X	X	X
85	02062500	Roanoke (Staunton) River at Brookneal, VA	BR	6254		X	X	
86	02064000	Falling River near Naruna, VA	PM	448		X	X	
87	02065500	Cub Creek at Phoenix, VA	PM	253	X	X	X	
88	02070000	North Mayo River near Spencer, VA	PM	280		X	X	
89	02072000	Smith River near Philpott, VA	BR	559		X	X	
90	02073000	Smith River at Martinsville, VA	PM	984		X	X	
91	02074500	Sandy River near Danville, VA	PM	290		X	X	
92	02075045	Dan River STP near Danville, VA	BR	5451		X	X	
93	02077000	Banister River at Halifax, VA	PM	1417	X	X	X	
94	02079640	Allen Creek near Boydton, VA	PM	139	X	X	X	

95	03165500	New River at Ivanhoe, VA	BR	3470		X	X	
96	03167000	Reed Creek at Grahams Forge, VA	VR	668	X	X	X	X
97	03168000	New River at Allisonia, VA	BR	5703		X	X	
98	03170000	Little River at Graysontown, VA	BR	800	X			
99	03171000	New River at Radford, VA	BR	7117		X	X	
100	03173000	Walker Creek at Bane, VA	VR	774	X	X	X	
101	03175500	Wolf Creek near Narrows, VA	VR	578		X	X	
102	03207800	Levisa Fork at Big Rock, VA	AP	769	X			
103	03208500	Russell Fork at Haysi, VA	AP	741	X	X	X	
104	03473000	S F Holston River near Damascus, VA	BR	785	X			
105	03475000	M F Holston River near Meadowview, VA	VR	533	X	X	X	
106	03488000	N F Holston River near Saltville, VA	VR	572	X			
107	03524000	Clinch River at Cleveland, VA	VR	1380	X	X	X	X
108	03531500	Powell River near Jonesville, VA	VR	826	X	X	X	X

EXPLANATION

Watershed abbreviated name/location index—See table 2 for additional descriptions; S F, South Fork; N F, North Fork; WV, West Virginia; F, Front; C, Cootes; Ck, Creek; Hwy, Highway; Bowl, Bowling; G, Green; R, Rocky; M F, Middle Fork; NE, Northeast; NW, Northwest; Shen, Shenandoah; Steph, Stephens; Jack, Jackson; Ruckrsvl, Ruckersville

1. Nanticoke, Delaware
2. Hogue Creek
3. Opequon, Steph City
4. Opequon, Berryville
5. Faye Spring
6. Dry Marsh Run
7. Opequon, WV
8. North River
9. Middle River
10. South R, Waynesboro
11. South R, Harriston
12. S F Shen, Luray
13. Gooney Run
14. S F Shen, F Royal
15. N F Shen, C Store
16. Linville Creek
17. Smith Creek
18. N F Shen, Mt Jack
19. N F Shen, Strasburg
20. Cedar Ck, Winchester
21. Cedar Ck, Hwy 11
22. Passage Creek
23. Crooked Creek
24. Manassas Run
25. Spout Run
26. Goose, Middleburg
27. Broad Run
28. Difficult Run
29. NE Anacostia

30. NW Anacostia
31. Cedar Run
32. Mattawoman
33. Aquia Creek
34. Hazel River
35. Rappahannock
36. Rapidan, Ruckrsvl
37. Robinson
38. Rapidan, Culpeper
39. Piscataway Creek
40. Dragon Swamp
41. North Anna River
42. Little River
43. South Anna River
44. Pamunkey River
45. Cohoke Mill Creek
46. Mattaponi, Bowl G
47. Mattaponi, Beulah
48. Jackson, Bacova
49. Back Creek
50. Dunlap Creek
51. Jackson, Covington
52. Potts Creek
53. Bullpasture River
54. Cowpasture River
55. James, Lick Run
56. Johns Creek

57. Craig Creek
58. Calfpasture River
59. Maury, Rockbridge
60. Maury, Buena Vista
61. James, Holcomb
62. James, Bent Creek
63. Hardware River
64. N F Rivanna River
65. Appomattox, Farmville
66. Appomattox, Mattoax
67. Deep Creek, Mannboro
68. Chickahominy
69. Nottoway, Rawlings
70. Nottoway, S Creek
71. Stony Creek, Dinwiddie
72. Blackwater, Dendron
73. Blackwater, Franklin
74. Meherrin, Lawrenceville
75. Meherrin, Emporia

76. S F Roanoke River
77. Roanoke, Lafayette
78. Roanoke, Roanoke
79. Roanoke, Niagara
80. Blackwater, R Mount
81. Goose, Bunker Hill
82. Goose, Huddleston
83. Big Otter, Bedford
84. Big Otter, Evington
85. Roanoke, Brookneal

86. Falling River, Naruna
87. Cub Creek, Phenix
88. North Mayo, Spencer
89. Smith, Philpott
90. Smith, Martinsville
91. Sandy, Danville
92. Dan RIver, Danville
93. Banister, Halifax
94. Allen Creek, Boydton
95. New River, Ivanhoe

96. Reed Creek, Grahams Forge
97. New River, Allisonia
98. Little, Graysontown
99. New River, Radford
100. Walker Creek, Bane
101. Wolf, Narrows
102. Levisa Fork, Big Rock
103. Russell Fork, Haysi
104. S F Holston, Damascus
105. M F Holston, Meadowview
106. N F Holston, Saltville
107. Clinch River, Cleveland
108. Powell River, Jonesville

Base modified from U.S. Geological Survey digital data, Krstolic, 2006

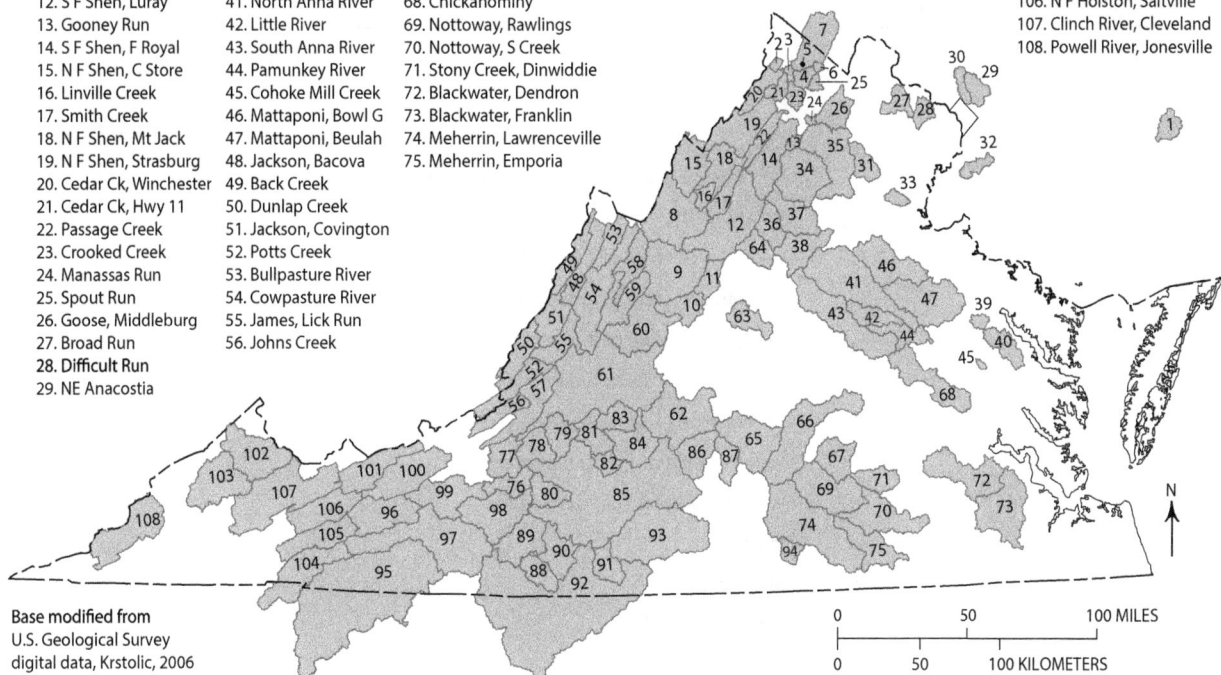

Figure 3: Names and locations of watersheds referenced in this study.

the steady state assumption, a number of equations can be written that represent the balance of mass moving into and/or out of the watershed. A total steady-state water balance across the watershed can be written as:

$$P - Q_s/A + U_i/A = ET_{vz} + ET_{rp} + U_o/A \qquad (1)$$

Where P is the average rate of precipitation, [L/t],

Q_s is the average rate of stream flow out of the watershed, [L^3/t],

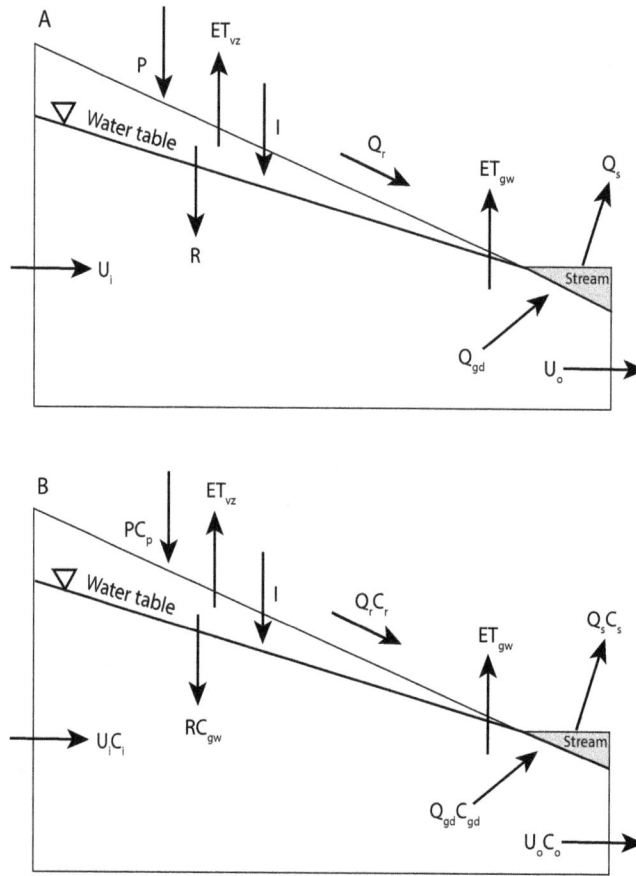

NOT TO SCALE

EXPLANATION

Components of budget

C_{gd}	Concentration of solute in groundwater discharge
C_{gw}	Concentration of solute in groundwater
C_i	Concentration of solute in incoming underflow
C_o	Concentration of solute in outgoing underflow
C_p	Concentration of solute in precipitation
C_r	Concentration of solute in surface runoff
C_s	Concentration of solute in stream
ET_{gw}	Evapotranspiration from groundwater
ET_{vz}	Evapotranspiration from vadose zone
I	Infiltration
P	Precipitation
Q_{gd}	Groundwater discharge
Q_r	Surface runoff
Q_s	Streamflow
R	Recharge
U_i	Underflow coming into watershed
U_o	Underflow going out of watershed

Figure 4: An idealized watershed showing components of the (A) water, and (B) conservative-solute budget.

A is the area of the watershed, $[L^2]$,

U_i is the average rate of groundwater underflow into the watershed, $[L^3/t]$,

U_o is the average rate of groundwater underflow out of the watershed, $[L^3/t]$,

ET_{vz} is the average rate of evapotranspiration from the soil or vadose zone, if distributed across the entire area of the watershed, $[L/t]$,

ET_{rp} is the average rate of evapotranspiration directly from groundwater near the stream in the riparian zone, if distributed across the entire area of the watershed, $[L/t]$,

L is the dimension of length, and

t is the dimension of time.

A similar equation can be written for the concentration of a conservative solute:

$$PC_p - Q_s C_s / A + U_i C_i / A = ET_{vz} C_{vz} + ET_{rp} C_{rp} + U_o C_o / A \qquad (2)$$

where C_p is the average concentration of the solute in precipitation, $[M/L^3]$,

C_s is the average concentration of the solute in the stream at the outflow point, $[M/L^3]$,

C_i is the average concentration of the solute in the groundwater flowing into the watershed, $[M^3/t]$,

C_{vz} is the average concentration of the solute in the water evapotranspiring from the vadose zone, $[M/L^3]$,

C_{rp} is the average concentration of the solute in the water evapotranspiring from the riparian zone, $[M/L^3]$,

C_o is the average concentration of the solute in the groundwater flowing out of the watershed, $[M/L^3]$, and [M] is the dimension of mass.

Because C_{vz} and C_{rp} are virtually zero, and the value $(U_i - U_o)$ is assumed to be negligible, equations 1 and 2 reduce to:

$$P - Q_s / A = ET_{vz} + ET_{rp} = ET \qquad (3)$$

where ET is the total evapotranspiration, $[L/t]$, and

$$PC_p A = Q_s C_s \qquad (4)$$

At this point, equation 4 assumes there is no source of solute from the land surface or subsurface mineral dissolution, but these sources are accounted for later when estimating C_r, the average concentration of the solute in the surface runoff. Other portions of the hydrologic budget can also be incorporated into mass balance equations, including those that represent water and solute budgets for the vadose zone:

$$R = I - ET_{vz} \qquad (5)$$

where R is the annual average rate of recharge to the water table, $[L/t]$, and I is the average rate of infiltration at the land surface, equal to $P - Q_r$, $[L/t]$, where Q_r is the surface runoff, and

$$RC_{gw} = I C_p \qquad (6)$$

where C_{gw} is the average concentration of the solute in the groundwater $[M/L^3]$. The thirty-year time over which the ET is represented allows for the change in storage in the unsaturated zone to be neglected. Equation 5 assumes that evaporation from ponded surface water is negligible, and that data were not collected from watersheds with substantial impounded surface water bodies. Equation 6 is often

used in arid environments to estimate recharge based on the amount of precipitation and the ratio of the chloride in precipitation to that in groundwater, with the assumption that Q_r at the site location is zero [27]. Additional equations can be written for the stream water balance:

$$Q_s = Q_{gd} + Q_r \qquad (7)$$

where Q_{gd} is the average annual groundwater discharge, or base flow, to the stream network, $[L^3/t]$; the water balance relating base flow and groundwater recharge:

$$Q_{gd} / A = R - ET_{rp} \qquad (8)$$

the stream solute balance:

$$Q_s C_s = Q_{gd} C_{gd} + Q_r C_r \qquad (9)$$

where C_{gd} is the concentration of the solute in the groundwater discharge to the stream, $[M/L^3]$; and C_r is the concentration of the solute in the runoff, $[M/L^3]$; and by applying a solute balance to equation 8, a solute relation between groundwater and base flow:

$$RC_{gw} A = Q_{gd} C_{gd} \qquad (10)$$

Q_{gd}/A is often referred to as the effective recharge and R as the total recharge [20]. In this study, the term "recharge" is used to mean total recharge.

Some of these budget components can be estimated from existing data, but some would be very difficult to estimate with available data; still other components could be calculated based on the known values and the above equations if all of the other values were known. In this study, available data was used to estimate precipitation, P, and average stream flow, Q_s. Evapotranspiration was then estimated using mass balance equation 3. By combining the stream balance equations 7 and 9, another equation can be obtained:

$$Q_r = Q_s \left(C_{gd} - C_s \right) / \left(C_{gd} - C_r \right) \qquad (11)$$

That represents the fraction of stream flow that is from surface runoff as a ratio of the concentrations in the stream and groundwater discharge, otherwise known as a chemical hydrograph separation. This equation can apply to the average concentrations over a long time period, or continuous concentrations measured over a short period of time. An 18-month time period between March 2007 and August 2008 was used during this study to estimate the fraction of surface runoff in watersheds. The average groundwater discharge component of stream flow was then calculated using water balance eq 7. To do this, the concentrations of C_s, C_{gd}, and C_r were estimated. The first two could be estimated from chemical hydrographs, but the latter had to be estimated independently. The value of C_p might help in estimating Cr, but obtaining precipitation samples in sufficient quantities over a wide expanse such as Virginia is difficult, and the assumption would have to be made that the solute in the stream originated only from precipitation—not a very good assumption in most localities. Instead, bounds were placed on C_r by envisioning two different end-member processes by which solutes in the streams might have originated. In one process, it is assumed that no solutes in the stream water originate by mineral dissolution in the subsurface, but rather are either originally present in the precipitation or originate by minerals (fertilizer, road salt, etc.) that dissolve into the precipitation on the land surface. Then mass balance equation 4 can be rewritten as:

$$C_r = Q_s C_s / PA \qquad (12)$$

This first assumption leads to a second assumption—that the solute concentrations of the surface runoff and infiltration are equal.

Based on this latter assumption, the only reason the solute in the stream is more concentrated than that in the precipitation is because evapotranspiration in the watershed removed water but not solute molecules in the soil zone. The second end-member process that can explain solute concentrations in streams is the opposite of the first-that virtually all of the solute in the stream was derived from subsurface mineral reactions, and that C_r is that of rainwater, C_p. In most watersheds, the conditions are likely to lie somewhere between these two end-member processes, so in this study we made calculations assuming both end members, and then also estimated the fraction of the stream solute that originates from the subsurface. In many watersheds, the calculations of C_r and Q_r based on the two end-member assumptions were not substantially different.

The final hydrologic budget component to estimate is recharge (R) to the water table. To estimate recharge, another component had to be estimated—either C_{gw} so that either equation 6 or 10 could be used to calculate recharge, or ET_{rp} so that either equation 5 or 8 could be used for the calculation. It is difficult to estimate C_{gw} because not enough wells with water-quality data are usually available to obtain a good statistical average. ET_{rp} is not easy to estimate, but the value is relatively small compared to the other components, so a substantial error in the ET_{rp} estimate is not likely to translate into a substantial error in the recharge estimate. This relatively small value of ET_{rp} relative to ETvz is supported by the fact that in the Piedmont and Valley and Ridge Provinces the depth to the water table is greater than one meter in all regions that are not immediately adjacent to streams [7].

Riparian evapotranspiration estimation

Estimates of ET_{rp} were obtained by using the seasonal difference between the values of C_{gd}. Most watersheds show a substantial difference in C_{gd}, with values being highest in late summer and early fall and lowest in late winter and early spring. This can be attributed to the presence of riparian ET during the summer and its absence during the winter. If the riparian zone has a chance to flush out over a number of months, then in late winter, $C_{gdw} = C_{gw}$. If this is the case then equations 8 and 10 can be rewritten as:

$$ET_{rps} = \left(Q_{gd}/A\right)\left(\left(C_{gds}/C_{gdw}\right) - 1\right) \tag{13}$$

where ET_{rps} is the riparian ET rate during the summer, [L/t], and

C_{gds} is the average concentration of the groundwater discharge during late summer, [M/t], and

C_{gdw} is the average concentration of the groundwater discharge during late winter, [M/t].

It was assumed that the summer riparian ET rate occurs for about one third of the year, with a small to negligible rate operating the remainder of the year. The equation for the estimated watershed mean-annual riparian ET calculation becomes that calculated for the summer (equation 13) divided by three:

$$ET_{rp} = \left(Q_{gd}/A\right)\left(\left(C_{gds}/C_{gdw}\right) - 1\right)/3 \tag{14}$$

One can observe from equations 13 and 14 that if there is no seasonal fluctuation in the concentration of discharging groundwater ($C_{gdw} = C_{gds}$), the riparian evapotranspiration would equal zero. Our estimates of riparian ET using equation 14 yielded values similar to other estimates [14,28] in the Mid-Atlantic region (see later in Riparian ET section), and were small compared to the magnitude of recharge and groundwater discharge. Using these values of ET_{rp}, equation 3 was used to compute values for ET_{vz}. Equation 5 was then used to calculate recharge for the watersheds by reducing infiltration by the amount

of vadose-zone evapotranspiration. According to the water balance, equation 8 could also be used to calculate recharge by adding the riparian ET to the base flow, and the resulting value would be the same.

Total evapotranspiration estimation

Total evapotranspiration for the watersheds of interest was estimated by subtracting stream flow from total precipitation using eq 3 [29]. A total of 60 watersheds were selected (Table 2) that met the criteria of complete flow record availability between 1971 and 2000. These dates were chosen because precipitation data were available from the PRISM climate database [30] as mean rates for that time interval for the entire Commonwealth of Virginia. Average flow rates from that time period were obtained from the USGS National Water Information System (NWIS) database. The assumption was made that for a long period of record, such as 30 years, three components of flux out of each watershed were negligible compared to the total flow of water: (1) water-use withdrawals, (2) the net underflow through the basin, and (3) change in storage of water within the watershed. All three components are believed to be small in Virginia for nearly all of the watersheds of interest. The magnitude of water-use withdrawals are discussed toward the end of this article, and found to be relatively small. Net underflow was suspected to be substantial in only a few localized karst regions of the Valley and Ridge province; those watersheds were excluded. Watersheds with substantial surface-water impounds were not used.

Once the total evapotranspiration for each watershed was estimated, the values were related to the precipitation and temperature data from the PRISM climate database. A multiple-regression equation was created that related the mean total evapotranspiration rate of each watershed to the precipitation rate, the mean maximum daily temperature, and the mean daily minimum temperature. All PRISM climate data averaged for the 1971-2000 data period were available as a raster grid for the entire Commonwealth on 800-meter spacing. A geographical information system was used to calculate an average temperature and precipitation value for each watershed. Evapotranspiration is known to be a function of climatic variables and, in this situation, the calculated evapotranspiration data correlated well with a multiple-regression equation of the form:

$$ET = aP + bT_{max} + c\,T_{min} + d \tag{15}$$

where T_{max} and T_{min} are the mean daily maximum and minimum temperatures, respectively, and a, b, c, and d are coefficients estimated by the regression, and have the values 0.370, 0.957, −0.383, and −34.277, respectively. The regression had an R^2 value of 0.844 and a slope of 0.91. Land cover data were also considered as a potential variable in the regression, but it did not substantially improve the regression and therefore was not included in the final equation [7]. For the remainder of Virginia, equation 15 was used to estimate total evapotranspiration by locality, along with a correction for percent impervious surface

Chemical hydrograph separation

The components of stream flow-surface runoff and base flow-are represented in the hydrologic budget in equations 7 and 11. We use the term base flow to represent groundwater discharge. Numerous studies have measured the concentrations of various solutes and isotopes during storm events to separate the hydrograph components of surface runoff and groundwater discharge since the 1970s [31,32]. This classical chemical hydrograph separation approach requires collecting and analyzing individual water samples frequently, and so is labor intensive and costly for long periods of time. This high cost precluded using this approach because of the large scale of this study. As an alternative, specific conductance (SC), which has been demonstrated to be effective for chemical hydrograph separation [33], was chosen as a proxy for total solute concentration in the stream. Even with the costs of the instrumentation and its maintenance, this latter approach proved to be very cost effective because data could be collected multiple times per hour (usually every 15 minutes) continuously for 18 months.

Instrumentation was installed on 75 streams (and one spring) across Virginia at real-time gaging sites (Table 2) for SC. Data were transferred to spreadsheets where both stream flow and SC could be plotted together [7]. The SC of the base-flow component was estimated by visual inspection of the SC data. A value for the base SC was estimated at the beginning of each month and the daily values were then interpolated in between these values. Drops in the SC measurements during high-flow peaks were assumed to be from sudden inflows of surface runoff or subsurface storm flow, and conversely, time periods long after high-flow peaks were assumed to contain little surface runoff component. On occasion, there was observed high-frequency variability in SC during low-flow periods that was attributed to causes other than rainfall. The base SC was often estimated to fall in the average range of this SC, and given that the percentage of flow occurring during these periods was low, the base-flow calculations were relatively insensitive to the base-SC estimate during those times. From this knowledge, the continuous SC of the base-flow component could be estimated and plotted. The surface-runoff (Q_r) and base-flow (Q_{gd}) components were then calculated for each time interval using equations 7 and 11 for two end-members, depending on the assumed value of C_r. A SC value of 15 microsiemens per centimeter ($\mu S/cm$) was used for one end-member and a value calculated using equation 12 was used for the other end-member. SC of rainwater was not measured directly in the study area, but rather the former value was used to represent the SC of average rainwater [34]. Data collection began in March 2007 and continued for 18 months, through August 2008.

During 2007-2008, water samples were also collected at approximately six-week intervals (during normal gage maintenance visits) from 90 stream gage sites and analyzed for SC and anion concentrations of chloride (Cl), sulfate, and nitrate [7]. Chloride tends to be the most conservative ion in the subsurface for most regions [34] and was therefore used as an indicator of the component of the dissolved salts that originated at the land surface. By using the Cl/SC ratio, the fraction of salts that were dissolved at the land surface, versus that dissolved by subsurface mineral dissolution could be estimated. A ratio of zero indicated zero salts from the land surface. To obtain the ratio that would likely represent zero salts from mineral dissolution, a situation was chosen in which land-surface salts would completely dominate the stream chemistry signal. Road salt runoff after a heavy winter road salting event was chosen to determine this ratio. The Difficult Run watershed in Fairfax County, Virginia, was sampled at 24 locations in January 2009, following a small rain event that followed a period of heavy road salting. A plot of chloride concentration versus SC for the Difficult Run samples and all of the other watershed samples [7] revealed that a Cl/SC ratio of about 0.33 was observed for all of the samples with a SC of greater than 1,000 $\mu S/cm$ (heavy road salt content). This ratio is characteristic of a stream that has 100 percent surface salts and virtually no mineral dissolution component. Conversely, many streams had a ratio below 0.03, indicating a low average surface-salt composition.

The mean specific conductance of the streams measured in Virginia is a reflection, in large part, of the solubility of minerals in the soils and rocks through which the groundwater passes [34]. Watersheds in the

Valley and Ridge Province had the highest mean SC values; especially the watersheds that were underlain by carbonate rocks, frequently have mean SC values in excess of 300 $\mu S/cm$ [7]. Conversely, watersheds in the Blue Ridge and Coastal Plain Provinces frequently had mean SC values less than 100 $\mu S/cm$ because of the relative abundance of quartz sand and lack of soluble minerals in the soils and rocks. Many of the watersheds that had groundwater-discharge SC values consistently well below 100 $\mu S/cm$ were too difficult to interpret; this is because the precipitation event did not create a signal that was substantially different than the random noise in the SC signal that was present during the measurement period. A second major reason why some watershed SC values could not be interpreted was because some streams had a substantial volume of water impounded upstream in a reservoir. These reservoirs controlled the flow in the downstream reaches and at the gage such that the natural response of the flow and SC to the precipitation events was muted. Watersheds with low SC or impounded water were not used for base-flow calculations, even though some of these watersheds were initially instrumented. Out of the 75 streams instrumented, only data from 48 were used for base-flow calculations, but historical SC and flow data from an additional 4 streams on the coastal plain of Maryland and Delaware were also used.

Regression analysis

In order to estimate the hydrologic budget components for all of Virginia, the results from the watersheds analysis were transferred to other localities using two regression equations. The first equation was that used to estimate total evapotranspiration, described previously. The second equation expressed the fraction of the precipitation that results in surface runoff as a function of landscape characteristics of the watersheds. The same landscape characteristics of Virginia localities (counties and independent cities) could then be put into the regression equation to obtain the surface-runoff-fraction component for each locality. Precipitation and temperatures for each locality were obtained from the PRISM climate database, and the evapotranspiration was obtained using that data in the ET regression equation developed from the watershed data. Surface runoff and ET were also adjusted for impervious surface (as described below). Riparian ET for the localities was estimated with a regression equation of percent marsh in the landscape based on temperature and topographic slope. With these estimates of surface-runoff-fraction and total and riparian ET for each locality, recharge and net-groundwater-discharge components were calculated by mass balance (Table 1).

A variety of different landscape characteristics were evaluated for correlation with the watershed estimates of surface runoff, including the physiographic province, land cover, rock type, median topographic slope, mean soil permeability, and percent impervious surface. After examination of each of these factors in the regression equation it was concluded that physiographic province, rock type, and percent impervious surface were capable of explaining much of the variability in the runoff between watersheds, and that topographic slope, soil permeability, and land cover were only capable of improving the fit by a very small insignificant amounts. There was also substantial amount of cross-correlation between these factors, for example between rock type and soil permeability and between land cover and topographic slope. Only a few watersheds had substantial percentages of impervious surface, which was not enough to determine the contribution to runoff implicitly in the regression. However, previous investigations [35] on the role of impervious surface on runoff have indicated an average of 29 percent increase in runoff for areas with 50 percent impervious surface. This ratio of surface runoff to impervious surface was applied to the regression estimate of surface runoff, and did improve the fit in the few watersheds that had substantial impervious surface cover. The same study that indicated the increase in surface runoff indicated a 38 percent decrease in ET for areas with 50 percent impervious surface. This percent of ET decline was also applied to the regression estimate of for the localities as a function of the climatic variables. These two effects of impervious surface were negligible in most of the counties, but substantial in the independent cities that had relatively high percentages of impervious surface.

Results

Estimates of hydrologic budget components

The components of the hydrologic budgets were first calculated for the watersheds based on the stream flow, climatic data, and chemical hydrograph separations in the watersheds (Table 1). These watershed results were used to create regression equations that described total ET and the mean annual surface runoff as a function of rock type and physiographic province. The components of the hydrologic budgets for all the localities were then calculated based on the climatic data for the localities, the regression equations for ET and surface runoff, and the water balance equations. Estimates of surface runoff and recharge may be particularly useful for water managers.

The hydrologic budget components were estimated for a number of watersheds across Virginia as an average annual rate in centimeters per year during the period 1971-2000. The precipitation was estimated by using the PRISM data directly without any additional interpretation. Mean annual precipitation rates for the watersheds used for the ET and chemical hydrograph separation calculations range from less than 100 cm/yr in the watersheds in the Shenandoah Valley to more than 125 cm/yr in some high-elevation watersheds in the Blue Ridge, Valley and Ridge, and Appalachian Plateau Provinces.

Total evapotranspiration

Total evapotranspiration was calculated for the watersheds using the water (mass) balance approach described earlier in sections 2.1 and 2.3 of this article in which the mean annual stream flow from 1971-2000 is subtracted from the mean annual precipitation of the same period multiplied by the watershed area. Results indicate that mean annual total ET rates in the watersheds evaluated in Virginia range from 60

cm/yr in some of the higher elevation watersheds in western Virginia to 80 cm/yr in some of the wetter and warmer watersheds in southwestern and southern Virginia (Figure 5). This range of values is very similar to that of potential ET estimated across Virginia at weather stations by the University of Virginia Climatology Center (http://climate.virginia.edu/va_pet_prec_diff.htm). Expressed as a percentage of precipitation, the ET rates for the watersheds range from less than 60 percent in some of the higher elevation watersheds in western and southwestern Virginia to more than 70 percent in some of the warmer watersheds in southern Virginia. When the ET rates for these watersheds were related to the mean annual precipitation and minimum and maximum daily temperature for the same watersheds, a regression (equation 15) was developed that contained four parameters. Different forms of the regression equation were fit to the data but a standard error of regression analysis indicated that four parameters were optimal for estimating the ET. A plot of the ET calculated using the water balance versus that estimated by the regression (equation 15) (Figure 6) indicates a relatively good fit (R^2=0.844, slope=0.91) and that ET in Virginia is controlled predominantly by variations in climate.

Chemical hydrograph separation

Hydrographs and records of specific conductance during the same period were obtained and plotted for 100 watersheds across the region [7]. In addition to the 75 watersheds instrumented with real time specific-conductance probes during this study, 25 watersheds that had historical specific conductance records were also examined. Three of these watersheds were from Maryland, one was from Delaware, and one was instrumented in Opequon Creek at Martinsburg, West Virginia. The watersheds in Maryland and Delaware were added as additional information for the Coastal Plain Province, as there were only two watershed records from the Virginia Coastal Plain that proved to be useful for chemical hydrograph separation.

Base flow

Base flow in 52 watersheds was estimated using the chemical hydrograph separation method described in section 2.4 of this article. Specific conductance was measured at the watersheds for a period of approximately 18 months between March 2007 and August 2008. One challenge in estimating a long-term mean base flow for a watershed is the assumption that this 18-month period represents average long-

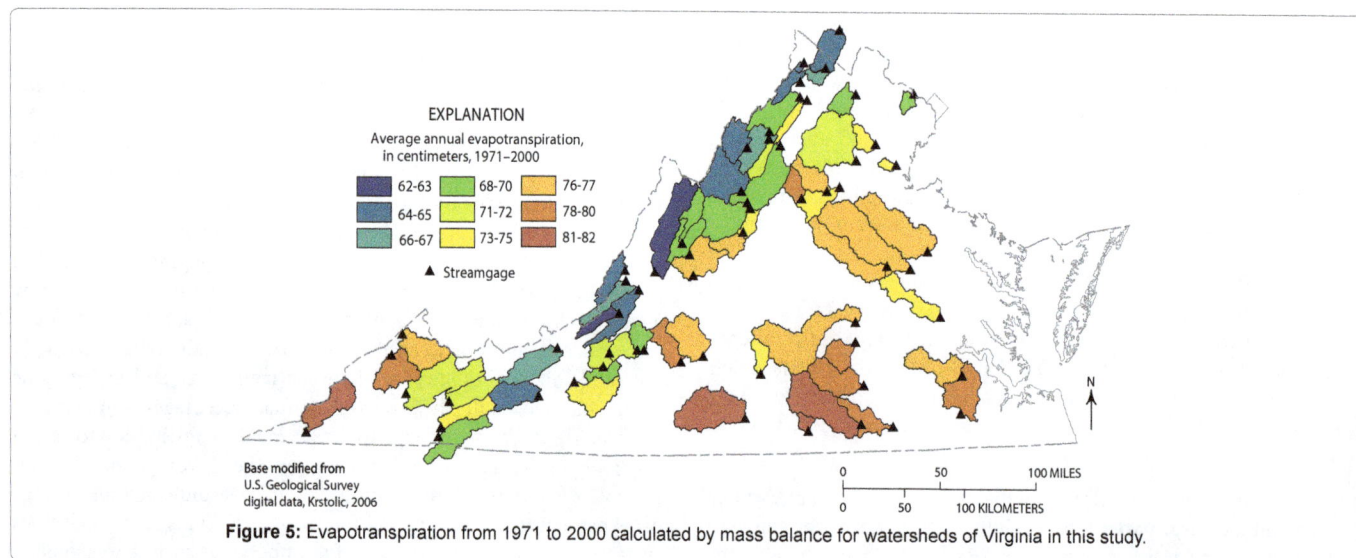

Figure 5: Evapotranspiration from 1971 to 2000 calculated by mass balance for watersheds of Virginia in this study.

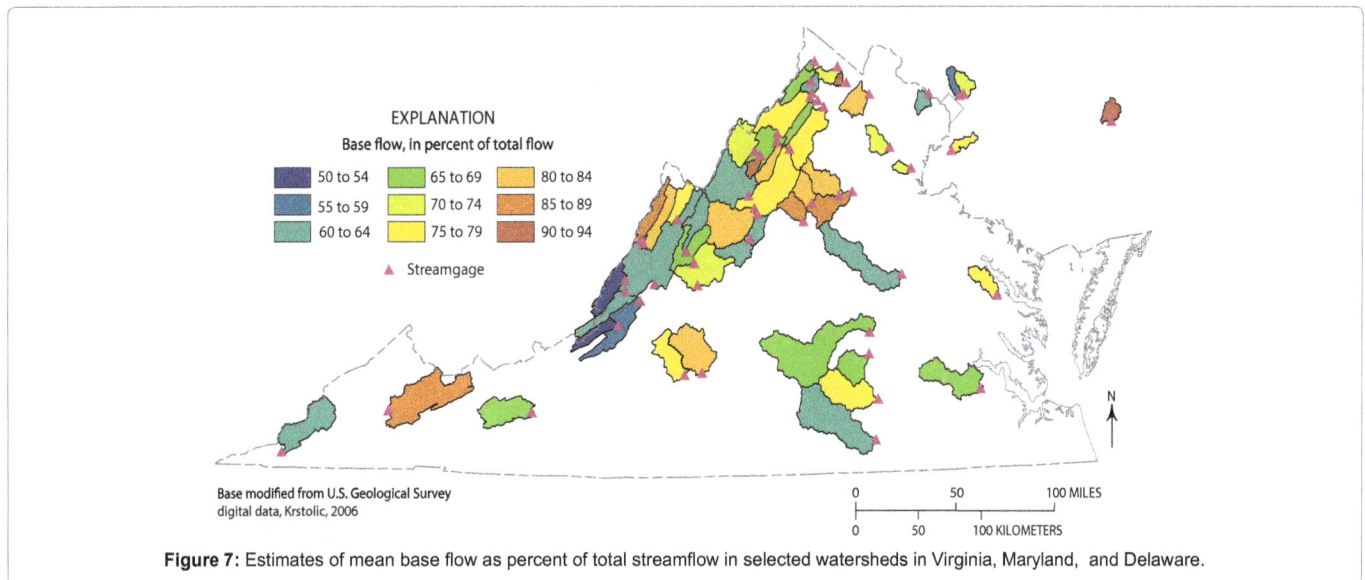

Figure 6: Comparison of evapotranspiration calculated by mass balance versus that estimated through the regression equation.

where BF equals the base-flow fraction and Q equals the mean monthly stream flow in cubic feet per second.

The long-term past monthly flows (Q) for each watershed were compiled and ranked by flow magnitude and input into equation 16 to obtain a flow-weighted, long-term, adjusted mean base flow. A long-term-adjustment ratio was then calculated by dividing the long-term adjusted mean base flow by the observed mean base flow. These long-term adjustment ratios were multiplied by estimates of the base flow that assumed the origin of the specific conductance was either from surface salts or subsurface mineral dissolution (as described earlier in section 2.4 of this article). An average base flow was then calculated from the two end-members based on a weighting term that is a function of the SC/Cl ratio.

Results of the base-flow analyses demonstrated a substantial difference in base-flow indices across Virginia (Figure 7). The base-flow index is the percentage of the mean annual stream flow that is base flow over the entire period of record, which in this study includes the long-term adjustments. The Valley and Ridge carbonate rocks consistently yield base-flow indices of over 90, whereas the Valley and Ridge siliciclastic rocks consistently yield values between 60 and 70 percent. The Piedmont watersheds also yield values typically between 60 and 70 percent, and the Blue Ridge watersheds yield values typically between 80 and 85 percent. The results revealed that the average base flow using this chemical separation was 72% of stream flow, as compared to 61% using a graphical separation technique. The latter value is typical of those presented in earlier studies for this region [16]. This primary finding led to the development of the regression equation for surface runoff as a percent of precipitation that was predominantly a function of the physiographic province and rock type (described earlier in section 2.5 of this paper). The range of base-flow indices in the individual watersheds ranged from under 60 percent in some of the siliciclastic rocks of western Virginia to more than 90 percent in some of the carbonate watersheds of the Shenandoah Valley. The sandy coastal plain watershed in Delaware also yielded a value over 90 percent.

Surface runoff

The long-term mean surface-runoff component of the hydrologic budget of each watershed was calculated by subtracting the long-

term flow conditions for the watershed. Upon examination of stream-flow records it was determined that a substantial number of the watersheds had flow conditions during the 18-month period of record that did not adequately represent long-term mean conditions. These watersheds were in a period of drought (mostly in southern Virginia) during that time, yielding higher than usual base-flow fractions and lower than usual surface-runoff fractions. To overcome this problem, base-flow estimates were adjusted to be consistent with long-term mean flow conditions. To accomplish this the monthly flow for each watershed was plotted versus its base-flow calculation [7] Log-linear curves of the form:

$$BF = a \ln (Q) + b\,Q + c \qquad (16)$$

were then fit to these data yielding the parameters a, b, and c,

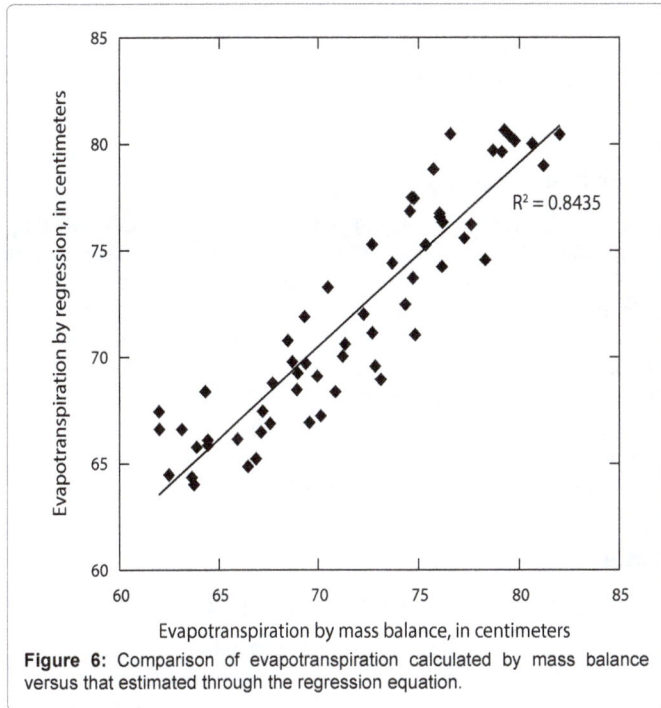

Figure 7: Estimates of mean base flow as percent of total streamflow in selected watersheds in Virginia, Maryland, and Delaware.

term base-flow component (base flow) from the total stream flow. The surface runoff values for the different watersheds across Virginia range from 5 cm/yr or less in the Valley and Ridge Province carbonate rocks and the Blue Ridge Province to 20 or more cm/yr in some of the siliciclastic rocks of the Valley and Ridge Province (Figure 8). The regression equation described in section 2.4 was used to estimate the surface runoff as a percentage of the precipitation depending on the physiographic province and rock type within that province (Table 3). In order for the regression to reflect only natural surfaces, an adjustment was made to calculate a "natural runoff" whereby the percent impervious surface was subtracted from the percent run off. When the regression was later applied to the localities, the effect of impervious surfaces was reintroduced, as described in section 2.5. The estimated percent of precipitation estimated to end up as surface runoff varied between approximately 4 and 16 percent.

Riparian evapotranspiration

Riparian Evapotranspiration, ET_{rp}, was calculated for each of the watersheds in which the chemical hydrograph method was employed, using the seasonal difference in specific conductance (eq 14). The values calculated for

ET_{rp} ranged from less than 1.2 cm/yr to more than 10 cm/yr (Figure 9). Estimates of ET_{rp} from an earlier investigation based on a combination of graphical hydrograph separation methods [16] also yielded a similar distribution of values of ET_{rp} for watersheds in Virginia (Figure 9).

Groundwater recharge

The mean recharge rate for a watershed can be calculated by subtracting the mean rate of vadose zone ET from the mean rate of infiltration (Eq 4). In our situation we have calculated a total ET and a riparian ET and the vadose zone ET is the latter subtracted from the former. Also the infiltration is the surface runoff subtracted from the precipitation, and given that we have values now for the latter two we could calculate the infiltration and then the recharge. As this analysis produces a closed hydrologic budget, the recharge can also be calculated by adding the groundwater discharge and the riparian ET with identical results. The calculated recharge rates for the various watersheds ranged between 20 and 45 cm/yr.

Estimates for localities

In order to apply the ET regression equations to the localities,

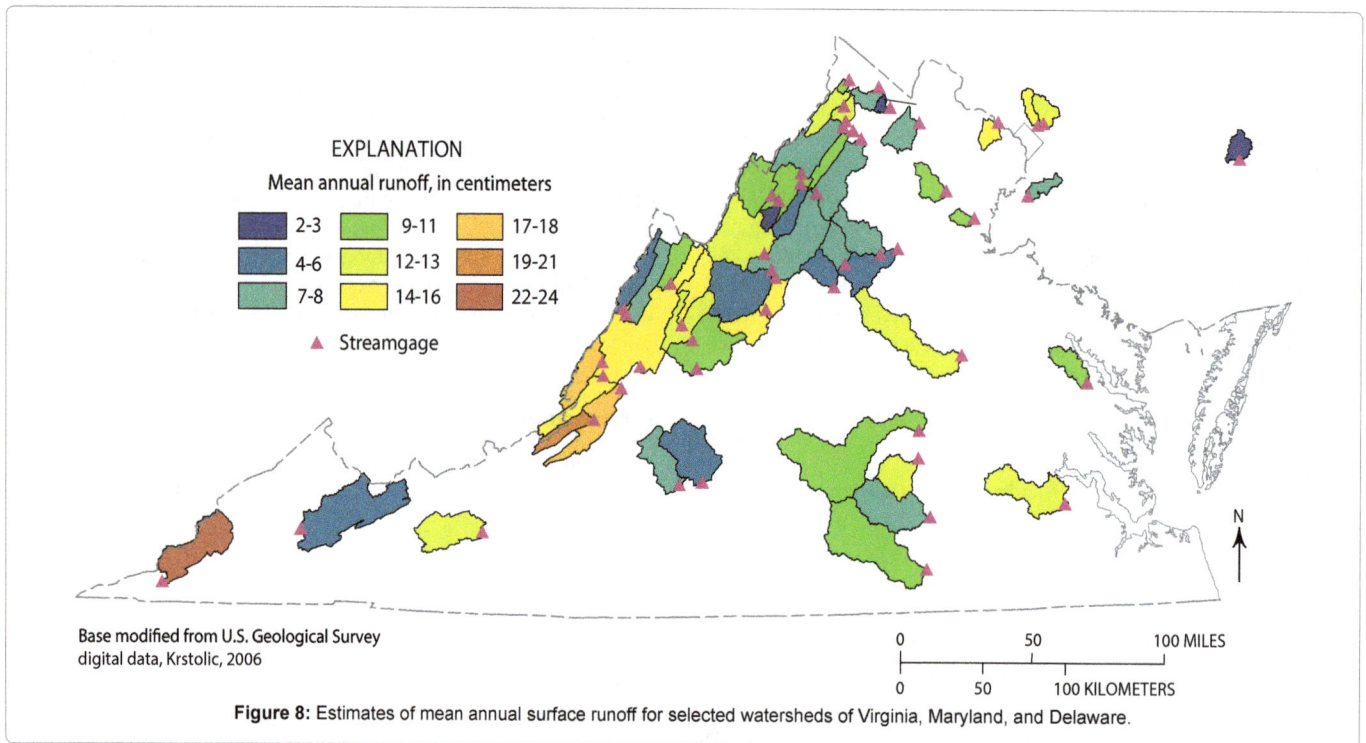

Figure 8: Estimates of mean annual surface runoff for selected watersheds of Virginia, Maryland, and Delaware.

Physiographic Province	Regression equations with linear constants and rock type variables*	Regression parameter values representing percent runoff from each corresponding rock type							
		a	b	c	d	e	f	g	h
Blue Ridge	R = aMV+bMS+cPL+dMB	1.0**	2.8	7.1	13.1	---	---	---	---
Coastal Plain	R = aFG+bMG+cCG	11.0	7.5	4.1	---	---	---	---	---
Piedmont	R = aNW+bSE+cMB+dCG	10.5	8.9	13.1	4.1	---	---	---	---
Valley and Ridge	R = aCD+bCOL+cOD+dOS +eSSL+fDS+gMS+hAP	19.6	1.0**	4.6	8.1	2.8	24.4	11.2	17.8

*MV=fraction metavolcanics, MS=fraction metasediments, PL=fraction plutonic, MB=fraction Mesozoic Basin, FG=fraction fine-grained sediment, MG=fraction mixed-grained sediment, CG=coarse-grained sediment, NW=fraction northwestern zone, SE=fraction southeastern zone, CD=fraction Cambrian dolomostones, COL=fraction Cambrian-Ordovician limestones, OD=fraction Ordovician Dolostones, OS=fraction Ordovician siliciclastics, SSL=fraction Silurian siliciclastics and limestones, S=fraction Devonian siliciclastics, MS=fraction Mississippian siliciclastics, AP=fraction Appalachian Plateau siliciclastics, R=percent of precipitation that runs off, see table 10 for the fractions of these rock types in the watersheds. **Values of 1.0 were assigned when the regression attempted to fit a value below zero.

Table 3: Runoff regession equations and their parameter values.

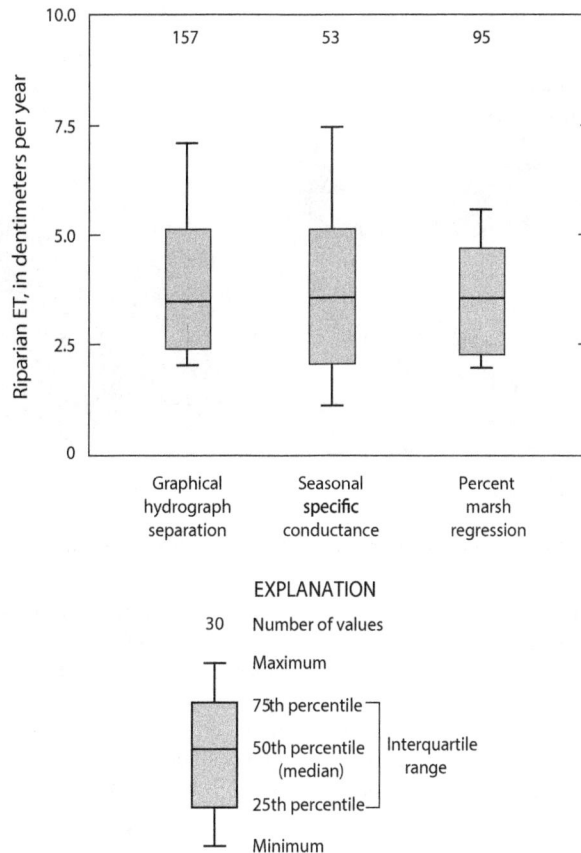

Figure 9: Comparison of different methods for estimating riparian evapotranspiration (ET).

certain climatic and land cover (marsh) variables were first needed for each locality. The climatic variables needed included the mean annual temperature, the mean annual precipitation, the mean daily maximum temperature, the mean daily minimum temperature, and the mean difference in daily temperature. In addition, the percentage of physiographic province and rock types in each county were required in order to apply the regression used to calculate the percent of precipitation that becomes surface runoff [7]. Resulting hydrologic budget components for the localities include precipitation, total ET, riparian ET, surface runoff, infiltration, recharge, net groundwater outflow, and net total outflow (Table 2).

Total evapotranspiration

The total ET for the localities of Virginia was estimated by the climate regression (eq 15) and the values thus reflect the local climatic conditions of each locality (Figure 10). The lowest values are 62 cm/yr or less in some of the far western and northern counties; these include Highland and Frederick Counties in the extreme north and west and Fairfax County in the northeast. The latter is relatively low because of the relatively high amount of impervious surface in the County. Many of the independent cities also have estimated total evapotranspiration of 62 cm/yr or lower because of the relatively high amounts of impervious surface (Figure 10). The highest evapotranspiration values are > 80 cm/yr and occur typically in the warmest counties in the southern region of Virginia. Lee and Patrick Counties, in southwestern Virginia, also have relatively high ET rates because of their high mean annual precipitation rates. Another useful way to express ET is by its relation to P, or as the ratio of ET to P. This is the fraction of precipitation that is evaporated

or transpired. For independent cities, this estimate is typically less than 55 percent and between 55 and 60 percent in southwestern Virginia. The value for Fairfax County is also in the latter range because of the relatively high amount of impervious surface, and the Atlantic coastal counties of Accomac, Northampton, and Virginia Beach are also in this range because of the effect of higher humidity near the ocean. The areas with the highest ratios of ET/P (above 66 percent) are the warmest counties in southern and south-central Virginia. Shenandoah County in the north is also in this upper range because of the relatively low mean annual precipitation rate. The values of ET estimated in this study agree reasonably well with other regional estimate of ET in Virginia [36].

Riparian evapotranspiration

Use of the seasonal SC estimates to estimate ET_{rp} on a local basis proved difficult because there was not an obvious spatial trend in the data. Therefore a third method was used in which three factors—the amount of riparian vegetation present, the mean annual air temperature, and the topographic relief—were used to estimate the ET_{rp}. The first factor was an indicator of the amount of riparian seepage present, and was represented by the percent marsh (or wetland) in the locality in the National Land Cover Database. The second factor related to the intensity of the total ET in the watershed, and the third factor represented the relative width of the floodplains likely to occur in the locality. By including the temperature and slope rather than using the percent marsh alone, a more consistently varying estimate of ET_{rp} was developed across Virginia. A correlation was established ($R^2=0.6031$)

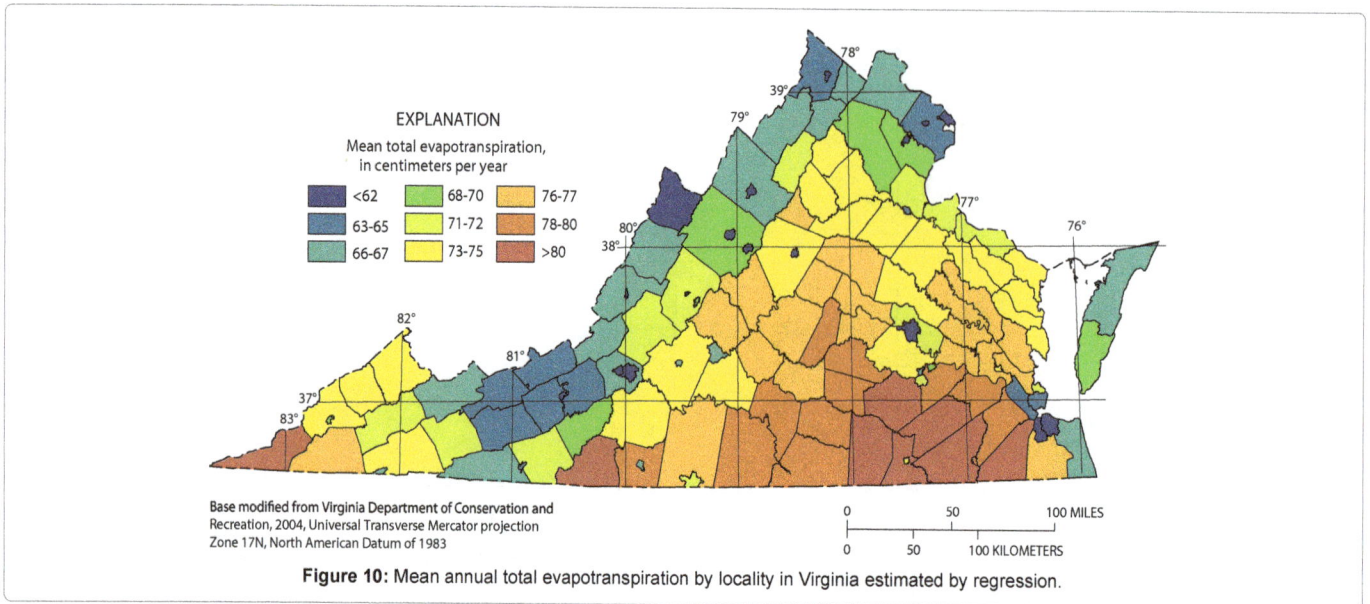

Figure 10: Mean annual total evapotranspiration by locality in Virginia estimated by regression.

between the slope and temperature in each of the 134 localities and the fraction of land cover that is marsh, using the relation:

$$\text{Log (FM)} = 0.167*T -0.067*S-11.085, \tag{17}$$

where FM is the fraction of land cover that is marsh, T is the mean air temperature (°F), and S is the topographic grade (dimensionless) (Figure 11). The riparian ET was then calculated using the formula:

$$\text{ET}_{rp} = -0.115*PS/\log(FM), \tag{18}$$

where PS is the fraction of pervious surface in the locality. The constant in this equation was adjusted such that the mean ET_{rp} of the localities was the same as that obtained for the watersheds in the other two estimates. This method also created a range of ET_{rp} similar to that produced by the other two methods (Figure 9). The uncertainty in the estimate of ET_{rp} for any given locality is relatively high compared to the magnitude of ET_{rp}, but given that the magnitude of ET_{rp} is small relative to other budget components, such as the total ET and the groundwater discharge, the effect of this uncertainty on the estimate of total recharge (which is calculated by adding the ET_{rp} to the base flow, or effective recharge) is relatively small.

The values estimated using equations 17 and 18 are strongly affected by the mean annual air temperature and topographic relief present in each locality (Figure 12). The values represent an estimate of the mean annual riparian ET for the entire area of the locality (not the local ET rate in the riparian zone itself). The lowest values are less than 2.5 cm/yr and are consistently found across the Valley and Ridge Province. Counties in the Blue Ridge province and vicinity of Washington D. C. have values that range 2.5 to 3.4 cm/yr. Values in the Piedmont Province and the northern counties of the Coastal Plain range between 3.5 and 5.7 cm/yr, whereas values in southeastern Virginia and the Tidewater area are between 5.8 and 7.4 cm/yr. These values all have an uncertainty associated with them that we estimate to be plus or minus 2 cm/yr, based on the range of values that have been estimated by other methods (Figure 9).

Net total outflow

The equivalent of total stream flow for a locality is the net total outflow (Table 1), which was calculated by subtracting the estimated

mean annual total ET from the mean annual precipitation. This term has also been referred to as the available precipitation, because it is the fraction of precipitation that is available in terms of surface water or groundwater. Results indicate that the net total outflow varies from about 30 to 50 cm in the Shenandoah Valley and Piedmont of central and southern Virginia to over 50 cm in the mountains of southwestern Virginia and the tidal regions of southeastern Virginia.

Surface runoff

Surface-runoff regression equations were used to predict the ratio of surface runoff to precipitation based on the physiographic province and bedrock type (Figure 13). Surface runoff rates in cm/yr for the localities were obtained by multiplying the mean annual precipitation rate by the runoff ratio. The percent of precipitation that rapidly runs off is estimated to range between 6 and 39 percent, based on locality in Virginia (Figure 14). Values less than 10 percent occur typically in the Blue Ridge Province or sections of the Coastal Plain where sandy soils are prevalent. Values greater than 20 percent occur in the Appalachian Plateau in southwestern Virginia, and in independent cities where there is a relatively large fraction of impervious surface. The mean annual values of surface runoff are controlled partly by the fraction of precipitation that runs off. Values of 8 to 10 cm/yr occur typically in the Blue Ridge or Coastal Plain. The carbonate rocks in the Shenandoah can produce similarly low values because precipitation is also relatively low there. Values of 22 cm/yr or greater occur typically in the Appalachian Plateau and in the independent cities.

Net groundwater discharge

The term, base flow, was used for the watersheds to indicate the groundwater discharge from that watershed to the stream, with the assumption that the groundwater discharge across the watershed divide was negligible. Counties and cities, however, have political boundaries that frequently do not align with subsurface watershed boundaries, resulting in the potential for substantial discharge of groundwater from those localities that is not base flow to streams. For example, in some small independent cities there are no prominent streams, and in some counties along the Chesapeake Bay much of the groundwater may discharge directly to the bay or coastal marshes. Both the inflow and outflow of ground water across non-stream locality boundaries may

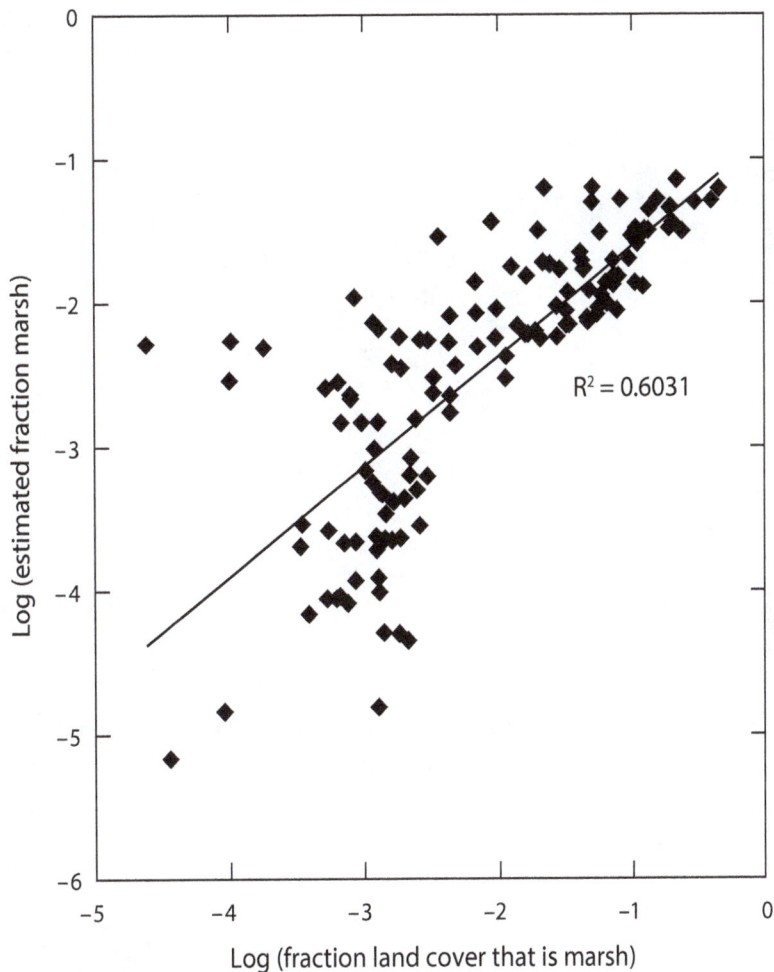

Figure 11: Comparison of fraction of locality land cover that is marsh and that estimated using temperature and topographic slope.

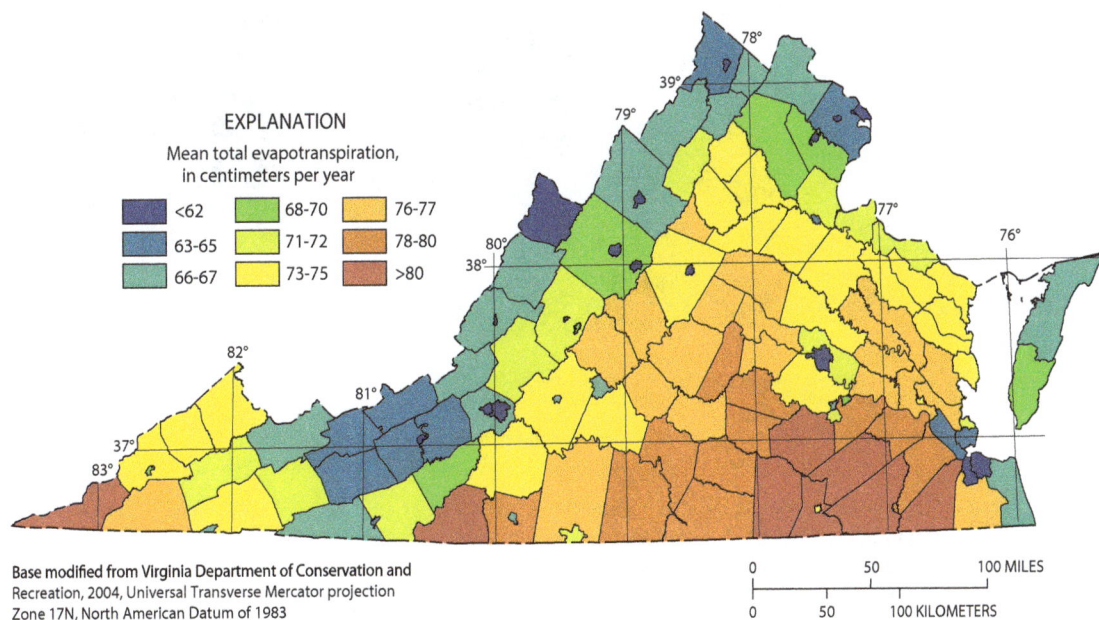

Figure 12: Estimates of mean annual riparian ET in Virginia from 1971 to 2000 by locality.

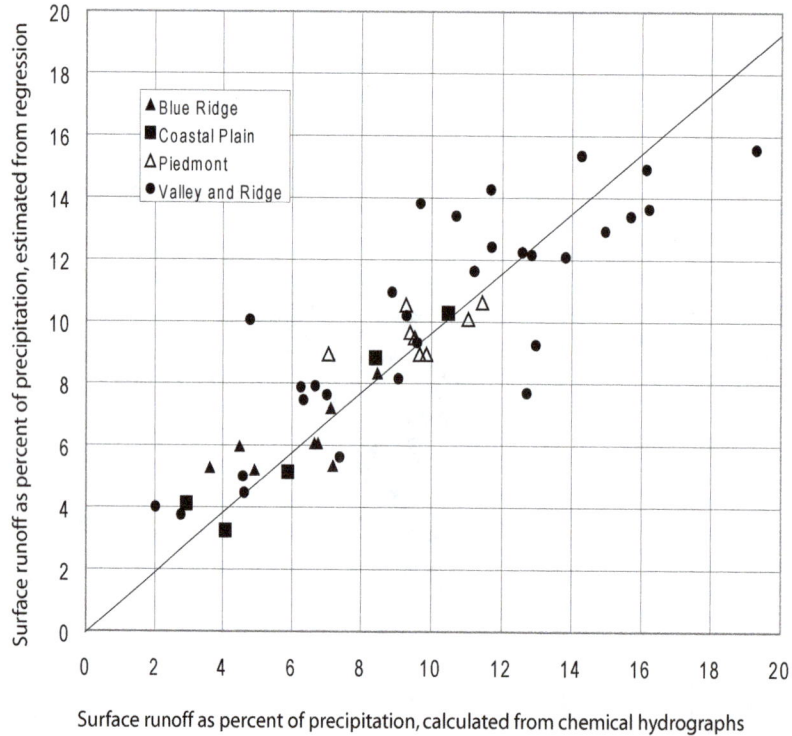

Figure 13: Comparison of calculated surface runoff versus that estimated by the regression.

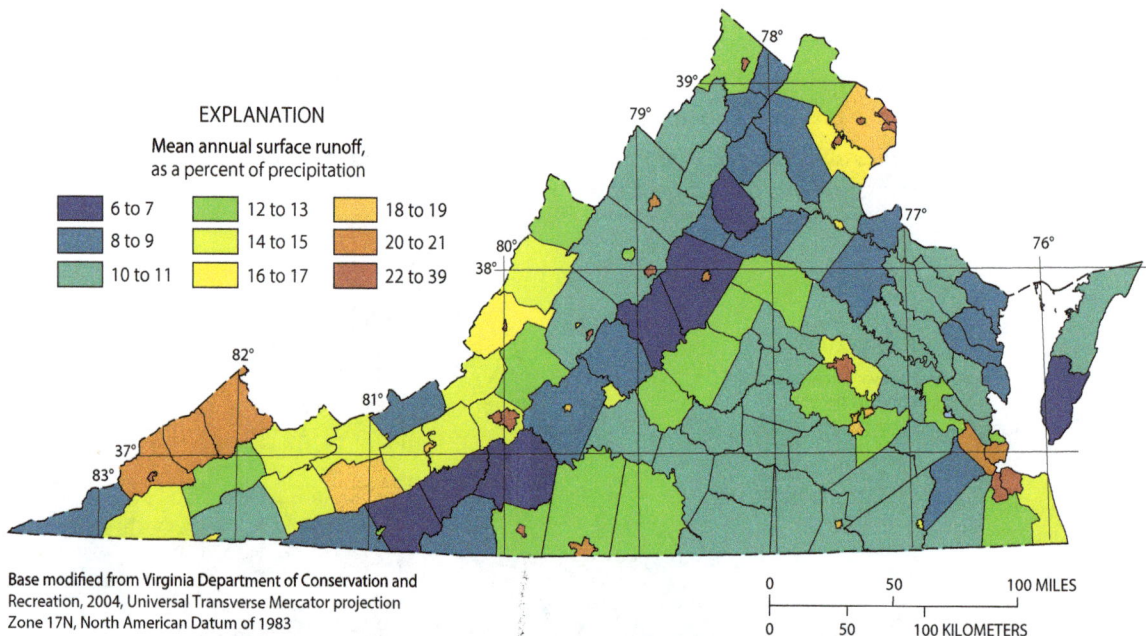

Base modified from Virginia Department of Conservation and
Recreation, 2004, Universal Transverse Mercator projection
Zone 17N, North American Datum of 1983

Figure 14: Estimates of surface runoff as a percent of precipitation in Virginia from 1971-2000 by locality.

be substantial, but only the net discharge (groundwater outflow minus inflow) is created by recharge within the locality, and is of concern in this study. Therefore, when describing discharge of groundwater from localities, the term "net groundwater discharge" is used rather than "base flow", although much of that discharge may actually occur as base flow. The estimated net groundwater discharge for the localities

is calculated by subtracting the estimated surface runoff from the net total outflow.

The net groundwater discharge for the localities varies from less than 22 cm/yr to approximately 40 cm/yr. Low values (<22 cm/yr) occur in the regions of western Virginia where precipitation is low or surface runoff is high, and in the Piedmont Province where total ET is relatively high. Alternatively, high values (>30 cm/yr) occur in the Blue Ridge Province where precipitation is high and surface runoff is low, and in counties where precipitation is high, such as Lee and Patrick Counties of southwestern Virginia. Another way to evaluate the net groundwater discharge is to estimate its value as a percentage of the net total outflow from a locality. The remainder of the net total outflow is by shallow rapid runoff processes. The percentage of net total outflow that is net groundwater discharge is the equivalent of a base-flow index for a watershed (Figure 15). The areas where the percent net groundwater discharge is low (less than 60 percent) are typically in areas of high surface runoff (the Appalachian Plateau and areas with a highly impervious surface). The areas where this value is high (75 percent or greater) are those with low surface runoff (the sandy soil regions of the Blue Ridge Province and Coastal Plain).

Infiltration

This means annual infiltration rate is calculated for the localities by subtracting the surface runoff from the precipitation. For this difference to represent actual infiltration, evaporation from ponded surface water must be negligible, which we believe to be the case for most localities. For localities where may not be the case (where there are large volumes of impounded water), this term includes the evaporation from ponded surface water. The rate is lowest (<95 cm/yr) typically in the Valley and Ridge Province and in areas of high impervious surface. The rate is highest (>105 cm/yr) in areas of high precipitation or sandy soil (such as the Blue Ridge Province). A large fraction of infiltration is subsequently lost to vadose ET; the remainder is groundwater recharge.

Groundwater recharge

The recharge rate to groundwater is important when planning for long-term groundwater resource use in any region. The first process that leads to groundwater recharge is the infiltration of rainfall into the ground. The recharge for the localities was calculated by subtracting the vadose zone ET from the infiltration. The vadose zone ET is defined here as the total ET minus the riparian ET. The exact equivalent

value for recharge can be arrived at by adding the riparian ET to the groundwater discharge. The localities with the lowest mean recharge rates (<25 cm/yr) are those in western Virginia in the Valley and Ridge or Appalachian Plateau where siliciclastic bedrock is present (Figure 16). The localities with the highest recharge (>35 cm/yr) are in the Blue Ridge Province, or where precipitation is high, or where ET is relatively low (the coastal localities).

Uncertainties in estimates

There are many uncertainties inherent in a study such as this one. First, the locality estimates included in this article are averages over each locality, and actual values may vary significantly within a locality based on the character of the local bedrock, land cover, and topography. The averages are also long-term mean estimates, and actual values of many of the components can vary significantly from year to year, and even more so from month to month, based on temporal variations in precipitation and air temperature. For example, recent studies in the Shenandoah Valley of Virginia have shown that groundwater recharge rates can vary significantly with annual precipitation, resulting in recharge rates which differ by a factor of two or more for dry versus wet years, and for valleys versus ridge tops [28,37,38].

Additionally, each component of the hydrologic budget that was measured or estimated from existing data is no more accurate than the assumptions that went into interpreting those measurements or data. Therefore, the precipitation data that was obtained from the PRISM climate group is limited to the accuracy of those data that are based on algorithms that interpolate precipitation data at stations throughout Virginia and attempt, for example, to account for changes in elevation. Watershed ET estimates were based on the assumption that long-term precipitation minus stream flow equals ET, and locality estimates were based on the ET regression derived from the watershed ET values and climatic factors. Therefore, individual ET averages for localities may vary by a few centimeters (associated with potential error in the watershed ET and regression). There are two uncertainties inherent in the surface runoff estimates: (1) the assumptions in the

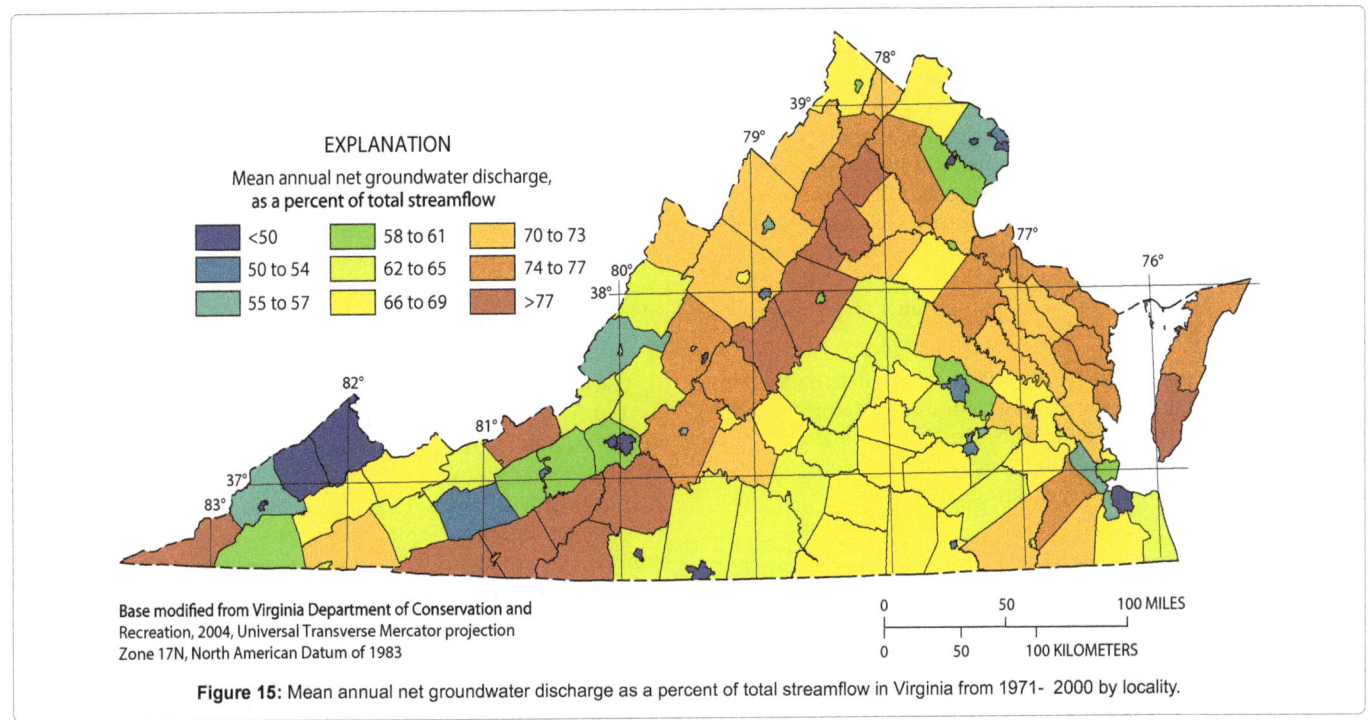

Figure 15: Mean annual net groundwater discharge as a percent of total streamflow in Virginia from 1971- 2000 by locality.

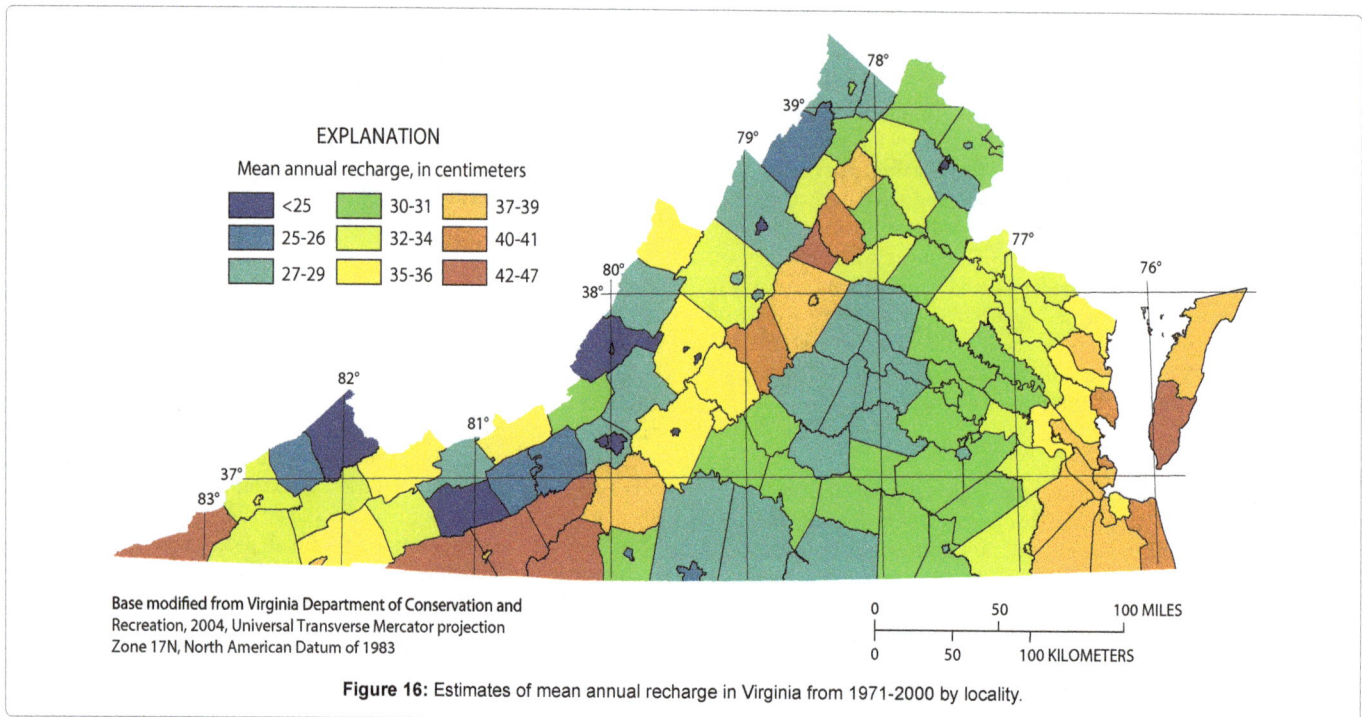

Figure 16: Estimates of mean annual recharge in Virginia from 1971-2000 by locality.

chemical hydrograph separation technique, and (2) uncertainty in the regression that estimates surface runoff based on province and rock type. Although the chemical hydrograph method provides additional information in comparison to that of standard graphical methods, estimates are made during the analysis, such as the base-flow specific conductance that is estimated by visual inspection. Also, recharge is assumed to be not so rapid that ET does not intercept the infiltration; this may not be the case in every type of terrain. The regression can easily include an uncertainty of 2 cm/yr, and 5 cm/yr in the Valley and Ridge Province. Given that the other components, such as recharge, are estimated by combining other components, these errors are potentially cumulative. The estimates of ET and surface runoff in the regions with high impervious surface (many of the independent cities) have been adjusted based on a somewhat general observation of the behavior of ET and runoff in such areas [35]. Different impervious surfaces can impart very different hydrological effects, and scaling up these behaviors into more regional estimates of system response can often be critical [39]. Thus the estimates for many of the independent cities have a higher degree of uncertainty than those localities with low percentages of impervious surface. Withdrawals of water for human use were not included for in this study, as the magnitude of the natural fluxes to the withdrawals [7] showed that the latter are usually quite small relative to the former.

Overall, given the relative reliability of the precipitation data [40], the agreement of the ET estimates with other recent estimates [26], and the long history of streamgaging by the U. S. Geological Survey, we believe the values of the budget components for the localities determined in this study are very useful estimates.

Conclusions

A study was undertaken to estimate the components of the hydrologic cycle for watersheds and localities (counties and independent cities) across Virginia. The components were estimated as long-term mean annual fluxes for each watershed or locality because such values are often needed by water-resource planners. The actual

values can, or course, vary greatly in time and space within localities. Flux estimates of components of the hydrologic cycle were made by creating water and solute budgets in which the various components balanced. The water and solute balance approach was combined with regression equations that were developed based on climatic and land surface characteristics. Mean annual precipitation was estimated for watersheds using the PRISM climate data from 1971-2000. Mean annual total evapotranspiration (ET) was estimated for watersheds by subtracting the long-term mean annual stream flow from the area of the watershed multiplied by the long-term mean annual precipitation. Surface runoff and base flow for the watersheds were estimated by using chemical hydrograph separation on real-time stream flow records for approximately 18 months during March 2007 through August 2008. These separations were performed using specific conductance. The results of the separation revealed that the average base flow using this chemical separation was 72% of stream flow, as compared to 61% using a graphical separation technique. This difference is consistent with previous chemical hydrograph studies, but is the first time this has been demonstrated to be consistent on a large scale and with a large number of watersheds. Riparian ET for the watersheds was estimated by comparing the mean summer versus mean winter specific conductance values of the base flows. Infiltration and recharge for the watersheds were calculated using the water balance assumption.

Mean annual precipitation for each locality was estimated using the PRISM climate data from 1971-2000. Mean annual total ET for the localities was calculated using a regression equation based on precipitation, the mean minimum daily temperature, the mean maximum daily temperature, and how these parameters varied with the ET values calculated for the watersheds. The surface runoff for the localities was estimated as a percent of precipitation by developing a regression equation, based on the relative area within any given physiographic province or rock type. Parameters for this equation were calculated by fitting these land characteristics to the surface runoff percentages observed in the watersheds. Net total outflow for the localities was estimated by subtracting the total ET from the precipitation. Net groundwater discharge for the localities was estimated by subtracting the surface runoff from the total net outflow. Riparian ET for the localities was estimated from a regression that estimated the percent marsh based on mean air temperature and topographic slope. Infiltration for the localities was estimated by subtracting surface runoff from precipitation. Recharge for the localities was calculated by adding the riparian ET to the net groundwater discharge.

The following estimates were made for the component fluxes across Virginia. As an annual long-term average for all of Virginia, 113 cm of precipitation falls on the land surface, of which 16 cm runs off the surface into streams, with the remaining 97 cm infiltrating into the soil zone. After infiltration, 65 cm evapotranspires from the vadose zone, leaving 32 cm to recharge the groundwater system at the water table. This groundwater migrates to the stream valleys where 4 cm evapotranspires in the riparian zone and the remaining 28 cm discharges to the stream. The 28 cm in the stream joins the 16 cm of surface runoff to result in 44 cm of mean annual stream flow. This stream flow plus the 69 cm of total ET balance the 113 cm of precipitation. Dividing the 28 cm of groundwater discharge by 44 cm of total stream flow indicates that 64 percent of stream flow is groundwater discharge on average.

The methods used in this study could easily be used in other regions of the United States or the world where (1) streams have been gaged for the last few decades, (2) there is plentiful climate data from the last few decades to estimate long-term average ET, and (3) specific conductance probes can be installed in the streams. In the western United States lack of continuous stream flow in many arid and semi-arid regions might make the implementation of this approach more difficult. In the eastern United States the physiographic provinces that are present in Virginia also extend north and south along most of the Atlantic coastline. Thus the base-flow and surface-runoff regressions might be able to be applied even without installing additional specific conductance probes. Alternatively, graphical hydrograph separation could be used in place of the more costly specific conductance approach. This study provides one example of how a water census could be developed for the United States or other countries where long-term climate and stream flow data sets exist.

Acknowledgments

Many individuals worked to install specific conductance probes, and collect and process stream flow and chemistry data during this study, including Gary K. Speiran, Roger M. Moberg, Jr., Donal C. Hayes, and George E. Harlow, Jr. This study would not have been possible without their tireless contributions in the field.

References

1. McFarland ER, Bruce TS (2006) The Virginia Coastal Plain hydrogeology framework. US Geol Surv Prof Paper 1731.

2. Heywood CE, Pope JP (2009) Simulation of groundwater flow in the Coastal Plain aquifer system of Virginia. US Geol Surv Scient Invest Rep 2009-5039.

3. Sanford WE, Pope JP, Nelms D (2009) Simulation of groundwater-level and salinity changes in the Eastern Shore, Virginia. US Geol Surv Scient Invest Rep 2009-5066.

4. Sanford WE, Nelms DL, Pope JP, Selnick DL (2012) Quantifying components of the hydrologic cycle in Virginia using chemical hydrograph separation and multiple regression analysis. US Geol Surv Scient Invest Rep 2011-5198.

5. McFarland ER (2010) Groundwater-quality data and regional trends in the Virginia Coastal Plain. US Geoll Surv Prof Paper 1772: 1906-2007.

6. Scanlon B, Healy R, Cook P (2002) Choosing appropriate techniques for quantifying ground-water recharge. Hydrogeol J 10: 18-39.

7. Healy RW, Scanlon BR (2010) Estimating groundwater recharge. Cambridge Univ Press, Cambridge.

8. Hufschmidt PW, Pilling JR, Oliver D, Hopkins HT, Ponton J, et al. (1981) Hydrology of Area 16, Eastern Coal Province, Virginia-Tennessee. US Geol Surv Open-File Rep 81-204.

9. Larson JD, Powel JD (1986) Hydrology and effects of mining in the upper Russell Fork basin, Buchanan and Dickenson Counties, Virginia. US Geol Surv Water-Resour Invest Rep 85-4238.

10. Rogers SM, Hufschmidt PW (1980) Quality of surface water in the coal-mining area of   southwestern Virginia. US Geol Surv Open-File Rep 80-769.

11. Rogers SM, Powell JD (1983) Quality of ground water in southern Buchanan County, Virginia. US Geol Surv Water-Resour Invest Rep 82-4022.

12. Powell JD, Larson JD (1985) Relation between ground-water quality and mineralogy in the coalproducing Norton Formation of Buchanan County, Virginia. US Geol Surv Water-Supply Paper 2274.

13. Harlow GE Jr, LeCain GD (1993) Hydraulic Characteristics of, and ground-water flow in, coal-bearing rocks of Southwestern Virginia. US Geol Surv Water Supply Paper 2388.

14. Swain LA, Mesko TO, Hollyday EF (2004) Summary of the hydrogeology of the Valley and Ridge, Blue Ridge, and Piedmont physiographic provinces in the Eastern United States. US Geol Surv Prof Paper 1422-A.

15. Hollyday EF, Hileman GE (1996) Hydrogeologic terranes and potential yield of water to wells in the Valley and Ridge physiographic province in the eastern and southeastern United States. US Geol Surv Prof Paper 1422-C.

16. Rutledge AT, Mesko TO (1996) Estimated hydrologic characteristics of shallow aquifer systems in the Valley and Ridge, the Blue Ridge, and the Piedmont physiographic provinces based on analysis of streamflow recession and base flow. US Geol Surv Prof Paper 1422-B.

17. Nelms DL, Harlow GE, Hayes DC (1997) Base-flow characteristics of streams in the Valley and Ridge, the Blue Ridge, and the Piedmont physiographic provinces of Virginia. U.S Geol Surv Water Supply Paper 2457.

18. Hayes DC (1991) Low-flow characteristics of streams in Virginia. US Geol Surv Water-Supply Paper 2374.

19. Richardson DL (1994) Ground-water discharge from the Coastal Plain of Virginia. US Geol Surv Water-Resources Invest Report 93-4191.

20. Lorenz DL, Delin GN (2007) A regression model to estimate regional ground water recharge: Ground Water 45: 196-208.

21. Risser DW, Gburek WJ, Folmar GJ (2005) Comparison of methods for estimating ground-water recharge and base flow at a small watershed underlain by fractured bedrock in the eastern United States. US Geol Surv Scient Invest Rep 2005-5038.

22. Ator SW, Denver JM, Krantz DE, Newell WL, Martucci SK (2005) A surficial hydrogeologic framework for the mid-Atlantic Coastal Plain. US Geol Surv Prof Paper 1680.

23. Hack JT (1982) Physiographic division and differential uplift in the Piedmont and Blue Ridge. US Geol Surv Prof Paper 1265.

24. Trapp H Jr, Horn MA (1997) Ground water atlas of the United States: Delaware, Maryland, New Jersey, North Carolina, Pennsylvania, Virginia, West Virginia. US Geol Surv Hydrol Atlas HA 730-L.

25. Senay GB, Leake S, Nagler PL, Artan G, Dickinson J, et al. (2011) Estimating basin scale evapotranspiration (ET) by water balance and remote sensing methods. Hydrol Proc 25: 2037-4049.

26. Sanford WE, Selnick DL (2013) Estimation of evapotranspiration across the conterminous United States using a regression with climate and land-cover data. J Amer Water Resour Assoc 49: 217-230.

27. Wood WW, Sanford WE (1995) Chemical and isotopic methods for quantifying groundwater recharge in a regional, semi-arid environment. Ground Water 33: 458-468.

28. Nelms DL, Moberg RM Jr (2010) Preliminary assessment of the hydrogeology and groundwater availability in the metamorphic and siliciclastic fractured-rock aquifer systems of Warren County, Virginia. US Geol Surv Scient Invest Rep 2010-5190.

29. Daniel JF (1976) Estimating groundwater evapotranspiration from stream-flow records. Water Resour Res 12: 360-364.

30. Daly C, Halbleib M, Smith JI, Gibson WP, Doggett MK, et al. (2008) Physiographically sensitive mapping of climatological temperature and

precipitation across the conterminous United States. Int J Climatol 28: 2031-2064.

31. Sklash MG, Farvolden RN (1979) The role of groundwater is storm runoff. J Hydrol 43: 45-65.

32. Hooper RP, Shoemaker CA (1986) A comparison of chemical and isotopic tracers: Water Resour Res 22: 1444-1454.

33. Stewart M, Cimino J, Ross M (2007) Calibration of base flow separation methods with streamflow conductivity. Ground Water 45: 17-27.

34. Hem JD (1970) Study and interpretation of the chemical characteristics of natural water (2d ed.). US Geol Surv Water-Supply Paper 1473.

35. Briel LI (1997) Water quality in the Appalachian Valley and Ridge, the Blue Ridge, and the Piedmont physiographic provinces, eastern United States. US Geol Surv Prof Paper 1422-D.

36. Lull HW, Sopper WE (1969) Hydrologic effects from urbanization of forested watersheds in the Northeast. US Department of Agriculture Forest Service Research Paper NE-146.

37. Harlow GE Jr, Orndorff RC, Nelms DL, Weary DJ, Moberg RM (2005) Hydrogeology and ground-water availability in the carbonate aquifer system of Frederick County, Virginia. US Geol Surv Scient Invest Report 2005-5161.

38. Nelms DL, Moberg RM Jr (2010) Hydrogeology and groundwater availability in Clarke County, Virginia. US Geol Surv Scient Invest Rep 2010-5112.

39. Mejia AI, Moglen GE (2010) Impact of the spatial distribution of imperviousness on the hydrologic response of an urbanizing basin. Hydrol Process 24: 3359-3373.

40. Fenneman NM, Johnson DW (1946) Physical Divisions of the United States: US Geol Survey, 1 sheet, scale 1: 7,000,000.

15

Extreme Weather and Flood Forecasting and Modelling for Eastern Tana Sub Basin, Upper Blue Nile Basin, Ethiopia

Ayenew Desalegn[1*], Solomon Demissie[2] and Seifu Admassu[2]

[1]Department of Meteorology and Hydrology, Institute of Technology, Arba Minch University, Arba Minch, Ethiopia
[2]School of Civil and Water Resources Engineering, Bahir Dar Institute of Technology, Bahir Dar University, Bahir Dar, Ethiopia

Abstract

River flood is a natural disaster that occurs each year in the Fogera floodplain causing enormous damage to the human life and property. Overflow of Ribb and Gummara rivers and backwater effects from Lake Tana has affected and displaced thousands of people since 2006. Heavy rainfall for a number of days in the upper stream part of the catchment caused the river to spill and to inundate the floodplain. Three models were used for this research; the numerical weather prediction model (WRF), physical based semi distributed hydrological model SWAT and the LISFLOOD-FP 1D/2D flood inundation hydrodynamic model to forecast the extreme weather, flood and flood modeling. Daily rainfall, maximum and minimum temperature for the forecasted period ranges from 0 to 95.8 mm, 18°C to 28°C and 9°C to 18°C, respectively. The maximum forecasted flow at Ribb and Gummara Rivers have 141 m³/s and 185 m³/s respectively. The flood extent of the forecasted period is 32 km²; depth ranges 0.01 m to 3.5 m; and velocity ranges from 0 to 2.375 m/s. This technique has shown to be an effective way of flood forecasting and modeling. Integrating Rainfall Runoff model with hydrodynamic model provides thus good alternative for flood forecasting and modeling.

Keywords: SWAT; LISFLOOD; WRF; Extreme weather; Forecasting and modeling

Introduction

Weather-related disasters are increasing in intensity and are expected to increase with climate change [1]. Approximately 70% of all disasters occurring in the world are related to hydro-meteorological event [2]. Death and destruction due to flooding continue to be all too common phenomena throughout the world; and affecting millions of people annually, which is about a third of all natural disasters throughout the world and are responsible for more than half of the fatalities [3].

Scientists agreed that changes in the earth`s climate will hit developing countries like Ethiopia first and the hardest because their economic are strongly dependent on crude forms of natural resources and their economic structure is less flexible to adjust to such drastic changes [4]. In Ethiopia, floods are common and occurring throughout the country with varying time and magnitude. Flood disasters are caused by rivers overflow or burst their banks and inundate to downstream flood plain land; particularly large scale flooding (riverine flooding) in the country is common in the low land flat parts due to high intensity of rainfall from highland parts [5].

As recently as 2006, flooding occurred in almost all parts of the country and devastate the entire country of which Lake Tana remains one of these areas regularly inundated. In spite of the recurrent flood problem, the existing disaster management mechanism has primarily focused on strengthening rescue and relief arrangements during and after flood disasters. Little work has been done in scientific context on minimizing the incidence and extent of flood damage; but need to forecast the extreme weather as well as extreme weather related disasters.

Hence, it is essential to forecast and model the occurrence of extreme weather related disasters to secure human life and property. Therefore, the objective of this study is to forecast extreme weather and flood, and evaluate the applicability of integrating WRF-SWAT-LISFLOOD-FP models to forecast flooding in Fogera floodplain, Easter Tana sub basin.

Materials and Methods

Study area

The study has conducted in the upper Blue Nile part of Ethiopia in Amahara Region, South Gondar Zone. Geographically the area is located between 10° 57′ and 12° 47′N and 36° 38′ and 38° 14′E (Figure 1). It has an aerial extent of about 4174.33 km² drained by Ribb and Gummara Rivers; which is nearly 600 km away from Addis Ababa. Different geographic futures like flood plain, high mountainous land with cold weather (Guna Mountain), Plateau, and rivers characterize it. The basins topography ranges from 1783 m near to Lake Tana up to 4089 m above mean see level on Guna Mountain. The climate is tropical highland monsoon where the seasonal rainfall distribution is controlled by the movement of the inter-tropical convergence zone and moist air from the Atlantic and Indian Ocean in the summer (June-September) [6]. The northward and southward movement of the Inter-Tropical Convergence Zone (ITCZ) controls the seasonal distribution of rainfall. Moist air masses have driven from the Atlantic and Indian Oceans during summer (June-September). During the rest of the year the ITCZ shifts southwards and dry conditions persists in the region between October and May.

The data set

Time series daily rainfall and temperature data for the selected

*Corresponding author: Ayenew Desalegn, Department of Meteorology and Hydrology, Institute of Technology, Arba Minch University, Arba Minch, Ethiopia
E-mail: aye.desalegn@gmail.com

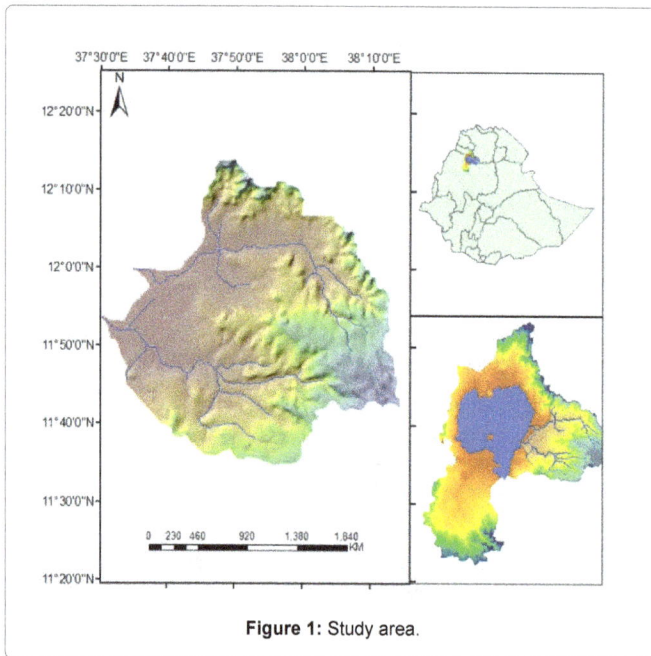

Figure 1: Study area.

stations from 1951-2014 were obtained from National Meteorological Agency of Ethiopia (NMA). The other variables evapotranspiration, solar radiation, wind speed and relative humidity, have simulated from SWAT weather Generator. Similarly, daily stream flow data of Ribb and Gummara rivers for the years of 1973 to 20014 have obtained from Ethiopian ministry of water Irrigation and Energy.

Spatial resolution of 30 × 30 m land use image has downloaded from landsat8 OPL sensor with 169 Path and 52 Row for 01/02/2014 and reclassified using supervised maximum likelihood land use classification method using GIS technique. In addition, soil data has extracted from Blue Nile Basin soil data (Soill90) obtained from Ministry of Water Irrigation and Energy of Ethiopia (MoWIE). River cross section data of Ribb and Gummara Rivers and Survey data for Fogera flood Plain have obtained from Tana Sub Basin Office (TaSBO) which has collected by MoWIE. The rivers width has also obtained by digitizing from ESRi high-resolution world imagery base map of resolution 1 m and better of resolutions (15 cm and 60 cm) on ArcGIS map window.

Extreme weather forecasting

Extreme weather has forecasted for the entire period of August 20, 2006-Sepetember 10, 2006 using a numerical weather prediction WRF model. To forecast the extreme weather nested three domains [Ethiopia (45 km), Northern part (15 km), and fogera (5 km)] resolution were selected by assuming that 1-degree ~ 111 km around equator. The model handled three domains at the same nest level (no overlapping nests), and/or three nest levels (telescoping). The nesting ratio for the WRF-ARW is three and the grid spacing of a nest was 1/3 of its parent.

The initial and lateral boundary, meteorological and terrestrial gridded data that used to run the WRF-ARW model has downloaded from Global Forecasting System (GFS) that has produced by the National Centers for Environmental Prediction (NCEP) and it has updated for every six hours. The Real-data has interpolated to run the NWP using WRF Pre-processing System (WPS). The WRF Model (ARW dynamical cores) was initialized numerical integration programs for real data processing. The output of the model WRF-ARW

has processed on WRF-post processing and visualized using Grid Analysis and Display System (GRADS). The output of the WRF model, weather data, has processed for SWAT model input. From the output of the model few parameter has selected and used as SWAT model impute (Figure 2).

Runoff forecasting

A conceptual, physically based, continuous SWAT model has employed to simulate stream flow. The SWAT (Soil and Water Assessment Tool) model was developed by the USDA (U.S. Department of Agriculture), ARS (Agriculture Research Service) and represents a continuation of roughly 40 years of modeling efforts [7]. SWAT is a public domain watershed scale model developed to predict the effects of land management on water, sediment, nutrients, pesticides and agricultural chemicals in small to large complex basins [8]. It is a physically based, semi-distributed parameter model with a robust hydrologic and pollution element that has successfully employed in a number of watersheds. Widely known application of SWAT model is simulating hydrology of a watershed, water quality, sediment yield and plant growth in relation to watershed management practices.

However, it has also applied for flow forecasting. The soil and water assessment tool (SWAT) can forecast the flow of a watershed but it is performed lower than Artificial Neural Network (ANN) models [9]. Hydrological modeling of a SWAT has used in flash flood forecasting with the application of three days weather forecast from the NWP (Numerical Weather Prediction) and the data from the NWP can be used with the SWAT model and provides relatively sound results [10]. Predicting flood hazard areas and damage reduction by flood inundation and the sediments using SWAT and HEC-RAS [11,12].

The major components of SWAT are climate, hydrology, erosion, land cover and plant growth, nutrients, pesticides and land management. The SWAT has used to simulate the hydrologic processes of the study watershed. Simulations of the hydrology of a watershed can separated into two major divisions. The first division is the land phase of the hydrologic cycle and the second division is the water or routing phase of the cycle [10].

For Land phase, the hydrologic cycle has based on the water balance equation:

$$SW_t = SW_0 + \sum_{I=0}^{t}(R_t - Q_{surf} - E_a - W_{seep} - Q_{gw}) \qquad (1)$$

Where, SW_t=final water content (mm)

SW_0=initial water content in time I (mm)

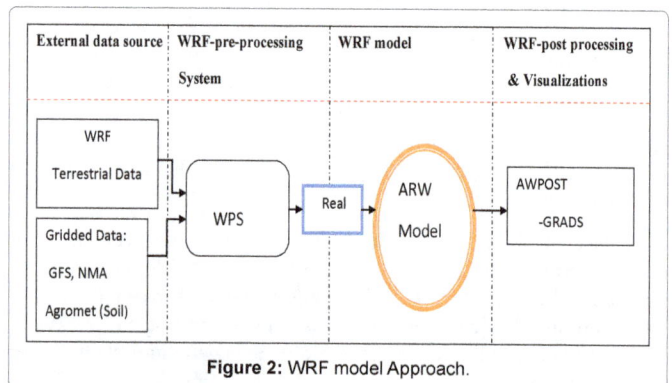

Figure 2: WRF model Approach.

t=time (in days, months, or years)

R_I=amount of rainfall in time I (mm)

Q_{surf}=amount of surface runoff in time I (mm)

E_a=amount of evapotranspiration in time I (mm)

W_{seep}=amount of water entering the vadose zone from the soil profile in time I (mm)

Q_{gw}=amount of return or base flow in time I (mm).

Surface runoff: Also known as overland flow, the part of the rainfall, infiltration excesses rainfall flowing along the slopes. SWAT uses the Soil Conservations Service (SCS) Curve Number (CN) method to calculate surface runoff. Surface runoff can express using the equation 2:

$$Q_{surf} = \frac{\left(R_{day} - I_a\right)^2}{R_{day} - I_a + S} \text{ and } I_a = 0.2 \times S \qquad (2)$$

Where, S=soil storage or retention

R_{day}=daily precipitation

I_a=initial surface abstraction that includes surface storage, interception and infiltration to moist soil surface up to runoff generation, all in mm water.

Soil storage or retention volume has expressed in terms of curve number CN (equation 3)

$$S = 25.4\left(\frac{1000}{CN} - 10\right) \qquad (3)$$

By substituting I_a and S in equation 2, surface runoff is expressed as:

$$Q_{surf} = \frac{\left(R_{day} - 0.2S\right)^2}{R_{day} + 0.8S} \qquad (4)$$

Surface runoff will occur when the amount of rainfall amount exceeds the initial abstraction and infiltration to the root zone of the soil. For a reason, CN is a function of land-use, soil and antecedent soil moisture content. These functional relationship and CN values can be obtained in the SWAT manual and user guide [13].

Before forecasting, the model has calibrated and validated using observed flow data. From the available data, 2 years (1994-1996) for warm-up, 9 years (1996-2004) for calibration and 5 years (2005-2009) for validation have used. Model calibration was performed using the Sequential Uncertainty Fitting version 2 (SUFI-2) interface of SWAT-CUP. SWAT-CUP is a separate calibration and uncertainty program developed by Abbaspour. It is a commonly used procedure for calibration and uncertainty analysis. Setegn et al. [14,15] compared different procedures and found SUFI-2 is better that gives good results even at smallest number of runs as compared to other procedures. The performance of model was evaluated using dimensionless Nash-Sutcliffe Efficiency (NSE) (Nash and Sutcliffe) and coefficient of determination (R²).

Flood modeling and forecasting

Among the most widely used hydraulic models LISFLOOD model has selected for this research. LISFLOOD is a distributed, raster- based; combination rainfall-runoff and hydrodynamic model embedded in a dynamic GIS environment [16-18], and has been developed for the simulation of hydrological processes and floods in European drainage basins. It is a flexible tool, which is capable of simulating hydrological processes on a wide range of spatial and temporal scales, maintaining high resolution even when simulating large catchment areas.

LISFLOOD-FP (Flood Plain) raster based inundation model. It shows a 2D/3D simulation of the floodplain has inundated and runs at a time step of seconds. The inputs for this module are, a high resolution DEM including all the topographic detail of features inside the modeled area considered necessary to produce probable flood inundation prediction, the rivers hydrograph, a map of flood source areas, and the outputs of the Flood Simulation model [17].

The LISFLOOD-FP, one of the modules of LISFLOOD, includes a number of numerical schemes (solvers) which simulate the propagation of flood waves along channels and across floodplains using the shallow water equations. The choice of numerical scheme is depend on the characteristics of the system has to be modeled, requirements of time for execution and type of data available. The momentum and continuity equations for the 1D full shallow water equations have given below (equations 5,6 respectively):

$$\frac{\partial Qx}{\partial t} + \frac{\partial}{\partial x}\left(\frac{Q_x^2}{A}\right) + gA\frac{\partial\left(h + z\right)}{\partial x} + \frac{gQ_x^2 n^2}{R^{4/3}A} = 0 \qquad (5)$$

$$\frac{\partial A}{\partial x} + \frac{\partial Qx}{\partial x} = 0 \qquad (6)$$

Where, Q_x=volumetric flow rate in the x Cartesian direction

A=cross sectional area of flow

h=water depth

z=bed elevation

g=gravity

n=Manning's coefficient of friction

R=hydraulic radius

t=time

x=distance in the x Cartesian direction.

Floodplain flow solvers: LISFLOOD Roe: The "Roe" solver includes all of the terms in the full shallow water equations was selected for this research [19]. The method has based on the Godunov approach and uses an approximate Riemann solver by Roe based on the TRENT model presented in Villanueva and Wright. The explicit descritisation is first order in space on a raster grid. It solves the full shallow water equations with a shock-capturing scheme. LISFLOOD-Roe uses a point wise friction based on the Manning´s equation, while the domain boundary/internal boundary (wall) uses the ghost cell approach. The stability of this approach has approximated by the Courant-Friedrichs-Levy (CFL) condition for shallow water models.

$$\frac{\partial Q}{\partial t} + \frac{\partial}{\partial x}\left(\frac{Q^2}{A}\right) + gA\frac{\partial\left(h_t + z\right)}{\partial x} + \frac{gQ^t n^2}{R^{4/3}A} = 0 \qquad (7)$$

Where, Q=discharge

t=time

A=cross section area

g=gravitational constant

h_t=water free surface height

z=bed elevation

n=Manning's coefficient

R=Hydraulic Radius.

Channel flow solvers: The "diffusive" channel flow solver has selected for this research uses the 1D diffusive wave equations and it includes the water slope term, which is able to predict backwater effects. Using the 1D-channel solvers, once channel water depth reaches bank full ight, water has routed onto adjacent floodplain cells has distributed as per the chosen floodplain solver. There is no transfer of momentum between the channel and floodplain, only mass. The 1D diffusive solvers assume that the in-channel flow component can be represented using a diffusive 1D wave equation with the channel geometry simplified to a rectangle. The 1D diffusive channel flow solver assumes that the channel to be wide and shallow, so the wetted perimeter is approximated by the channel width such that the lateral friction is neglected.

Results and Discussion

Extreme weather forecasting

The extreme weather for the study area has forecasted using a numerical weather prediction model WRF-ARW from 20 August 2006-10 September 2006. The weather parameters have forecasted at a six-hour time step and converted to daily for SWAT model input. Air temperature, wind speed at two meter, solar radiation, relative humidity, precipitation, geopotential height, sea surface temperature and Surface temperature were among the outputs of the WRF model. Precipitation and temperature of the output parameters have selected for SWAT model input to forecast the flood. The result in Figure 3 shows that Eastern Tana Sub basin has subjected to intense and heavy rains during the selected period. The developments of intensive weather events that invade Eastern Tana sub basin during 20 August 2006-7 September 2006, have characterized by "exceptional and extremely heavy rainfall," which affected almost all part of the Eastern Tana Sub Basin.

The forecasted rainfall of the selected station has obtained from the WRF output gridded data. Unfortunately, the selected stations point has no the same coordinate with girded point. Hence, the forecasted rainfall for the station points has obtained from the neighboring gridded points using regression method. The cumulative of forecasted rainfall is similar with the cumulative of the observed. The forecasted daily rainfall for the forecasted period ranges from 0 to 95.8 mm. A very intense and heavy rain has occurred during 25th of the days almost all over the entire sub basin. Even though the WRF model captures rainfall climatology in the study area both in space and time basin, there is variations of the forecasted rainfall for selected stations. The maximum rainfall has recorded in the upper part of the sub basin climatological stations, which are D/Tabor, Lewaye, K/Dnigay and M/Eyesus for the entire period. The maximum daily rainfall has observed at K/Dnigay station about 95.8 mm followed by Lewaye and D/Tabor stations about 60 mm and 40 mm respectively. In the meanwhile, the minimum daily rainfall has recorded in the lower part of the study area stations, which were Yifag, Wanzaye (Figure 4).

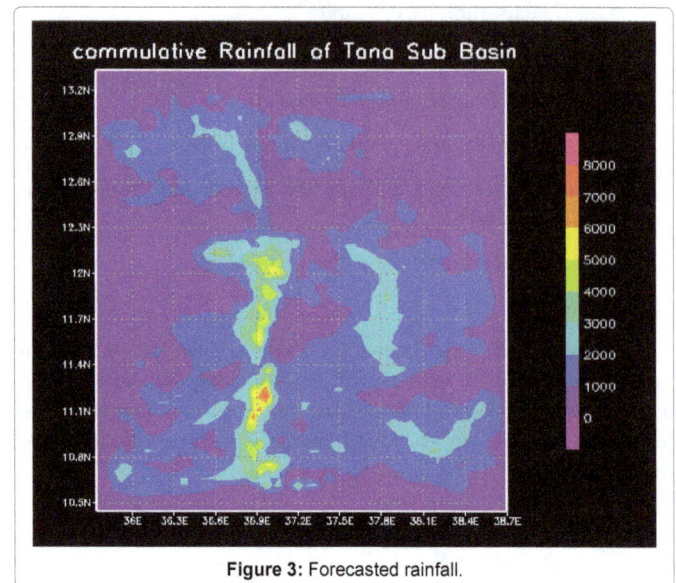

Figure 3: Forecasted rainfall.

Similarly, the forecasted temperature for the station points has obtained from the neighboring gridded points using regression method (Figure 5). As can be seen Figure 6, the spatial variation of average temperature over the Tana Sub basin.

The maximum forecasted air temperature for the selected period of the entire sub basin ranges from 18°C to 28°C. Generally, the WRF model has well forecasted the maximum temperature compared with the observed data for the sub basin.

Flow modelling and forecasting

Hydrological model calibration and validation: The calibration and validation of the model was a key factor in reducing the uncertainty and increasing confidence in its predicative abilities, which makes the application an effective model. Information on the sensitivity analysis, calibration and validation of multivariable SWAT models was provided to assist watershed modelers in developing their models to achieve their watershed management goals [20]. SWAT simulation has executed for the 1994-2009 period to provide two-years for an initialization period. Calibration of SWAT has performed for 1994-2004, while 2005-2009 have used as the validation years.

The goodness of fit of the model is evaluated using coefficient of determination (R^2) and Nash-Sutcliffe Efficiency (NSE). It has found that the model has strong predictive capability as shown in Table 1. Statistical model efficiency criteria fulfilled the requirement of $R^2>0.6$ and NSE>0.5 which is recommended by SWAT developer [21]. This showed the model parameters represent the processes occurring in the watershed to the best of their ability given available data and can used to predict watershed response for various outputs (Figures 7 and 8).

Flow forecasting: The forecasted weather data using NWP-WRF model has used as input for SWAT model. The simulated value has considered as forecasted flow. It has also found that the simulated flow rate using NWP-WRF data was lower than the observations for both watersheds for consecutive five days from 27 August 2006-1 September 2006. This was because the rainfall from the NWP-WRF model was lower than the measured rainfall. In Summary, the simulated flow rates for the rivers using data rom NWP-WRF were higher than the observations flow at Ribb River and lower than at Gummara River. The maximum forecasted flow at Ribb was 141 cm but the maximum

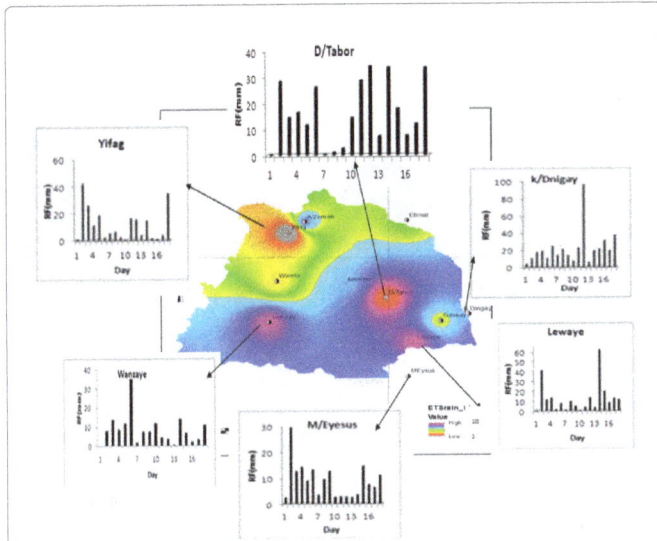

Figure 4: Forecasted rainfall for selected stations of Eastern Tana Sub Basin.

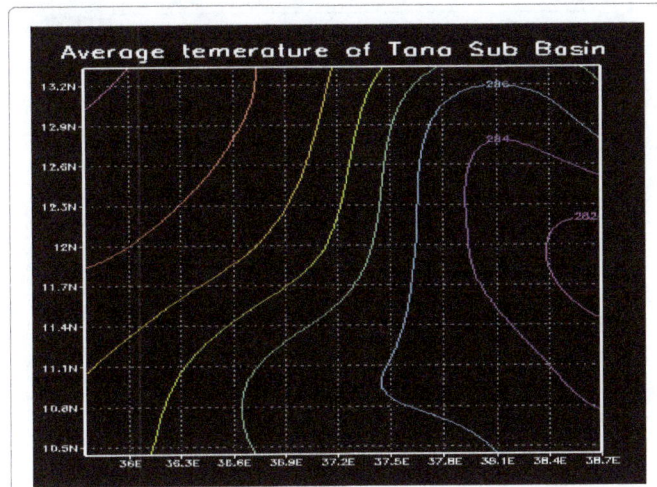

Figure 5: Forecasted Temperature for Eastern Tana Sub basin.

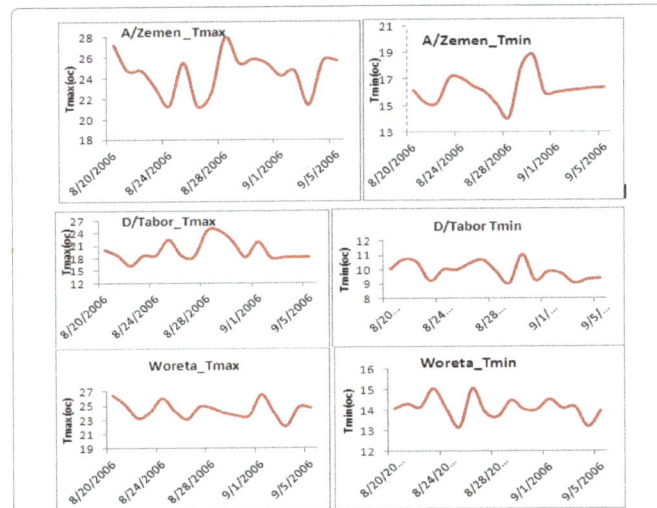

Figure 6: Forecasted Temperature for the selected stations Eastern Tana Sub Basin.

Watershed	Calibration		Validation	
	NSE	R²	NSE	R²
Gummara	0.75	0.77	0.73	0.74
Ribb	0.72	0.73	0.67	0.76

Table 1: SWAT Model Calibration.

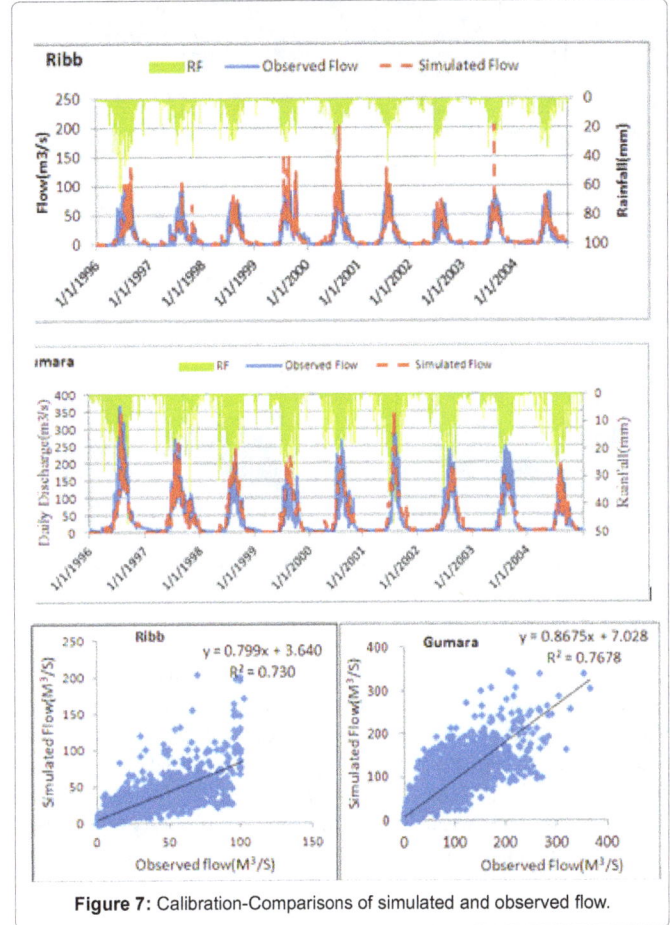

Figure 7: Calibration-Comparisons of simulated and observed flow.

observed flow was 93 cm. Similarly, for Gummara, the maximum forecasted and observed flow was 185 cm but the maximum observed flow was 206 cm (Figure 9).

Flood modeling

Both upstream and downstream boundary conditions have given for the diffusive channel solver. The upstream boundary is the forecasted flow rate at gauging site of Ribb and Gummara rivers; and the downstream boundary condition is the Lake Tana water level. The advantage of the diffusive channel solver over Kinematic solver is that the tributaries have handled automatically by LISFLOOD-FP. To simulate a dynamic flood wave both upstream and downstream time varying boundary condition (QVAR and HVAR) have used.

The forecasted flood extent for the design period 20 August 2006-10 September 2006 has computed by integrating the hydrology model (SWAT) and a hydrodynamics model (LISFLOOD). The output from SWAT that is a hydrograph used as upper boundary for LISFLOOD model and the Lake level interims of elevation used as lower boundary. Therefore, the LISFLOOD computes the flood extent on accounts of the boundary conditions, the rivers width, and river cross-section and

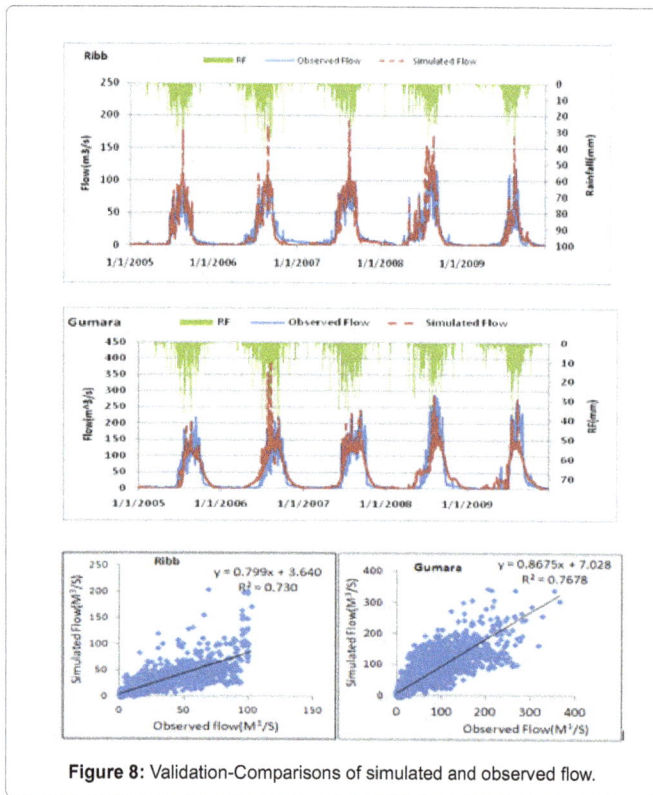

Figure 8: Validation-Comparisons of simulated and observed flow.

Manning's friction coefficient.

The flood extent obtained from LISFLOOD-FP has processed on GIS environment. The extent of flood for the forecasted period is 329 Km^2. The flood depth ranges from 0.01 m to 3.5 m and the maximum depth is at the rivers. The flood velocity for the forecasted period ranges from 0 to 2.375 m/s. The model has not accounted the rainfall over the flood plain and the small rivers those are not tributary of the main rivers (Ribb and Gummara) (Figure 10). This might be under estimate the flood extent.

Flood model verification

The goodness of fit between the created flood map from flood model and the flood map extracted from the satellite images has assessed by the measure of Relative Error (RE) and F-statistics (F). As shown the Figure 11 indicates that the inundation area of the extracted flood images from the satellite is 259.7 Km^2 and predicted flood inundation area is 256.9 Km^2. The area of overlapping portion of the two flood inundations is 236.55 Km^2 with RE of 0.01 and F-statistics of 84.47%. This shows that the compared areas of the flood inundation are similar to each other but they are not geospatially similar. As can be seen in Figure 11 the satellite image shows more flooded area in the side of Ribb River but the forecasted flood area is more in the Gummara riverside. This seems rescannable because the satellite image also accounts the logged water over the area due to rainfall and other tributaries. Near to the Ribb, river and center of the flood plain there are tributaries, which are causing flood.

The goodness of fit between the created flood map from flood model and the flood map extracted from the satellite images in Table 2 shows that the model has well fitted.

Where, $RE = \dfrac{|A_o - A_p|}{A_p}$ (8)

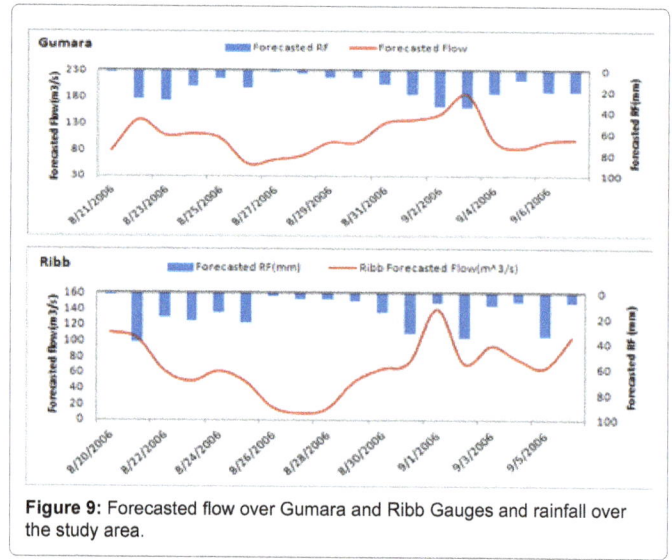

Figure 9: Forecasted flow over Gumara and Ribb Gauges and rainfall over the study area.

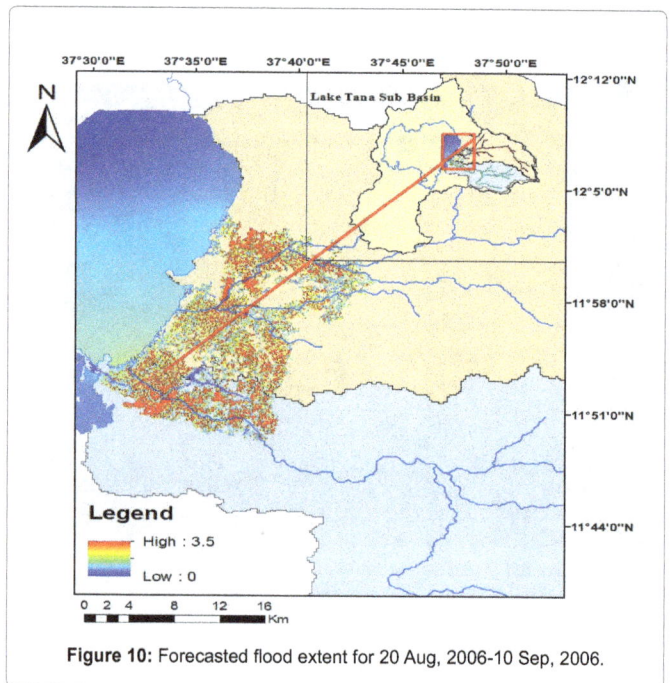

Figure 10: Forecasted flood extent for 20 Aug, 2006-10 Sep, 2006.

$$F = \left(\frac{A_{op}}{A_o + A_p - A_{op}} \right) \times 100 \qquad (9)$$

A_o indicates the inundation area of the extracted flood images from

the satellite

A_p refers to the predicted flood inundation area

A_{op} represents the intersection of A_o and A_p.

Conclusion and Recommendations

Flooding is the main challenge natural hazard in Eastern Tana

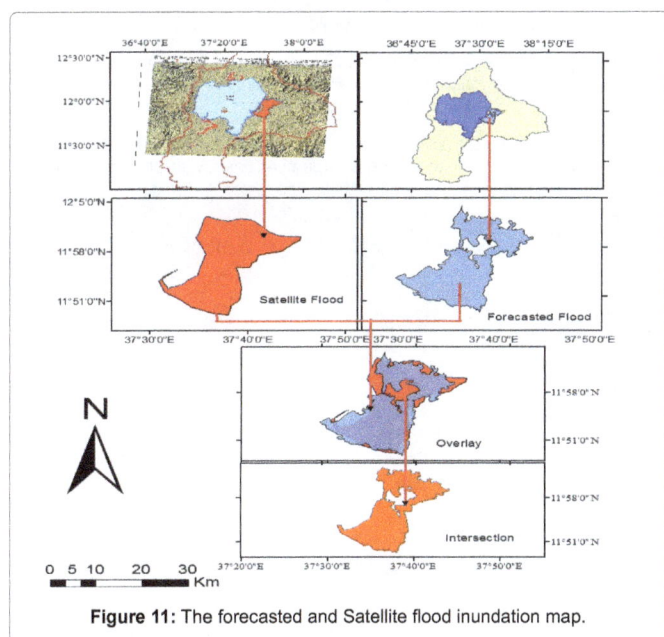

Figure 11: The forecasted and Satellite flood inundation map.

Year	RE	F
2006	0.011	84.47

Table 2: Measure of goodness of fit of the satellite image and forecasted flood extent.

sub Basin that affects human life and property badly. Thousands of people are displacing each year and the people in the area are always a misery due to flood. The Ethiopian government has taken an operation measure to control the impact of flood but the measure has abused by the farmers that they love the flood to get fertile soil.

The extreme weather over the study area is controlled by the movement of the Inter-Tropical Convergence Zone (ITCZ) position and orientation, monsoon trough, Low level jet (Somali Jet), Southern hemisphere high pressures, Southerly (cross equatorial) moisture flows, Strengthening frequency of Tropical Easterly Jet (TEJ), ENSO events and seasonal rainfall features Wet/dry summer (La Nina/El Nino).

Flooding has estimated by integrating three models (WRF-SWAT-LISFLOOD) and the approach that gave a good result. Even though the main source of flood was included in the flood model domain during flood modeling, few sources have left (the rainfall and the tributaries over the floodplain). This will a little bit underestimated the estimated flood. This work is the first work in Ethiopia and the method can extend for other regions with similar/different climates.

The ongoing rapid land use change and expansion of agricultural area in this study area will have negative effects on the runoff properties. To attenuate the occurrence of flood on Fogera flood plain, a better land use management system is required, which can impede the unregulated conversion from one land use to another land use. To avoid future flood disasters, flood early warning and forecasting system, flood management and flood mitigation plans are need to be able to react quickly to areas affected by flooding. Flood monitoring system is required to assess, on a continuous basis, the areas affected by floods and to have emergency measures plan to reduce the damage of exceptional floods. Also, need to aware the community/ farmers about the effect of flood strongly to solve problems related to dyke breaking.

Further investigations should consider on the possibility of flood

forecasting and modeling with including other events in the area. Future works are needed on establishing early warning system by considering the outputs of this study.

Acknowledgements

I would like to thank the Blue Nile Water Institute for funding this research and providing necessary facilities. First, I would like to express my sincere gratitude to my advisors Dr. Solomon Demissie and Dr. Seifu Admassu for the continuous support of my study and research, for their patience, motivation, enthusiasm, and immense knowledge. Their guidance helped me in all the time of research and writing of this thesis. I could not have imagined having a better advisor for my research. My sincere thanks also go to Dr. Essayas Kaba, Mr. Fasikaw Atanaw and Mr. Mamaru Moges for their guidance and help on land use classification, guiding on writ up of my thesis under difficult condition and crowded time.

References

1. Parry M (2007) Climate Change 2007: impacts, adaptation and vulnerability. Contribution of Working Group II to the fourth assessment report of the Intergovernmental Panel on Climate Change, Cambridge University Press.

2. Barrientos HG, Swain A (2014) Linking flood management to integrated water resource management in guatemala: A critical review. International Journal of Water Governance 4: 53-74.

3. Berz G (2000) Flood disasters: lessons from the past-worries for the future. Proc Inst Civ Eng Water Marit Energy 142: 3-8.

4. Bryan E, Deressa T, Gbetibouo G, Ringler C (2009) Adaptation to climate change in Ethiopia and South Africa: Options and constraints. Environ. Sci. Policy 12: 413-426.

5. Deressa TT, Hassan RM, Ringler C, Alemu T, Yesuf M (2009) Determinants of farmers' choice of adaptation methods to climate change in the Nile Basin of Ethiopia. Global environmental change 19: 248-255.

6. Kebede S, Travi Y, Alemayehu T, Marc V (2006) Water balance of Lake Tana and its sensitivity to fluctuations in rainfall, Blue Nile basin, Ethiopia. Journal of hydrology 16: 233-247.

7. Williams JR, Arnold JG, Kiniry JR, Gassman PW, Green CH (2008) History of model development at Temple, Texas. Hydrological sciences journal 53: 948-960.

8. Arnold J, Srinivasan RS, Muttiah RS, Williams JR (1998) Large area hydrologic modeling and assessment part I: Model development. Journal of the American Water Resources Association (JAWRA) 34: 73-89.

9. Demirel MC, Venancio A, Kahya E (2009) Flow forecast by SWAT model and ANN in Pracana basin, Portugal. Advances in Engineering Software 40: 467-473.

10. Wangpimool W, Pongput K, Supriyasilp T, Kamol PN, Sakolnakhon S, et al. (2013) Hydrological evaluation with swat model and numerical weather prediction for flash flood warning system in Thailand. Journal of Earth Science and Engineering 3: 349.

11. Rivera S, Hernandez A, Ramsey RD, Suarez G (2007) Predicting flood hazard areas: a swat and HEC-RAS simulations conducted in Aguan river basin of Honduras, Central America. Paper presented at ASPRS 2007 Annual Conference.

12. Jung CG, Joh HK, Yu YS, Park JY, Kim SJ (2012) Study on damage reduction by flood inundation and the sediments by SWAT and HEC-RAS modeling of flow dynamics with watershed hydrology-for 27 july 2011 heavy storm event at GonjiamCheon watershed. Journal of the Korean Society of Agricultural Engineers 54: 87-94.

13. Neitsch SL, Arnold JG, Kiniry JR, Williams JR (2011) Soil and water assessment tool theoretical documentation version 2009. Texas Water Resources Institute Technical Report No. 406, Texas Water Resources Institute.

14. Setegn SG, Srinivasan R, Dargahi B (2008) Hydrological modelling in the Lake Tana Basin, Ethiopia using SWAT model. The Open Hydrology Journal 2: 49-62.

15. Yang J, Peter R, Abbaspour KC, Xia J, Yang H (2008) Comparing uncertainty analysis techniques for a SWAT application to the Chaohe Basin in China. Journal of Hydrology 358: 1-23.

16. Roo APJD, Wesseling C, Deursen VV (2000) Physically based river basin modelling within a GIS: the LISFLOOD model. Hydrological Processes 14: 1981-1992.

17. Roo AD, Gouweleeuw B, Pozo JT, Sattler K (2003) Development of a European flood forecasting system. International Journal of River Basin Management 1: 49-59.

18. Roo APJD (1999) LISFLOOD: a rainfall-runoff model for large river basins to assess the influence of land use changes on flood risk in RIBAMOD: river basin modelling, management and flood mitigation, concerted action by European Communities, pp: 349-358.

19. Trigg MA, Wilson MD, Bates PD, Horritt MS, Alsdorf DE, et al. (2009) Amazon flood wave hydraulics. Journal of Hydrology 374: 92-105.

20. White KL, Chaubey I (2005) Sensitivity analysis, calibration, and validations for a multisite and multivariable SWAT model. Journal of the American Water Resources Association (JAWRA) 41: 1077-1089.

21. Santhi C, Arnold JG, Williams JR, Dugas WA, Srinivasan R, et al. (2001) Validation of the SWAT model on a large river basin with point and nonpoint sources. Journal of the American Water Resources Association (JAWRA) 37: 1169-1188.

16

Snyder Unit Hydrograph and GIS for Estimation of Flood for Un-Gauged Catchments in Lower Tapi Basin, India

Sudhakar BS[1], Anupam KS[2] and Akshay OJ[3]

[1]Institute of Science, Nirma University Science & Technology, Ahmedabad-382481, India
[2]Institute of Engineering & Technology, J.K Lakshmipat Universitsy, Jaipur-302026, India
[3]Department of Civil Engineering, Pandit Dindayal Petroleum University, Gandhinagar, India

Abstract

In the flood prone catchments, it is needful to estimate the discharge, standard lag time, time of peak, and flood response of each watershed in the basin. The SUH method offers considerable advantage over others, and thus, has been chosen for estimation of flood response, contribution of flooding potential, percentage of flood volume for 25 sub-watersheds. The discretion of sub-watershed for estimation of peak discharge, time of peak, alternate lag time, and width of SUH at 50% and 75% of peak found to offer advantages over other methods. This paper considers Snyder Unit Hydrograph (SUH) with GIS based spatial database for calculating discharge at Lower Tapi Basin (LTB). The hydrological parameters of each sub-watershed such as river length, length of centroid, spatial area, land use, lateral slope, and terrain and soil factors have been extracted from GIS database. The geo-data has been combined with topographical maps to produce a digital elevation model (DEM) of 50 m cell size. The analysis for all 25 sub-watersheds exhibit that 35.07 m^3/s and 4.55 m^3/s and 13.23 hours and 4.33 hours have been highest and lowest peak flow and time of peak respectively. The SUH model has been validated for peak discharge at a gauge site Amli (E73^{0}23' N21^{0}23') where discharge data were collected during 2010 and 2011 monsoon. A comparison between measured and SUH modelled discharge shows good fit within a mean variability range of 5-7%. The SUH methods ability to estimate hydrological parameters including peak flow discharge shows wider replication for un-gauged catchments.

Keywords: Hydrological modelling, Discharge estimation, Snyder unit hydrograph, GIS

Introduction

In the flood prone catchments it is needful to calculate peak flood discharge from each watershed. It is desirous to estimate lag time, time to peak, and flood response of each watershed, as above parameters affect the channel flow and peak flood formation. Therefore, reliable estimation of them is of prime importance and more so for ungauged catchments. The recent International Association of Hydrological Sciences (IAHS) initiative on prediction in ungauged basins (PUB) has opened opportunities to carry-out research in data poor or ungauged basin. In India most of the watersheds below 500 km² are ungauged or sufficient hydrological data is not available [1-3]. Lower Tapi basin a geographical area of 1998 km² and a river length of 106 km between Ukai dam and Hazira is one among ungauged catchments. The basin has been receiving periodic floods occurred every 3-4 years interval. The recent flood in the basin has been during 6-14 August 2006 causing more than INR 22,000 crore (~US$ 4.5 billion) economic loss and 300 people being killed.

Traditional techniques for design flood estimation use historical rainfall-runoff data and unit hydrographs derived from them and to overcome such difficulties, the use of physically based rainfall-runoff estimation methods such as the geomorphological instantaneous unit hydrograph (GIUH) have evolved [4]. Several methods on peak flood discharge and associated parameters estimation have been suggested in the literature for ungauged basins. However, the parameter reliability between various methods varies to a large extent and none found to be suitable universally.

Sharma et al. [5] have estimated flooding potential watershed using SCS-CN method for identification of watershed, for a part of lower Tapi Basin. The data calculation gives good results for rainfall runoff modelling and suggested that the method may be good tool

for runoff estimation for lower Tapi basin and un-gauged catchment like Varekhadi catchment. Sherman [6] proposed an advance theory of unit hydrograph for estimating surface runoff in gauged basins [4]. This theory has been considered an important contribution to the field of hydrology in deriving the flood hydrograph. However, UH theory has limitation on precise runoff prediction due to limiting assumptions. Snyder [7] developed a set of empirical equation for synthetic unit hydrograph (SUH) in large number of catchments in Appalachian Highland of eastern United State [5]. The SUH method has better acceptability for ungauged basins unlike the Sherman's Unit Hydrograph method. Literature supports that SUH method has applications for watersheds having large variability ranging from 25 km² to 25,000 km².

Hoffmeister et al. [8] in their research work, developed a synthetic unit hydrograph for an un-gauged basin in New Zealand [6]. The authors tested three different methods viz. Snyder method, Common's dimensionless method, SCS dimensionless hydrograph for catchment each represent dominant hydrological and physiographic characteristics of that region. Their research results indicate that Snyder's UH method gives best results as compared to later ones. In other study, Wayal et

*Corresponding author: Sudhakar BS, Institute of Science, Nirma University Science & Technology, Ahmedabad- 382481, India
E-mail: sudhakar.sharma@nirmauni.ac.in

al. [9] has derived empirical equations for ungauged catchment based on snyder's relation on SUH in India [7]. The study was carried out in parts of Krishna and Pennar river basin in South India. The synthetic relations derived by Wayal [9] can be applied only under watersheds having similar topographical and climatologically regions like Krishna and Pennar. Later Adebayo [10] developed unit hydrograph and compared the performance of Snyder's method, SCS method, Gray's method for eight sub-watersheds in south-west Nigeria [8]. The authors suggested use of SCS-method since topographic, climatic and basin properties at daily time scale were not available. The authors found SCS-method suitable as compared to others Abid et al. [11] compared the runoff hydrographs estimated by the SCS and Snyder UH models with the observed runoff hydrographs in Kasilian watershed [9]. The authors observed that the calculated runoff hydrographs by these models have good fitness with observed runoff hydrographs. Limantara [12] has been of the view that SUH could become the source of some important information that is necessary for the reliable of hydraulic structures [10]. In his paper he analyzed the design flood hydrograph through the uses of areal rain data inputs. Therefore, this paper uses SUH method to estimate peak flood discharge for partial gauged site in LTB catchment in India.

Study area description

Tapi is the second largest westward draining inter-state river in India after mighty River Narmada. The basin finds its outlet in the Arabian Sea after passing Surat city in Gujarat that is bounded on the three sides by the hill ranges. The Tapi River is divided in three zones, viz. Upper Tapi basin, Middle Tapi Basin, and Lower Tapi Basin (LTB). The portion between Ukai Dam to Arabian Sea has been considered as LTB, mainly occupying Surat and Hazira twin city along with tens of small towns and villages along the river course. The lower tapi basin extends over an area of 1998 km². The Surat and Hazira twin cities are downstream of Ukai Dam almost 106 km distance and are affected by recurrence floods at regular intervals.

The Geographic Location of study area is Longitude 72˚42' to 73˚40'and Latitude 21˚08' to 21˚30' (Figure 1). LTB receive an average annual rainfall of 1376 mm, and these heavy downpours result into devastating floods and water loggings mainly between Ukai dam and Hazira town downstream. The major crops grown in the study area are cotton and maize followed by Soybean. The land use prevailing in the study area is mixed forest, agriculture land, rural and urban settlements. The topography of Surat is gently sloping and flat. Therefore, it can be stated that study area has multiple problems in flood formation. This necessitates the need for monitoring and solutions that are simple, based on Remote Sensing and GIS, require use of Hydrological Modelling. The main reasons for flooding in Surat depend on heavy rainfall and discharge due to high water levels from Ukai dam. Therefore, the flood problems of the river system are inundation due to over flowing of the banks. High tide during certain period also play significant role in flood formation for this 106 km stretch.

Methodology

The research methodology used for estimation of discharge using SUH method consist of three steps viz. geo-database development, estimation of hydrological parameters and field data collection (Figure 2). The detailed description on each step has been given below.

Geo-database development

The geo-database for LTB has been created using topological maps, satellite remote sensing images and field surveys using GPS. Topographical maps at 1:50000 scales were collected, geo-referenced, and digitized for themes such as contours, level points, streams, and watershed boundary. Based on information obtained from maps, attribute properties to various themes have been assigned. The geo-data base on above listed themes has been cross-checked with field and attributes were revised. It was decided to carry out engineering survey for almost 327 cross-sections across the Tapi River. A digital elevation model (DEM) for 50 m cell size and 2.5 m vertical accuracy for LTB has been generated (Figure 3). The accuracy assessment carried on DEM for selected locations shows a good fit and has been in coherence with actual elevation values. The DEM has been considered as basis for delineation of sub-watershed boundary, geographical areas and longitudinal slopes along river stretches. The 25 sub-watersheds for LTB has been delineated using hydrological model software BASIN (EPA, 2007) on DEM (Figure 4). A threshold area of 10 km² has been considered for delineation of sub-watershed boundary. The other hydrological parameters such as watershed area, river length, and

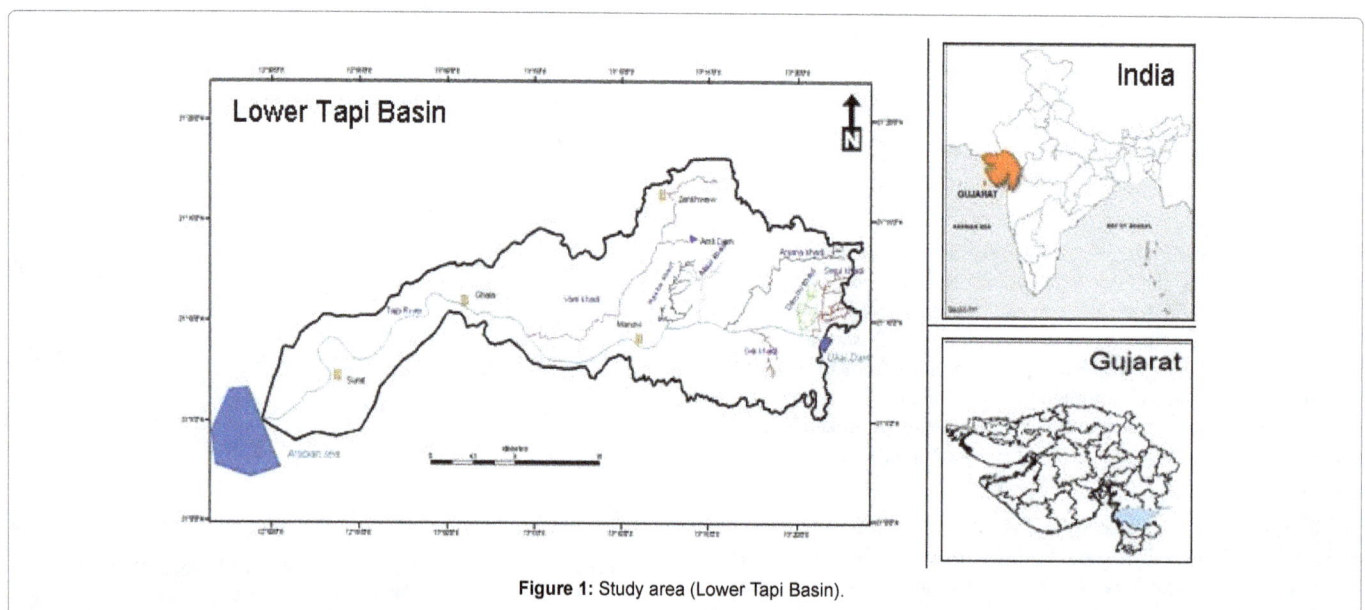

Figure 1: Study area (Lower Tapi Basin).

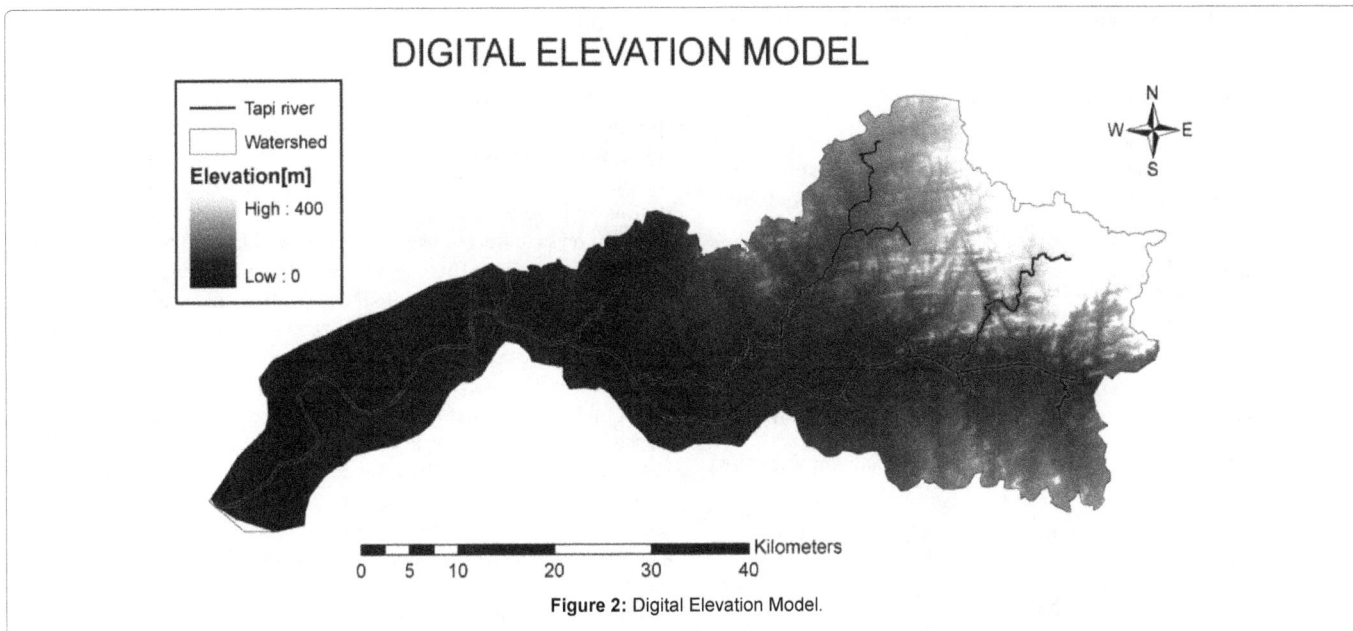

Figure 2: Digital Elevation Model.

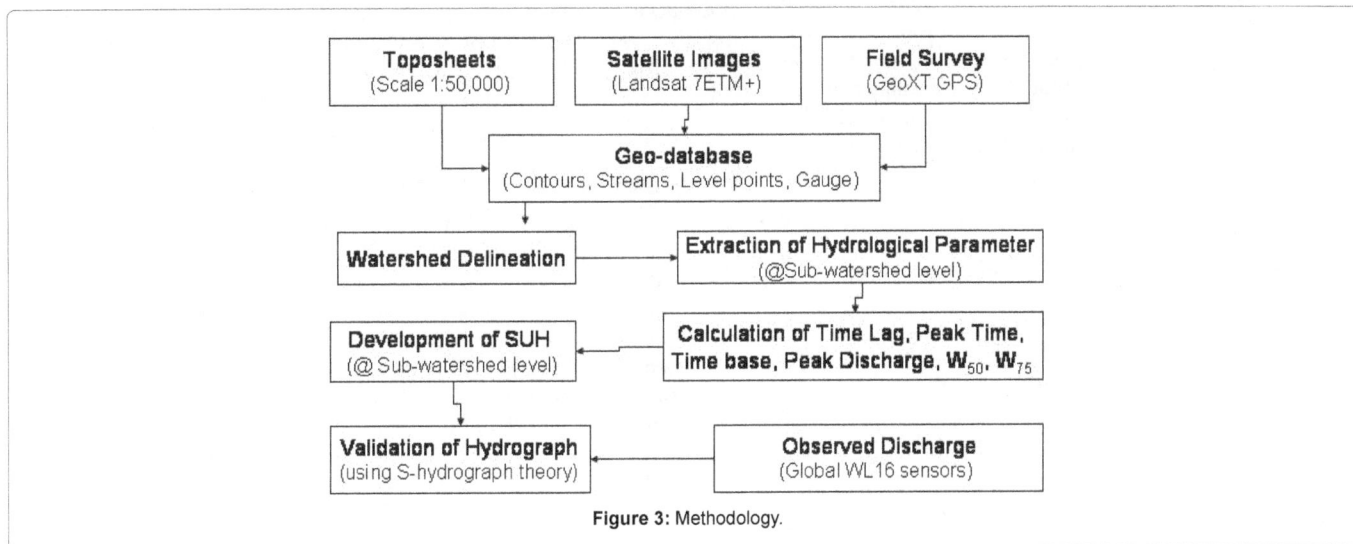

Figure 3: Methodology.

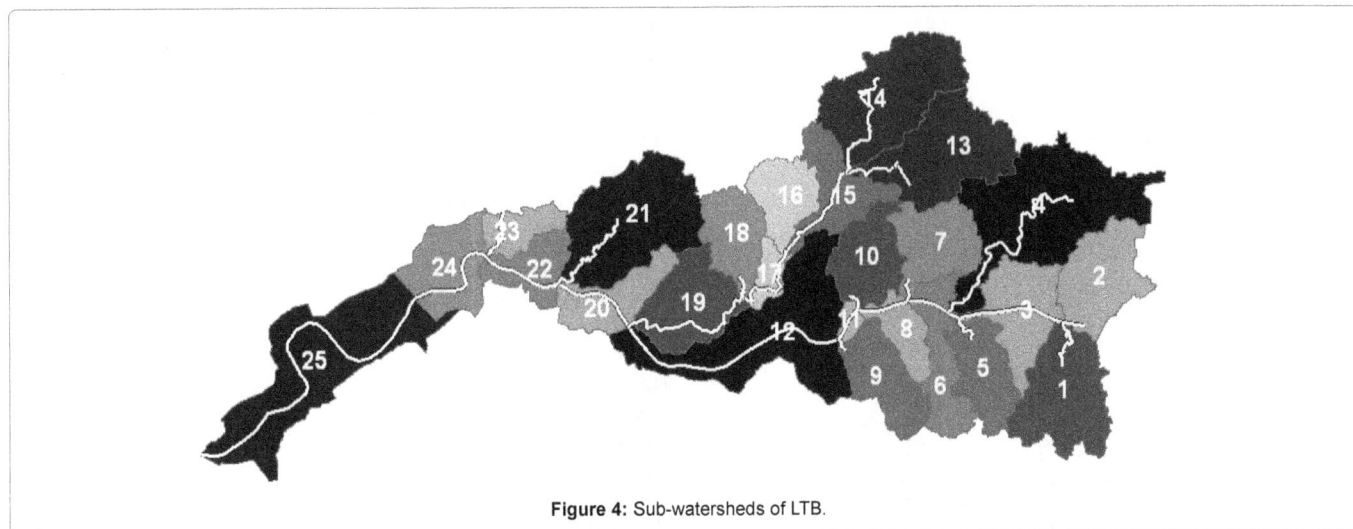

Figure 4: Sub-watersheds of LTB.

length of centroid have been derived using measurement tool in Arc-GIS.

Estimation of hydrological parameter

The following methodology has been used for calculation of hydrological parameters resulting into Snyder unit hydrograph (SUH). The steps involved in the calculation have been detailed out from equation (1) to (8) below [11,12].

The Snyder standard lag time (T_{lag}) [in hours];

$$T_{lag} = C_t \left(L \times L_{ca} \right)^{0.3} \tag{1}$$

Where the terms,

C_t = Lag Coefficient [1.2- 2.2] dependent upon basin properties

L = main channel length from basin outlet to upstream watershed boundary [km]

L_{ca} = main channel length from outlet to a point opposite the Centre of gravity [km]

The duration of UH (T_d) [in hours];

$$T_d = \frac{T_{lag}}{5.5} \tag{2}$$

As the term T_d is variable for each watershed and depends on shape and size. We propose to prepare SUH for each sub-watershed having common time duration of 1 hour. Therefore, above calculated time duration (T_d) is not desired duration. The alternative lag time [$T_{lag.alt}$] can be computed using equation (3);

$$T_{lag.alt} = T_{lag} + 0.25 \left(T_{da} - T_d \right) \tag{3}$$

Where, T_d = Previously Calculated duration [in hours], T_{da} = New desired duration [in hours]

Time of Peak (T_p) [in hours]:

$$T_p = \frac{T_d}{2} + T_{lag,alt} \tag{4}$$

The terms T_d and $T_{lag.alt}$ have been obtained from equation (2) and equation (3).

Peak Discharge (Q_p) [in m³/s] from each sub-watershed is calculated using below equation:

$$Q_p = \frac{2.78 \times A \times C_p}{T_{lag,lag}} \tag{5}$$

Where, A= area of sub-watershed [in km²], C_p= Peak flow coefficient [0.5- 0.7] that dependent upon basin characteristics.

Time base of the UH [in days] is calculated as follows:

$$T_{base} = 3 + \frac{T_{lag,alt}}{8} \tag{6}$$

Furthermore, Snyder proposed that shape of Unit Hydrograph is very important which has been approximated using the widths W_{50} and W_{75} at 50% and 75% of the peak discharge. The UH width at W_{50} and W_{75} can be calculated based on equations (7) and (8) below:

$$W_{50} = 756 Q_p - 1.081 \tag{7}$$

$$W_{75} = 450 Q_p - 1.081 \tag{8}$$

Hence, it is possible to develop the shape, standard time lag, time base, time of peak, and peak discharge of each sub-watershed.

Field data collection

The elevation data points in some parts of LTB have been limited; therefore additional field surveys using Trimble Geo-XT global positioning system (GPS) have been conducted. The GPS has been used as DGPS and expected to have horizontal and vertical accuracy of 1m and 3 m respectively. The accuracy check between GPS elevation points and top o-sheet elevation points shows good fit. However, the field surveys conducted using GPS has been limited to sub-watersheds having flat terrains. These elevation data points have been integrated in Arc-GIS to develop accurate DEM, as discussed under geo-database section above. The water depths in river channel have been measured using WL-16 U automatic water level sensors procured from Global Water USA. It was decided to have four automatic water level sensors of 25 m cable length fitted with data logger of capacity 81,800 records. The sensor has 0.1 mm water level measurement accuracy and can record 10 reading per second. The time duration of 30 sec for monsoon year 2010 and 120 sec for monsoon year 2011 were considered in our field based water-level data measurements. The water-level measurement stations in LTB are Kathore, Visdalia, Ghodsamba and Amli dam. Later, the discharge data from Amli station has been considered for validation of hydrological model. The water-level data from all four discharge stations have been imported in notebook using dedicated software. The output data has standard spread sheet format as *.csv (comma separated by values) and output is acquired in excel format using window-XP compatible water level logger software. These data were used for calculating actual discharge using Stage-discharge curve which were later compared with predicted peak flood discharge.

Results and Discussions

As discussed earlier, the entire LTB consist of 25 sub-watersheds having average geographical area of 65.7 km². The sub-watersheds identified are part of river tributaries such as Serul khadi, Damini khadi, Gal khadi, Anjana khadi, Mau-khadi, Rakha khadi, and Vare khadi. The entire database has been developed in Arc-GIS and has been analysed using two spatial data analysis tools viz. project management and measurement tool respectively. The sub-watershed parameters such as geographical area, main river channel length, and length of centroid from watershed outlet were calculated. The attribute data table obtained after sub-watershed analysis along with hydrological parameters has been depicted in Table 1 below. The DEM of LTB at 50 m cell size has been prepared using topo-maps and GPS survey measurements, as shown in Figure 3. DEM has mean elevation of 83.86 m with 71.02 as SD value. Maximum and minimum elevation value of DEM is 400 m and 0 m respectively. The z-value of each DEM cell has been used to delineate sub-watersheds using BASIN 4.0 software from US-EPA. The slope analysis of 25 sub-watersheds shows a mean value of 26.28%. The sub-watersheds 2, 4, 7 and 14 situated in head watersheds have very high slope in the range of 40-45.0%. However, the sub-watersheds 24 and 25 situated near sea coast have flat slope and their value has been found in the range of 2-2.5%.

The process of estimating the sub-watershed SUH can be described in following steps; (i) Calculate standard lag time using equation (1) based on main channel length, channel length from outlet to centroid point, (ii) Calculate the duration of UH, (iii) Calculate the alternative lag time using equation (3) for 1-hr UH. Thereafter, (iv) time of Peak (Tp) and peak discharge can be calculated using equation (4) and equation (5). In next step (v) the time base (T_{base}) is estimated using equations (6) that can considered important in Unit Hydrograph theory. Finally, (vi) the shape of SUH is very important which is determined by approximating width at 50% of peak discharge (W_{50})

Watershed	L (km)	Lc (km)	Area (km)	Ave Slope (%)	T_p (hr)	Q_p (m³/s)
1	7.33	10.52	89.27	29.11	9.85	18.22
2	2.35	10.44	67.88	43.70	7.78	17.49
3	11.68	6.61	66.16	29.71	7.80	17.50
4	26.72	19.56	160.78	45.57	13.23	23.93
5	3.75	9.56	51.35	30.39	8.86	11.76
6	4.73	10.25	50.66	27.73	6.86	15.39
7	2.68	7.14	57.67	41.58	8.27	14.25
8	5.59	5.94	29.82	15.66	7.84	7.81
9	1.51	8.39	56.29	23.88	8.65	13.24
10	1.98	9.75	55.59	29.13	9.65	11.60
11	3.49	2.34	6.77	7.80	4.33	35.07
12	25.62	18.00	150.52	19.09	12.77	23.26
13	6.22	10.38	9.27	33.65	10.38	17.87
14	14.65	10.97	112.64	40.13	10.99	20.42
15	11.15	10.03	50.99	33.18	8.65	12.01
16	0.11	5.17	41.96	36.61	7.26	11.97
17	9.60	5.75	15.54	19.75	7.10	4.55
18	2.76	7.88	47.26	29.75	8.44	11.41
19	16.57	10.55	68.04	17.77	9.52	14.40
20	9.62	7.65	46.45	23.34	7.68	12.45
21	13.64	14.21	121.36	29.62	11.96	20.12
22	8.55	4.86	37.98	17.74	6.59	12.08
23	6.13	5.63	27.11	26.78	7.90	7.04
24	10.45	9.07	60.27	2.70	8.23	14.97
25	25.32	15.80	161.54	2.77	12.27	26.05

Table 1: Sub-Watershed Parameters Involved in SUH.

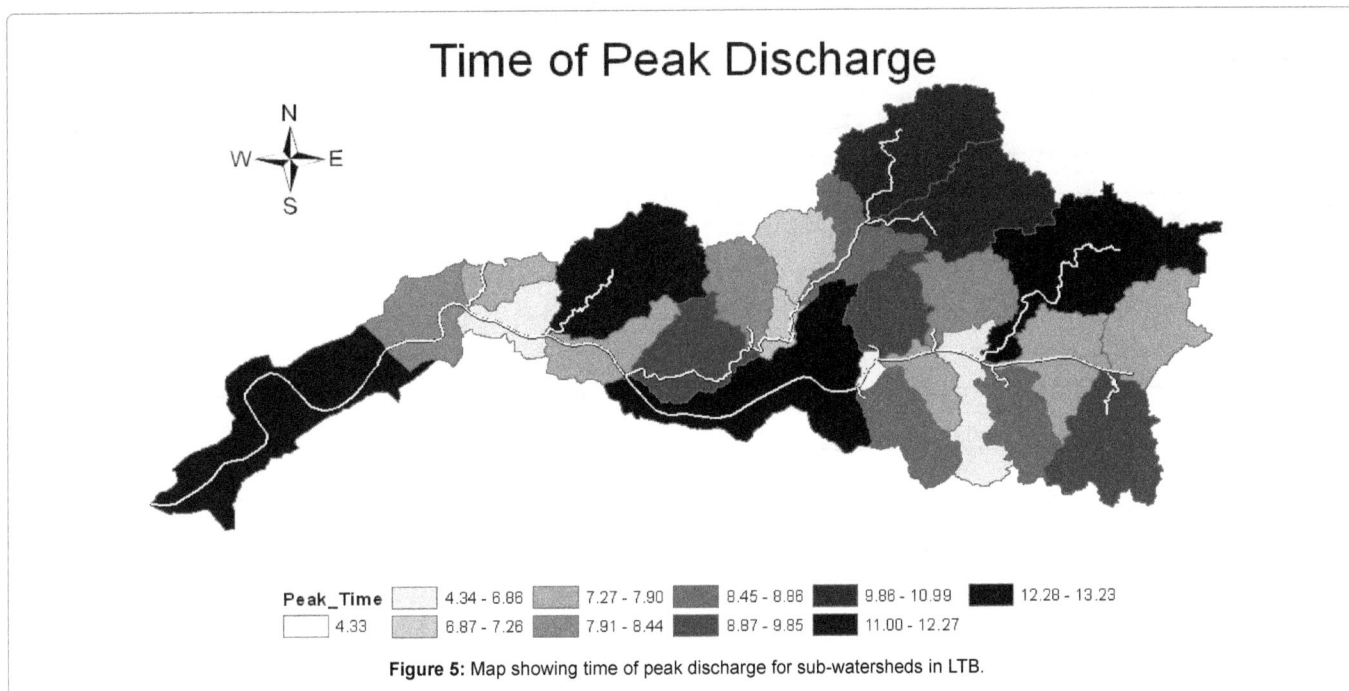

Figure 5: Map showing time of peak discharge for sub-watersheds in LTB.

and 75 % of the peak discharge (W_{75}) using the equation (7) and (8). Figure 4 showing time of peak (Tp) shown in Figure 5, however other hydrological parameters are given in Table 1.

The systematic analysis on steps (i) to (vi) as discussed above reveals the following outcome. The value of peak discharge increases when standard lag time decreases, and peak discharge increases when sub-watershed area increases. The sub-watershed level discrete analysis presents that sub-watershed 4, 12, 21, 25 have slow response time due to larger watershed area. The sub-watershed 11 has quickest response time due to smaller geographical area and shortest river channel length. Other sub-watersheds like 2, 3, 6, 8, 16, 17, 20 & 22 have quick response time due to smaller geographical area and shorter river channel length. In calculating the peak discharge as per equation (5), the sub-watershed

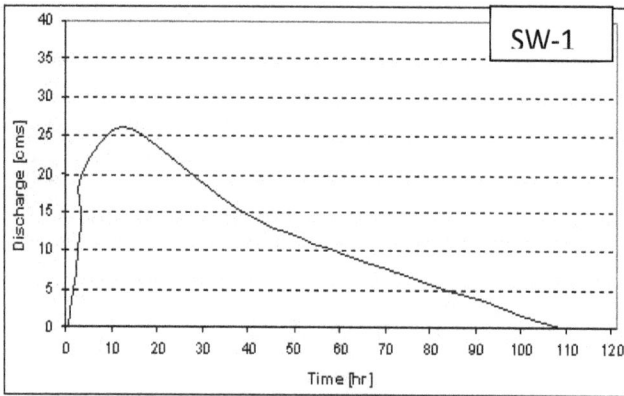

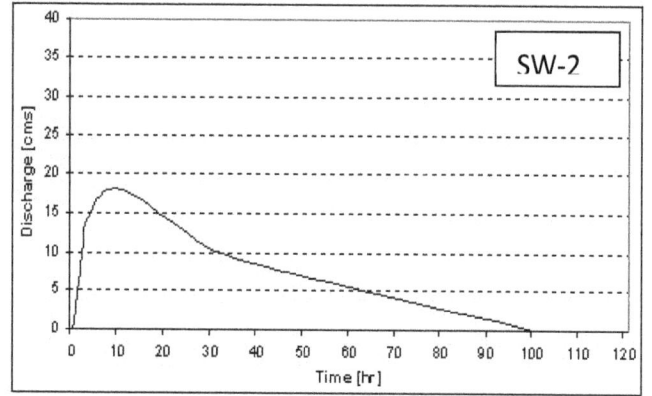

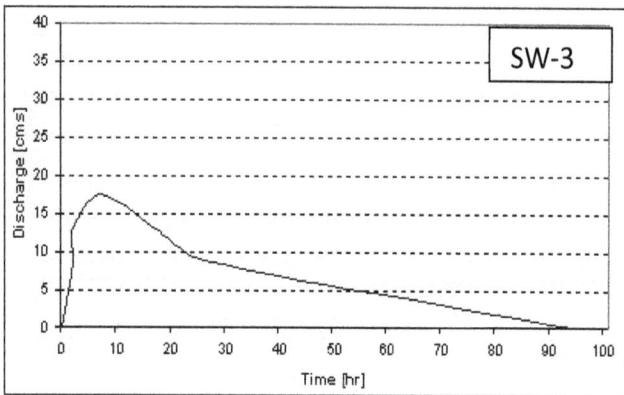

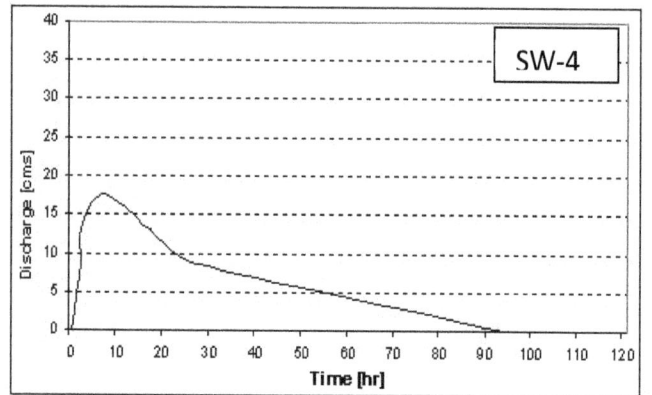

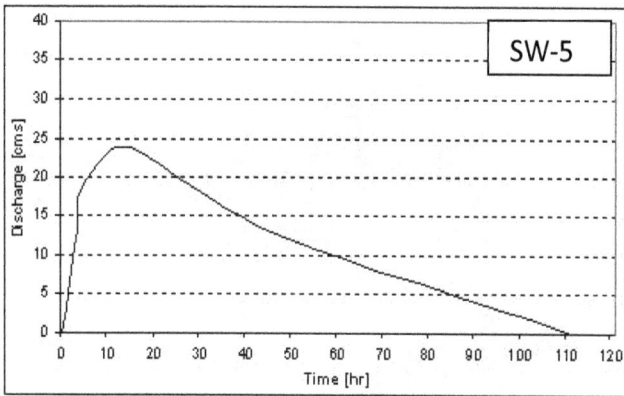

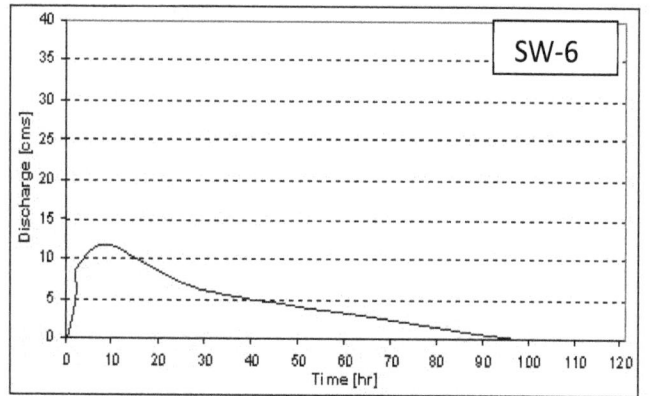

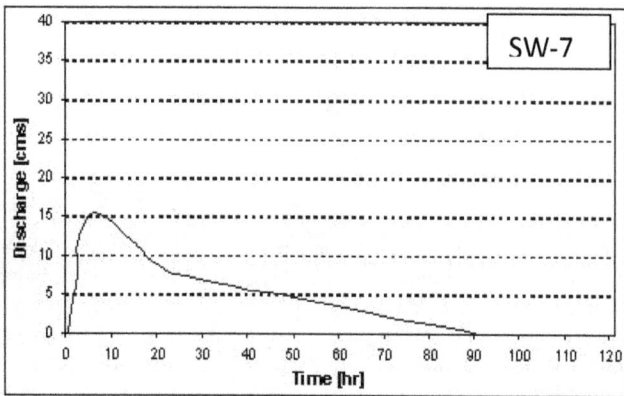

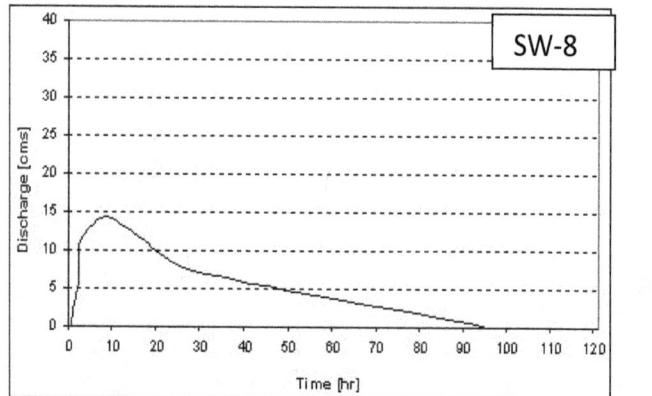

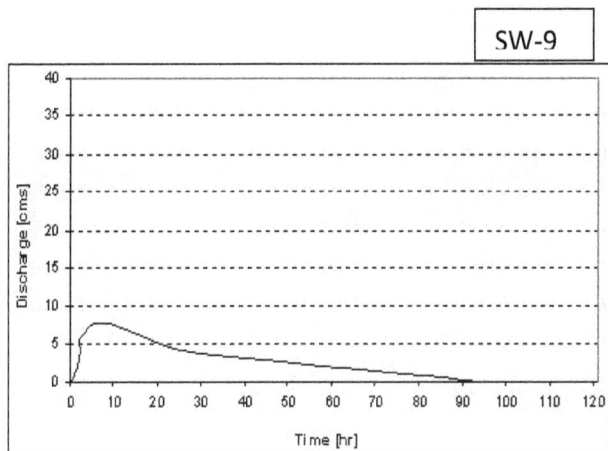

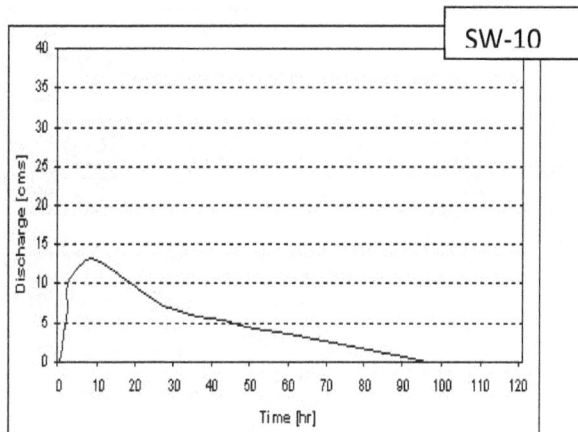

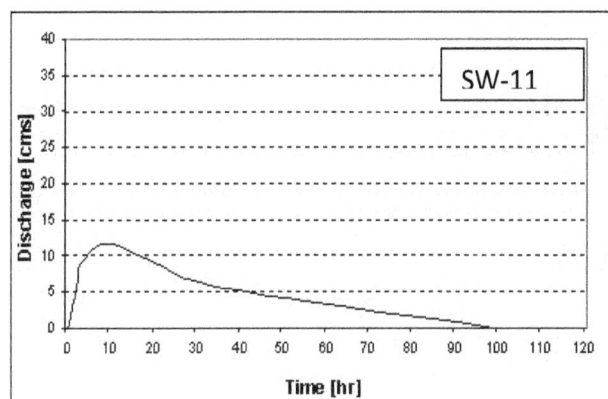

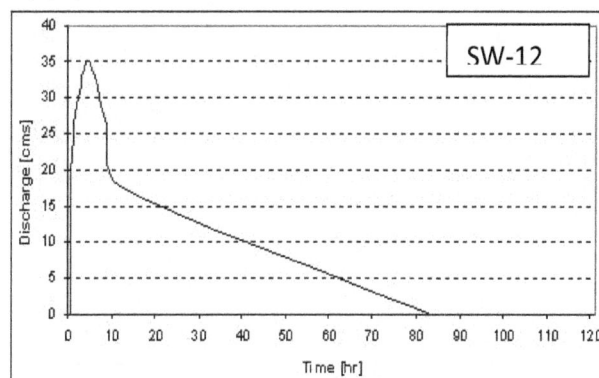

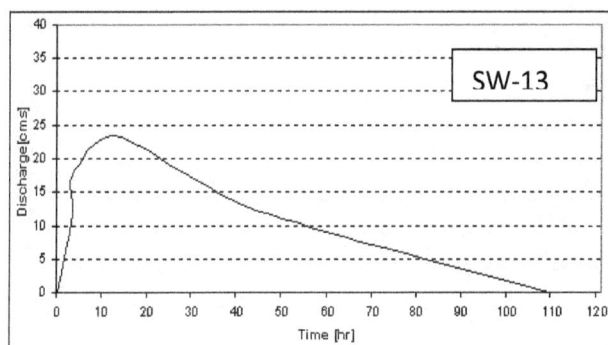

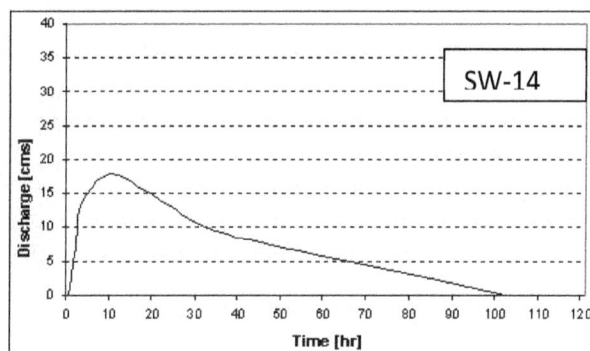

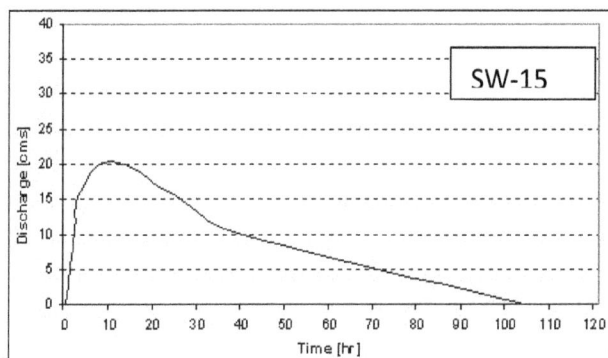

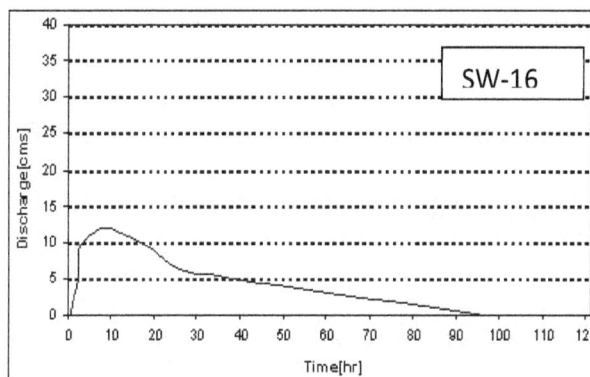

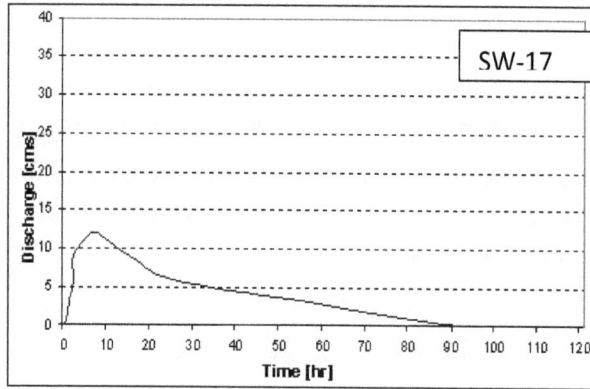

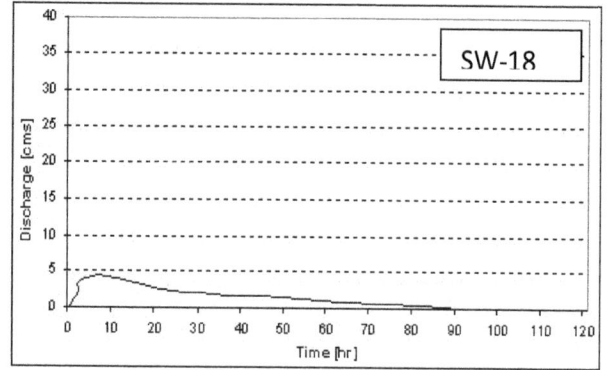

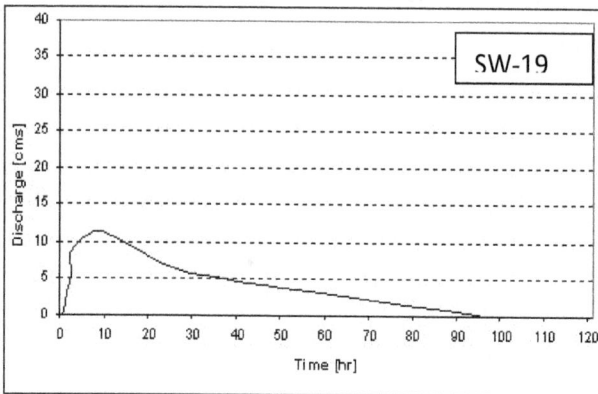

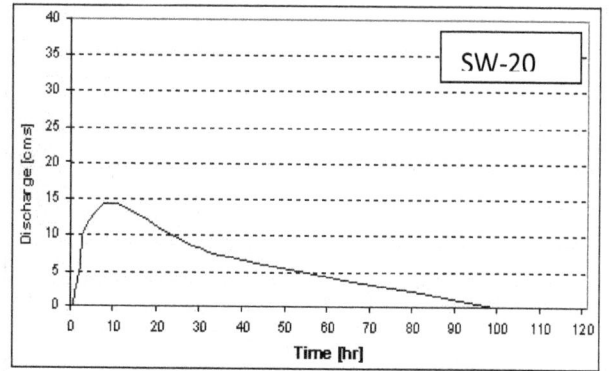

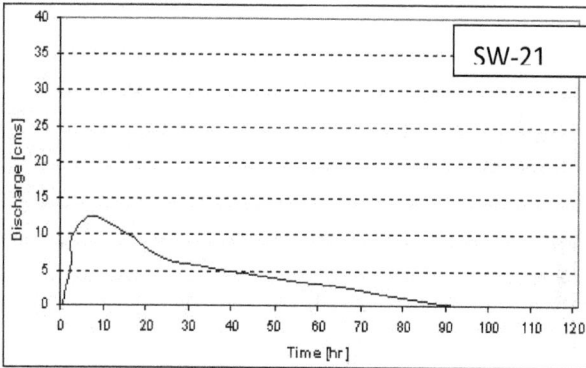

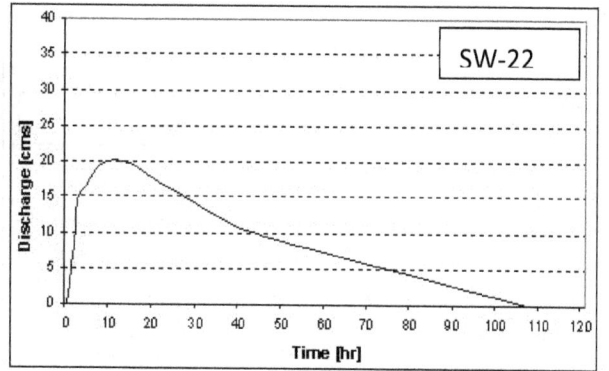

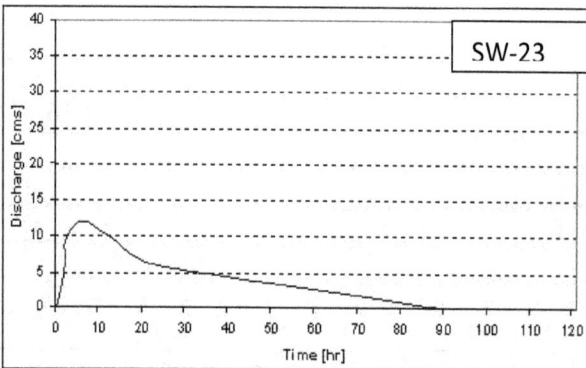

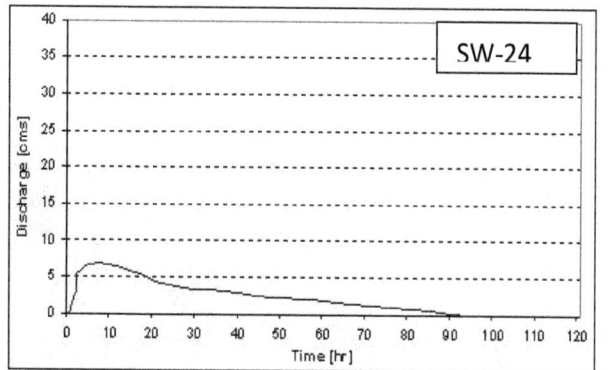

Figure 6: Unit Hydrograph of 25 Sub-Watersheds.

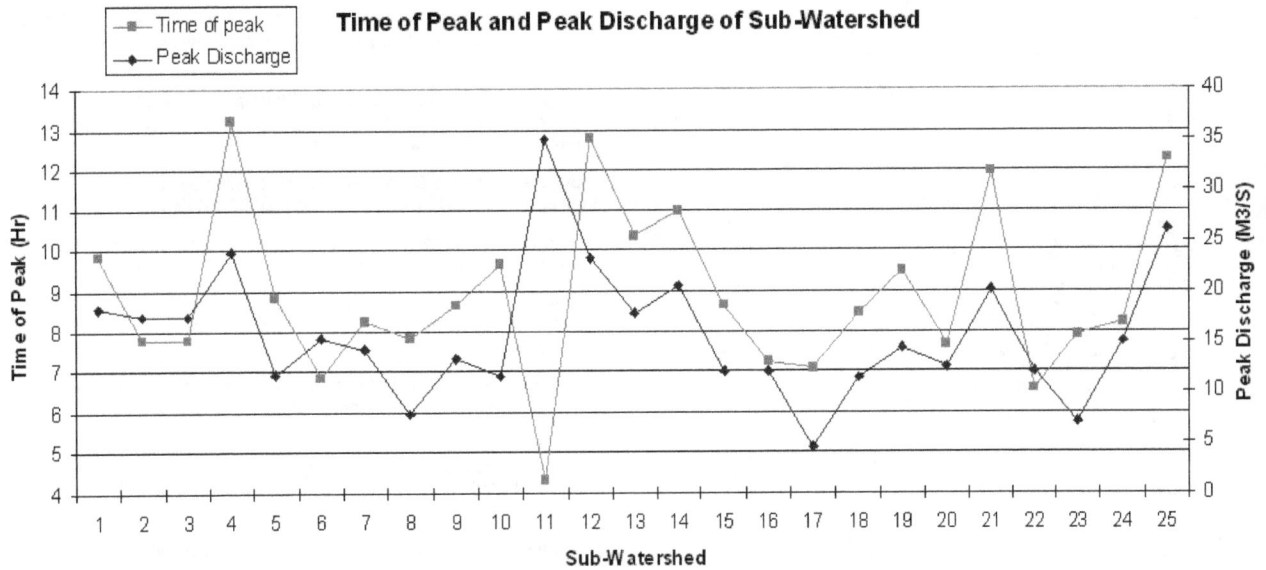

Figure 7: Time of Peak and Peak Discharge of 25 Sub-Watersheds.

S.N	Duration	Rainfall (cm)	Discharge (predicted) M³/s	Discharge (measured) m³/s	Difference	% Error
1	July 4-5, 2010	4.7	80	75	5	6.67
2	July10-11, 2010	5.5	94	86	8	9.32
3	August 8-9, 2010	11.1	191	159	32	16.75
4	August 18-19, 2010	3.4	57	54	3	5.55
5	August 24-25, 2010	7.9	133	118	15	12.71
6	August 26-27, 2010	6.1	104	96	8	8.33
7	August 30-31, 2010	3.0	51	47	4	8.51
8	Sep 1-2, 2010	2.9	40	37	3	8.11
9	Sep 6-7, 2010	9.6	159	132	27	20.45
10	Sep 15-16, 2010	4	68	59	9	8.67

Table 2: Amli Sub-watershed (13) Validation for 2-day Event.

11 gives highest peak discharge due to quick response and smallest geographical area. Sub-watersheds 4, 12, 14, 21 and 25 also gives high discharge value while sub-watersheds 7, 8 and 23 gives low discharge value. The sub-watershed 25 has highest area of 161.54 km², highest discharge of 26.05 m³/s, and time of peak as 12.27 hours (Figure 7).

The unit response from all the sub-watersheds as UH has been given in Figure 6. Later, the discharge from UH for few events has been calculated using summation hydrograph (S-hydrograph). Thereafter, the discharge hydrograph for a gauge site Amli has been considered for validation. The observed discharge from WL16 automatic water level sensor and SUH derived discharge have been compared for model validation (Table 2). The peak discharge and runoff volume can be considered an important parameter while deciding the non-structural measures for flood modelling and flood forecasting.

Flooding potential of each sub-watershed was estimated using SUH. The value of Volume and surface runoff were calculated for extracting percentage flood contribution for each sub-watershed which is shown in Table 3. It was found that sub-watershed number 4, 12 and 25 are high flooding potential having percentage contribution 9.24, 8.61 and 9.24 respectively. It was also observed that sub-watershed number 8, 17 and 23 are low flooding potential having percentage contribution 1.66, 0.86 and 1.50 respectively. Flooding potential depends on peak discharge and area of watershed which is clearly reflected in results found out.

Conclusion

The SUH method has been chosen for estimation the flood response, contribution of flooding potential, percentage of flood volume for 25 sub-watersheds in LTB. The Geo-spatial database for LTB has been developed and sub-watershed parameters such as river length, length of centroid, spatial area, land use, lateral slope, and terrain and soil factors for estimating discharge. The geo-database developed in Arc-GIS environment from topo-sheets (1:50,000), satellite remote sensing images and GPS have been combined and kept in a common database. A DEM of 50 m cell size has been produced and sub-watershed characterisation was completed. The analysis for all 25 sub-watersheds exhibit that 35.07 m³/s and 4.55 m³/s and 13.23 hours and 4.33 hours have been highest and lowest peak flows and time to peak respectively. The SUH model has been validated for peak discharge at Amli gauge site where discharge data were collected using water level sensors WL-16U during monsoon season of the year 2010 and 2011. Measured discharge and SUH estimated discharge shows good fit within a mean variability range of 5-7%. SUH methods ability to estimate hydrological parameters including peak flow discharge shows wider replication for un-gauged catchments in LTB. Snyder unit hydrograph method could become a good source of information for flood related issues. It gives peak flow estimation with time of peak at a various sub-watersheds is of vital importance in flood forecast as it is useful in computing flood discharge for various rainfall event.

ID	Qp [m³/s]	Tp [hr]	Volume [m³]	Runoff [%]	Contribution [%]
1	18.22	9	1147708.45	50	5.02
2	17.49	7	849901.78	51	3.72
3	17.48	7	786496.15	47	3.44
4	23.93	13	2110347.34	52	9.24
5	11.76	8	677286.57	52	2.97
6	15.39	6	637061.88	50	2.79
7	14.25	8	743751.79	51	3.26
8	7.81	7	379515.89	50	1.66
9	13.24	8	714865.61	50	3.13
10	11.6	9	709826.27	50	3.11
11	35.07	4	757411.99	44	3.32
12	23.26	12	1967536.20	52	8.61
13	17.87	10	1189984.87	51	5.21
14	20.42	11	1470045.89	52	6.44
15	12.05	8	607239.82	47	2.66
16	11.97	7	538578.88	51	2.36
17	4.55	7	196534.05	50	0.86
18	11.41	8	616058.65	58	2.70
19	14.4	9	881163.65	51	3.86
20	12.45	7	582583.07	50	2.55
21	20.12	12	1593293.62	52	6.98
22	12.08	6	478304.84	50	2.09
23	7.04	8	342098.83	50	1.50
24	14.97	8	754388.39	49	3.30
25	26.05	12	2109771.42	52	9.24

Table 3: Volume and surface runoff were calculated for extracting percentage flood contribution for each sub-watershed.

References

1. Singh AK, Sharma AK (2009) GIS and a remote sensing based approach for urban flood-plain mapping for the Tapi catchments, India, IAHS Publ 331: 389-394.

2. Singh AK, Sharma S, Vakharia U, Sharma AK (2011) Estimating hydrological parameters in ungauged basin using field measurements and GIS. Proceedings ISRS2011- Bhopal, 9-11 Oct, 2011.

3. Singh AK, Sharma S, Jain AO (2011) Mapping and predication of surface run-off using SCS-CN method. Proceedings ISG2011 Ajmer.

4. Bhaskar N, Parida B, Nayak A (1997) Flood Estimation for Ungauged Catchments Using the GIUH 123: 4.

5. Sharma SB, Singh AK (2014) Assessment of the Flood Potential on a Lower Tapi Basin Tributary using SCS-CN Method integrated with Remote Sensing & GIS data. J Geogr Nat Disast 4: 2-7.

6. Sherman LK (1932) The relation of hydrographs of runoff to size and character of drainage-basins, Eos, Transactions American Geophysical Union 13: 332-339.

7. Sudhakar Sharma B, Anupam Singh K (2014) Lower Tapi Basin Tributary using SCS-CN Method integrated with Remote Sensing & GIS data. J Geogr Nat Disast 4: 128.

8. Snyder FF (1938) Synthetic unit graphs. Trans Americans Geophysics. Un. 19.

9. Hoffmeister G, Weisman RN (1977) Accuracy of synthetic hydrographs derived from representative basin. Hydrological Sciences 22: 297-312.

10. Wayal AS, Parmeswaran PV, Ameta NK (2008) Derivation of unit hydrographs for ungauged catchments, First International Conference on Emerging Trends in Engineering and Technology, pp. 1029-1034.

11. Adebayo WS, Solomon OB, Ayanniyi MA, Sikiru FO (2009) Evaluation of synthetic unit hydrograph method for development of design storm hydrographs for rivers in South-West- Nigeria. Journal of American Science 5: 23-32.

12. Adib A, Salarijazi M, Najafpour K (2010) Evaluation of synthetic outlet runoff assessment models. J Appl Sci environ Manage 14: 13-18.

17

Comparison of Seepage Simulation in a Saline Environment below an Estuary Using MODFLOW and SEAWAT

Wissam Al-Taliby*, Ashok Pandit and Howell Heck

Department of Civil Engineering, Florida Institute of Technology, Melbourne, FL 32901, United States of America

Abstract

This paper compares the results produced by MODFLOW, a constant-density model, to results produced by SEAWAT, a variable-density model, to investigate the feasibility of using MODFLOW in a saline environment below an estuary known as the Indian River Lagoon. The comparison was conducted over sixteen numerical simulation cases at different conditions of estuarine salinity C_L, hydraulic conductivity anisotropy ratio K_r, and water table elevations on the freshwater boundaries in a two-dimensional vertical domain. The use of MODFLOW at the study site under the calibrated K_r distribution ranging from 1000-20,000 was found to accurately match the field-measured and SEAWAT-simulated results with a remarkable increase in accuracy at higher groundwater elevations. The study determined a critical value of K_r of 1000 above which, MODFLOW simulations of the variable-density problem produced results that agreed well with those produced by SEAWAT. However, MODFLOW starts to produce significant errors with K_r below the critical value and hence, it should not be used for simulating variable-density environments when $K_r < 1000$. The amount of submarine groundwater discharge (SGD) predicted by either model, and also MODFLOW accuracy in predicting the SGD are directly proportional to the head difference between the groundwater divide elevation and the lagoon water surface, but to a lower extent, are inversely proportional to C_L.

Keywords: Coastal aquifer; MODFLOW; SEAWAT; Indian River lagoon; Estuary; Submarine groundwater discharge

Introduction

The presence of high concentrations of dissolved solids in groundwater alters the fluid density and may result in spatial density variations within the flow system. Significant fluid density gradients can substantially affect the groundwater flow patterns introducing thereby mathematical and numerical complexities for simulating such density-dependent flow systems [1]. Examples of such variable-density environments are: saltwater intrusion [1-7], Submarine Groundwater Discharge (SGD) [8,9], aquifer storage and recovery [10-13], brine migration [14], coastal wetland hydrology [15], injection of liquid waste in deep saline aquifers [16], and disposal of radioactive waste in salt formations [17,18]. Numerical modeling of variable-density groundwater flow and transport environments such as in saline environments, where the physics of flow and transport are density-driven, typically relies on the use of variable-density numerical models that incorporate the relationship between fluid density and solute concentration by iteratively solving the flow and transport governing equations. These models are also termed as coupled models. An example of a variable-density coupled model is SEAWAT [19-21] which was developed by the U.S. Geological Survey (USGS). SEAWAT couples a modified version of MODFLOW [22-24], as a flow simulator for solving the flow equation, and MT3DMS [25] as a solute-transport simulator for solving the mass transport equation [8]. One of the coupling schemes in SEAWAT is termed the implicit approach which is adopted in this paper. An implicit coupling incorporates iterative solution of the flow and transport equations at each time step until the density difference is within a specified value [19]. On the other hand, in situations where spatial density variation is so small that it can be considered to be negligible, a constant-density model such as MODFLOW, also developed by the USGS, may also be used in variable-density environments to solve the flow equation when the modeler's main objective is to simulate the flow but not the transport conditions.

While it is clearly preferable and generally accepted that a variable-density coupled flow and transport model best simulates the physical situation in a saline groundwater environment, there may be reasons why a modeler may wish to use a constant-density model in a variable-density environment. Reasons why modelers would prefer to use a constant-density model such as MODFLOW over a variable-density coupled model such as SEAWAT, even in a saline or variable-density environment, could be: (a) constant-density model simulations require remarkably smaller computation times compared to a variable-density coupled model, and thus, an improved computational efficiency is achieved especially if numerous simulations are required for calibration and validation, b) high level- accuracy simulation results may not always be required in some problems where dropping some important physics to get faster results that are reasonably accurate for making some groundwater management decisions is more practical, c) the ability of the constant-density simulation, like a MODFLOW simulation to be followed by multiple transport codes simulations such as MT3DMS [25] and RT3D [26], in case the model is to be used later for transport simulations, d) familiarity of groundwater modeling community with the constant-density model [19,27], and e) variable-density coupled models require a larger number of parametric values such as values for molecular diffusion and dispersivity which leads to a higher level of uncertainty.

Comparisons of results obtained by variable-density or coupled models and constant-density or uncoupled models in saline environments have been reported in past several studies. Some of these comparison studies have been conducted in the context of the Henry

---

*Corresponding author: Wissam Al-Taliby, Postdoctoral Research Associate, Department of Civil Engineering, Florida Institute of Technology, Melbourne, FL 32901, United States of America, E-mail: waltaliby2011@my.fit.edu

problem [28] or on laboratory scale physical models [29-33] while other studies used large scale models [6,34-36]. Researchers have found that under some saline conditions, a constant-density model is capable of producing results similar to a variable-density model. For example, Simpson et al. [29] solved the Henry's problem using both variable-density and constant-density models and found that the predicted location of the 0.5 isochlor, under steady state conditions, were quite similar although the constant-density isochlor did not advance as far into the aquifer as the variable-density isochlor. However, they also found that the matching of the predicted steady state 0.5 isochlors by the two models was much closer when the recharge rate, q, was increased to 2q and 4q. Finally, Simpson et al. [29] noted that the variable-density solution reached the steady state condition much sooner than the constant-density solution indicating that a constant-density model may not be suitable if transient solutions are required. Simpson et al. [30] found that the discrepancies between results obtained from variable-density and constant-density models were more significant for a modified Henry's problem where the freshwater recharge rate of the standard Henry's problem was halved. Dentz et al. [31] compared variable-density and constant-density analytical and numerical solutions of the Henry problem over a range of two dimensionless groups, the coupling parameter, defined as the ratio of buoyancy flux and viscous forces, and the Péclet number, defined as the ratio of advective and dispersive effects. Dentz et al. [31] concluded that variable-density and constant-density solutions resulted in similar flow patterns when the coupling parameter was far lower than 1 (this corresponds to higher freshwater recharge rate) at moderate Péclet numbers. Goswami et al. [32], in an experimental and numerical analysis, using variable-density and constant-density (i.e., coupled and uncoupled) SEAWAT simulations, of a laboratory-scale porous media tank, made similar conclusions to those of Simpson and Clement [29,30]. Abarca et al. [33] solved the Henry problem in its standard diffusive form and a modified dispersive form for both variable-density and constant-density conditions. Abarca et al. [33] concluded that, unless the diffusion coefficient of the standard form is reduced by a factor of 10, or the standard form is replaced by the proposed diffusive form, constant-density solution would not result in dramatic changes in the simulated freshwater hydraulic heads or concentration distributions compared to variable-density solution.

Arlai et al. [34] found that both the density-dependent coupled model and the density- independent uncoupled model produced basically similar plume migrations in a Bangkok aquifer. Arlai et al. [34] hypothesized that this was due to groundwater pumping playing a major role in plume migration compared to the role played by density effects. However, in a subsequent paper, Arlai et al. [35] showed that the predicted steady state saline concentrations, in a 1000 by 100 m vertical plane of a saturated coastal aquifer, were much different when using a density-dependent model than a density-independent model, and that the density-dependent model results were closer to the predicted Ghyben-Herzberg interface. Motz et al. [6] conducted multiple numerical experiments on a two-dimensional vertical section of a theoretical coastal aquifer problem with freshwater recharge boundary on the upstream and a seacoast boundary on the downstream using MODFLOW and SEAWAT. Motz et al. [6] found that MODFLOW could closely match the hydraulic heads and fluxes simulated by SEAWAT on the freshwater side of the aquifer when the coastal boundary was represented by specified freshwater hydraulic heads. Subsequently, Motz et al. [36] compared coupled (density-dependent) and uncoupled (density-independent) SEAWAT solutions of saltwater intrusion and seepage circulation on the seacoast boundary

of the same system described in Motz et al. [6]. In that comparison study, Motz et al. [36] observed that the density-independent solution produced similar results of saltwater intrusion and seepage circulation to that produced by the density-dependent solution when the ratio of freshwater recharge rate to the density-driven flux was increased. Thus, depending on the boundary conditions and the aquifer properties of a specific saline environment problem, it is reasonable to conclude that there may be situations when a constant-density model can replicate the results of a variable-density coupled model in simulating saline environments although that is not always the case.

The studies discussed above showed through, numerical, analytical, or non-dimensional formulation of variable-density problems, that aquifer anisotropy ratio of hydraulic conductivity K_r, and the advective effect of the regional freshwater recharge are amongst the most significant factors controlling the mixed convection ratio (ratio of buoyancy forces to advective forces) on which, density effects depend substantially in saline environments. Sensitivity analyses conducted in this paper also demonstrated the importance of these parameters. Thus, these two parameters, in addition to the salt concentration which is the main source of density-variation, are therefore expected to affect the accuracy of a constant-density solution of variable-density problems where they may increase or dampen the density effects. Also, it is generally known that the density variation has limited effect on horizontal flow and is mostly observed when the flow in a saline environment is vertical, the case that occurs at high K_r. Therefore, it is expected that a constant-density solution would resemble a variable-density solution if there is a relatively small vertical flow component which is likely to happen if K_r is large. However, no critical K_r value, where the two solutions become similar, has been defined in the literature especially for a real-world saline environment. It is also unknown if there would still be some vertical flow component at that critical K_r, i.e., if the two models would produce similar results even if there is a vertical flow component.

This paper uses a real-world, two-dimensional transect in the vertical plane below the Indian River Lagoon (IRL), an estuary located on the east-central coast of Florida, to determine if a constant- density model such as MODFLOW and a variable-density model such as SEAWAT, that are using the same calibrated distribution of K_r can produce similar results and if they can match 1) measured freshwater hydraulic head contours in the unconfined aquifer below the estuary, 2) groundwater flow directions in the unconfined aquifer, 3) total submarine groundwater discharge (SGD) into the estuary from the adjacent unconfined aquifer, and 4) spatial distribution of SGD below the estuary. The investigation then extends to determine: a) the critical K_r below which, results of constant-density and variable-density models become significantly different, b) if there is still a vertical flow component occur at the critical K_r, c) how the IRL salinity C_L can affect the amount of SGD into the lagoon and the accuracy of MODFLOW in predicting that amount, and d) how the MODFLOW accuracy improves by increased regional groundwater table elevations on the freshwater boundaries under the calibrated K_r and high C_L. The constant-density and variable-density modeling results are also compared under both field measured and wide ranges of C_L and water table elevations at freshwater boundaries using the calibrated K_r values. Thus, this study provides guidance to modelers, regarding values of the anisotropy ratio, at which constant-density models can be used for solving real-world saline environment problems and efficiently overcome the computational burden associated with variable-density models. The paper also provides further guidance by quantifying the error resulting from dropping the physics from the variable-density problem by approximating it as a constant-density problem. It is well-known that constant-density model simulations take less computation time than variable-density model simulations. However, the authors are not aware of studies that have compared and quantified the stark contrast between the computation times used by the two models in a real- world variable density environment. Thus, the computation time difference has been identified and discussed in this study as well. The MODFLOW version used in this study was MODFLOW-2000 [23]. This study also uses SEAWAT-2000 [20], which couples MODFLOW-2000 and MT3DMS [25] using the implicit finite difference scheme.

Description of Study Area

The study site for this comparison is a transect in the Indian River Lagoon (IRL) which is termed as the Palm Bay transect. The IRL is a coastal estuary that extends over approximately 250 km on the east coast of Florida (Figure 1). The width of the lagoon varies in the range of 0.8-8.0 km, while its depth ranges from 1 to 3 m. The IRL is connected to the Atlantic Ocean via the Ponce de Leon inlet, Jupiter inlet, Sebastian inlet, Fort Pierce inlet, and the St. Lucie inlet. Everywhere else, it is separated from the ocean by a coastal island strip known as the Barrier Island. The IRL is underlain by an unconfined aquifer which consists mainly of sand and shells with some silt, sandy clay, and clay. The Hawthorn formation, which is a confining impervious layer, underlies the unconfined aquifer and consists mainly of marl and clay [37]. The majority of groundwater seepage into the IRL comes from the watershed on the west side known as the Mainland although some seepage also comes from the Barrier Island.

As shown in Figure 2, the Palm Bay transect is oriented perpendicular to the IRL shoreline, and extends from the groundwater

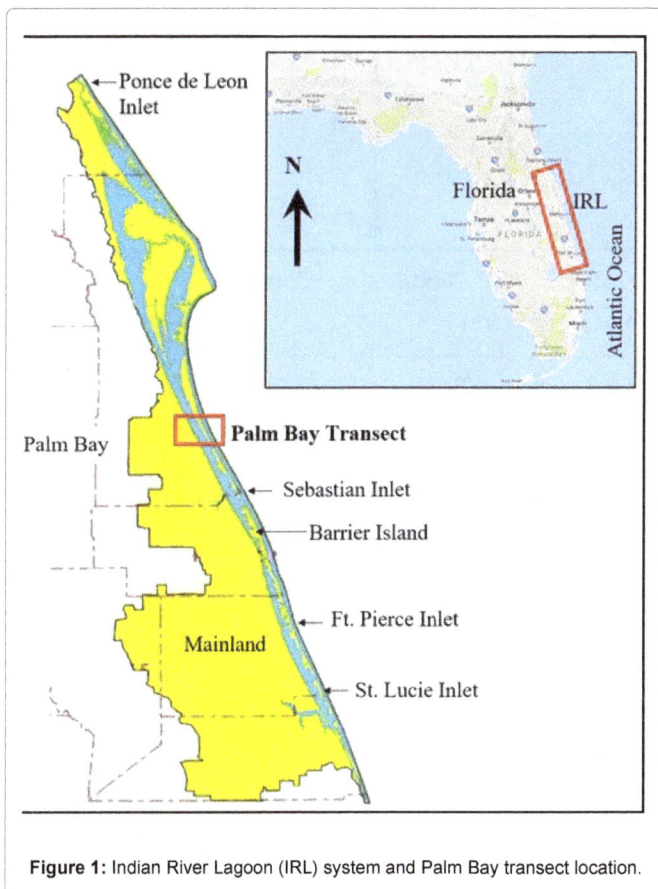

Figure 1: Indian River Lagoon (IRL) system and Palm Bay transect location.

divide on the Mainland to the Atlantic Ocean passing through the IRL and the Barrier Island. As a result of the mixing of salt water from the ocean through the inlets, and freshwater received through rainfall, groundwater, non-point runoff and point runoff through canals and rivers draining an approximately 3575 km² [38], the IRL waters are generally brackish and have a salinity in the annual range of about 10.8 to 37.8 g/L at the study site. This corresponds to a normalized salinity range of 0.3-1.05 based on ocean salinity of 36 g/L.

Data Collection

Initially, the horizontal domain of the transect was determined by locating the groundwater divide (Figure 2) using multiple monitoring wells installed on the Mainland. Relevant field data is measured to determine appropriate boundary conditions, and for the calibration of the numerical models. Water table elevations were measured in several wells installed on the Mainland and Barrier Island to obtain water surface profiles which were used as boundary conditions in the numerical models. The lagoon bed profile at the transect location shown in Figure 3 was determined by traversing the lagoon cross section by a boat and measuring the bed depth under the water surface at approximately 50 locations. Piezometric heads and groundwater salinity are simultaneously measured at eight stations situated across the IRL transect as shown in Figures 2 and 3. Three of these stations are located on the lagoon shores and are termed WS1, WS2, and ES, while five stations are somewhat uniformly spaced along the lagoon and are labelled S1 to S5. Nested shallow and deep piezometers were installed at each of the eight lagoon stations as depicted in Figure 3. The depth of the shallow locations ranges from 0.5 to 1.5 m while that of the deep locations ranges from 1.5 to 6.1 m below the IRL bed. Piezometers were made of 1.9 cm diameter PVC pipes. A 30 cm long, 1.9 cm diameter well screen made of slotted PVC pipe with #10 slot size, was attached to the end of each piezometer. Mid- screen locations range from (0.5-4.31 m) and (1.72-6.82 m) below the National Geodetic Vertical Datum of 1929 (NGVD 29) for the shallow and deep piezometers, respectively.

Piezometers were driven into the aquifer using a jetting technique in which, a 3.175 cm diameter PVC casing was jetted into the sediment to the desired depth by a 1.49 Kw centrifugal pump. Once the desired depth is reached, a piezometer was inserted into the casing. The casing was then pulled out of the sediment slowly and sand was then packed around the screen and then the annular space around the piezometer was sealed by bentonite clay chips. The depth of the clay seal from the lagoon bed was about 180 cm and 30 cm for the deep and the shallow piezometers, respectively. The water that was introduced into the wells during the jetting phase was removed by wells development. In order to protect the piezometers from potential sabotage, shallow piezometers were terminated approximately 5 cm above the lagoon bed while the deep piezometers were terminated within 60 cm from the lagoon water surface. The top of each piezometer was threaded and fitted with an O-ring seal where a cap was screwed to create a leak-proof seal. Whenever measurements were taken, the cap was removed and a PVC extension pipe was screwed into the top of the piezometer so that it extends above the water surface.

Groundwater salinity was measured by extracting samples from the deep and shallow piezometers, and reading the salinity with a YSI salinity meter, Model 85. The salinity meter converts conductivity readings to salinity based on ASTM algorithms found in ASTM Designation D1125-82, and has a measurement range of 0 to 80 parts per thousands (ppt), a resolution of 0.1 ppt, and an accuracy of 0.5 percent. Lagoon water salinity was also measured at each of eight lagoon stations at the same time as the other measurements. The lagoon

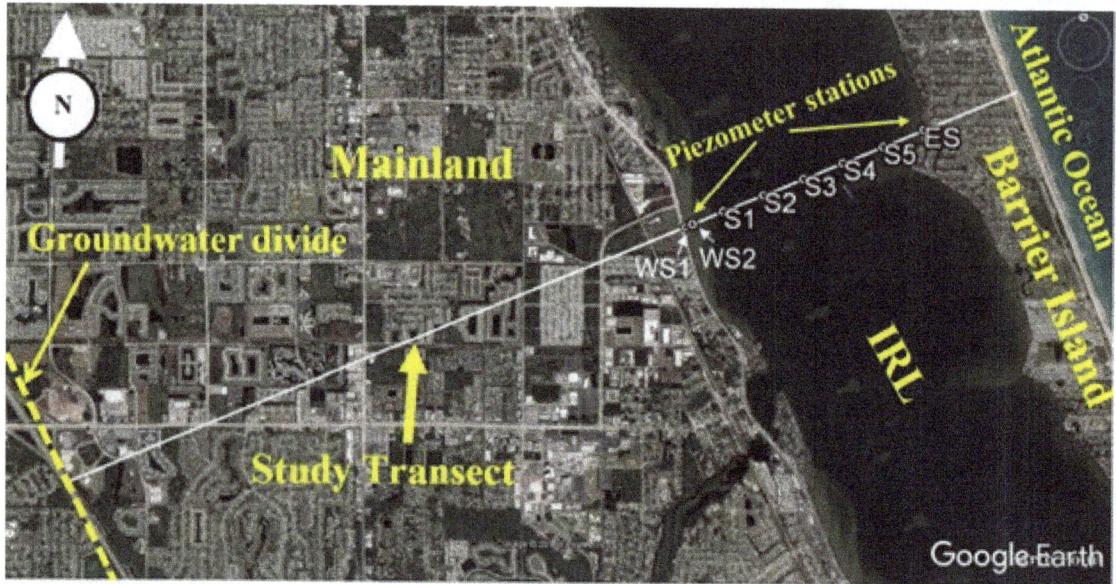

Figure 2: Domain and orientation of the Palm Bay transect.

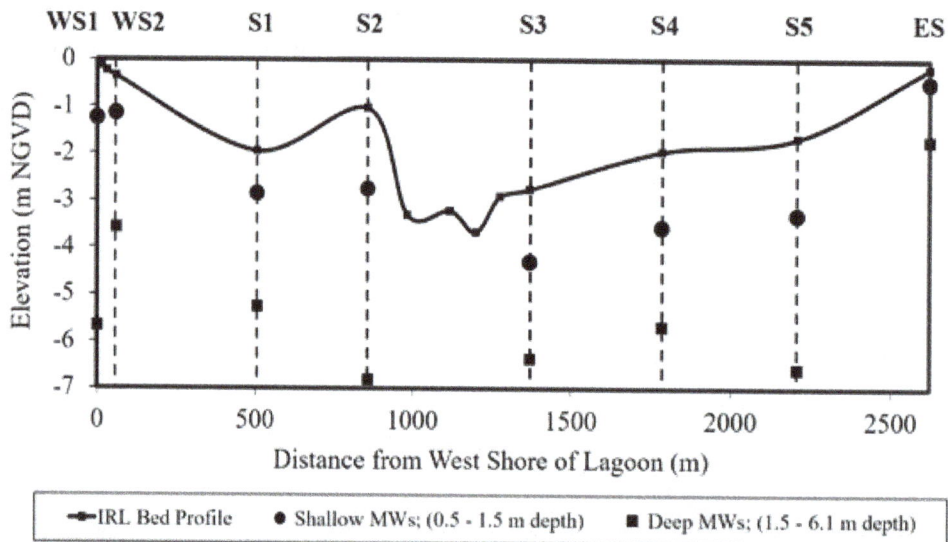

Figure 3: The IRL bed profile and locations of monitoring wells used for the measurement of groundwater hydraulic heads and salinity at the Palm Bay transect.

water surface elevation from the NGVD 29 was established during each sampling event based on three bench marks installed in the study area.

The horizontal hydraulic conductivity K_h at the stations S1 through S5 in addition to WS1 and ES, at both shallow and deep depths, was estimated in the laboratory by obtaining grain size distribution curves and using the Hazen's equation, $K=C(d_{10})^2$, where d10 is the effective diameter, and C is an empirical constant=0.01. Soil samples used for the K_h analyses were extracted at these stations from shallow locations ranging in depth of 0 to 0.5 m and deep locations ranging in depth of 1.4 to 1.9 m. An average K_h of about 30 m/day is obtained for both deep and shallow locations. A sampling event on any given day, comprised of measuring the water table elevations in the Mainland and Barrier Island wells, the piezometric heads and groundwater salinity

in the offshore and onshore piezometers, and the IRL water surface elevation and salinity. This study uses field data of three sampling events conducted on May, August, and September. Table 1 gives the measured elevations of water table at the locations of the groundwater divides on the mainland and the Barrier Island as well as the measured elevations of the IRL water surface for the three sampling events. Values of the head difference ΔH between the groundwater divide and the IRL surface are also shown in Table 1.

Numerical Modeling

The details regarding the MODFLOW and SEAWAT numerical modeling set up including models domain and discretization, initial and boundary conditions, and calibration and aquifer parameters are presented subsequently.

Sampling event	Groundwater divide elevation (m above NGVD29)		IRL water surface elevation (m above NGVD29)	ΔH (m)
	Mainland	Barrier Island		
May	7.518	0.366	0.183	7.335
August	8.637	0.387	0.259	8.378
September	8.829	0.591	0.576	8.253

Table 1: Measured Elevations of Groundwater Divides on Mainland and Barrier Island and Water Surface of IRL at Different Sampling Events.

Models domain and discretization

The model domain for both MODFLOW and SEAWAT is a two-dimensional vertical cross- section extending horizontally over a distance of 9,740 m from the groundwater divide at the Mainland to the Atlantic Ocean (Figure 4). The left-hand boundary of the domain is the groundwater divide at the Mainland, while the right-hand boundary is the Atlantic Ocean. The width of the IRL at this transect is 2.61 km. The groundwater divide on the Mainland is located at 6.23 km from the west shore of the IRL, while the Atlantic Ocean is 0.9 km from the east shore. The bottom boundary is the top of the Hawthorn Formation, which is assumed to be horizontal, impermeable, and is at a depth of approximately 33.5 m at the Palm Bay transect location. The model domain was discretized into one row, 76 columns and 22 layers (Figure 4). The single row of the finite difference mesh was arbitrarily set with a width of 1 m. The columns spacing ranged from 15 to 300 m, while layers spacing ranged from 0.3 to 6 m. The exact finite difference mesh shown in Figure 4 is used identically in both MODFLOW and SEAWAT models.

Initial and boundary conditions

Initial condition for the hydraulic heads is 33.5 m (i.e., the total depth of the model domain) at all internal nodes for both MODFLOW and SEAWAT, while the saltwater concentrations are set to zero at all internal nodes for SEAWAT. The boundary conditions are described in the following section.

MODFLOW boundary conditions: When setting up the boundary conditions for MODFLOW, the mainland and Barrier Island freshwater boundaries (AB and CD) shown in Figure 4 were modeled as constant head boundaries (Dirichlet boundaries). The constant head values of AB and CD boundaries were calculated from the functions $f_1(x)$ and $f_2(x)$ obtained from the field measured water table elevations in the Mainland and Barrier Island, respectively. Each of these functions is a polynomial equation of a trend line obtained from statistical regression between the head values measured at each monitoring well and its distance from the lagoon shoreline. Both of the Hawthorn formation (EF) and the groundwater divide (FA) boundaries were modeled as no flow boundaries (Neumann boundaries). Motz et al. [6] found that specifying equivalent freshwater heads at the brackish water boundaries in a constant-density model yielded results closest to those obtained by a variable-density coupled model. Representing the boundary conditions at the saline boundaries in constant-density models by specified freshwater hydraulic heads was also used by Simpson and Clement [29-31,36]. Therefore, in MODFLOW, the hydraulic heads at the lagoon bed boundary (BC) and the ocean boundary (DE) were converted to freshwater hydraulic heads using equations 2 and 4, respectively, and these boundaries were also treated as constant head boundaries. This is equivalent to coupling the flow and transport only at the saltwater boundaries (i.e., the lagoon and the ocean) while ignoring the coupling everywhere else inside the model domain. The boundary conditions used in MODFLOW can be mathematically described in equations 1 to 6 below:

Boundary AB: $h = f_1(x)$ (1)

Boundary BC: $h = h_f = z + \left[\left(1 + \left(\dfrac{\rho_s - \rho_f}{\rho_f} \right) C_L \right) (h_s - z) \right]$ (2)

Boundary CD: $h = f_2(x)$ (3)

Boundary DE: $h = h_f = z + \left[\left(1 + \left(\dfrac{\rho_s - \rho_f}{\rho_f} \right) C_s \right) (h_s - z) \right]$ (4)

Boundary EF: $\partial h / \partial z = 0$ (5)

Boundary FA: $\partial h / \partial x = 0$ (6)

where h is the hydraulic head specified in the boundary conditions, h_f is the equivalent freshwater hydraulic head, $f_1(x)$ and $f_2(x)$ are functions obtained by respectively measuring water table elevations on the Mainland and Barrier Island, z is the elevation of the nodes from the top of the Hawthorn Formation, C_L and C_s are the measured concentration of the lagoon and seawater in the form of normalized salinity, respectively, ρ_s is the saltwater density, ρ_f is the freshwater density, and h_s is the height of the piezometric surface above the Hawthorn formation.

SEAWAT boundary conditions: SEAWAT model requires the boundary conditions for both hydraulic heads and saltwater concentrations. The hydraulic head boundary conditions used for SEAWAT are identical to those used for MODFLOW in equations 1-6. However, the conversion into freshwater hydraulic heads at the IRL and Ocean boundaries BC and DE in equations 2 and 4, respectively, is done internally in SEAWAT using the specified concentration at each boundary [21]. The saltwater concentration boundary conditions for SEAWAT can be mathematically described as:

Boundary AB: C=0   (7)

Boundary BC: $C = C_L$ (8)

Boundary CD: C=0   (9)

Boundary DE: $C = C_s$ (10)

Boundary EF: $\partial C / \partial z = 0$ (11)

Boundary FA: $\partial C / \partial x = 0$ (12)

Model calibration and aquifer parameters

The vertical hydraulic conductivity of the aquifer below the estuary at the study transect was calibrated using both statistical and visual methods. In the statistical method, the model-predicted and field-measured freshwater hydraulic heads nodal values at the shallow and deep locations shown in Figure 3 were compared by estimating the Root-Men Squared Error (RMSE), Nash-Sutcliffe Efficiency (NSE) index, and testing the null hypothesis using the two-sided test. In the visual comparison method, the calibration was conducted by visually matching the predicted and measured freshwater head equipotential lines below the study transect. Model calibration resulted in a RMSE of 0.05 m, a NSE of 0.96, a two-sided t-test not rejecting the null hypothesis that the means of the measured and predicted values are equal, and a good visual comparison of the measured and predicted freshwater head equipotential lines.

The calibrated values of the aquifer vertical hydraulic conductivity ranged from 0.0015 to 0.03 m/day with a predominant value of

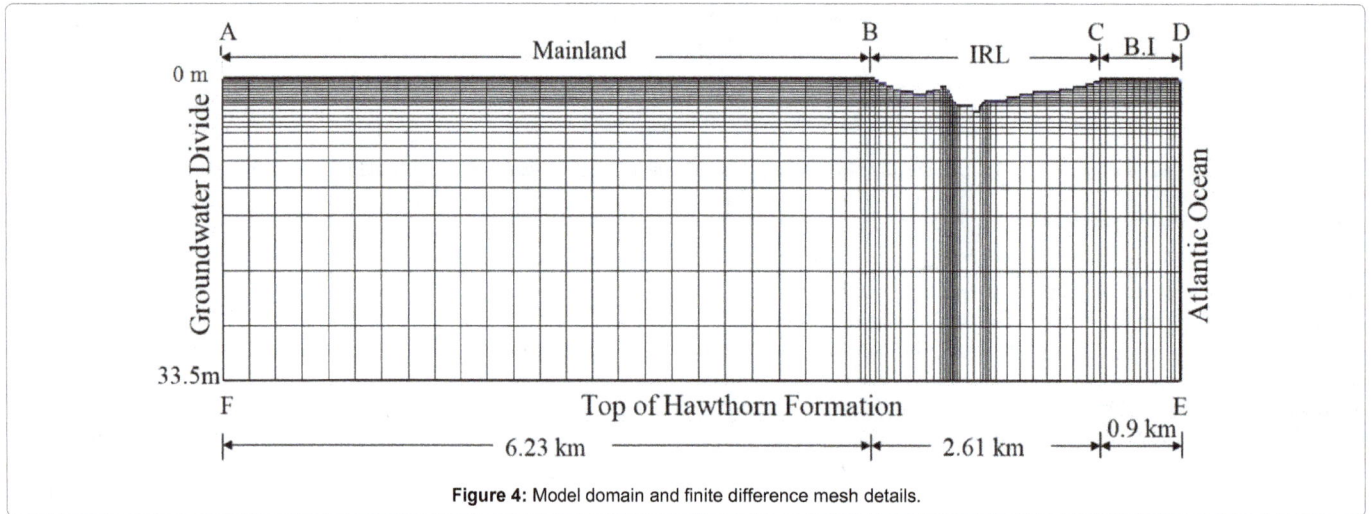

Figure 4: Model domain and finite difference mesh details.

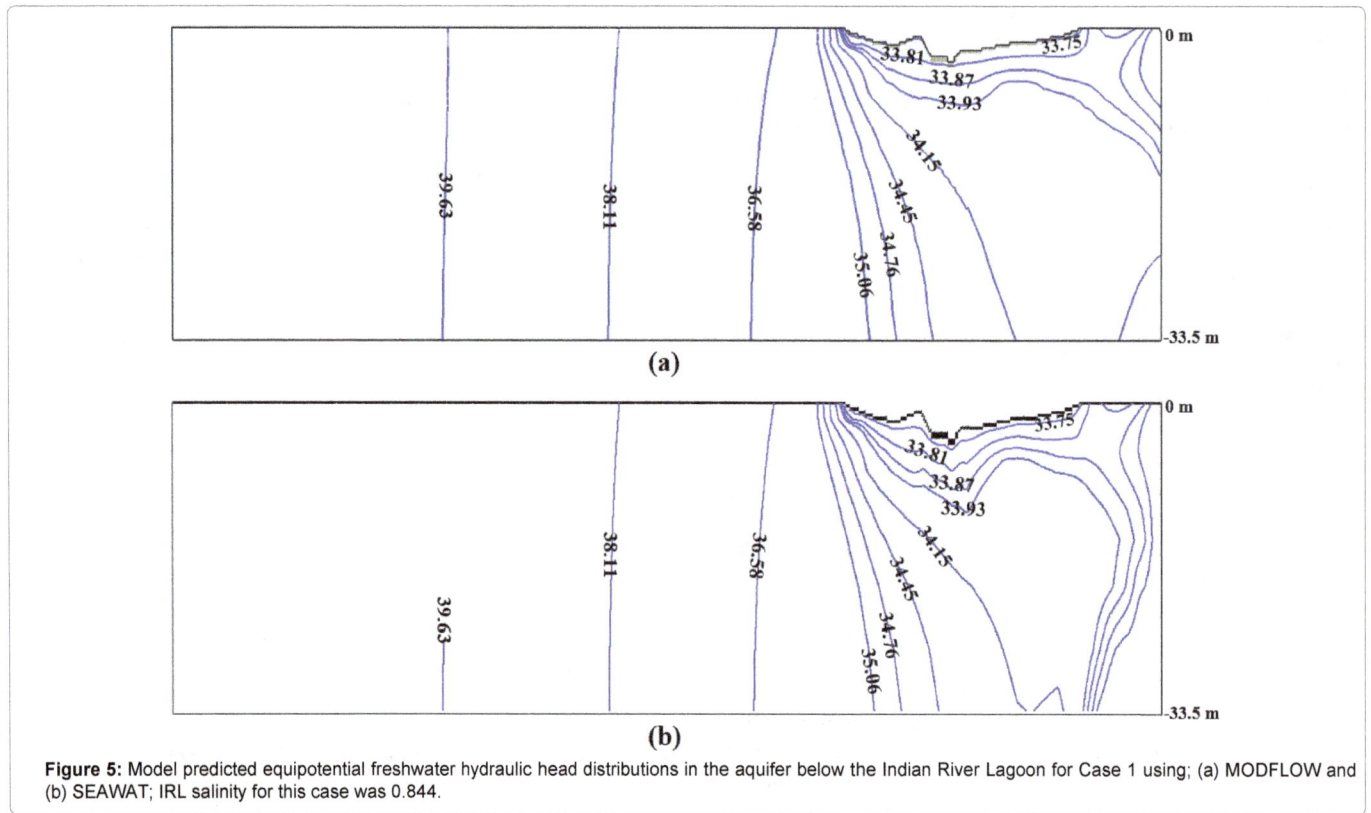

(a)

(b)

Figure 5: Model predicted equipotential freshwater hydraulic head distributions in the aquifer below the Indian River Lagoon for Case 1 using; (a) MODFLOW and (b) SEAWAT; IRL salinity for this case was 0.844.

0.015 m/day. The value of the horizontal hydraulic conductivity was predominantly 30 m/day. This range of hydraulic conductivity values leads to K_r ranging from 1000-20,000 with a predominant value of 2000. The K_r is defined in this paper as the ratio of horizontal and vertical hydraulic conductivities (K_h/K_v). The input data and aquifer properties used in the MODFLOW and SEAWAT models in this study are listed in Table 2. The K_h was estimated from Hazen's equation as described previously. The K_v and lagoon salinity are variable as discussed in the study. Freshwater and seawater normalized salinities and densities are constants. Multiple dispersivity values were investigated and found not to affect the predicted seepage values.

Results and Discussion

Results produced by MODFLOW and SEAWAT models were compared for sixteen cases described in Table 3. In presenting the SGD results of these cases, the term, relative error, implies the difference between the two sets of model results with the SEAWAT results assumed to be the "true" results. The results of Cases 1 to 3 compare the outputs of both models using the calibrated distribution of K_r ranging from 1000-20,000 at different boundary conditions measured during field sampling events conducted on May, August, and September. The respective lagoon normalized salinity C_L in the three cases was 0.844, 0.583 and 0.306 and the measured groundwater elevations

Data or Aquifer Property	Value
Horizontal hydraulic conductivity, K_h (m/day)	30
Vertical hydraulic conductivity, K_V (m/day)	variable (see Table 3)
Porosity, n	0.3
Specific Storage, S_s (m^{-1})	0.00001
Specific yield, S_y	0.01
Longitudinal dispersivity, α_L (m)	30
Transverse dispersivity, α_T (m)	3
Diffusion coefficient, D_m (m^2/d)	0
Seawater normalized salinity, C_s	1
Freshwater normalized salinity, C_f	0
Lagoon water normalized salinity, C_L	variable (see Table 3)
Density of seawater, ρ_s (kg/m^3)	1025
Density of freshwater, ρ_f (kg/m^3)	1000

Table 2: Input Data for Models.

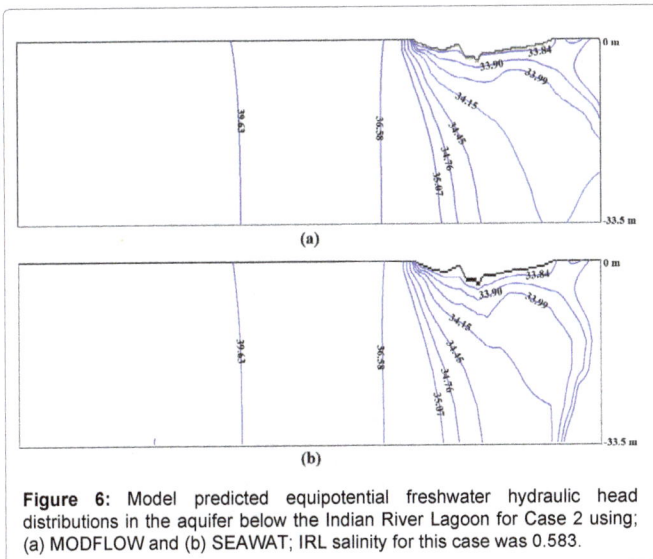

Figure 6: Model predicted equipotential freshwater hydraulic head distributions in the aquifer below the Indian River Lagoon for Case 2 using; (a) MODFLOW and (b) SEAWAT; IRL salinity for this case was 0.583.

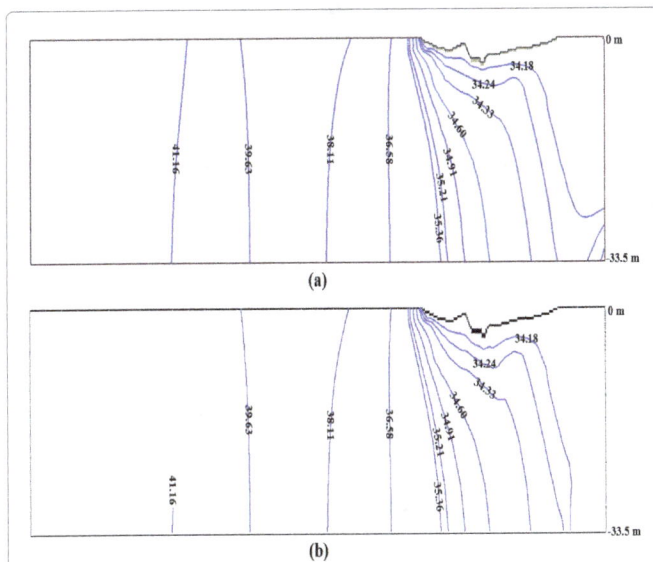

Figure 7: Model predicted equipotential freshwater hydraulic head distributions in the aquifer below the Indian River Lagoon for Case 3 using; (a) MODFLOW and (b) SEAWAT; IRL salinity for this case was 0.306.

on the boundaries in addition to the IRL water surface elevations are presented in Table 1. In the Cases 4 through 8 (Table 3), C_L was arbitrarily increased from 0.9 to 1.05 to determine how MODFLOW results compare with SEAWAT results under higher IRL salinity conditions. Cases 4 through 8 utilize the same Mainland and Barrier Island boundary conditions and the same IRL water surface elevation measured on May and used in Case 1. Cases 4 through 8 also utilize the calibrated K_r distribution, hence the aquifer K_r for these cases also varies from 1000-20,000 as in Cases 1 to 3. Cases 9 through 13 tested the accuracy of MODFLOW at decreasing K_r under relatively high saline condition as the lagoon salinity was kept at 0.9. In these cases, the K_r was changed from 100,000 to 10 while the boundary conditions and lagoon salinity were kept identical to those used for Case 4. Case 14 is identical to Case 12 with the exception that the C_L was reduced from 0.9 to 0.3. This numerical experiment was conducted to determine the accuracy of MODFLOW at low anisotropy ratio and at low lagoon salinity. Cases 15 and 16 are also identical to Case 1 with the exception that the water table elevations used for the boundary conditions of the Mainland and the Barrier Island cells were increased by 5 percent in Case 15 and 10 percent in Case 16 to examine the effect of higher water table elevation. In Case 15, the water table elevations at the water table divides on the Mainland and the Barrier Island were 2.05 m and 1.7 m, respectively higher than in Case 1. Also, in Case 16, the respective elevations on the Mainland and the Barrier Island were 4.1 m and 3.4 m higher than in Case 1.

Cases No. 1 to 3

The following results are compared for Cases 1 to 3: a) freshwater hydraulic head distributions below the IRL, b) flow directions in the form of velocity vectors, c) total SGD into the IRL from the underlying aquifer, and d) spatial distribution of SGD below the IRL. Figures 5-7 compare the calibrated MODFLOW and SEAWAT model-simulated freshwater hydraulic head distributions in the entire modeling domain for Cases 1, 2, and 3, respectively. Even though the shape and magnitude of the equipotential freshwater hydraulic heads are fairly different for the three cases, both MODFLOW and SEAWAT predicted almost identical distributions. Inspecting the comparisons at the two saline boundaries of the models, i.e., the lagoon and Ocean boundaries, (Figures 5-7), it can be seen that the equipotential lines predicted by the two models show some discrepancy on the bottom half portion of the sea boundary where salt intrusion is predominant, while they seem to be fairly similar on the top portion. On the lagoon boundary, the two distributions are very similar for all three cases except that MODFLOW is always predicting the contours locations a little higher than their actual locations predicted by SEAWAT. However, this discrepancy in the contour locations below the estuary seems to become very minor for Case 3 (Figure 7), where the IRL salinity is the lowest and the groundwater elevation on the Mainland (Table 1) is the highest, compared to the first two cases. In general, the variable-density effects predominating the sea side of the model, do not seem to have significant effect on the accuracy of predicting the equipotential lines below the estuary especially in Case 3.

A comparison of the model predicted velocity vectors by MODFLOW and SEAWAT in Case 1 is shown in Figure 8. In general, both models predicted very similar flow patterns in that: a) the groundwater flows upward into the IRL, and b) there is recirculation of the ocean water from the lower part of the aquifer adjacent to the ocean boundary back into the ocean at the upper region of the aquifer. The meteoric groundwater discharge (MGWD) originating from the Barrier Island splits into two directions with a portion going to the

Case no.	Anisotropy ratio Kr	IRL salinity CL	Boundary conditions	Total SGD into IRL		
				SEAWAT (m³/day/m)	MODFLOW (m³/day/m)	Relative error (%)
1	1000-20,000	0.844	May	1.790 × 10⁻⁴	1.960 × 10⁻⁴	9.4
2	1000-20,000	0.583	August	2.170 × 10⁻⁴	2.250 × 10⁻⁴	3.9
3	1000-20,000	0.306	September	1.820 × 10⁻⁴	1.880 × 10⁻⁴	3.2
4	1000-20,000	0.9	May	1.777 × 10⁻⁴	1.957 × 10⁻⁴	10.1
5	1000-20,000	0.95	May	1.764 × 10⁻⁴	1.953 × 10⁻⁴	10.7
6	1000-20,000	1	May	1.751 × 10⁻⁴	1.949 × 10⁻⁴	11.3
7	1000-20,000	1.025	May	1.747 × 10⁻⁴	1.947 × 10⁻⁴	11.5
8	1000-20,000	1.05	May	1.744 × 10⁻⁴	1.946 × 10⁻⁴	11.6
9	1,00,000	0.9	May	4.120 × 10⁻⁵	4.340 × 10⁻⁵	5.3
10	10,000	0.9	May	1.070 × 10⁻⁴	1.170 × 10⁻⁴	9.1
11	1,000	0.9	May	2.080 × 10⁻⁴	2.380 × 10⁻⁴	14.6
12	100	0.9	May	2.590 × 10⁻⁴	3.570 × 10⁻⁴	38.2
13	10	0.9	May	2.190 × 10⁻⁴	5.020 × 10⁻⁴	129.4
14	100	0.3	May	3.170 × 10⁻⁴	3.530 × 10⁻⁴	11.2
15	1000-20,000	0.844	(a)	4.730 × 10⁻⁴	4.950 × 10⁻⁴	4.5
16	1000-20,000	0.844	(b)	7.680 × 10⁻⁴	7.930 × 10⁻⁴	3.3

Note: SGD in m³/day/m is the total SGD in m³/day per m of transect width; (a) and (b) imply that the groundwater elevations collected on May were increased by 5% and 10% in Cases 15 and 16, respectively.

Table 3: Description and Results of Numerical Modeling Cases.

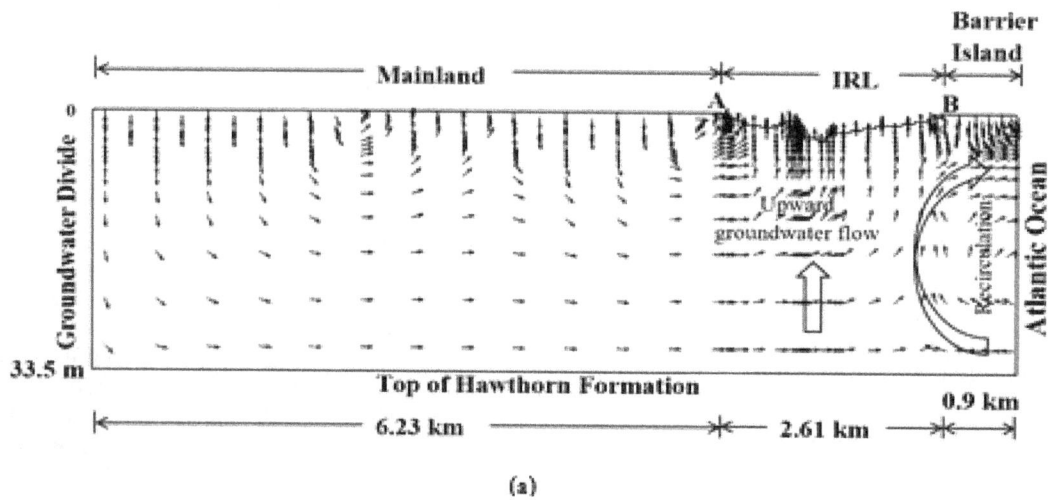

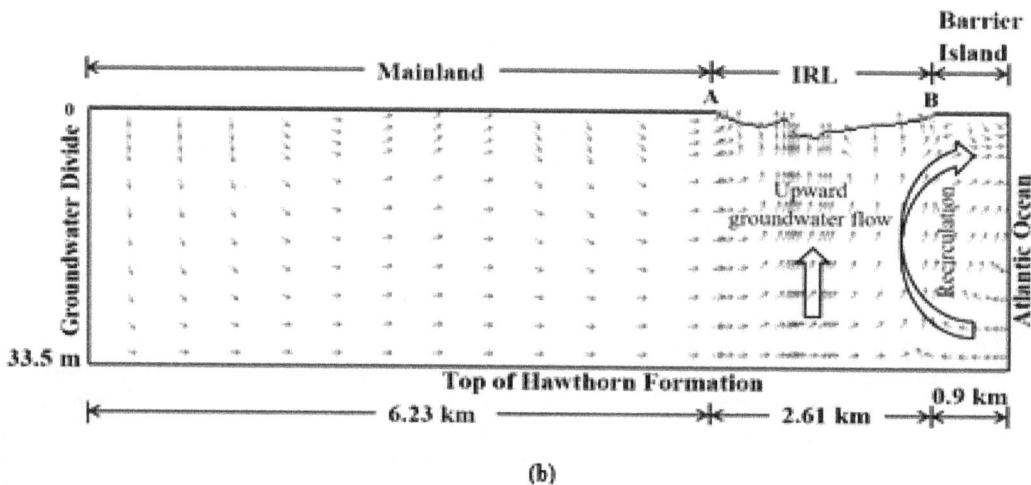

Figure 8: Groundwater flow direction at the Palm Bay Transect for Case 1 using (a) MODFLOW (b) SEAWAT, (arrows are not to scale).

IRL and the rest going to the Atlantic Ocean. Both models predict that the flow direction below the lagoon is mostly oriented upward towards the lagoon even though the vertical hydraulic conductivity K_v is significantly small. The simulated results of Cases 2 and 3 are not shown as they were very similar to the patterns of groundwater flow shown in Figure 8.

The values of the predicted total SGD into the IRL transect, over the three cases, are shown in Table 3, and the relative error in the MODFLOW results ranged from 3.2 to 9.4 percent. In all these cases, MODFLOW predicted slightly higher SGD with the highest error (9.4 percent) occurring in Case 1 when the IRL salinity was relatively higher and the groundwater elevation on the Mainland (Table 1) was relatively lower. Regardless of the difference in the predicted SGD, both models predicted that the SGD into the IRL is the highest on August (Case 2) followed by September (Case 3) then May (Case 1). However, referring to the measured groundwater elevations on the Mainland, it can be noticed that those elevations were the highest on September followed by August and then on May. Therefore, one may expect that the highest SGD into the IRL would follow that sequence. However, the measured IRL water surface elevations shown in Table 1, seem to also increase with increasing Mainland groundwater elevation. Therefore, higher mounts of rainfall on the boundaries do not necessarily produce a significantly higher SGD since the rainfall not only increases the water table elevation but also elevates the water level in the estuary thereby maintaining the hydraulic gradient. Infact, the predicted SGD values seem to be proportional to the head differences ΔH between the groundwater divide on the mainland and the IRL water surface (Table 1).

A comparison of the predicted spatial distribution of the SGD, from the west shore to the east shore of the IRL transect for the first three cases, is illustrated in Figures 9-11, respectively. It can be seen that the patterns predicted by MODFLOW and SEAWAT are nearly identical. Both models predict higher seepage on the west shore side as most groundwater seepage comes from of the Mainland.

Figures 9-11 don't only show that both constant-density and variable-density models agree very well, but also show that the SGD is not just a near-shore phenomenon but extends all the way up to 1600 m from the shoreline. This result indicates that the SGD into the IRL can occur at much greater distances away from the shoreline than predicted by previous studies (Martin et al.). Also, there can be high and low SGD producing zones within a few meters of each other due to a sudden variation in the vertical hydraulic conductivity K_v values. For example, there is a sharp decline, followed by a sudden rise, in the SGD flux within 400 m from the west shore. This sharp variation in SGD into the IRL can be missed in seepage meter studies unless seepage meters are placed within 10 m (or even closer) of each other. Although Cases 1 to 3 share the same K_r distribution, the IRL salinity and the groundwater elevations on the boundaries and the IRL water surface elevations were different. Therefore, it cannot be decided at this point if the results of these cases presented in Table 3 and Figures 5-11 were mostly affected by the changing C_L, the boundary conditions, or both. Therefore, in the next phase (Cases 4-8) of this investigation, comparisons were accomplished at a changing C_L while keeping the K_r and the boundary conditions constant as in Case 1.

Cases No. 4 to 8

The SGD values predicted by both MODFLOW and SEAWAT for Cases 4 through 8 are shown in Table 3 and these results in addition to the results of Case 1 indicate that the accuracy of MODFLOW in predicting the SGD decreases slightly with increasing IRL salinity.

As a result, MODFLOW accuracy in predicting the SGD spatial distribution, which depends directly on the SGD value, would also decrease slightly with increasing salinity. The results also indicate that the highest relative error (11.6%) occurs when the normalized lagoon salinity has its highest value of 1.05 (Case 8). The amount of SGD into the IRL predicted by both models seems to be inversely proportional to the lagoonal water salinity. However, changing the IRL salinity for these cases seems not to change the amount of seepage as much as noticed between Cases 1 to 3. It can be concluded that the amount of SGD is primarily affected by the hydraulic gradient present between the IRL and the groundwater levels on the boundaries and is affected, to a lot lower level, by the IRL salinity. Although not presented here, the equipotential freshwater hydraulic head distributions predicted by MOFLOW and SEAWAT for Cases 4 to 8 did not change significantly with increasing IRL salinity and were very similar to those shown in Figure 5 for Case 1. These results also indicate that MODFLOW-predicted freshwater hydraulic head distributions stay similar to those predicted by SEAWAT even if the IRL salinity increases significantly although the error in predicting SGD increases slightly. Therefore, it can be concluded that the shape and magnitude of the freshwater hydraulic head distributions on one hand, and the agreement of those distributions in shape and magnitude between the two models depend primarily on the Mainland groundwater divide elevation and not the IRL salinity for the cases having the same K_r distribution. This explains why MODLOW-predicted freshwater hydraulic head distribution for Case 3 (Figure 7), which has the highest groundwater divide elevation at the Mainland (Table 1), was more similar to those of SEAWAT compared to Figures 5 and 6.

Cases No. 9 to 13

It can be seen from the SGD values shown in Table 3 that the relative error is less than 10% when K_r is 10,000 or higher (Cases 9 and 10) and less than 15% where K_r is 1000 or higher. However, when K_r is lowered below 1000, the relative error becomes significantly higher and is as high as 129.4 percent when K_r is equal to 10 (Case 13). The large difference in the predicted SGD values is also reflected in the freshwater hydraulic head contours predicted by the two models as shown in Figure 12 which compares the results for Case 12. The comparison of the freshwater hydraulic head contours of Case 13, which has the highest relative error of 129.2 percent, although not shown here, was even worse than that shown in Figure 12. At those low K_r values the predicted flow directions below the lagoon were mostly vertical either upward or downward. Downward vertical flow gives rise to the saline estuarine water overlaying the aquifer with a salinity as high as 0.9 to penetrate deeper into the aquifer while the upward vertical flow drifts more seawater up into the aquifer. Those vertical flow patterns make the variable-density effects below the estuary to be predominant over the advective effects causing the constant-density MODFLOW model to fail at this point. Inspecting Figure 13 that presents the SGD results of Cases 9 to 13 shows that a K_r of 1000 (Case 11) looks to be the critical value below which, the use of the constant-density MODFLOW model results in significant loss of accuracy. A comparison of flow directions in a selected portion of the model domain extending horizontally from the west shore to the east shore of the lagoon and vertically from the lagoon surface down to an arbitrarily selected depth of -9 m NGVD 29 is shown in Figure 14. It can be seen from Figure 14 that MODFLOW is still predicting very similar flow directions to those produced by SEAWAT at the critical K_r of 1000 although there is still a vertical flow component below the lagoon.

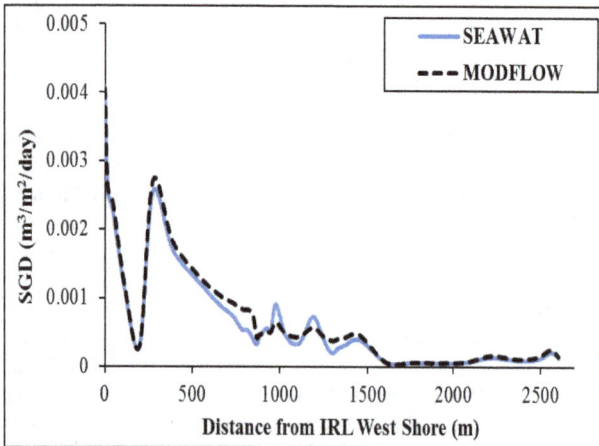

Figure 9: Spatial distribution of submarine groundwater discharge (SGD) to the Indian River Lagoon for Case 1.

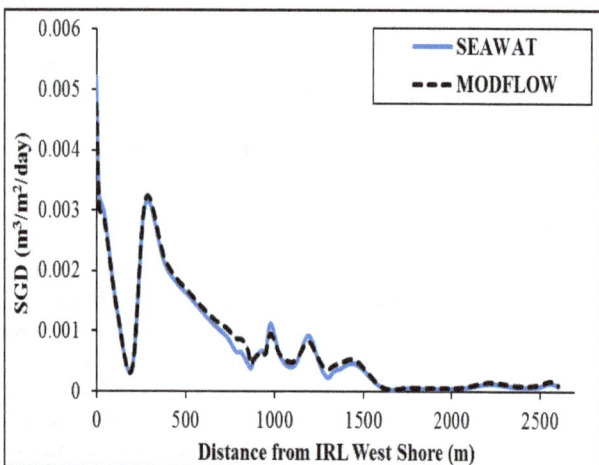

Figure 10: Spatial distribution of submarine groundwater discharge (SGD) to the Indian River Lagoon for Case 2.

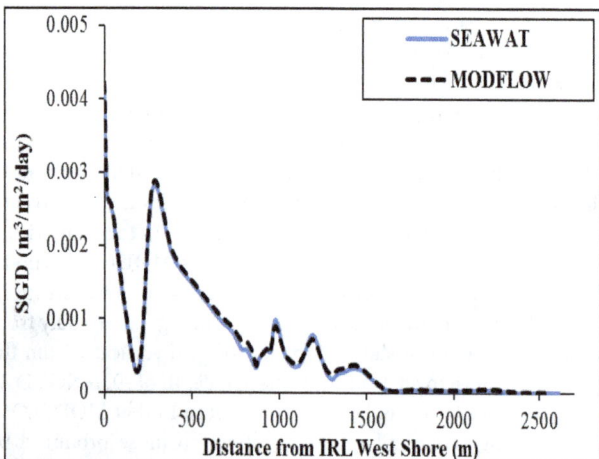

Figure 11: Spatial distribution of submarine groundwater discharge (SGD) to the Indian River Lagoon for Case 3.

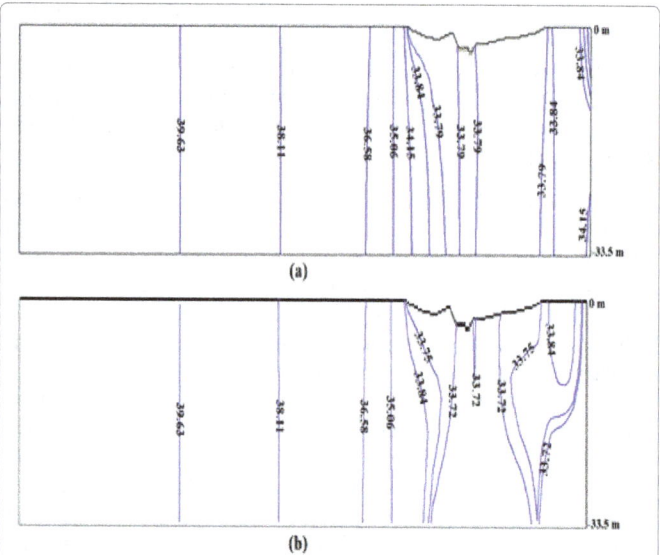

Figure 12: Model predicted equipotential freshwater hydraulic head distributions in the aquifer below the Indian River Lagoon for Case 12 using; a) MODFLOW and b) SEAWAT; IRL salinity for this case is 0.9.

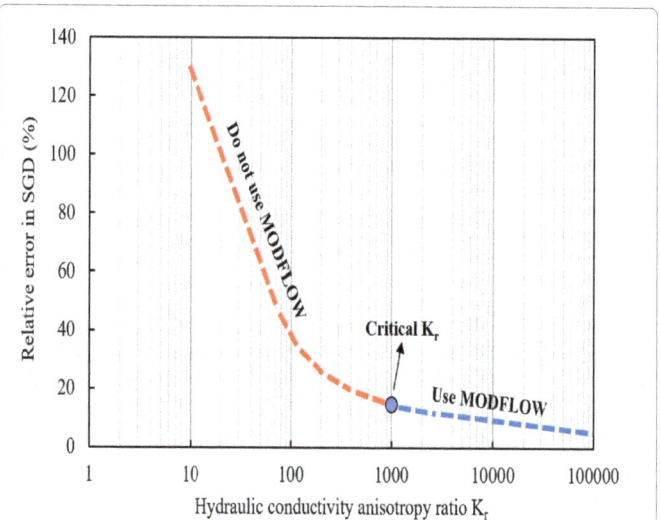

Figure 13: Results of Cases 9 to 13 showing where it is safe to use MODFLOW and where it is not and the critical K_r value of 1000 in between.

Case No. 14

Case 14 is a repetition of Case 12 with the exception that the C_L is 0.3 instead of 0.9 while K_r was kept at 100. Results of this case show that although reducing the IRL salinity from 0.9 to a value closer to freshwater salinity resulted in reducing the relative error in SGD prediction from 38.2 to 11.2 percent (Table 3), MODFLOW fails to produce accurate freshwater hydraulic head distribution below the lagoon at such a low K_r (Figure 15). It is obvious that modeling this system based on a constant-density assumption under low K_r conditions regardless of C_L value is not accurate at all.

Cases No. 15 and 16

These cases assess the degree of improvement in the MODFLOW accuracy in predicting the SGD values and the freshwater hydraulic head equipotential lines if Case 1 was re-run with the same K_r and C_L but with higher groundwater elevations on the Mainland and Barrier Island boundaries. The freshwater hydraulic head equipotential lines predicted by the two models for Cases 15 and 16 are shown in Figures 16 and 17, respectively. These figures as well as the SGD values shown in Table 3 indicated that the accuracy of MODFLOW improves substantially if the water table elevations are higher. Increasing the water table elevations by 5% in Case 15 and 10% in Case 16 resulted in reducing the error in estimating the SGD from 9.4% in Case 1 to 4.5% and 3.3% in Cases 15 and 16, respectively. The drop of the relative error from 9.4% into 4.5% and 3.3% is equivalent to an increase in MODFLOW accuracy by 52% and 65%, respectively.

Model computation time

The computation times for all cases described in Table 3 were approximately 0.07 seconds for MODFLOW and approximately 564 seconds for SEAWAT. Analysis conducted by refining the finite difference mesh to 258 columns and 110 layers compared to 76 columns and 22 layers in the original model showed that the discrepancy between the computation times for the two models increased even more than that. MODFLOW required approximately 1.5 seconds while SEAWAT required approximately 12,600 seconds (3.5 hours) for simulating the results of Case 1. The computation times reported here are clock times of PC with an Intel Core i3-2310M 2.10 GHz CPU. These computation times indicate that, depending on the mesh size, a constant-density model may be faster by a factor of 8000 or more than a coupled model in solving identical problems. This difference in computation times

is likely to become even higher for complex three-dimensional (3-D) models.

Summary and Conclusions

This paper investigates the accuracy of results produced by MODFLOW, a constant-density model, in a saline environment below an estuary known as the Indian River Lagoon (IRL), by comparing its results with those of SEAWAT, which is a variable-density coupled model. The results that were compared included: 1) measured freshwater hydraulic head contours in the unconfined aquifer below the estuary, 2) groundwater flow directions in the unconfined aquifer, 3) total submarine groundwater discharge (SGD) into the estuary from the adjacent unconfined aquifer, and 4) spatial distribution of SGD below the estuary. The comparison was conducted over sixteen numerical experiments at different conditions of estuarine salinity C_L, hydraulic conductivity anisotropy ratio K_r, and water table elevations on the freshwater boundaries.

The results presented in this paper showed that: a) the use of MODFLOW for modeling the IRL at the study site under its calibrated K_r range of 1000-20,000 was satisfactory and accurate to within approximately 3 to 9% regardless of the IRL salinity and groundwater elevations on the boundaries with an increase in its accuracy by about 52% and 65% by increasing the measured groundwater elevations by 5% and 10%, respectively, b) results produced by MODFLOW can be in close agreement with those obtained by SEAWAT if K_r is greater than a critical value of 1000 regardless of the lagoon salinity, the conditions

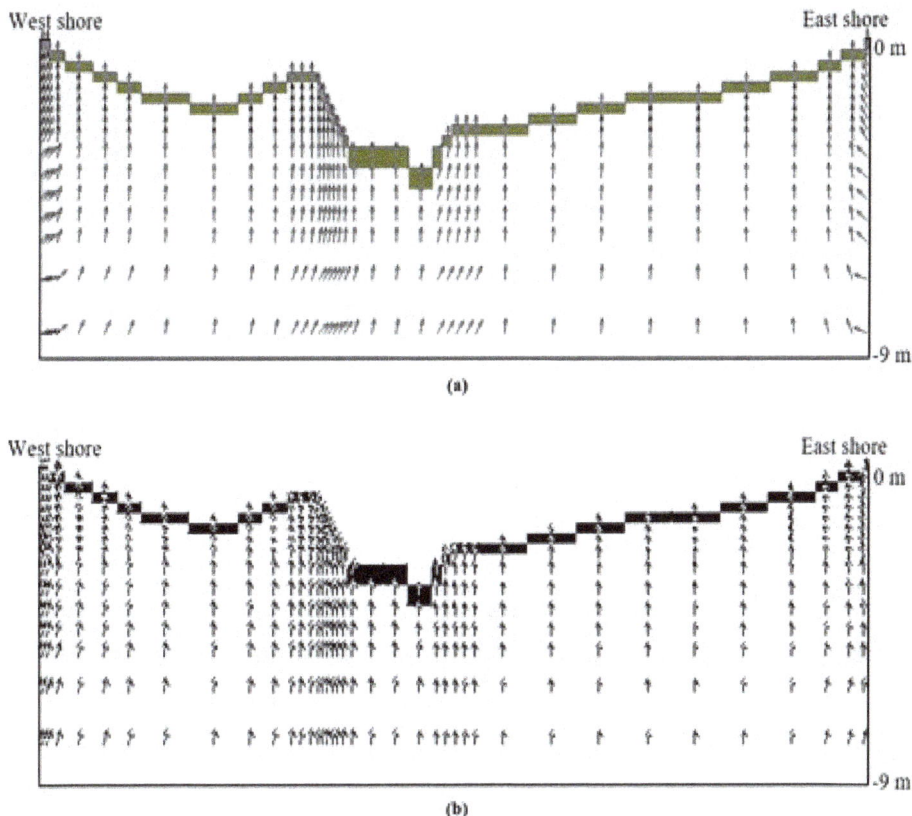

Figure 14: Flow directions below the IRL transect profile up to a depth of 9 m NGVD29 at the critical Kr of 1000 predicted in Case 11 showing that the predicted flow directions using (a) MODFLOW and (b) SEAWAT are very similar and that the flow below the lagoon is still vertical at the critical Kr.

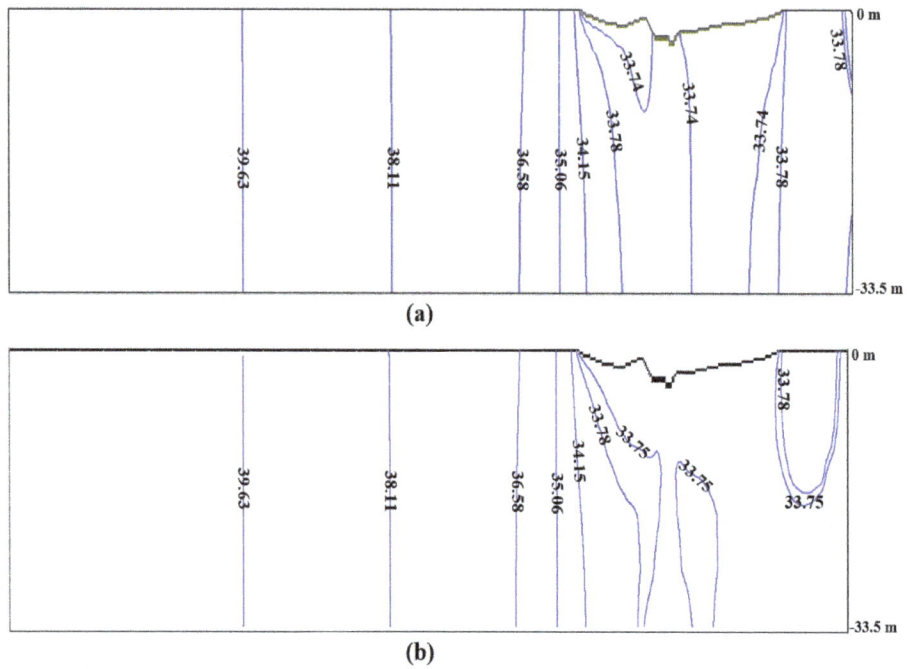

Figure 15: Model predicted equipotential freshwater hydraulic head distributions in the aquifer below the Indian River Lagoon for Case 14 using; a) MODFLOW and b) SEAWAT; IRL salinity for this case is 0.3.

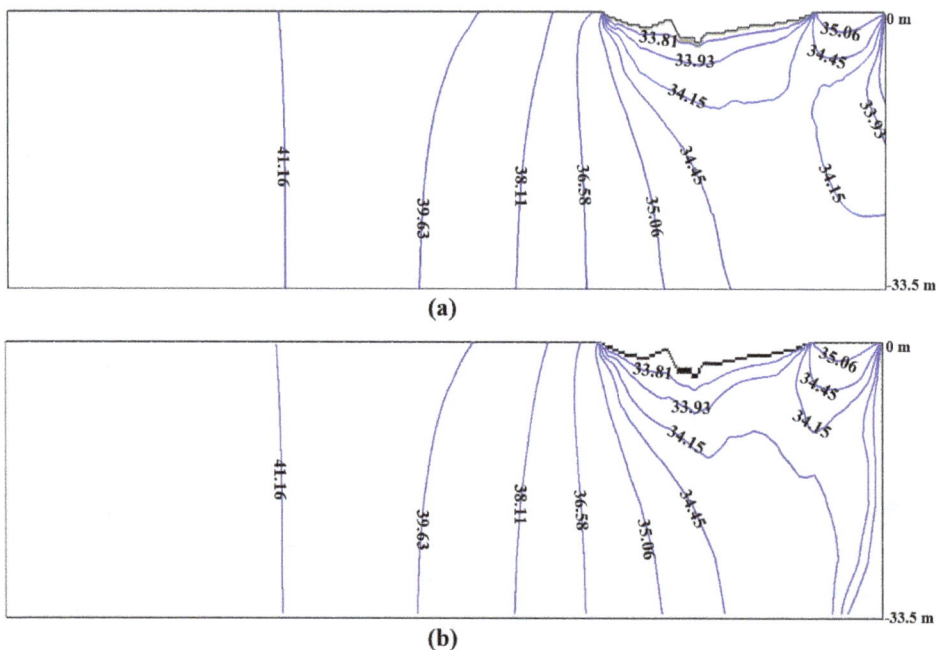

Figure 16: Model predicted equipotential freshwater hydraulic head distributions in the aquifer below the Indian River Lagoon for Case 15 using; a) MODFLOW and b) SEAWAT; IRL salinity for this case is 0.844.

under which, MODFLOW produced results within less than 15% to those predicted by SEAWAT, c) MODFLOW should probably not be used in saline environments if K_r is less than 1000 under any conditions even when lagoon salinity is low, d) there is still vertical flow component predominating below the lagoon even at the critical K_r of 1000, e) the amount of SGD predicted by either model and also the MODFLOW

accuracy in predicting the SGD are directly proportional to the head difference between the groundwater divide elevation and the lagoon water surface, but, to a lower extent, are inversely proportional to C_l, f) both MODFLOW and SEAWAT predicted that the SGD can occur at much greater distances away from the shoreline than predicted by previous studies, g) both MODFLOW and SEAWAT agreed very well

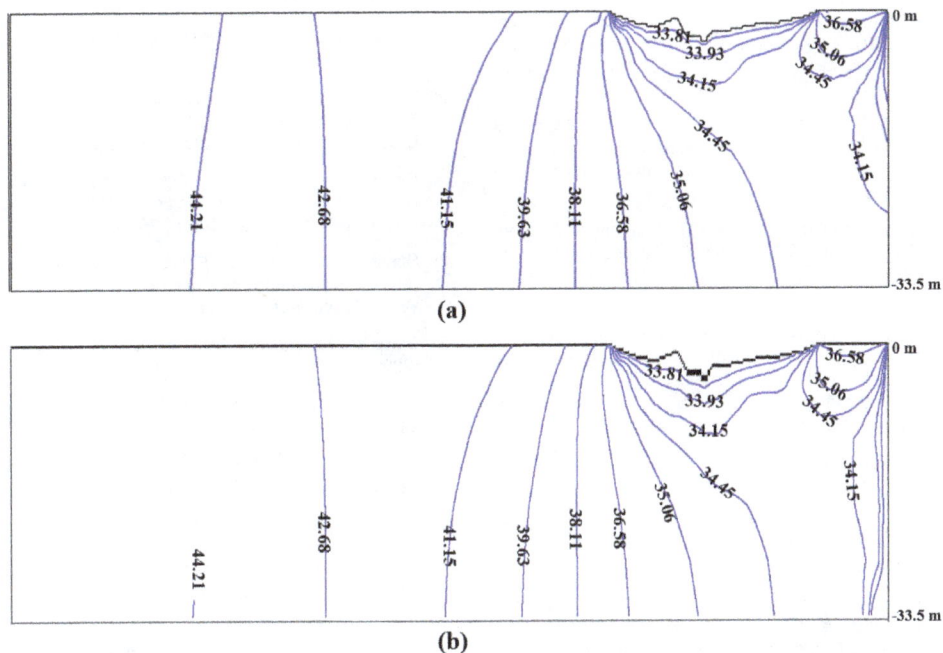

Figure 17: Model predicted equipotential freshwater hydraulic head distributions in the aquifer below the Indian River Lagoon for Case 16 using; a) MODFLOW and b) SEAWAT; IRL salinity for this case is 0.844.

in showing high and low SGD producing zones across the transect and both models showed that depending on the K_v value, those zones can occur within a few meters of each other and may be missed by seepage meter studies, and h) for the two meshes used in the analyses, MODFLOW was faster than SEAWAT by a factor of greater than 8000 and this discrepancy in computation times becomes even more significant as the mesh is refined.

In summary, under certain conditions, constant-density models such as MODFLOW could be a viable option for modeling saline environments, particularly estuarine environments, if the primary reason for the modeling is to determine flow rates and not saltwater transport, as there is a vast discrepancy in the computation times needed by the two models to solve identical problems. Of course, it is advisable that modelers compare results for their boundary value problem with both variable-density and constant-density models prior to deciding if the constant-density model is feasible or not.

Acknowledgments

We greatly appreciate the support provided by the St. Johns River Water Management District (SJRWMD) for funding the work of this research.

References

1. Paniconi C, Khlaifi I, Lecca G, Giacomelli A, Tarhouni J (2001) A modelling study of seawater intrusion in the Korba Coastal Plain, Tunisia. Physics and Chemistry of the Earth, Part B: Hydrology, Oceans and Atmosphere 26: 345-351.

2. Shoemaker WB, Edwards KM (2003) Potential for saltwater intrusion into the lower Tamiami aquifer near Bonita Springs, southwestern Florida. Water Resources Investigations Report 03- 4262, U.S. Geological Survey, Tallahassee, FL.

3. Dausman A, Langevin CD (2005) Movement of the saltwater interface in the surficial aquifer system in response to hydrologic stresses and water-management practices, Broward County, Florida. Scientific Investigations Report 2004-5256, U.S. Geological Survey, Reston, Virginia.

4. Lin J, Snodsmith JB, Zheng C, Wu J (2009) A modeling study of seawater intrusion in Alabama Gulf Coast, USA. Environ Geol 57: 119-130.

5. El-Bihery M (2009) Groundwater flow modeling of quaternary aquifer Ras Sudr, Egypt. Environ Geol 58: 1095-1105.

6. Motz L, Sedighi A (2009) Representing the coastal boundary condition in regional groundwater flow models. J Hydrol Eng 14: 821-831.

7. Ding F, Yamashita T, Lee HS, Pan J (2014) A modelling study of seawater intrusion in the liao dong bay coastal plain, china. Journal of Marine Science and Technology (Taiwan) 22: 103-115.

8. Langevin CD (2003) Simulation of submarine ground water discharge to a marine estuary: Biscayne Bay, Florida. Ground Water 41: 758-771.

9. Li X, Hu BX, Burnett WC, Santos IR, Chanton JP (2009) Submarine ground water discharge driven by tidal pumping in a heterogeneous aquifer. Ground Water 47: 558-568.

10. Ward JD, Simmons CT, Dillon PJ (2007) A theoretical analysis of mixed convection in aquifer storage and recovery: How important are density effects. J Hydrol 343: 169-186.

11. Ward JD, Simmons CT, Dillon PJ, Pavelic P (2009) Integrated assessment of lateral flow, density effects and dispersion in aquifer storage and recovery. J Hydrol 370: 83-99.

12. Minsley BJ, Ajo-Franklin J, Mukhopadhyay A, Morgan FD (2011) Hydrogeophysical methods for analyzing aquifer storage and recovery systems. Ground Water 49: 250-269.

13. Zuurbier KG, Zaadnoordijk WJ, Stuyfzand PJ (2014) How multiple partially penetrating wells improve the freshwater recovery of coastal aquifer storage and recovery (ASR) systems: A field and modeling study. J Hydrol 509: 430-441.

14. Shafer JM, Brantley DT, Waddell MG (2010) Variable-density flow and transport simulation of wellbore brine displacement. Ground Water 48: 122-130.

15. Langevin C, Swain E, Wolfert M (2005) Simulation of integrated surface-water/ground-water flow and salinity for a coastal wetland and adjacent estuary. J Hydrol 314: 212-234.

16. Baidariko EA, Pozdniakov SP (2011) Simulation of liquid waste buoyancy in a deep heterogeneous aquifer. Water Resources 38: 972-981.

17. Rumynin VG, Mironenko VA, Sindalovsky LN, Boronina AV, Konosavsky PK, et al. (2000) Conceptual and numerical modelling of density induced migration of radioactive contaminants at the Lake Karachai waste disposal site. Proceedings of the ModelCARE'99 Conference, September 20, 1999 - September 23, IAHS Press, Zurich, Switz, pp: 405-411.

18. Fraser Harris AP, McDermott CI, Kolditz O, Haszeldine RS (2015) Modelling groundwater flow changes due to thermal effects of radioactive waste disposal at a hypothetical repository site near Sellafield, UK. Environmental Earth Sciences 74: 1589-1602.

19. Guo W, Langevin CD (2002) User's Guide to SEAWAT: A Computer Program for Simulation of Three-Dimensional Variable-Density Ground-Water Flow. Techniques of Water-Resources Investigations 6-A7, U.S. Geological Survey, Tallahassee, FL.

20. Langevin CD, Shoemaker WB, Guo W (2003) MODFLOW-2000, the U.S. Geological Survey modular ground-water model: Documentation of the SEAWAT-2000 version with the variable- density flow processes (VDF) and the integrated MT3DMS Transport Processes (IMT). USGS Open-File Rep. 03-426, U.S. Geological Survey, Tallahassee, FL.

21. Langevin CD, Guo W (2006) MODFLOW/MT3DMS-based simulation of variable density ground water flow and transport. Ground Water 44: 339–351.

22. McDonald MG, Harbaugh AW (1988) A Modular Three-Dimensional Finite-Difference Ground- Water Flow Model. Techniques of Water-Resources Investigations 6-A1, U.S. Geological Survey, Reston, Virginia.

23. Harbaugh AW, Banta ER, Hill MC, McDonald MG (2000) MODFLOW-2000, the U.S. Geological Survey Modular Ground-Water Model: User Guide to Modularization Concepts and the Ground-Water Flow Process. USGS Open-File Rep. 00-92, U.S. Geological Survey, Reston, Virginia.

24. Harbaugh AW (2005) MODFLOW-2005, the U.S. Geological Survey Modular Ground-Water Model-The Ground-Water Flow Process. USGS Techniques and Methods 6-A16, U.S. Geological Survey, Reston, Virginia.

25. Zheng C, Wang P (1999) MT3DMS-a modular three-dimensional multispecies transport model for simulation of advection, dispersion, and chemical reaction of contaminants in ground-water systems: Documentation and user's guide. Jacksonville, Florida. Contact Report SERDP-99-1, U.S. Army Corps of Engineers.

26. Clement TP (1997) RT3D-A Modular Computer Code for Simulating Reactive Multi-Species Transport in 3-Dimensional Groundwater Aquifers. Pacific Northwest National Laboratory, Richland, Washington.

27. Hill MC, Poeter E, Zheng C, Doherty J (2003) MODFLOW 2001 and other modeling Odysseys. Ground Water 41: 113-113.

28. Henry HR (1964) Effects of dispersion on salt encroachment in coastal aquifers. Sea water in coastal aquifers. Geological Survey Water Supply Paper 1613-C, U.S. Geological Survey, Washington, DC, pp: 70-84.

29. Simpson MJ, Clement TP (2003) Theoretical analysis of the worthiness of Henry and Elder problems as benchmarks of density-dependent groundwater flow models. Adv Water Resour 26: 7-31.

30. Simpson MJ, Clement TP (2004) Improving the worthiness of the Henry problem as a benchmark for density-dependent groundwater flow models. Water Resour Res 40: 1-11.

31. Dentz M, Tartakovsky DM, Abarca E, Guadagnini A, Sánchez-Vila X, et al. (2006) Variable- density flow in porous media. J Fluid Mech 561: 209-235.

32. Goswami RR, Clement TP (2007) Laboratory-scale investigation of saltwater intrusion dynamics. Water Resour Res 43: W04418.

33. Abarca E, Carrera J, Sánchez-Vila X, Dentz M (2007) Anisotropic dispersive Henry problem. Adv Water Resour 30: 913-926.

34. Arlai P, Koch M (2007) Need for density-dependent flow and transport modeling of horizontal seawater and vertical saltwater intrusion in the Bangkok multilayer-aquifer system. Proceedings of the 12th National Convention on Civil Engineering, Phitsanulok, Thailand, May 2-4.

35. Arlai P, Koch M (2009) The importance of density-dependent flow and solute transport modeling to simulate seawater intrusion into a coastal aquifer system. Proceedings of the International Symposium on Efficient Groundwater Resources Management (IGS-TH 2009), Bangkok, Thailand, February 16-21.

36. Motz L, Sedighi A (2013) Saltwater intrusion and recirculation of seawater at a coastal boundary. J Hydrol Eng 18: 10-18.

37. Brown DW, Kenner WE, Crooks JW, Foster JB (1962) Water resources of Brevard County, Florida. Report of Investigations 28, U.S. Geological Survey, Tallahassee, FL.

38. Martin JB, Cable JE, Smith C, Roy M, Cherrier J (2007) Magnitudes of submarine groundwater discharge from marine and terrestrial sources: Indian River Lagoon, Florida. Water Resour Res 43: W05440.

18

Simulation of Hydro Climatological Impacts Caused by Climate Change: The Case of Hare Watershed, Southern Rift Valley of Ethiopia

Biniyam Yisehak Menna*

Department of Meteorology and Hydrology, College of Natural Sciences, Arba Minch University, Arba Minch, Ethiopia

Abstract

Ethiopia will be more vulnerable to climate change. Because of the less flexibility to adjust the economic structure and being largely dependent on agriculture, the impact of climate change has far reaching implication in Ethiopia. Simulation models of watershed hydrology and water quality are extensively used for water resources planning and management. The study aims to Simulate Hydro Climatological impacts caused by Climate Change: the case of Hare Watershed, Southern Rift Valley of Ethiopia. In the study the daily data values of rainfall and discharge for the current period of 1980-2006 were used. Historical Representative Concentration Pathway (RCPs) data of precipitation and temperature were used to extract raw climate variables. The raw RCPs data were corrected using a bias correction method. The downscaled climate data such as, RCP4.5 and RCP8.5 scenarios was used for the future period assessment. Soil water assessment tool (SWAT) models were used to Simulate Hydro Climatological impacts caused by Climate Change. Calibration and validation of the model output were performed by comparing predicted streamflow with corresponding measurements from the Hare river outlet for the periods 1991-2002 for calibration and 2003-2006 for validation. The models' calibration results show a good agreement with the observed flow with the coefficient of determination is 0.85 and a Nash Sutcliffe efficiency is 0.73. The result of mean monthly percentage changes of climate variables from the baseline period were used to simulate future projections of stream flow. Stream flow projections for future time periods showed that mean monthly stream flow may increase by 12.2, 8.0, and 13.9% at 2020s, 2050s, and 2080s, respectively, from the baseline period for RCP4.5 scenario, whereas for RCP8.5 scenario, it will be expected to increase by 7.3, 13.4, and 15.4% for 2020s, 2040s, and 2080s, respectively. The model simulations considered only future climate change scenarios assuming all spatial data constant. But change in land use scenarios other climate variables will also contribute some impacts on future stream flow.

Keywords: Climate change; Climate projection; RCPs; Streamflow; Bias correction; Hare watershed

Introduction

Climate change refers to any systematic change in the long-term statistics of climate elements (such as temperature, pressure, or winds) sustained over several decades or longer time periods. Climate change describes changes in the global temperature over time (i.e., increase in global temperature or global warming) and its consequences on other climatic variables, such as pressure, humidity, wind etc. Observations that delineate how global temperature has increased in the past shows, the global average surface temperature have increased by 0.74°C/Century [1].

The impact of climate change on water resources are the most crucial research agenda in worldwide level [2]. This change in climate causes a significant impact on the water resource by disturbing the normal hydrological processes. Future change in overall flow magnitude, variability and timing of the main flow event are among the most frequently cited hydrological issues [3,4].

The IPCC findings indicate that developing countries, such as Ethiopia, will be more vulnerable to climate change. It may have far reaching implications to Ethiopia for various reasons, mainly as its economy largely depends on agriculture and low adoptive coping. A large part of the country is arid and semiarid, and is highly prone to drought and desertification. Climate change and its impacts are, therefore, a case for concern to Ethiopia. Hence, assessing vulnerability to climate change impact mapping and preparing adaptation options as part of the national program is very crucial for the country [5].

In spite of the fact that the impact of different climate change scenarios is projected at global scale, the exact type, and magnitude of the impact at catchment scale is not investigated in most parts of the world. Hence, identifying local impact of climate change at watershed level is quite important. The economy of Ethiopia mainly depends on agriculture, and this in turn largely depends on available water resources. Given that a large part of the country is arid and semiarid and highly prone to drought and desertification, this represents a significant risk. Also, the country has a fragile highland ecosystem that is currently under stress due to increasing population pressure. Hare watershed is one of tributaries of Lake Abaya and high competition for irrigation water used among upstream and downstream irrigation sites. In addition, it can be considered as representative watershed where there is high landscape and climatic zone different with in short distance [6].

Therefore, the effect of climate change on water availability (with respect to water resource analysis, management, and policy formulation in the country) in the Hare watershed have not been adequately addressed. Hence, it is necessary to improve our understanding of problems involved to the change in climate.

In this climate change impact study, soil water assessment tool (SWAT) model was used. The framework of this approach is, firstly representative concentration pathway (RCPs) and then downscale or

*****Corresponding author:** Biniyam Yisehak Menna, Department of Meteorology and Hydrology, College of Natural Sciences, Arba Minch University, PO Box 21, Arba Minch, Ethiopia, E-mail: b1n1y21@gmail.com

correct RCPs data. Next feed the downscaled or corrected RCPs data into calibrated and validated hydrological models. After that, simulate streamflow for three future periods considering the climate change scenario. Finally, carry out hydro climatological impacts caused by climate change.

Study Area and Dataset

Study area

The study area, Hare River watershed, is located in the Abaya-Chamo sub-basin of the southern Ethiopian Rift Valley and drains to Lake Abaya, which is the second largest lake of the country. The watershed is situated between 37° 27⊠ and 37° 37⊠ Eastern longitude and 6° 03⊠ and 6° 18⊠ Northern latitude and has a land area of 153 km². The topography of the study area is generally increasing in elevation from the downstream to the upstream. The middle reach of the watershed is mainly covered by steep slopes characterized through abrupt faults [7].

The climate of the Hare watershed ranges from tropical to alpine due to its great difference in altitude and topographical elevation. The average annual temperature is 23°C and 14°C, and mean annual rainfall are 750 mm and 1300 mm at the lowland and highland respectively. Generally, about 55.56% is Dega (Humid), 22.84% is Woyna Dega (Sub-humid), 11.73% is Wurch (Alpine) and 9.25% is Kola (Sub-arid). The lower watershed area is characterized by dry Kola while the middle part of the watershed is characterized by moist woyna-dega and much of the area in the northern part is dominated by Dega and the tip is Wurch. The main and small rainy seasons at Hare watershed occur from April-May and September and October respectively. The spatial rainfall distribution at Hare watershed indicates that major increase of rainfall takes place with an increase in elevation from 1180 m up to 3,480 m above sea level (a.s.l.) [6] (Figure 1).

Dataset meteorological data

Required long year daily precipitation data were collected from three meteorological stations such as, Arba Minch, Chencha, and Mirab Abaya. Daily maximum and minimum temperature data were collected from Arba Minch station. The historical weather data for above three station were obtained from National Meteorological Service Agency (NMSA) from 1980 to 2006 (Figure 2).

Downscaled RCPs data

The Intergovernmental Panel for Climate Change (IPCC) has published projections of future climate change scenarios in a series of reports. There was a fundamental change between the fourth and fifth assessment reports (AR4 and AR5) [8-10] and in order to reflect such differences as well as model variability, the study looked at two future scenarios. The first scenario considers what the future climate will be under conditions with a representative concentration path (RCP) that assumes that radiative forcing will stabilize at 8.5 W/m² in 2100 (RCP8.5); the second less extreme scenario assumes that radiative forcing will stabilize at 4.5 W/m² in 2100 (RCP4.5).

Downscaled rainfall, and average, minimum and maximum temperatures for the period 1951-2100 have been obtained from CORDEX-Ethiopia database. The data correspond to three RCP scenarios- RCP2.6, RCP4.5, RCP6 and RCP8.5. In order to best conduct a future climate change study, RCP 4.5 and RCP 8.5 forced scenarios were selected from 2010 to 2100 for 3 climate stations, and downscaled to the same climate stations which were used for SWAT model for the hydro Climatological simulations (Table 1).

Hydrological data

Stream flow was used for calibrating and validating the SWAT model simulation. Daily stream flow data were obtained from Ministry of Water Irrigation and Energy (MWIE) for Hare River from 1980 to 2006.

Spatial data

Digital elevation model (DEM), land use/land cover, and soil are the three spatial data inputs required by SWAT model. DEM describes the elevation of any point in a given area at a specific spatial resolution as a digital file. DEM is one of the essential inputs required by SWAT: (1) to delineate the watershed into a number of sub-watersheds or sub-basins and (2) to analyze the drainage pattern of the watershed, slope, stream length, width of channel within the watershed. The DEM was obtained from USGS website with a resolution of 30 m by 30 m.

The land use map of the study area ware downloaded from USGS website with path-169 and row-056, 20 January 2000 satellite image. It has 8-bands, and with special resolution of 30 m × 30 m processed in ArcGIS was used for the hydrologic study input data. The land use of the area has reclassified based on the available topographic map, Arial photography and satellite images. The reclassification of land use map was done to represent the land use according to the specific land cover types such as type of Forest land, Water body, Cultivated land, Bush land and Bare land.

SWAT model requires different soil textural and physico-chemical properties such as soil texture, available water content, hydraulic conductivity, bulk density and organic car-bon content for different layers (up to 3 layers) of each soil type. These data were obtained mainly from the following sources: Ministry of Agriculture (MoA) and Africa CD-ROM (Food and Agriculture Organization of the United Nations [11], Major Soils of the world CD-ROM [12], Digital Soil Map of the World and Derived Soil Properties CD-ROM [13], Proper-ties and Management of Soils of the Tropics CD-ROM [14] (Figure 3).

Methodology

This study, applying bias correction method for downscaled climate variables (precipitation and temperature) is the first and basic step for this impact assessment to be described. Stream flow modeling with SWAT is the second step in the methodology. Finally, climate change impact study using hydrological model is the subject to be discussed.

Bias correction method of downscaled climate data

Often, the downscaled RCPs data cannot be directly used for impact assessment as the computed variables may differ systematically from the observed ones. Bias correction is therefore applied to compensate for any tendency to overestimate or underestimate the mean of downscaled variables. Bias correction factors are computed from the statistics of observed and simulated variables [15]. Two bias correction methods were tried in this study. First, the nonlinear bias correction method proposed by Nader et al. [16] and the second method called "delta approach". The formulas used for rainfall and temperature bias correction are indicated in Equations 1 and 2. Corrections factors were computed for each month.

Precipitation correction

In the bias correction technique, nonlinear correction each daily

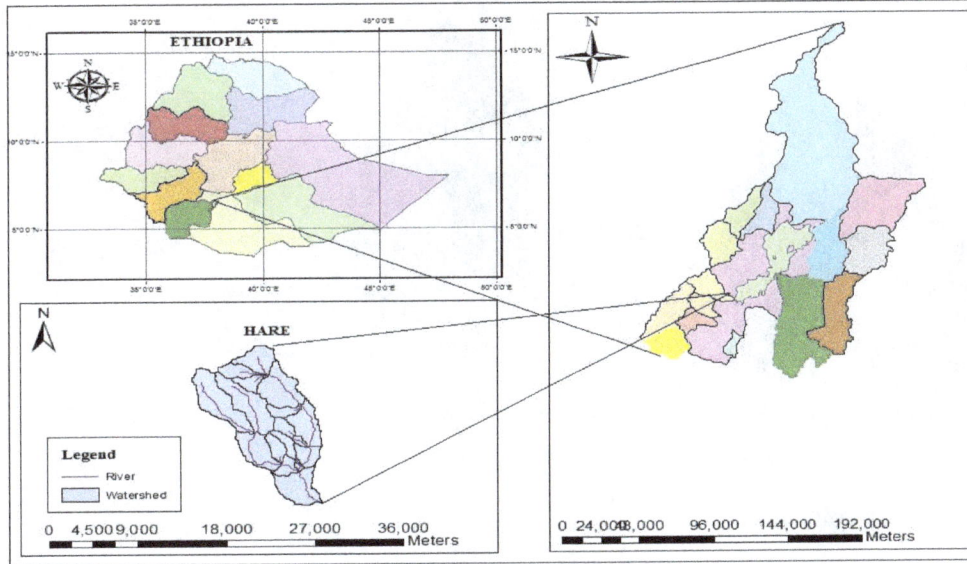

Figure 1: Location of Hare Watershed.

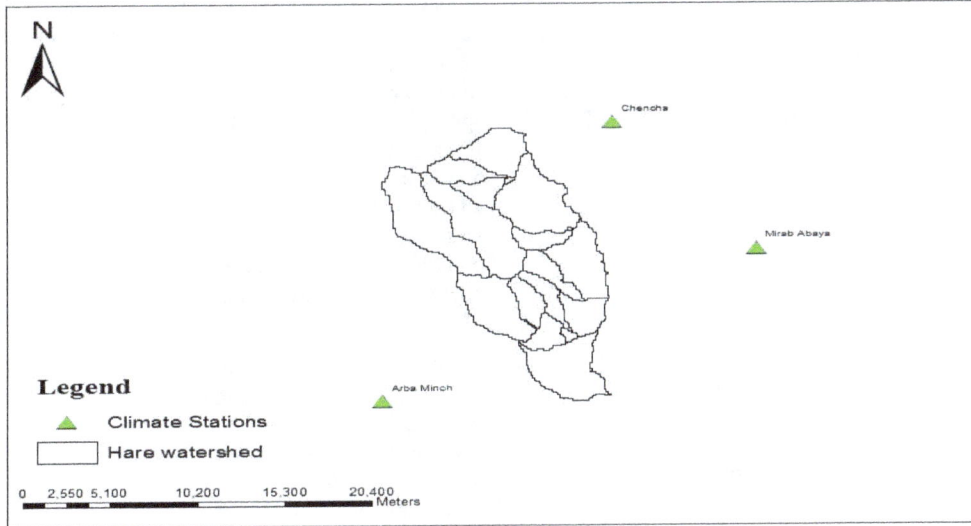

Figure 2: Location of climate station of Hare watershed.

Stations	Grid Code	Corresponding output for the study area	
		Latitude	Longitude
Arba Minch	GP111207	5.72	37.4
Mirab Abaya	GP111208	6.16	37.4
Chencha	GP111209	6.6	37.4

Table 1: Location of CORDEX-Ethiopia output grids for the study area [35].

precipitation amount P is transformed to a corrected P' a power transformation equation have been used.

$$P' = aP^b \qquad (1)$$

Where P' is the simulated data in the projection period, and a and b are the parameters obtained from calibration in the baseline period and subsequently applied to the projection period. They are determined by matching the mean and coefficient of variation (CV) of simulated data with that of observed data [16].

Temperature correction

For temperature, monthly systematic biases were calculated for the baseline period by comparing RCPs outputs with the observations the monthly mean biases correction have been calculated according to the following Equation [17].

$$T_c = T_{om} + \frac{\delta_0}{\delta_r}\left(T_r - T_{rm}\right) \qquad (2)$$

Where; T_c is bias corrected future temperature, Tom is mean of observed temperature in base period, Trm is mean of RCPs temperature in base period and T_r is RCPs temperature of base period δ_r and δ_0 represent the standard deviation of the daily RCPs output and observations in the reference period respectively.

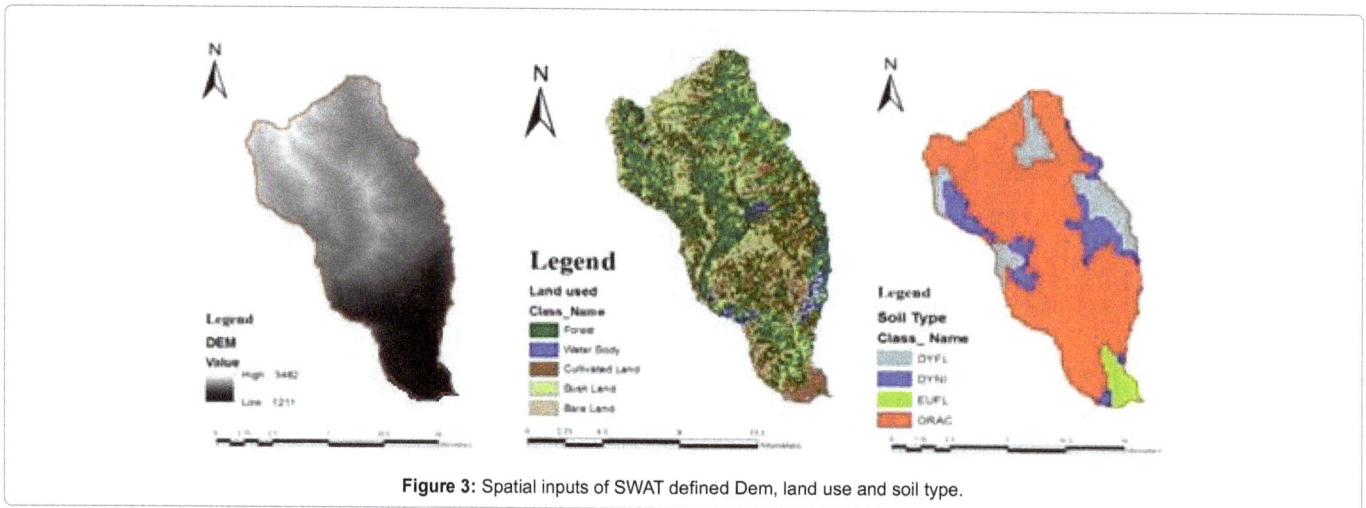

Figure 3: Spatial inputs of SWAT defined Dem, land use and soil type.

Bias correction performance evaluation

The performance of the bias correction method was evaluated using the root means square error (RMSE), the mean absolute error (MAE) and the Relative Error (RE) [18]. They are calculated by the following equations (3), (4) and (5).

$$RMSE = \sqrt{\frac{\sum_{i}^{N}\left(Y_i' - Y_i\right)^2}{N}} \tag{3}$$

$$MAE = \frac{1}{N}\sum_{i=1}^{N}\left|Y_i' - Y\right| \tag{4}$$

$$RE = \frac{\frac{1}{N}\sum_{i=1}^{N}\left(Y_i' - Y\right)}{Y_{mean}} \tag{5}$$

Where, Y_i is the observed value at time step i, Y_i is simulated value at time step i, Y_{mean} is the mean of observed values, and N is the number of observations.

Arc SWAT model approach

Watersheds can be subdivided into sub watersheds and further into hydrologic response units (HRUs) to account for differences in soils, land use, crops, topography, weather, etc. The model has a weather generator that generates daily values of precipitation, air temperature, solar radiation, wind speed, and relative humidity from statistical parameters derived from average monthly values. The model computes surface runoff volume either by using modified SCS curve number method or the Green and Ampt infiltration method. Flow is routed through the channel using a variable storage coefficient method or the Muskingum routing method. SWAT has three options for estimating potential evapotranspiration: Hargreaves, Priestley-Taylor, and Penman-Monteith. The model also includes controlled reservoir operation and groundwater flow model. The important equations used by the model are discussed below. The detailed and complete descriptions are given in the SWAT theoretical documentation. SWAT splits hydrological simulations of a watershed into two major phases: the land phase and the routing phase. The difference between the two lies on the fact that

water storage and its influence on flow rates is considered in channelized flow [19].

Weather generator

Lack of full and realistic long period climatic data is the problem of developing countries. Weather generators solve this problem by generating data having the same statistical properties as the observed ones [20]. SWAT requires daily values of precipitation, maximum and minimum temperature, solar radiation, relative humidity and wind speed. The climatic data collected from the three meteorological stations in the study area however, have too many missing data. As SWAT has a built-in weather generator called WGEN that is used to fill the gaps, all the missing values were filled with a missing data identifier, -99. The weather generator first independently generates precipitation for the day. Maximum temperature, minimum temperature, solar radiation and relative humidity are then generated based on the presence or absence of rain for the day. Finally, wind speed is generated independently [21].

For the sake of data generation, weather parameters were developed by using the weather parameter calculator WXPARM [22] and dew point temperature calculator DEW02 [23], which were downloaded from the SWAT website. The WXPARM program reads daily values of solar radiation (calculated from daily sunshine hours), maximum and minimum temperatures, precipitation, relative humidity, and wind speed data. It then calculates monthly daily averages and standard deviations of all variables as well as probability of wet and dry days, skew coefficient, and average number of precipitation days in the month. The DEW02 programs reads daily values of relative humidity, and maximum and minimum temperature values and calculates monthly average dew point temperatures [21].

Hydrological component of SWAT

The simulation of the hydrology of a watershed is done in two separate divisions. One is the land phase of the hydrological cycle that controls the amount of water, sediment, nutrient and pesticide loadings to the main channel in each sub-basin. Hydrological components simulated in land phase of the hydrological cycle are canopy storage, infiltration, redistribution, evapotranspiration, lateral subsurface flow, surface runoff, ponds, tributary channels and return flow. The second division is routing phase of the hydrologic cycle that can be defined as the movement of water, sediments, nutrients and organic chemicals through the channel network of the watershed to the outlet. In the land phase of hydrological cycle, SWAT simulates the hydrological cycle based on the water balance equation.

$$SW_t = SW_0 + \sum_{i=1}^{t}\left(R_{day} - Q_{surf} - E_a - W_{seep} - Q_{gw}\right)i \qquad (6)$$

In which SW_t is the final soil water content (mm), SW_0 is the initial soil water content on day i (mm), t is the time (days), R_{day} is the amount of precipitation on day i (mm), Q_{surf} is the amount of surface runoff on day i (mm), E_a is the amount of evapotranspiration on day i (mm), W_{seep} is the amount of water entering the vadose zone from the soil profile on day i (mm), and Q_{gw} is the amount of return flow on day i (mm).

Using the above equation, the soil moisture content for the given area is simulated. Since the soil moisture storage is the main concern of this study, the brief description of some of the key model components are provided in this thesis. More detailed descriptions of the different model components are listed in Sallis et al. [19]. Soil water may follow different paths of movement: vertically upward (plant uptake), vertically downward (percolation), or laterally-contributing to stream flow. The vertical movement as plant uptake removes the largest portion of water that enters the soil profile.

The amount of soil water is usually measured in terms of water content as percentage by volume or mass, or as soil water potential, this soil water content is highly depending on the water balance values given in equation 6. Mostly, taking the precipitation as source of soil water content and reduction of run off, actual evapotranspiration, and ground water from precipitation is result in availability of water in the soil. Therefore, SWAT model revealed quantitatively the value of soil water content (SW) depends on the above water balance values.

However, water content does not necessarily describe the availability of the water to the plants, more indicates how the water moves within the soil profile. The only information provided by water content is the relative amount of water in the soil.

Soil water dynamics can be thought of as comparable to a sponge. When a sponge is saturated by soaking it in water when it is lifted out of the water, any excess water will drop off it. This is equivalent to drainage from the macro pores in the soil. Once the sponge has stopped dripping it is at field capacity.

When the sponge is squeezed, it is easy to get the first half of the water out. This first squeeze is equivalent to draining the sponge to the stress point and the water is removed like the RAWC (readily available water-holding capacity). Squeezing the second half of the sponge out is much harder. This is like draining the sponge to permanent wilting point. The total water squeezed out of the sponge from when it stopped dripping is the TAWC (Total Available Water-Holding Capacity). But no matter how hard the sponge is squeezed there is no way to get all the water out of it. The water left is the equivalent to the hygroscopic water found in soil. This sponge analogy is similar to how plant roots find getting moisture from the soil. From field capacity to the stress point it is easy to get the water. From the stress point to the permanent wilting point plants find it much harder to draw water from the soil and their growth is stunted. Below the permanent wilting point no further water can be removed and the plant dies [21].

Percolation is the downward movement of water in the soil. SWAT calculates percolation for each soil layer in the profile. Water is allowed to percolate if only the water content exceeds the field capacity of that layer [19].

Surface runoff occurs whenever the rate of precipitation exceeds the rate of infiltration. SWAT offers two methods for estimating surface runoff: the SCS curve number procedure and the Green and Ampt infiltration method [24]. Using daily or sub daily rain-fall, SWAT simulates surface runoff volumes and peak runoff rates for each HRU. In this study, the SCS curve number method was used to estimate surface runoff because of the unavailability of sub daily data for Green and Ampt method.

Lateral flow is common in areas with high hydraulic conductivities in surface layers and an impermeable or semi-permeable layer at a shallow depth. Rainfall will percolate vertically up to the impermeable layer and develops a saturated zone stored above this layer. This is called a perched water table, which is the source of water for lateral subsurface flow. SWAT incorporates a kinematic storage model for subsurface flow [19].

The peak discharge or the peak surface runoff rate is the maximum volume flow rate passing a particular location during a storm event. SWAT calculates the peak runoff rate with a modified rational method. In rational method, it assumed that a rainfall of intensity I begins at time t=0 and continues indefinitely, the rate of runoff will increase until the time of concentration, $t=t_{conc}$. The modified rational method is mathematically expressed as:

$$q_{peak} = \frac{\alpha tc * Q_{surf} * Area}{3.6 * t_{conc}} \qquad (7)$$

Where, q_{peak} is the peak runoff rate (m³/s), αtc is the fraction of daily rainfall that occurs during the time of concentration, Q_{surf} is the surface runoff (mm), Area is the sub-basin area (km²), t_{conc} is the time of concentration (hr), and 3.6 is a conversion factor.

Potential evapotranspiration: There are many methods that are developed to estimate Potential Evapotranspiration (PET). Three methods are incorporated into SWAT: The Penman-Monteith method [25], the Priestley-Taylor method and the Hargreaves method [26]. For this study, Penman-Monteith method were used.

Groundwater: The simulation of groundwater is partitioned into two aquifer systems i.e., an unconfined aquifer (shallow) and a deep-confined aquifer in each sub basin. The unconfined aquifer contributes to flow in the main channel or reach of the sub basin. Water that enters the deep aquifer is assumed to contribute to stream flow outside the watershed [27]. In SWAT the water balance for a shallow aquifer is calculated with equation 8.

$$aq_{sh,i} = aq_{sh,i-1} + W_{rchrg} + Q_{gw} - W_{deep} - W_{pump,sh}$$

where: $aq_{sh,i}$ is the amount of water stored in the shallow aquifer on day i (mm), $aq_{sh,i-1}$ is the amount of water stored in the shallow aquifer on day i-1 (mm), W_{rchrg} is the amount of recharge entering the aquifer on day i (mm), Q_{gw} is the groundwater flow, or base flow, into the main channel on day i (mm), WR_{evap} is the amount of water moving into the soil zone in response to water deficiencies on day i (mm), W_{deep} is the amount of water percolating from the shallow aquifer into the deep aquifer on day i (mm), and $W_{pump,sh}$ is the amount of water removed from the shallow aquifer by pumping on day i (mm).

SWAT Model Setup

Watershed delineation

Watershed delineator tool in Arc SWAT allows the user to delineate the watershed and sub-basins using DEM. Flow direction and accumulation are the concepts behind to define the stream network of the DEM in SWAT. The monitoring point is added manually and the numbers of sub-basin are adjusted accordingly. Finally, the catchment area is delineated to be 153 km^2 for Hare River catchment, and 17 sub-basins are formed for the whole catchment.

HRU analysis

HRU analysis helps to load land use map and soil map and also incorporates classification of HRU into different slope classes. The land use map as well as soil map was overlapped 100% with the delineated watershed and 103 HRUs were formed for the study area. Spatial inputs of slope, soil, and land use were used to define the catchment. 30-meter resolution of USGS DEM topographic data was used for slope classification. Extracted topographic map with the catchment boundary shows an elevation range from 1211 to 3482 m amsl.

Weather data definition

Available meteorological records (1980-2006) (i.e., precipitation, minimum and maximum temperatures, relative humidity, and wind speed) and location of meteorological station are prepared based on Arc SWAT input format and integrated with the model using weather data input wizards. Arba Minch meteorological station data were used as weather generator for this study.

Sensitivity analysis

Sensitivity analysis is a technique of identifying the responsiveness of different parameters involving in the simulation of a hydrological process. For big hydrological models like SWAT, which involves a wide range of data and parameters in the simulation process, calibration is quite a cumbersome task. Even though, it is quite clear that the flow is largely affected by curve number, for example in the case of SCS curve number method, this is not sufficient enough to make calibration as little change in other parameters could also change the volumetric, spatial, and temporal trend of the simulated flow. Hence, sensitivity analysis is a method of minimizing the number of parameters to be used in the calibration step by making use of the most sensitive parameters largely controlling the behavior of the simulated process. This appreciably eases the overall calibration and validation process as well as reduces the time required for it. Besides, as Dutra et al. [28] indicated, it increases the accuracy of calibration by reducing uncertainty.

The sensitivity analysis was undertaken by using a built-in tool in SWAT that uses the Latin Hypercube One-factor-At-a-Time (LH-OAT). Details of this method are explained in Robinson et al. [29]. After the analysis, the mean relative sensitivity (MRS) of the parameters was used to rank the parameters, and their category of Dutra et al. [28] classification. He divided sensitivity was also defined based on the sensitivity into four classes as shown in Table 2. Eime et al. [4] indicated that there can high (0.20) be a significant variation of hydrological processes between individual watersheds. This, therefore, justified the need for the sensitivity analysis made in the study area. The analysis involved a total of 28 parameters. For the study area, the sensitivity analysis should be carried out for a period of twelve years, which included both calibration period (from January 1, 1993 to December 31, 2002) and the warm-up period (From January 1, 1991 to December 31, 1992).

Model calibration and validation

Calibration is tuning of model parameters based on checking results against observations to ensure the same response over time. This involves comparing the model results, generated with the use of historic meteorological data, to recorded stream flows. In this process,

model parameters varied until recorded flow patterns are accurately simulated. Model calibration of SWAT run can be divided in to several steps. Among these Water balance and stream flow generation are the most important part is also considered.

Ref. [19] distinguished three types of calibration methods: the manual trial-and-error method, automatic or numerical parameter optimization method; and a combination of both methods. According to the authors, the manual calibration is the most common and especially recommended in cases where a good graphical representation is strongly demanded for the application of more complicated models. However, it is very cumbersome, time consuming, and requires experience. Automatic calibration makes use of a numerical algorithm in the optimization of numerical objective functions. The method undertakes a large number of iterations until it finds the best parameters. The third method makes use of combination of the above two techniques regardless of which comes first. For this study, the first and the third approach was considered [30].

The manual calibration of this study was done based on the procedures recommended in SWAT user manual. Water balance calibration normally takes care of the overall flow volume and its distribution among the different hydrologic components, whereas temporal flow calibration is concerned about the flow time lag and the hydrograph shape [21]. For this case study also, as the soil moisture data is not available at the station, one of the water balance component with observed data, (stream flow) is used for calibration and validation purpose. The automatic calibration was done using Parameter Solution (Parasol) [4]. This method was chosen for its applicability to both simple and complex hydrological models.

Calibration for water balance and stream flow is first done for average annual conditions. Once the model is calibrated for average annual conditions, it can be repeated to monthly or daily records to fine-tune the calibration. Accordingly, the annual and monthly calibration was taken for the study area. Flow calibration was performed for a period of ten years from January 1, 1993 to December 31, 2002 using the sensitive parameters identified. However, flow was simulated for twelve years from January 1, 1991 to December 31, 2002, within which the first two year was considered as a warm up period. Flow validation was performed for a period of four years from January 1, 2003 to December 31, 2006.

The flow was calibrated manually using the observed flow gauged at the outlet of the watershed. First of all, the surface runoff flow components of gauged flow were balanced with that of the simulated flow.

Afterwards the adjusted flow was further calibrated temporally by making delicate adjustments to ensure best fitting of the simulated flow curves with the gauged flow curves. Manipulation of the parameter values were carried out within the allowable ranges recommended by SWAT developers. The factor of goodness fit can be quantified by the coefficient of determination (R^2) and Nash-Sutcliff efficiency (NSE) between the observations and the final best simulations. Coefficient of determination (R^2) and Nash-Sutcliffe coefficient (NSE) are calculated by:

$$R^2 = \frac{\left[\left(\sum_i Q_{m,i} - \bar{Q}_m\right)\left(Q_{s,j} - \bar{Q}_s\right)\right]^2}{\left(\sum_i Q_{mi,j} - \bar{Q}_m\right)^2 \sum_i \left(Q_{s,i} - \bar{Q}_s\right)^2} \tag{9}$$

$$NSE = 1 - \frac{\sum_i (Q_m - Q_s)i^2}{\sum_i (Q_{m,i} - \overline{Q}_m)^2} \qquad (10)$$

In which Q_m is the measured discharge, Q_s is the simulated discharge, $\overline{Q}_m$ is the average measured discharge and $\overline{Q}_s$ is the average simulated discharge.

Hydrologic impact of future climate change scenario

The main objective of downscaling is to generate a reliable estimation of meteorological variables corresponding to given scenario of the future climate so that these meteorological variables will be used as basis for different types of impact studies. Therefore, after calibration and validation of hydrological models with historical record, the next step in the investigation is to simulate river flows in the watershed corresponding to future climate conditions by using the downscaled precipitation and temperature in to SWAT model. Such simulation helps to identify the hydro climatological impacts caused by climate change.

The future climate variables that is downscaled precipitation and temperature found as an output from the RCPs and corrected RCPs were given as an input to the SWAT model. Then simulation results corresponding to each of downscaling scenario time period (current, 2020s, 2050, and 2080s) are analyzed for all month of the year.

Results and Discussion: Bias Correction Results

In this study, the bias correction method was applied to simulate climate variables: precipitation for three meteorological stations (Arba Minch, Chencha and Mirab Abaya) and maximum temperature and minimum temperature for Arba Minch station.

The results of the bias correction method were evaluated using residual plots (difference between the simulated and observed values) of precipitation and temperature in terms of mean. Standard model performance statistics tests were carried out by computing the root mean square error (RMSE), the mean absolute error (MAE) and the relative error (RE).

The monthly residuals of bias corrected climate variables (precipitation, maximum temperature, and minimum temperature) are presented in Figure 4. The figures show that a significant improvement is achieved by applying bias correction method on RCPs since the bias corrected results provided smaller monthly mean bias values than the raw RCPs results. In other words, the bias corrected results are closer to the observed values than the raw RCPs results at the three stations.

As shown in the Figure 4a, the mean monthly residuals of raw precipitation are between -83.1 mm and +62.8 mm and after bias correction the residuals are between -13.6 mm and +0.01 mm. Figure 4b shows that, the mean monthly residuals of raw precipitation are between -738.1 mm and +34.7 mm and after bias correction the residuals are between -11.7 mm and +17.7 mm. Figure 4c shows that, the mean monthly residuals of raw precipitation are between -49.8 mm and +56.1 mm and after bias correction the residuals are between -37.5 mm and +0.02 mm which indicates that significant improvement is achieved after bias correction.

Again, for Arba Minch station, the bias corrected mean daily maximum temperature and minimum temperature was compared with the raw mean daily maximum temperature and minimum temperature using residual plots.

As shown in the Figure 5a and 5b the mean daily residual plots of raw and bias corrected mean daily maximum temperature and minimum temperature. It is found that while the raw mean daily maximum temperature residuals are between -1.5°C and +1.4°C, after bias correction the range is between -1.1°C and +1°C in terms of mean and raw mean daily minimum temperature residuals are between -2.3°C and +3.4°C, after bias correction the range is between -1.4°C and +2.1°C in terms of mean.

The bias corrected results show that the corrected RCPs provided fewer errors than the raw RCP results. From Table 3, it can be seen that the models' performance (RMSE, MAE, and RE) for precipitation, maximum temperature, and minimum temperature was good and almost the same.

Future climate projection precipitation

The rainfall expected to experience a mean monthly increase by 6.40, 2.56, and 16.30% for RCP4.5 scenario at 2020s, 2050s, and 2080, respectively. The mean annual increase was repeated by RCP8.5 scenario with 8.56, 8.08, and 15.85% at 2020s, 2050s, and 2080s, respectively. These values show that increasing in rainfall is not uniform; instead, it differs from time period to time period

A mean monthly rainfall projection of Kiremt (Jun-September) increase by 1.79, 1.43, and 1.68% for RCP4.5 scenario and RCP8.5 scenario by 4.10, 3.96, and 12.52% at 2020s, 2050s, and 2080, respectively and Bega (October-January) shows increasing by 18.91, 21.46, and 58.84% for RCP4.5 scenarios and RCP8.5 scenarios by 31.52, 37.42, and 56.00% at 2020s, 2050, and 2080s, respectively. A mean monthly rainfall projection of Belg (February-May) decrease by 1.50, 15.20, and 16.77% for RCP4.5 scenario and RCP8.5 scenario by 9.94, 17.14, and 20.97% at 2020s, 2050s, and 2080, respectively. A rainfall projection of Kiremt (Jun-September) and Bega (October-January) shows increasing the two emission scenarios whereas Belg (February-May) projection shows a decreasing mean monthly rainfall for the two emission scenarios (Figure 6a and 6b).

Percentage change in monthly, seasonal, and annual precipitation for the period 2010-2099 generally increasing during the Kiremit (wet season=June–September) for the long-term future and also indicate a corresponding increase in precipitation for the Belg (less rainy season=February-May) for 2050s and 2080s [31].

Maximum temperature

The projected mean monthly maximum temperature shows increasing trend for all time periods by 0.01, 0.02, and 0.11°C for RCP4.5 scenario for 2020s, 2050s, and 2080, respectively. RCP8.5 scenario also shows an increase of mean monthly maximum temperature with 0.01, 0.03, and 0.11°C for 2020s, 2050s and 2080s, respectively. As compared to RCP8.5 scenario, RCP4.5 scenario is almost the same (Figure 7a and 7b).

Minimum temperature

The projected minimum temperature shows an increasing trend in all time periods. In this case, both the RCP4.5 and RCP8.5 emission scenarios predict the future minimum temperature in similar manner. For RCP4.5 scenario, mean monthly minimum temperature increases by 0.06, 0.07, and 0.12°C and for RCP8.5 scenario 0.32, 0.15, and 0.13°C for 2020s, 2050s, and 2080s, respectively. Mean monthly variation of minimum temperature is higher than maximum temperature (Figure 8a and 8b).

According to the Ethiopian National Meteorological Services

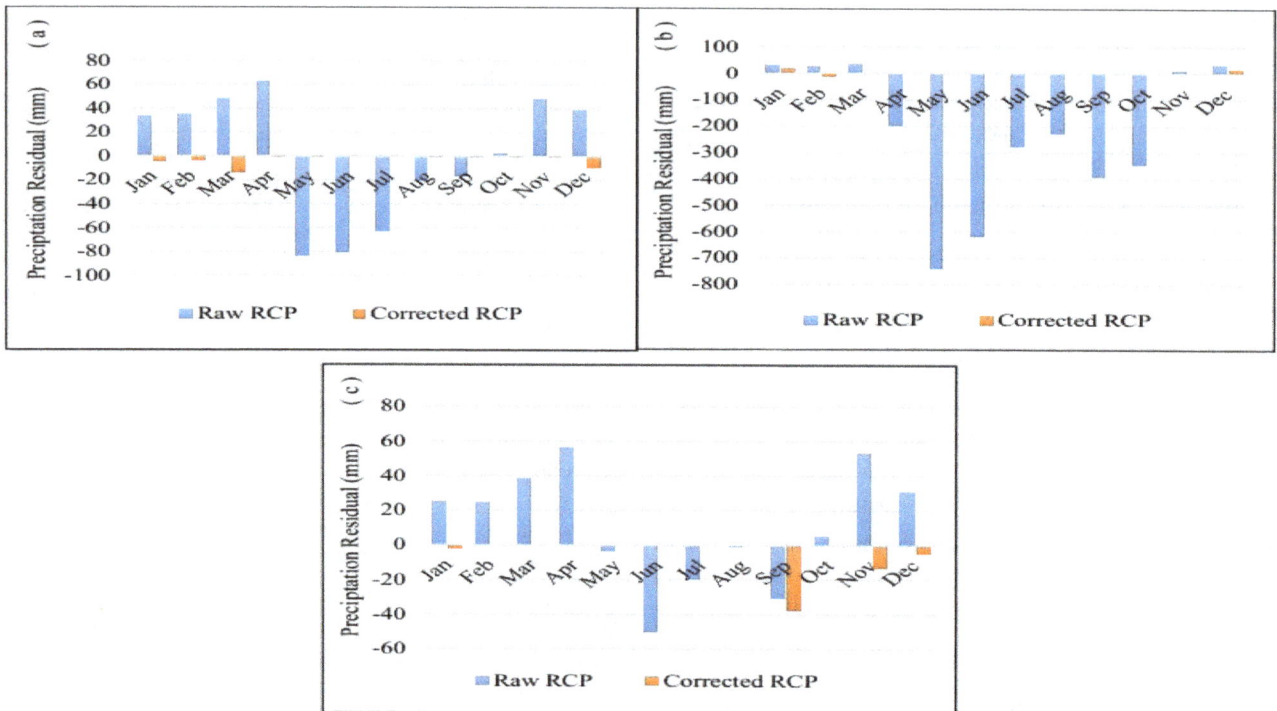

Figure 4: Mean monthly residuals of raw RCPs and corrected RCPs (1980-2006) a) Arba Minch, b) Chencha and c) Mirab Abaya.

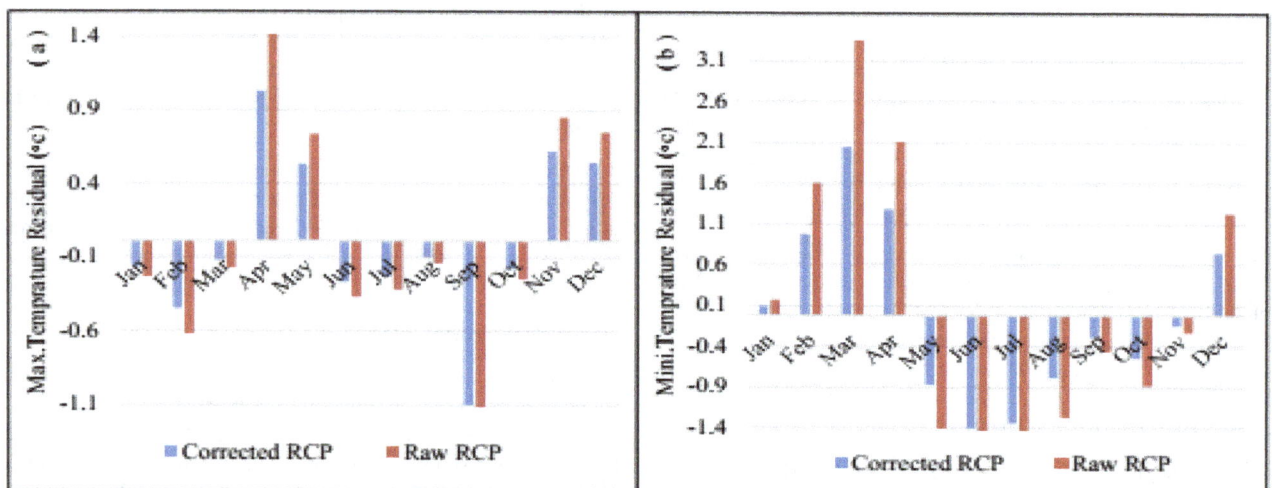

Figure 5: Mean monthly residuals temperature of raw RCPs and corrected RCPs (1980-2006) at Arba Minch.

Class	Index	Sensitivity
I	0.00≤/I/<0.05	Small to Negotiable
II	0.05≤/I/<0.2	Medium
III	0.2≤/I/<1	High
IV	/I/≥1	Very high

Table 2: Sensitivity Class for SWAT model Source: [28].

Agency [5] the study for 40 stations showed that there have been very warm and very cold years. However, the general trend showed there was an increase in temperature over the last 50 years. The study also noted that the minimum temperature is increasing at a higher rate than the maximum temperature.

Hydrological model calibration and validation

In this study, soil, and water assessment hydrological tool (SWAT) model were employed to simulate streamflows for Hare river watershed using observed and bias corrected precipitation and temperature data.

As mentioned in the methodology part, flow sensitivity analysis has been carried out for 26 hydrological parameters using SWAT sensitivity analysis. The most sensitive parameters considered for calibration were Curve number (CN2), Soil Evaporation Compensation factor (ESCO), Groundwater recession factor (ALPHA_BF), Soil Available Water Capacity (SOL_AWC), Soil depth (SOL_Z), Saturated hydraulic conductivity (SOL_K), Slope (SLOPE), Threshold depth of water in

Station	Evaluation	Precipitation	Temperature Max	Temperature Min
Arba Minch	RMSE	12.29	1.76	1.46
	MAE	1.69	0.0629	0.0672
	RE	0.0329	2.6×10^{-15}	0.0238
Chencha	RMSE	9.36	-	-
	MAE	1.49	-	-
	RE	-0.0156	-	-
Mirab Abaya	RMSE	10.55	-	-
	MAE	1.76	-	-
	RE	0.077	-	-

Table 3: Performance statistics of bias correction method.

the shallow aquifer required for return flow occur (GWqmn), Ground water delay (GW_DELAY), Average slope length (SLSUBBSN) and Plant evaporation compensation factor (EPCO). Result of sensitivity analysis was used to conduct the calibration of SWAT. The calibration was performed with the seven sensitive parameters of stream flow. After several iterations, fitted values for the seven parameters were gained (Table 4).

Performance of the best simulation of stream flow result which used these fitted parameter values for calibration and validation is shown in Table 5. Govindan et al. [30] stated that Nash–Sutcliff efficiency (NSE) values greater than or equal to 0.50 are considered adequate for SWAT model application. Hence, it is observed that SWAT exhibited strong performance in representing the hydrological conditions of the catchment. It can be seen from the flow hydrographs (Figure 9a and 9b) that the simulated flows well matched the observed flows except for peak values in the calibration period and low values in the validation period for both daily and monthly time steps.

Simulation of hydro climatological impacts caused by climate change

One of the main objectives of this study was simulation of hydro climatological impacts caused by climate change over the study area. Therefore, this section of result and discussion is one of the ultimate goals of this study. For this purpose, the changes of downscaled climate variables (temperature and precipitation) from the baseline climate were used to get 2020s, 2050s, and 2080s (each has 30 years daily climatic data including the baseline) time series data. SWAT simulation was run four times (for the baseline, 2020s, 2050s, and 2080s) by keeping constantly calibrated soil, crop, and slope parameters to quantify climate change impact only. Simulation results of stream flow for the three future time periods, 2020s, 2050s, and 2080s, were compared with the baseline period simulation.

Mean monthly stream flow may increase by 12.2, 8.0, and 13.9% for 2020s, 2050s, and 2080s, respectively, from the baseline stream flow for RCP4.5 scenario whereas for RCP8.5 scenario, the model shows an increase by 7.3, 13.4, and 15.4% for 2020s, 2040s, and 2080s, respectively (Figure 10a and 10b).

The same streamflow metrics used for assessing the effects land use change were also used for assessing the effects of climate change. Consistent with other research [8,32,33], the climate only scenarios had a larger effect on streamflow than land use change. In 2050s, mean annual streamflow is found to increase in all the scenarios. The maximum projected increase in discharge of 32.46% was found for

Rank	Parameter	SWAT default		Fitted Value
		Lower bound	Upper bound	
1	CN2	35	98	64 to 85
2	ESCO	0	1	9.5
3	ALPHA_BF	0	a	0.06
4	SOL_AWC	0	1	0.01 to 0.30
5	SOL_Z	0	3000	200 to 490
6	SOL_K	0	100	4.3 to 80
7	SLOPE	0	1	0.06 to 0.75
8	GWqmn	0	5000	50
9	GW_DELAY	0	50	20
10	SLSUBBSN	10	150	10 to 43
11	EPCO	0	1	0.85

Table 4: Calibrated fitted values of flow-sensitive parameters.

Criteria	Calibration (1991-2002)	Validation (2003-2006)
Coefficient of determination (R^2)	0.85	0.84
Nash-Sutcliffe efficiency (NSE)	0.73	0.77

Table 5: Calibrated model simulation performance.

GISSE2-H (RCP 4.5), ware as MIROC-ESMCHEM (RCP 8.5) and BCC-CSM1.1

(RCP 8.5) showed almost same increase in projected annual discharges which are 26.89% and 26.27% consecutively. In 2080s, the maximum projected increase in discharge of 47.44% was found for BCC- CSM1.1 (RCP 8.5). Other scenarios gave variable increase in discharge, 29.50%, 18.96%, 38.82% and 13.60% increase in mean annual streamflow were found for MIROC-ESMCHEM (RCP 8.5), HADGEM2- ES (RCP 8.5), GISS-E2-H (RCP 4.5) [34].

Conclusions and Recommendations

The bias correction for downscaled RCPs data, the raw RCPs data was corrected by a bias correction methods successfully as the simulated climate variables produced consistent results with the historical records. The residuals between corrected and historical monthly values were smaller than the residuals between raw RCP and historical monthly values of precipitation and temperature. The models' evaluation of performance for precipitation, maximum temperature, and minimum temperature was good and almost the same. The bias correction method the most accurate method for downscaled climate variables to reproducing the main features of the observed data.

The rainfall projection expected to experience a mean monthly increase by 6.40, 2.56, and 16.30% for RCP4.5 scenario at 2020s, 2050s, and 2080, respectively. The mean annual increase was repeated by RCP8.5 scenario with 8.56, 8.08, and 15.85% at 2020s, 2050s, and 2080s, respectively. For both RCPs scenarios, the maximum and minimum temperature projection expected increasing for all future period.

The simulations were done using the IPCC RCP4.5 and RCP8.5 emission scenario. The curve number, Soil Evaporation Compensation factor (ESCO), and Alfa base flow parameter showed a relatively very higher sensitivity. Soil water assessment tool (SWAT) models were well calibrated and validated using observed flow data as the coefficient determination was above 0.7 and Nash Sutcliffe efficiency index was above 0.5 for watershed.

Mean monthly percentage changes of climate variables from the baseline period were used to simulate future projections of stream

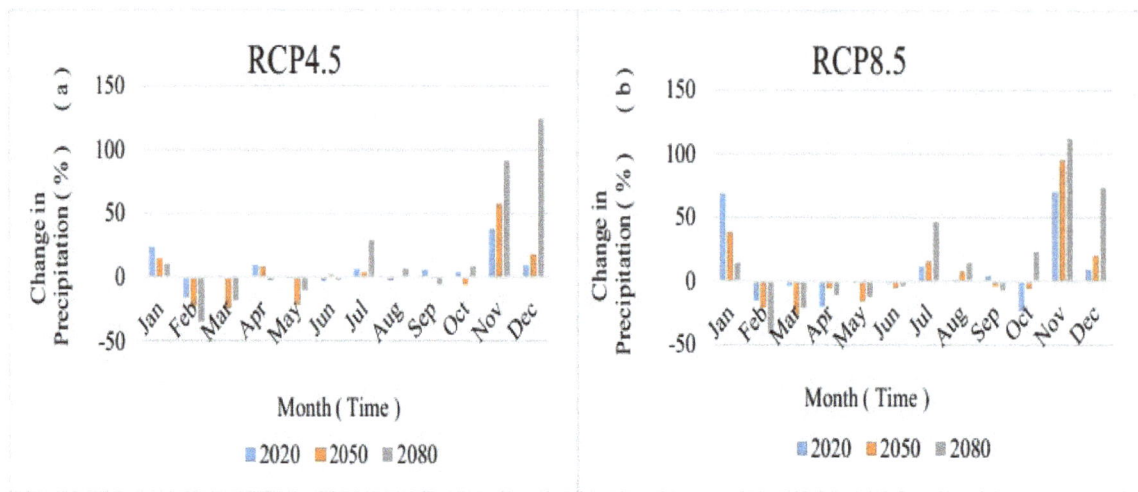

Figure 6: Future changes in mean monthly precipitation for a) RCP4.5 and b) RCP8.5 scenarios.

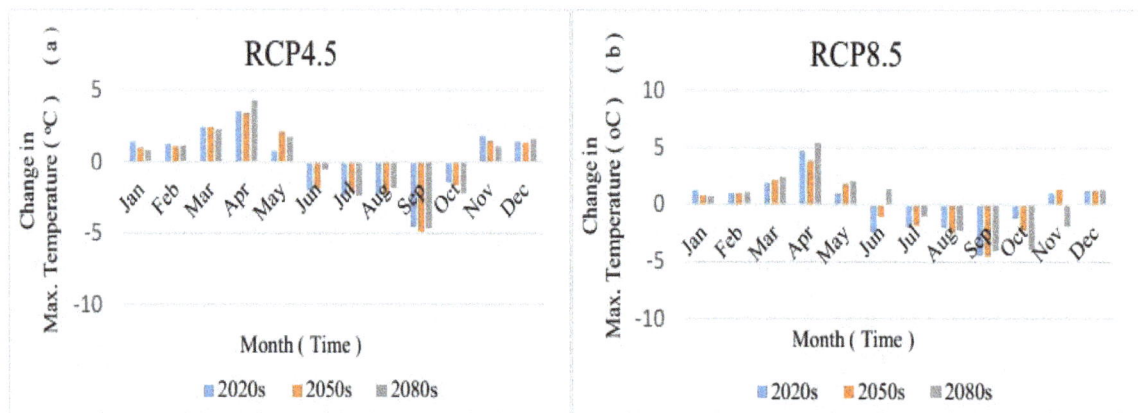

Figure 7: Future changes in mean monthly Maximum temperature for a RCP4.5 and b RCP8.5 scenarios.

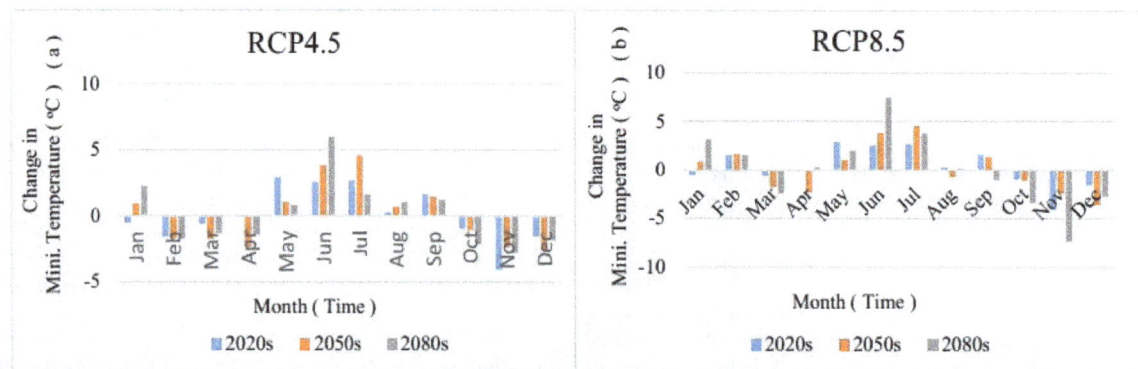

Figure 8: Future changes in mean monthly Minimum temperature for a RCP4.5 and b RCP8.5 scenarios.

flow. Stream flow projections for future time periods showed that mean monthly stream flow may increase by 12.2, 8.0, and 13.9% at 2020s, 2050s, and 2080s, respectively, from the baseline period for RCP4.5 scenario, whereas for RCP8.5 scenario, it will be expected to increase by 7.3, 13.4, and 15.4% for 2020s, 2040s, and 2080s, respectively.

Data quality and availability should be stressed much more while using distributed hydrological models. The applications SWAT models were very challenging and a lack of appropriate data was one of the biggest concerns throughout. Without proper data, model implementation is very difficult if not impossible.

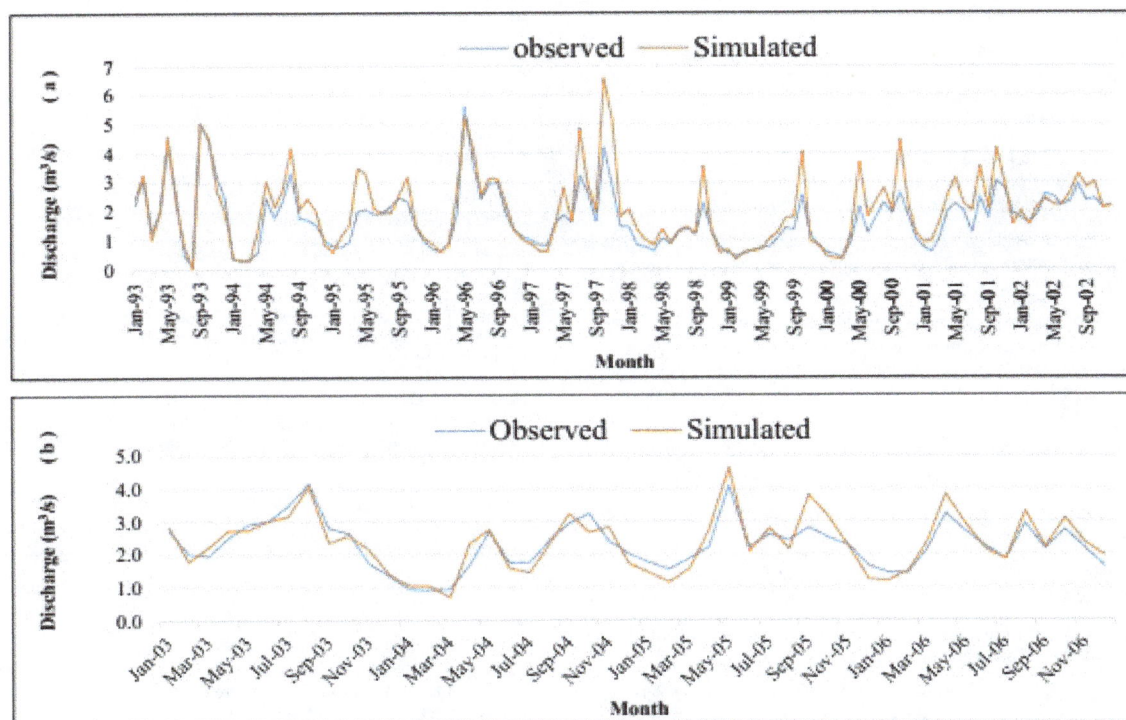

Figure 9: a) Calibration and b) Validation result of average monthly simulated and gauged flows.

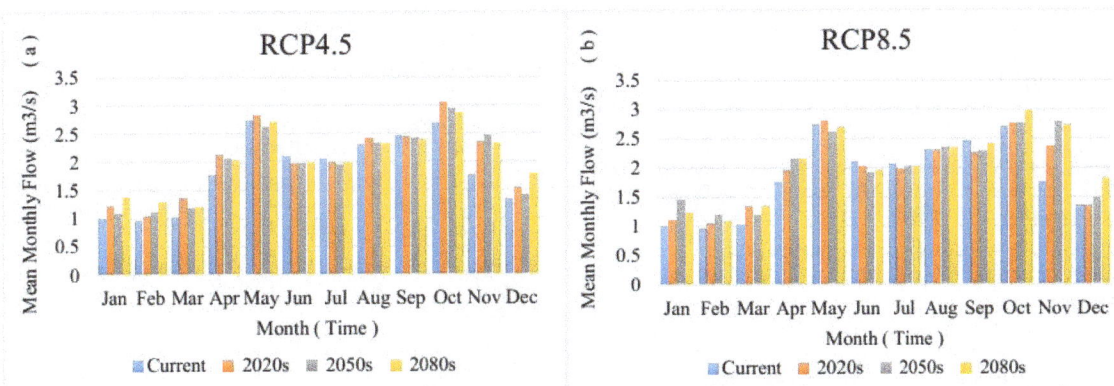

Figure 10: Mean monthly flow for a RCP4.5 and b RCP8.5 scenarios.

The model simulations considered only future climate change scenarios assuming all spatial data constant. But change in land use scenarios other climate variables will also contribute some impacts on future stream flow.

Uncertainty analysis is recommended to assess the uncertainty associated with the bias correction method, the hydrological models, and the RCPs. Finally, to link the study results to the sustainable development of the community, it is recommended to use the hydrological models' outputs in water resource management models to take the socio-economic aspects of the hydrological system [35-39].

Acknowledgements

I would also like to thank the Ministry of Water Irrigation and Energy, Ministry of Agriculture and National Meteorological Services Agency of Ethiopia for providing me discharge, soil and climate data respectively.

References

1. Salvi K, Kannan S, Ghosh S (2011) Statistical Downscaling and Bias Correction for Projections of Indian Rainfall and Temperature in Climate Change Studies. 2011 International Conference on Environmental and Computer Science 19: 7-11.

2. IPCC-TGICA (2007) General Guidelines on the Use of Scenario Data for Climate Impact and Adaptation Assessment. Version 2, Prepared by Carter TR on behalf of the Intergovernmental Panel on Climate Change, Task Group on Data and Scenario Support for Impact and Climate Assessment, p: 66.

3. Frederick KD (2002) Introduction. In: Water resource and climate change. Fredrick KD (ed.), Northampton MA, Edward Elgar Publishing, p: 514.

4. Van Griensven A, Meixner T, Grunwald S, Bishop T, Diluzio M, et al. (2006) A global sensitivity analysis tool for the parameters of multi-variable catchment models. Journal of Hydrology 324: 10-23.

5. NMSA (National Meteorological Service Agency) (2001) Initial National Communication of Ethiopia to the United Nations Framework convention on Climate Change (UNFCCC). National Meteorological Services Agency.

6. Kassa TM (2009) Watershed hydrological responses to changes in land use and land cover, and management practices at hare watershed, Ethiopia.

7. Förch G (2007) Assistance to Arba Minch Water Technology Institute (AWTI). A Brief History of German Project Support and its Impacts on the Ethiopian and German Partners. Proceedings of Lake Abaya Research Symposium (LARS), FWU Water Resources Publication, p: 6.

8. Caldwell PV, Sun G, McNulty SG, Cohen EC, Myers JM (2012) Impacts of impervious cover, water withdrawals, and climate change on river flows in the conterminous US. Hydrology and Earth System Sciences 16: 2839-2857.

9. Moss RH, Edmonds JA, Hibbard KA, Manning MR, Rose SK, et al. (2010) The Next Generation of Scenarios for Climate Change Research and Assessment. Nature 463: 747-756.

10. Stocker TF, Qin D, Plattner GK, Tignor M, Allen SK, et al. (2013) Climate Change 2013: The Physical Science Basis. Contribution of Working Group I to the Fifth Assessment Report of the Intergovernmental Panel on Climate Change. Cambridge University Press, Cambridge, United Kingdom and New York, NY, USA, p: 1535.

11. FAO (1998) The Soil and Terrain Database for northeastern Africa (CDROM) FAO, Rome.

12. FAO (Food and Agricultural Organization) (2002) Major Soils of the World. Land and Water Digital Media Series. FAO Soil Buletin No 19, Rome, Italy.

13. FAO (1995) Digital Soil Map of the World and Derived Soil Properties (CDROM) Food and Agriculture Organization of the United Nations, FAO.

14. Van Wambeke A (2003) Properties and management of soils of the tropics. FAO Land and Water Digital Media Series No. 24, FAO, Rome.

15. Gebrie GS, Engida AN (2015) Climate Modeling of the Impact of Climate Change on Sugarcane and Cotton for Project on 'a Climate Resilient Production of Cotton and Sugar in Ethiopia.

16. Leander R, Buishand TA (2007) Resampling of regional climate model output for the simulation of Extreme River flows. J Hydrol 332: 487-496.

17. Ho CK, Stephenson DB, Collins M, Ferro CAT, Brown SJ (2012) Calibration Strategies A Source of Additional Uncertainty in Climate Change Projections. Bulletin of the American Meteorological Society 93: 21-26.

18. Shamarokh A (2012) Hydrological Impact of Climate Change in Semi-Urban Watershed. MSc Thesis, Civil and Environmental Engineering (Bangladesh Univ of Engg. & Tech.).

19. Neitsch SL, Arnold JG, Kiniry JR, Williams JR (2005) Soil and Water Assessment Tool, Theoretical Documentation: Version 2005. Temple, TX. USDA Agricultural Research Service and Texas A & M Black Land Research Centre.

20. Danuso F (2002) Climak. A Stochastic Model for Weather Data Generation. Italian Journal of Agronomy 6: 57-71.

21. Yakob M (2009) Climate change impact assessment on soil water availability and crop yield in anjeni watershed Blue Nile basin. A Thesis Submitted to School of Graduate Studies Arba Minch University in Partial Fulfillment of the Requirement for the Degree of Master of Science in Meteorology.

22. Williams JR (1969) Flood routing with variable travel time or variable storage coefficients. Trans ASAE 12: 100-103.

23. Liersch S (2003) The Programs dew.exe and dew02.exe User's Manual. Berlin, p: 5.

24. Green WH, Ampt GA (1911) Studies on soil physics: 1. The flow of air and water through soils. J Agric Sci 4: 11-24.

25. Monteith JL (1965) Evaporation and the environment. In the State and Movement of Water in living Organisms. XIXth Symposium Soc For Exp Biol, Swansea, Cam-bridge University Press, pp: 205-234.

26. Hargreaves GL, Hargreaves GH, Riley JP (1985) Agricultural benefits for Senegal River basin. J Irrig and Drain Engr 111: 113-124.

27. Arnold JG, Allen PM, Bernhardt G (1993) A comprehensive surface-groundwater flow model. Journal of Hydrology 142: 47-69.

28. Lenhart T, Eckhardt K, Fohrer N, Frede HG (2002) Comparison of two different approaches of sensitivity analysis. Physics and Chemistry of the Earth 27: 645-654.

29. Huisman S, Griensven AV, Srinivasan R, Breuer L (2004) European SWAT School, Advanced Course, p: 112 (Institute for Landscape Ecology and Resource Management, University of Giessen, Hienrich-BuffRing 26, 35392 Giessen, Germany).

30. Santhi C, Arnold JG, Williams JR, Dugas WA, Srinivasan R, et al. (2001) Validation of the SWAT Model on a Large River Basin with Point and Non-Point Sources. Journal of the American Water Resources Association 37: 1169-1188.

31. Dile YT, Berndtsson R, Setegn SG (2013) Hydrological Response to Climate Change for Gilgel Abay River, in the Lake Tana Basin-Upper Blue Nile Basin of Ethiopia. PLoS ONE 8: e79296.

32. Lockaby G, Nagy C, Vose JM, Ford CR, Sun G, et al. (2013) Forests and water. In: The Southern Forest Futures Project: Technical Report. SRSGTR-178. USDA-Forest Service, Southern Research Station, Asheville, NC, USA.

33. Sun G, McNulty SG, Myers JA, Cohen EC (2008) Impacts of Multiple Stresses on Water Demand and Supply Across the Southeastern United States. Journal of the American Water Resources Association 44: 1441-1457.

34. Sarfaraz A (2015) Impact of climate change on future flow of Brahmaputra river basin using swat model. MSc Engineering Thesis, Department of Water Resources Engineering, Bangladesh University of Engineering and Technology (BUET), Dhaka.

35. Biniyam Y, Kemal A (2017) The Impacts of Climate Change on Rainfall and Flood Frequency: The Case of Hare Watershed, Southern Rift Valley of Ethiopia. J Earth Sci Clim Change 8: 383.

36. Chalise SR (1994) Mountain Environments and Climate Change in the Hindu Kush- Himalayas. In: Beniston M (ed.), Mountain Environments in Changing Climates, Routledge, London, UK, pp: 383-404.

37. Refsgaard JC, Storm B (1996) MIKESHE. In: Computer Models in Watershed Hydrology. Singh VJ (ed.), Highland Ranch, Colo: Water Resources Publications, pp: 809-846.

38. Ho CK, Stephenson DB, Collins M, Ferro CAT, Brown SJ (2015) Calibration Strategies a Source of Additional Uncertainty in Climate Change Projections. Bull Amer Met Soc 93: 21-26.

39. Wurbs RA, Muttiah RS, Felden F (2005) Incorporation of climate change in water availability modeling. Journal of Hydrologic Engineering 10: 375-385.

19

Long-term Changes in Annual Precipitation and Monsoon Seasonal Characteristics in Myanmar

Win Win Zin[1]* and Martine Rutten[2]

[1]*Yangon Technological University, Insein Rd, Yangon, Myanmar*
[2]*Delft University of Technology, Stevinweg 1, 2600 GA Delft, The Netherlands*

Abstract

Understanding the spatial and temporal precipitation patterns is essential for climate change and water resources management in Myanmar. In this study, trend detection was performed on historical precipitation series with a maximum length of 49 years. The Mann-Kendall non-parametric test and the Pettitt test were applied to detect trends and step changes in annual total precipitation, annual maximum precipitation and monsoon onset and withdrawal dates. Abrupt changes in trend detection were taken into account and trend analyses on the segments of records discriminated by change points were performed. For 17 out of 82 stations dominated significant trends in annual total precipitation were revealed when ignoring the presence of change points. Upward trends were mainly detected in the southern and coastal parts of the country and downward trends were mainly detected in the central part of the country. However, only two significant trends were detected in the subseries divided by change point. It is noticed that monsoon withdrawal dates have shifted to be early.

Keywords: Mann-Kendall; Myanmar; Precipitation; Pettitt; Trend

Introduction

Understanding precipitation patterns is vital for water resources management in Myanmar. People's well-being and income in this agriculturally dominated country particularly rely on the availability of sufficient water [1]. Flooding due to too much water is the most commonly occurring form of natural disaster in Myanmar, affecting yearly over 2,000,000 people [2]. Spatiotemporal variation in precipitation in Myanmar is large. The annual total average precipitation varies from 900 mm in the central to about 4600 mm in the south (See Figure 1). Most precipitation falls within the monsoon period from May till October. Changes in precipitation will cause changes in availability and excess of water and are important to anticipate in water resources management, particularly in a rapid developing county such as Myanmar.

There seems a direct influence of global warming on precipitation and changes in precipitation patterns are expected to occur globally [3]. In 2000, Sen et al. [4] reported, based on an analysis of station data over the period 1947 to 1970, no significant trends in precipitation in Myanmar. More recent research showed some indication of changes in precipitation patterns. Based on gridded precipitation products and reanalysis data [5] report doubling in May precipitation in Myanmar over the period from 1979 to 2010. According to climate projections Myanmar precipitation patterns are expected to change. The multimodel (of 20 CMIP3) mean suggests enhanced precipitation over Myanmar [6]. The mean annual precipitation in the Bago river basin is projected to increase by 90 mm to 115 mm by 2050 [7]. Yet, statistical evidence of changes occurring in station data in Myanmar is still missing.

We here present an analysis of trends in precipitation Myanmar based on a set of 82 station long term time series and thereby extend Sen et al. [4] work to check if changes can be detected if more recent data is included. We performed the trend analysis first by assuming linear trends and ignoring change points and second by considering change points. In addition, we present a trend analysis and change point analysis of monsoon onset and withdrawal. Finally, we discuss to which extent detected changes can be attributed to climate change and land use changes.

Materials and Methods

Precipitation time series of 82 stations were tested for trends in total annual precipitation and maximum annual precipitation. The Mann-Kendall non-parametric test, which is widely used [8-12] to detect trends, was applied. Trend analysis is often disturbed by the presence of change points. Therefore, the trend is analysis was followed by change point analysis as suggested by Villarini et al. [13]. The change per unit time was estimated by applying Sen's estimator of slope and the Pettitt test was done to detect step changes. In addition, trends in onset, withdrawal and duration of the monsoon were analyzed. We used non-parametric methods and removed serial correlation to avoid false trend detection due to persistence in and/or heave tailed probability density distributions of the time series.

Stations

The monitoring network used in this study (Figure 2) consists of 82 precipitation stations with a maximum sample size of 49 years spanning from 1967 to 2015. The monitoring network is assumed to reflect the meteorological conditions. The length of data set in this study suffices the minimum required length in searching evidence of climatic change

*Corresponding author: Win Win Zin, Department of Civil Engineering, Yangon Technological University, Insein Rd, Yangon, Myanmar
E-mail: winwinzin@ytu.edu.mm

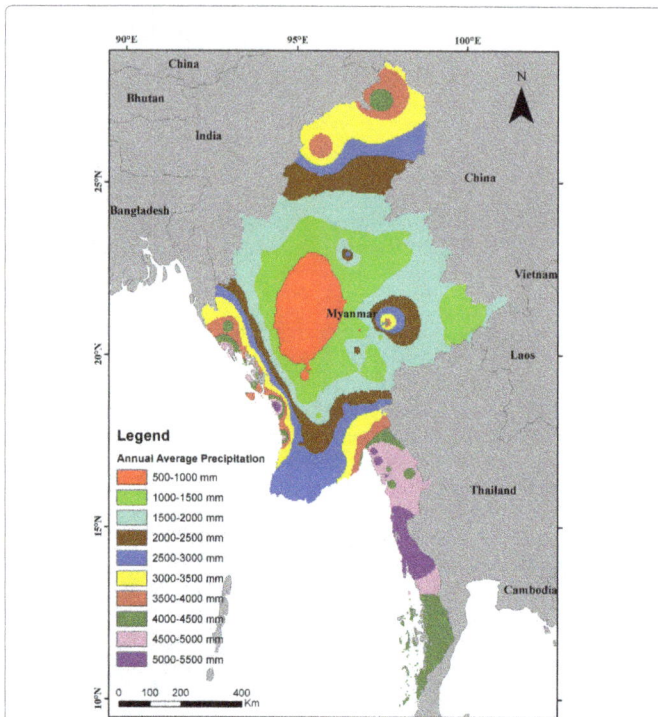

Figure 1: Mean annual precipitation of Myanmar. Isohyets are developed based on average annual precipitation of two decades (1990-2010).

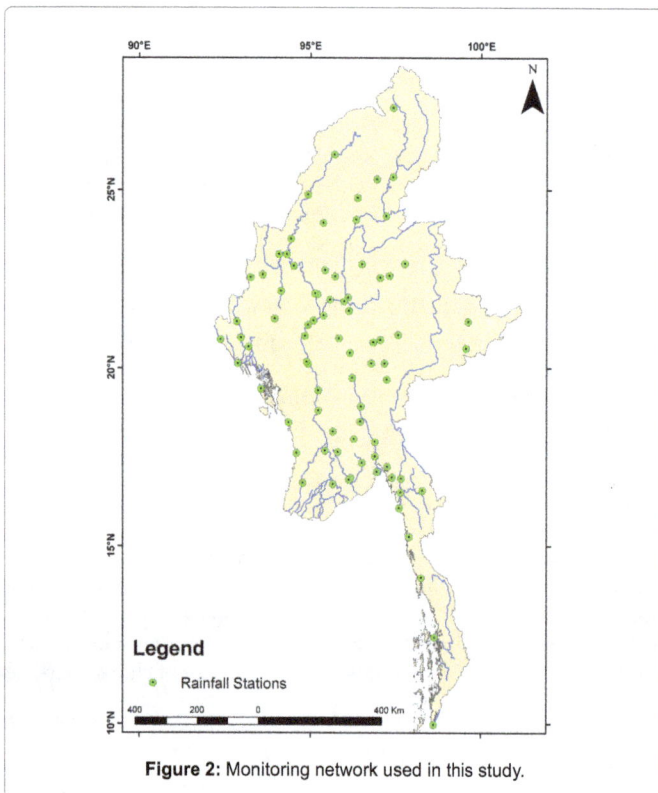

Figure 2: Monitoring network used in this study.

in hydro-climatic time series, which is 25 years [14]. More specifically, 60 out of 82 stations have a sample size between 41 and 49 years, and the remaining stations have a sample size between 30 and 40 years. Monsoon onset and withdrawal dates were collected from the department of Meteorology and Hydrology in Myanmar (DMH).

Pettitt test for change point analysis

As a nonparametric test that allows detection of changes in the mean (median) when the change point time is unknown, the Pettitt test [15] has been suggested by Villarini et al. [13] to analyze the change point. This test is based on a version of the Mann-Whitney statistic for testing whether the two samples $X_1, \ldots, X_t$ and $X_{t+1}, \ldots, X_T$ come from the same population. The test statistic K_T is defined as

$$K_T = \max \left| U_{t,T} \right| = \max_{1 \le t \le T} \left(K_T^+, K_T^- \right) \qquad (1)$$

Where $U_{t,T} = \sum_{i=1}^{t} \sum_{j=t+1}^{T} \mathrm{Sgn}\left(X_i - X_j \right)$ (2)

The p value of the test statistic is computed using the limiting distribution approximated by Pettitt [15], which is valid for continuous variables [13]. And the 95% confidence level was used to evaluate the significance of change points. The significant level is determined approximately by

$$p = \exp\left(\frac{-6k_T^2}{T^3 + T^2} \right) \qquad (3)$$

Mann-Kendall test for trend detection

The Mann–Kendall (MK) statistical test [16,17] is a non-parametric, rank based test for monotonic trend detection [18-20]. It is widely used and recommended by the World Meteorological Organization [20,21] for trend detection because it is more robust against false trend detection due to non-normally distributed time series than its parametric alternatives, yet nearly as powerful.

A critical assumption, when considering precipitation, is that elements in the time series are assumed to be independent. Several authors [22,23] investigated the sensitivity of the MK test to serial correlations and proposed procedures to reduce the effect of serial correlation. Long term persistence can induce a statistically significant trend even though no trend is present. We followed the procedures described in Partal et al. [8] and first removed any lag-1 serial correlations over 5% significance level before applying the Mann-Kendall test. The Mann-Kendall test returns the significance level of a monotonic trend. The magnitude of the trend was calculated with Sen's estimator, also following procedures described in Partal et al. [8].

Results on Annual Precipitation Analysis

Results on change point analysis

Thirteen stations show step change with a confidence over 95% in total annual total precipitation (Table 1) and twelve in annual maximum precipitation (Table 2). Four abrupt changes happened from 1981 to 1990, seven from 1991 to 2000, two from 2001 to 2010 in the analysis of annual total precipitation. Five abrupt changes happened from 1981 to 1990, five from 1991 to 2000, two from 2001 to 2010 in the analysis of annual maximum precipitation.

An abrupt change in annual total precipitation at Kyaukphyu was

Station	Pettitt test for change point			
	K_T	t	shift	p
Kalaewa	294	2001	+	0.01
Kawthaung	294	1993	+	0.014
Kyaikkame	180	2000	+	0.002
Kyaukme	185	1991	-	0.011
Kyaukphyu	368	1989	+	0.001
Loikaw	248	2000	+	0.033
Magway	336	1994	+	0.003
Maubin	256	1988	+	0.049
Myeik	266	1993	+	0.014
Myitkyina	254	1983	+	0.048
Naungcho	187	1999	+	0.011
Pinlaung	275	2001	+	0.019
Sittwe	261	1989	+	0.016

Table 1: Results of Pettitt test for annual total precipitation.

Station	Pettitt test for change point			
	K_T	t	shift	p
Bhamo	256	1989	+	0.043
Chauk	223	1995	-	0.034
Gwa	134	1992	+	0.028
Kawthaung	348	1981	+	0.002
Kyauktaw	160	1994	-	0.042
Maungdaw	199	2007	+	0.039
Mingaladon	268	2003	+	0.012
Minkin	317	1995	+	0.001
Myitkyina	281	1983	+	0.02
Pinlaung	293	1990	+	0.011
Thandwe	238	1987	+	0.047
Theinzayat	231	1993	+	0.009

Table 2: Results of Pettitt test for annual maximum precipitation.

detected to have occurred around 1989. The mean level significantly shifted upward from 4281 mm in the period (1967-1989) to 5013 mm in the period (1990-2015) as can be viewed in Figure 3. An abrupt change in annual total precipitation at Kawthaung was detected to have occurred around 1993. The mean value significantly shifted upward from 3889 mm in the period (1967-1993) to 4470 mm in the period (1994-2015) as can be viewed in Figure 4.

An abrupt change in annual total precipitation at Magway was detected to have occurred around 1994. The mean value significantly shifted upward from 749 mm in the period (1967-1994) to 947 mm in the period (1995-2015) as can be viewed in Figure 5.

An abrupt change in annual total precipitation at Sittwe was detected to have occurred around 1989. The mean value significantly shifted upward from 4423 mm in the period (1967-1989) to 5032 mm in the period (1990-2015) as can be viewed in Figure 6.

An abrupt change in annual maximum precipitation at Chauk as detected to have occurred around 1995. The mean value significantly shifted downward from 86 mm in the period (1967-1995) to 62 mm in the period (1996-2015) as can be viewed in Figure 7. An abrupt change in annual maximum precipitation at Myitkyina was detected to have occurred around 1983. The mean value significantly shifted upward from 108 mm in the period (1967-1983) to 141 mm in the period (1984-2015) as can be viewed in Figure 8.

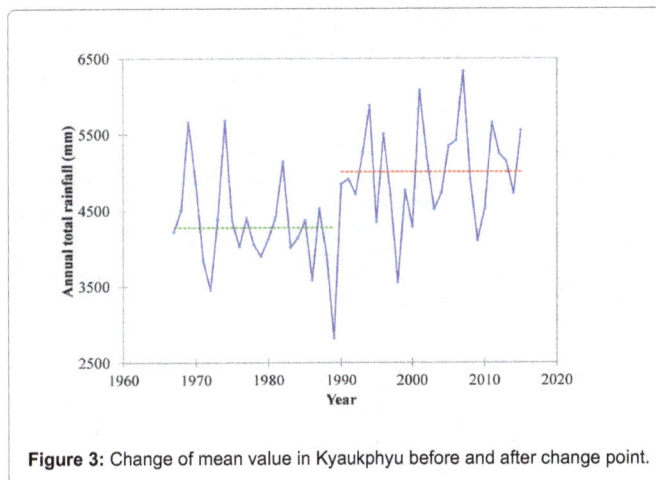

Figure 3: Change of mean value in Kyaukphyu before and after change point.

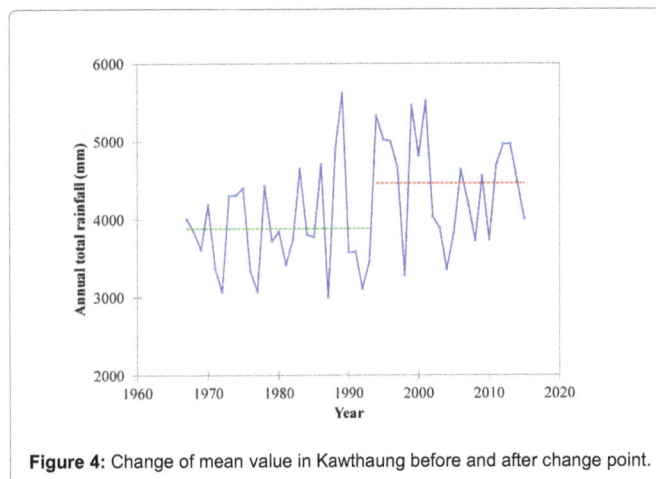

Figure 4: Change of mean value in Kawthaung before and after change point.

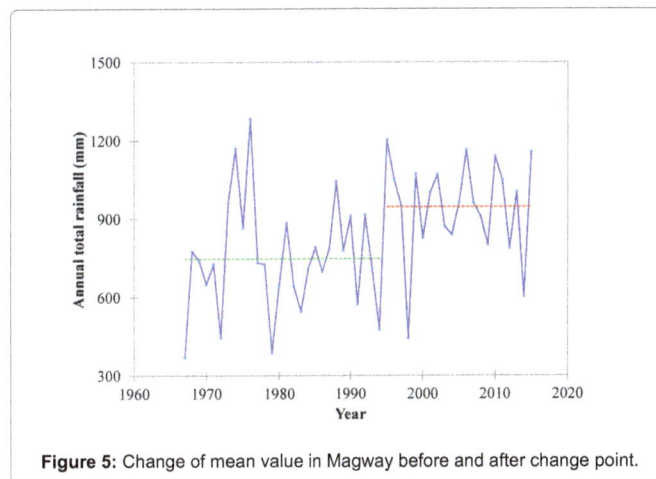

Figure 5: Change of mean value in Magway before and after change point.

Results on trend analysis

The Mann–Kendall method was applied to time series from the selected precipitation stations. All trend results in this research have been evaluated at 10% level of significance to ensure an effective exploration of the trend characteristics of the study region. Significance level indicates the trend's strength and Sen's slope estimator indicates the magnitude of the trend. We considered the lag 1 serial correlation before applying the Mann-Kendall test for the significance of the trend. 75 stations do not show significant serial correlation in annual total precipitation series and 80 stations do not show significant serial correlation in annual maximum precipitation series.

Table 3 shows results of trend analysis under monotonic trend assumption. Among 82 stations, the number of stations containing a positive trend in annual precipitation is 54 and the number containing a negative trend is 28.

Results of detected stations (significance level less than 10%) are described in Tables 4 and 5 respectively. Among 82 stations, 17 stations showed a significant trend, the remaining 65 stations showed no significant trend in annual total precipitation. Among the 17 stations with significant trends, fourteen stations (Kalaewa, Kawthaung, Kyaikkame, Kyaukphyu, Kyauktaw, Lashio, Magway, Maubin, Mawlamyine, MyaukU, Myeik, Myitkyina, Yangon, Zaungtu) showed an upward trend, three stations (Chauk, Lashio and Kyaukme) displayed downward trends. Some stations were noted to have particularly strong trends. Annual total precipitation at Magway, Kyaikkame and Kyaukphyu displayed an increasing trend at 1% significance level. The strongest increasing trend was detected at the Kyaikame station (35.2 mm per annual) in annual total precipitation. The strongest decreasing trend was detected at the Kyaukme station (7.5 mm per annual) in annual total precipitation.

Among 82 stations, 14 stations showed a significant trend, the remaining 68 stations showed no significant trend in annual maximum precipitation. Annual maximum precipitation at Kawthaung, Kyauktaw and Minkin displayed an increasing trend at 1% significance level. The strongest increasing trend was detected at the Kyauktaw station (3.0 mm per annual) in annual maximum precipitation.

Figures 9 and 10 show the spatial distribution of trends for annual total precipitation and annual maximum precipitation for the selected stations in Myanmar. The spatial distribution of the trends for the annual precipitation exhibited that downward trends were mainly detected in the central and eastern parts of the country, and adversely, in the southern and coastal parts of the country upward trends were detected at most of the stations.

Significant trends for annual total precipitation and annual maximum precipitation are shown in Figures 11 and 12 respectively.

Besides, trend has also been done for the stations with change points. Trend analysis was done separately for the subseries divided by the change points. Trends in annual total series and annual maximum series prior to and posterior to change points are shown in Tables 6 and 7 respectively. When change point are considered, only three significant trends were found in annual series.

The direction of the trends of subseries prior to and posterior to the change points was the same at five stations and adverse direction of trend was found in the subseries at the other eight stations in annual total precipitation series. At Kalaewa station, the subseries prior to the change point has decreasing trend and an increasing trend was seen after change point.

Series	General trends			Significant trend		
	Total	down	up	Total	down	up
Annual precipitation	82	28	54	17	3	14
Annual maximum precipitation	82	28	54	14	2	12

Table 3: Results of trend analysis under monotonic trend assumption.

Station	Mann-Kendall test			Sen's slope estimator (mm/year)		
	Z	Trend	Significance	Q	$Q_{95(L)}$	$Q_{95(U)}$
Chauk	-1.98	Down	**	-4.14	-8.16	-0.11
Kalaewa	1.78	up	*	4.57	-0.37	10.80
Kawthaung	2.16	Up	**	15.93	1.78	31.00
Kyaikkame	2.62	Up	***	35.26	11.44	59.99
Kyaukme	-2.05	Down	**	-7.44	-14.36	-0.27
Kyaukphyu	2.93	Up	***	19.64	8.33	33.57
Kyauktaw	2.49	Up	**	31.36	7.09	54.23
Lashio	-2.08	Down	*	-4.20	-8.01	0.64
Magway	2.77	Up	***	6.42	2.58	10.94
Maubin	2.24	Up	**	8.54	1.32	17.38
Mawlamyine	1.83	Up	*	14.26	-1.71	29.57
MyaukU	1.92	up	*	21.61	-0.37	42.22
Myeik	2.12	Up	**	11.13	2.00	19.58
Myitkyina	1.96	Up	**	7.23	0.13	14.08
Naungcho	1.89	Up	*	6.68	-0.24	11.69
Yangon	1.77	Up	*	5.63	-0.86	13.44
Zaungtu	2.38	Up	**	22.96	6.34	40.13

Table 4: Results of Mann-Kendall Test with Sen's slope estimator for annual total precipitation by ignoring the presence of change points.

* $\alpha = 0.1$ level of significance

** $\alpha = 0.05$ level of significance

*** $\alpha = 0.01$ level of significance

The direction of trends of subseries prior to and posterior to the change point was the same at six stations and adverse direction of trend was found in the subseries at the other four stations in annual maximum precipitation series.

Station	Mann-Kendall test			Sen's slope estimator (mm/year)		
	Z	Trend	Significance	Q	$Q_{95(L)}$	$Q_{95(U)}$
Bhamo	1.84	Up	*	0.68	0.05	1.53
Gwa	2.11	Up	**	2.71	0.29	5.18
Katha	1.92	Up	*	0.72	0	1.39
Kawthaung	2.58	Up	***	1.06	0.3	2.13
Kyaukphyu	1.86	Up	*	1.2	-0.04	2.55
Kyauktaw	3.59	Up	***	3.03	1.55	4.68
Maungdaw	2.36	Up	**	2.13	0.57	4.0
Mingaladon	2.44	Up	**	0.76	0.11	1.37
Minkin	2.81	Up	***	1.12	0.42	2.37
Phyu	-1.74	Down	*	-0.87	-1.8	0.17
Pinlaung	1.71	Up	*	0.35	-0.05	0.81
Pyinmana	-1.77	Down	*	-0.37	-0.89	0.03
Theinzayat	1.85	Up	*	1.0	-0.05	2.0
Thandwe	1.75	Up	*	1.23	-0.13	2.48

Table 5: Results of Mann-Kendall Test with Sen's slope estimator for annual maximum precipitation by ignoring presence of change points.

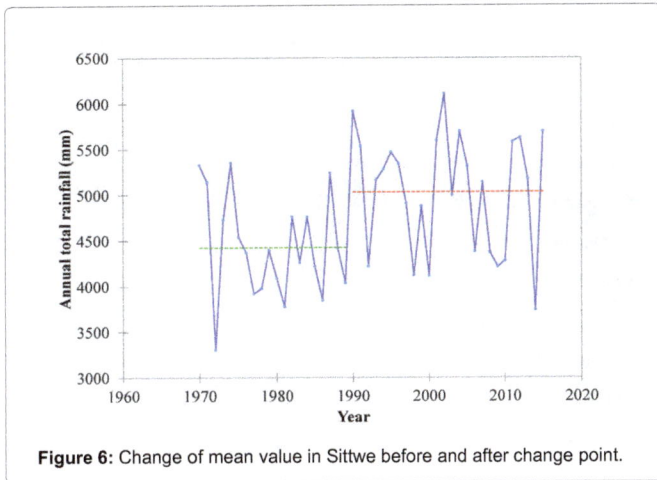

Figure 6: Change of mean value in Sittwe before and after change point.

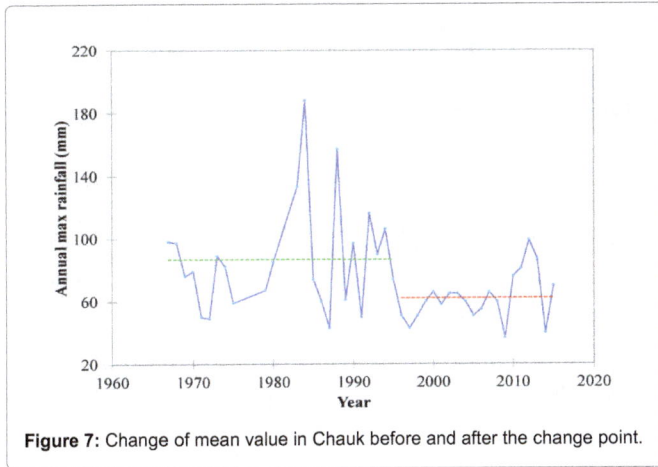

Figure 7: Change of mean value in Chauk before and after the change point.

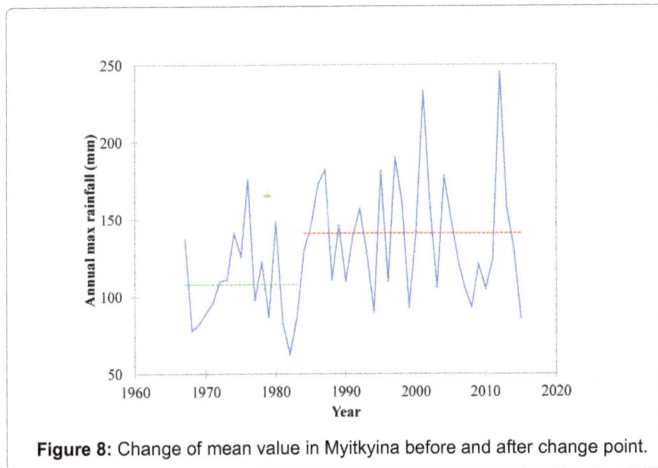

Figure 8: Change of mean value in Myitkyina before and after change point.

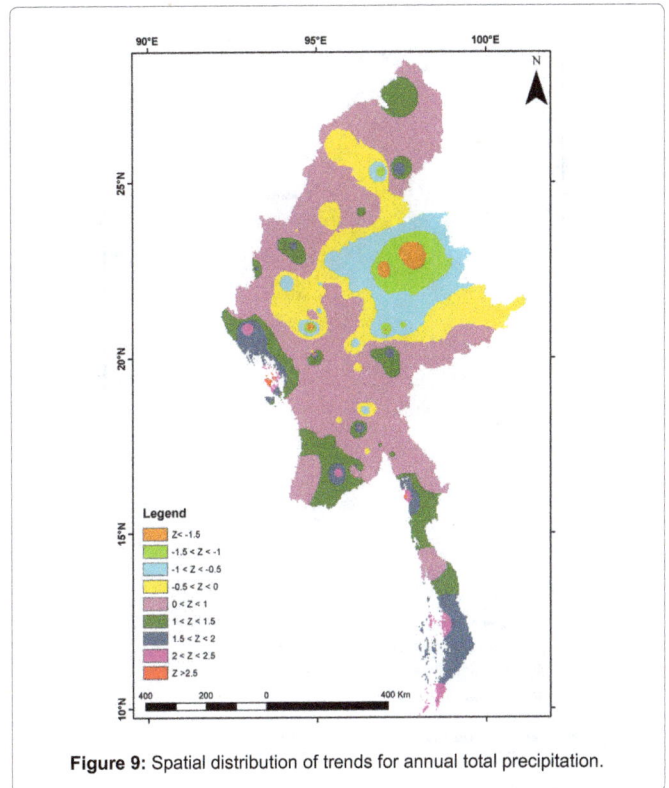

Figure 9: Spatial distribution of trends for annual total precipitation.

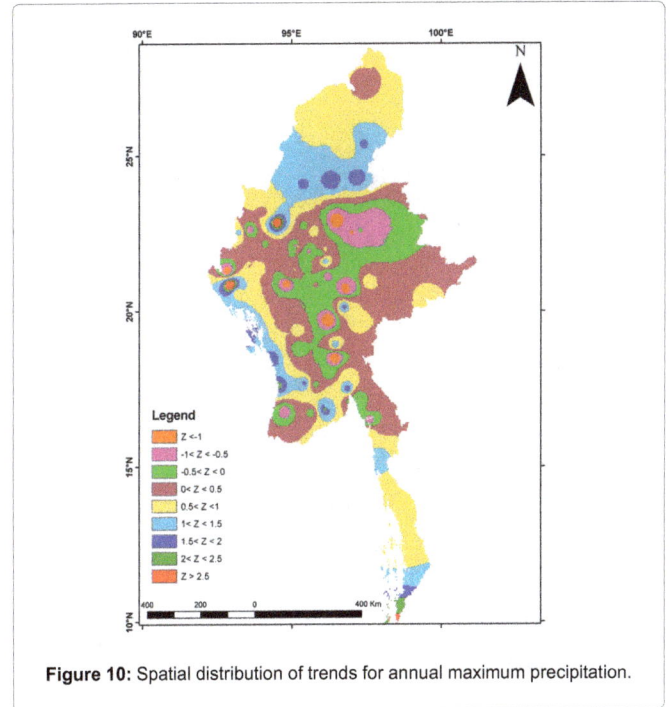

Figure 10: Spatial distribution of trends for annual maximum precipitation.

Results on southwest monsoon onset and withdrawal analysis

In addition to trends in annual total and annual maximum precipitation, trends in monsoon onset and withdrawal were investigated. Monsoon onset and withdrawal days were collected from the Department of Meteorology and Hydrology (DMH). Due to late onset and early withdrawal of the monsoon, the duration of monsoon has become shorter in last three decades.

Results of Mann-Kendall test for monsoon time series over the period 1967-2015 are shown in Table 8. Monsoon duration and withdrawal days displayed a significant decreasing trend.

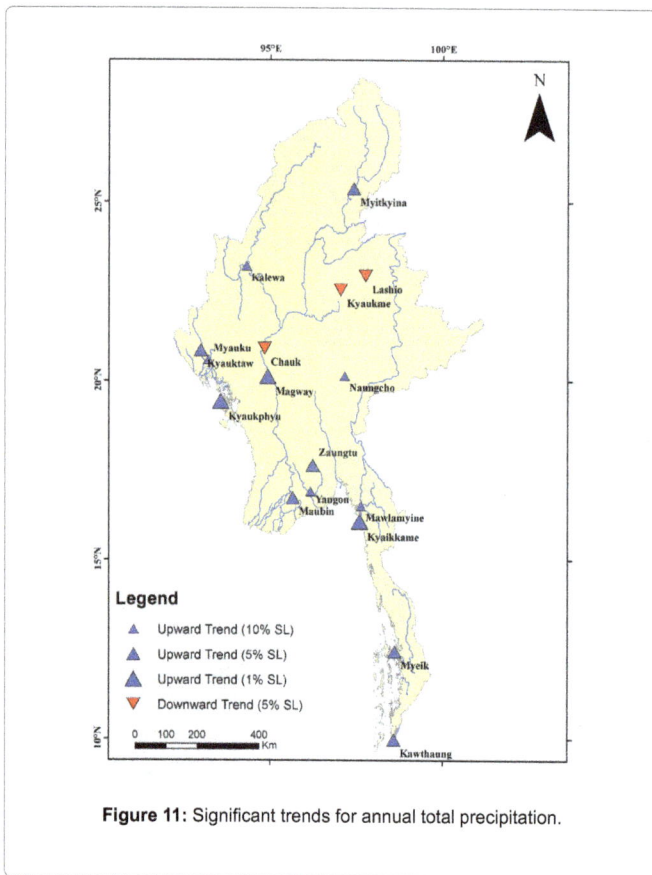

Figure 11: Significant trends for annual total precipitation.

Station	Change point	prior to change points		posterior to change points	
		Z	direction	Z	direction
Kalaewa	2001	**-1.73**	-	**1.68**	+
Kawthaung	1993	0.04	+	-0.69	-
Kyaikkame	2000	-0.14	-	0.2	+
Kyaukme	1991	1.03	+	-0.21	-
Kyaukphyu	1989	-1.37	-	1.17	+
Loikaw	2000	**-1.73**	-	0.88	+
Magway	1994	0.1	+	0	+
Maubin	1988	0	+	0.26	+
Myeik	1993	-1.31	-	0.79	+
Myitkyina	1983	-0.95	-	0.26	+
Naungcho	1999	-0.23	-	-0.77	-
Pinlaung	2001	-1.19	-	-0.11	-
Sittwe	1989	-1.01	-	-0.06	-

Table 6: Trends in annual total precipitation series prior to and posterior to change points (The bold values denote 90% confidence level)

Station	Change point	Prior to change points		Posterior to change points	
		Z	Direction	Z	Direction
Bhamo	1989	-0.61	-	0.02	+
Chauk	1995	0.3	+	1.55	+
Kawthaung	1981	0.2	+	-0.58	-
Kyauktaw	1994	1.15	+	0.88	+
Mingaladon	2003	-0.18	-	0.86	+
Minkin	1995	-0.82	-	**1.85**	+
Myitkyina	1983	-0.18	-	-0.02	-
Pinlaung	1990	-0.9	-	-0.72	-
Theinzayat	1993	-0.21	-	-0.87	-
Thandwe	1987	-0.76	-	-0.09	-

Table 7: Trends in annual maximum precipitation series prior to and posterior to change points (The bold values denote 90% confidence level)

The results of Pettitt test for monsoon time series over the period 1967-2015 are shown in Table 9. The monsoon duration was about 135 days before 1988 and 120 days after 1988. It was observed that monsoon duration and withdrawal days have shifted significantly according to the Pettit test (p=0.95). Moreover, it was noticed that monsoon withdrawal dates have shifted to be early. The change of mean value in monsoon withdrawal and monsoon duration are shown in Figures 13 and 14.

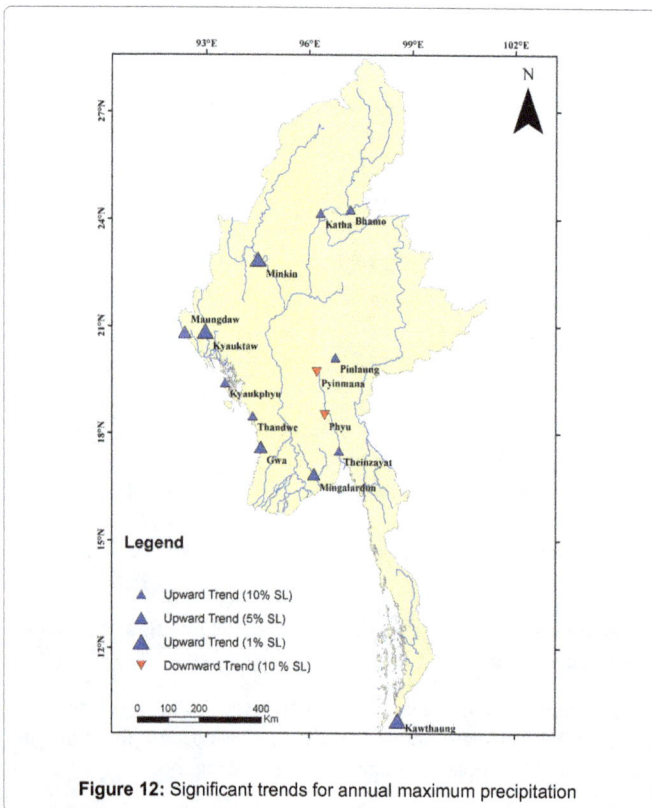

Figure 12: Significant trends for annual maximum precipitation

	Mann-Kendall test	
	Z	Significance
Monsoon duration	-2.29	5 % SL
Withdrawal days	-3.15	1 % SL
Onset days	1.16	No significant

Table 8: Results of Mann-Kendall test for monsoon analysis.

	K_T	t	shift	p
Monsoon duration	**359**	**1988**	-(shorter)	**0.0026**
Monsoon withdrawal	**458**	**1989**	- (earlier)	**0.0001**
Monsoon onset	191	1976		0.238

Table 9: Results of Pettitt test for monsoon time series (Data series with significant shifts at 0.05 significance level are shown in bold).

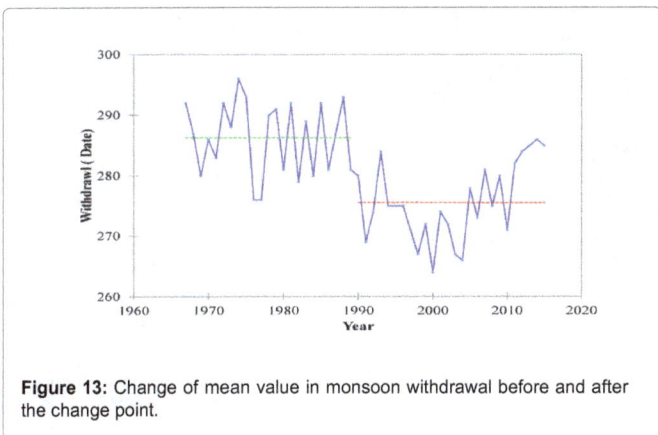

Figure 13: Change of mean value in monsoon withdrawal before and after the change point.

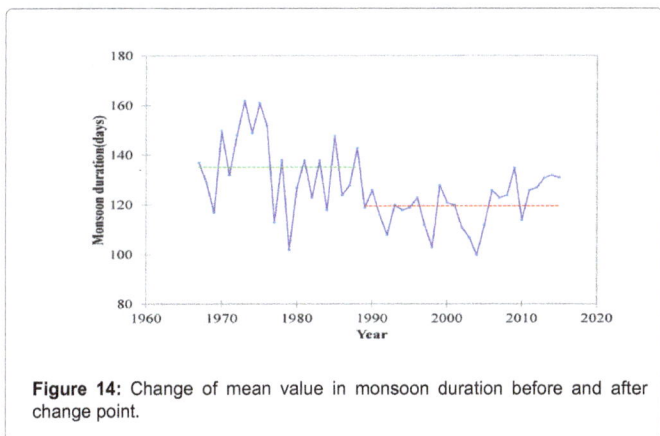

Figure 14: Change of mean value in monsoon duration before and after change point.

Discussion

We detected some significant changes in precipitation in Myanmar over the last few decades. Significant trends were obtained if the absence of change points was assumed; however, mostly no significant trends were obtained in the subseries divided by the change points. Thus, abrupt behavior of the time series should be taken into account in trend analysis, or else, results of trends could be misleading. Subseries after the change point analysis are considered short for the trend detection, hence the results of trend for the sub-series at some stations may not be robust.

According to Wang et al. [5], global warming has led to cyclone intensification particularly in the pre-monsoon season which has led to increase in overall precipitation. An intensification of the monsoon the shorter monsoon season is also reported by the Myanmar Department of Meteorology and Hydrology (DMH).

Wang et al. [5] hypothesize a regime shift from apparent mid latitude circulation regime to a tropical/monsoonal one in the pre-monsoon as an explanation for the sudden increase in precipitation in the 1990s. Changes might also be partly attributed to multi decadal modes of variability rather than one directional climate change. D'Arrigo et al. [24] show based on an analysis of teak ring data that there are significant decadal to multi decadal modes of variability that can be linked to monsoon precipitation. They suggest that these modes reflect the remote influence of the Pacific Decadal Oscillation and related regime shifts, and their impact on monsoon precipitation.

In addition to the external climate change factors mentioned above, land use changes in Myanmar may have caused significant changes in precipitation patterns. Feddema et al. [25] demonstrated the effect of land use changes on climate and specifically the Asian monsoon. Large scale land use changes could affect both strength and timing of the monsoon yet effects were different for the different change scenarios used. In Myanmar deforestation has been the major land use change so far. Over the period from 1990 to 2000 forests have declined by 0.3% annually [26] and rates have most probably been increasing since due to the faster economic development. The trend attribution and the relation between the observed precipitation trends and climate change should be addressed in future modeling and data analysis studies.

Conclusion

The paper presents an analysis of trends in precipitation in Myanmar based on long term historical station data. On the basis of annual precipitation at 82 stations in Myanmar, the long-term monotonic trend and abrupt changes have been investigated. This study performs a simultaneous evaluation of gradual and abrupt changes in annual precipitation in Myanmar using Mann–Kendall trend test and Pettitt change-point test. The application of a trend detection framework resulted in the identification of some significant trends under monotonic trend assumption. The results of Mann-Kendall analysis indicated that 17 stations have significant trend in annual total precipitation, 14 with upward trend and 3 with downward trend, the other 65 stations have no significant trend. Fourteen stations have significant trend in annual maximum precipitation, 12 with upward trend and 2 with downward trend, the other 68 stations have no significant trend.

The direction of precipitation trend was, in general, upward across the country though not statistically significant for all stations. Upward trends were mainly detected in the southern and coastal parts of the country and downward trend was mainly in the central part of the country. The shift analysis based on Pettitt test, performed in order to highlight the possible change points, showed probable change years in the decades 1980-1990 and 1990-2000 for annual precipitation.

There are 17 out of 82 stations that were dominated by significant trend in annual total precipitation series when ignoring the presence of change points. Only three significant trends were detected by considering change points in annual total precipitation series. There are 14 out of 82 stations that were dominated by significant trend in annual maximum precipitation series when ignoring the presence of change point.

Only one statistically significant trend was detected by considering change points in annual maximum precipitation series. Therefore, a study of monotonic trends in a time series regardless of the abrupt shifts in mean or change in direction of trends can lead to misleading conclusions.

In addition, a clear shift in the monsoon duration and withdrawal dates was detected. Onset dates were delayed and withdrawal dates were advanced leading to a shortening of the monsoon seasons. The trend attribution and the relation between the observed precipitation trends and climate change should be addressed in future studies with the inclusion of the influence of temperature.

Changes in precipitation lead to changes in runoff and water availability that can be critical in the fast-developing country Myanmar, especially in more dry regions. It is important to investigate present and probable future climatic change patterns and their impacts on water resources so that appropriate adaptation strategies may be implemented.

References

1. Taft L, Evers M (2016) A review of current and possible future human-water dynamics in Myanmar's river basins. Hydrological Earth System Science 20: 4913-4928.

2. Kar AK, Winn LL, Lohani AK, Goel NK (2011) Soft computing–based workable flood forecasting model for Ayeyarwady River Basin of Myanmar. Journal of Hydrologic Engineering 17: 807-822.

3. Trenberth KE (2011) Changes in precipitation with climate change. Climate Research 47: 123.

4. Sen Roy N, Kaur S (2000) Climatology of monsoon rains of Myanmar (Burma). Int J Climatol 20: 913-928.

5. Wang SY, Buckley BM, Yoon, JH, Fosu B (2013) Intensification of premonsoon tropical cyclones in the Bay of Bengal and its impacts on Myanmar. Journal of Geophysical Research: Atmospheres 118: 4373-4384.

6. Turner AG, Annamalai H (2012) Climate change and the South Asian summer monsoon. Nature Climate Change 2: 587-595.

7. Htut AY, Shrestha S, Nitivattananon V, Kawasaki A (2014) Forecasting Climate Change Scenarios in the Bago River Basin, Myanmar. J Earth Sci Clim Change 5: 228.

8. Partal T, Kahya E (2006) Trend analysis in Turkish precipitation data. Hydrological Processes 20: 2011-2026.

9. Li ZL, Xu ZX, Li JY (2008) Shift trend and step changes for runoff time series in the Shiyang River basin, northwest China. Hydrological Process 22: 4639-4646.

10. Hirsch RM, Slack JR, Smith RA (1982) Techniques of trend analysis for monthly water quality data. Water Resources Research 18: 107-121.

11. Xu ZX, Takeuchi K, Ishidaira H (2003) Monotonic trend and step changes in Japanese precipitation. Journal of Hydrology 279: 144-150.

12. Xu ZX, Takeuchi K, Ishidaira H, Li J (2005) Long-term trend analysis for precipitation in Asian Pacific FRIEND river basins. Hydrological Processes 19: 3517-3532.

13. Villarini G, Serinaldi F, Smith AJ, Krajewski FW (2009) On the stationarity of annual flood peaks in the continental United States during the 20th century. Water Resour Res 45: W08417.

14. Burn HB, Elnur MAH (2002) Detection of hydrologic trends and variability. Journal of Hydrology 255: 107-122.

15. Pettitt AN (1979) A non-parametric approach to the change-point problem. Appl Stat 28: 126-135.

16. Mann HB (1945) Nonparametric tests against trend. Econometrica 13: 245-259.

17. Kendall MG (1948) Rank Correlation Methods. Hafner, New York, USA.

18. Helsel DR, Hirsch RM (1992) Statistical Methods in Water Resources. Elsevier, Amsterdam.

19. Serrano A, Mateos VL, Garcia JA (1999) Trend analysis of monthly precipitation over the iberian peninsula for the period 1921-1995. Physics and Chemistry of the Earth, Part B: Hydrology, Oceans and Atmosphere. 24: 85-90.

20. Yue S, Pilon P, Phinney B, Cavadias G (2002) The influence of autocorrelation on the ability to detect trend in hydrological series. Hydrological Processes 16: 1807-1829.

21. Mitchell JM, Dzerdzeevskii B, Flohn H, Hofmeyr, WL, Lamb HH, et al. (1966) Climatic change. (Report of a working group of the Commission for Climatology). WMO Technical Note No. 79. World Meteorological Organization, p: 79.

22. Hirsch RM, Slack JR (1984) A nonparametric trend test for seasonal data with serial dependence. Water Resources Research 20: 727-732.

23. Kulkarni A, Von Storch H (1995) Monte-Carlo experiments on the effect of serial correlation on the Mann-Kendall test of trend. Meteorologische Zeitschrift 4: 82-85.

24. D'Arrigo R, Ummenhofer CC (2015) The climate of Myanmar: evidence for effects of the Pacific decadal oscillation. Int J Climatol 35: 634-640

25. Feddema JJ, Oleson KW, Bonan GB, Mearns LO, Buja LE, et al. (2005) The importance of land-cover change in simulating future climates. Science 310: 1674-1678.

26. Leimgruber P, Kelly DS, Steininger MK, Brunner J, Müller T, Songer M (2005) Forest cover change patterns in Myanmar (Burma) 1990-2000. Environmental Conservation 32: 356-364.

20

Determination the Origin of Mineralization in the Coastal Aquifer Northeast Tunisia by Isotopic Method

Mzoughi Aroua[1]*, Ben Hamouda Mohamed Fethi[2] and Bouhlel Salah[1]

[1]Laboratory of Mineral Resources and Environment, Faculty of Sciences of Tunis, University of Tunis El Manar, Tunis, Tunisia
[2]Hydrology and Isotopic Geochemistry Unit, CNSTN Sidi-Thabet Technology Center, Tunisia

Abstract

A geochemical and isotopic techniques were undertaken to characterize groundwater in Northeast Tunisia. Hydrogeochemical investigations demonstrated that groundwater can be classified into different water facies. The Ras Djebel-RafRaf aquifer showed a (Ca-Cl-SO_4) and (Na-Cl-NO_3) water type. Data inferred from ^{18}O and deuterium isotopes in groundwater samples indicated recharge with modern rainfall. Water characterized by lower $\delta^{18}O$ and δ^2H values is interpreted as recharged by non-evaporated rainfall originating from Mediterranean air masses from Mediterranean air masses at higher altitude. However, water with relatively enriched $\delta^{18}O$ and δ^2H contents is thought to reflect the occurrence of an evaporation process related to the long-term practice of flood irrigation.

Keywords: Hydrogeology; Geochemistry; Coastal aquifer; Tunisia

Introduction

The region of Bizerte (Figure 1) is situated in boundary Northen of Tunisia. It forms with the Cap Blanc the most advanced point of Africa. It is exposed to humid winds from the northwest and West, and receives an average rainfall of 570 mm/year [1], This coastal occupies an area of 50 Km². It is limited by Djebels bouchoucha, Touchela and Bab Banzart in NW, Djebels Ennadour and Demna in South, the Mediterranean Sea in the North and NE.

This zone is influenced by a sub-humid Mediterranean climate. The annual temperature is around 18° C. The ETP, about 1197 mm/year. This basin is drained by many wadis. The use of water is essential since the Ras Djebel is an agricultural region by excellence, in Fact the major part of population practise agricultural activities.

In result, this aquifer undergoes increased anthropic pressure through overexploitation and construction of two dams at Wadi Beni Ata and Wadi Shaab Eddoud.

Geology Setting

From a geologic standpoint, the main geologic features of this study area are [2,3].

(a) Upper Cretaceous, represented by the lower senonian and the lower campanian which is divided into two distinct entities: the marls senonian and the calcareous marl campanian.

(b) Lower Paleocene, represented by grey marls.

(c) Eocene, formed by limestone.

(d) Mio-Pliocene, formed by continental materials: Marls, clays, conglomerates, gypsum and sands.

(e) Quaternary, represented by the villafranchian, formed essentially by limestone.

Materials and Methods

For the evaluation of groundwater quality, 50 water samples were collected from the shallow aquifer of Ras Djebel plain during October (Figure 1), the groundwater samples were analysed for chemical and isotopic composition.

Measurement of temperature, electrical conductivity and pH were measured in the field. Major elements (SO_4^{2-}, Cl^-, NO_3^-, Ca^{2+}, Mg^{2+}, Na^+ and K^+) were analysed at the laboratory of National Center for Nuclear Sciences and Technologies. These elements are expressed in mg/L^{-1}. Salinity was analyzed by evaporating a specified volume of water sample in a graduated capsule for 24 h in an incubator at 105°C.

The dry residue concentration was determined by subtracting the final mass from the initial mass of the sample that was placed in the incubator. Ca^{2+} and Mg^{2+} were measured by Liquid Ion Chromatography while K^+ and Na^+ were measured by flame photometer.

The results of chemical analyses of the investigated groundwater samples are presented in Table 1.

Results and Discussion

Hydrochemistry

Physical-chemical parameters: The overall chemical characteristics of groundwater are presented in Table 1. The pH values range from 6, 88 to 8, 3 with an average value of 7, 3. The temperature values are between 20, 2°C and 25°C. The values measured on the surface water are influenced by the surface air temperatures.

The total dissolved solids and the Electrical Conductivity (EC) values range from 1, 03 g/L to 12, 96 and 0, 2 to 0, 7 µS/cm. The spatial distribution maps (Figure 2) shows that salinity increase in the direction of groundwater flow. The Electrical Conductivity (EC) is well correlated with the total dissolved solids (TDS).

The factor of correlation is in the order of 1.

Major ion chemistry:

---

*Corresponding author: Mzoughi Aroua, Unit of Geology and Applied Geochemistry (UR11ES16), Department of Geology, Faculty of Sciences of Tunis, University of Tunis El Manar, 2092 Tunis, Tunisia, E-mail: Arwa_mzoughi@yahoo.com

N°èch	pH	Cond	sec	Ca	Mg	Na	K	SO$_4$	Cl	NO$_3$	HCO$_3$
1	7.2	2.41	1567	220	40.32	328.9	8.19	324.52	479.25	23.3	351.54
2	6.9	5.37	3491	368	86.64	749.8	9.75	845.24	955.7	70.59	423.46
3	7.1	2.4	1560	180	32.88	280.6	28.47	333.26	323.1	97.4	572.26
4	7.2	2.15	1398	200	32.4	230	8.59	349.32	323.15	152.1	353.4
5	7.3	2.46	1599	142	62.16	331.2	8.97	467.31	370.23	86	523.9
6	7.7	2.4	1560	106	65.52	354.2	15.6	402.26	388.57	144.4	416.64
7	7.1	2.49	1619	156	48.48	299	30.81	354.42	455.79	90.87	407.34
8	8	2.92	1898	201	66.48	386.4	9.75	573.34	556.17	19.11	391.84
9	7.1	3.82	2483	137	104.64	616.4	7.02	492.12	753	40.17	701.84
10	7.5	4.04	2626	196	47.76	464.6	31.98	534.34	657.9	29.15	502.2
11	7.2	1.66	1079	89	28.32	149.5	72.15	202.16	238.48	49.92	221.34
12	7.2	3.56	2314	114	59.76	595.7	24.57	464.89	648.78	21.35	513.98
13	7	6.7	4355	319	67.44	1299.5	85.8	986.48	1567.7	41.24	544.36
14	7	5.37	3491	401	104.64	832.6	93.6	854.89	1327.5	12.68	564.82
15	7.2	2.93	1905	222	26.64	503.7	3.51	452.14	670.24	45.24	550.56
16	7.2	2.81	1827	260	30.96	411.7	2.34	335.22	660.9	47.58	412.3
17	7	3.36	2184	252	30.96	496.8	6.24	521.76	608.74	15.11	626.2
18	7.1	3.22	2093	178	66.72	397.9	31.98	254.72	632.97	48.26	574.12
19	7	3.25	2113	270	138.72	328.9	38.22	443.93	733.48	57.82	515.84
20	7.1	3.3	2145	250	104.88	368	23.79	459.82	806.07	54.21	493.52
21	6.9	2.97	1931	205	86.88	393.3	17.16	304.88	744.33	34.42	512.74
22	7	3.31	2152	263	119.04	342.7	7.41	350.4	843.33	53.24	550.56
23	7.3	1.44	936	66	39.36	138	18.33	160.01	147.61	73.22	288.3
24	7.2	3.4	2210	240	47.52	446.2	13.26	526.93	560.96	60.16	600.78
25	6.9	7.09	4609	542	207.84	984.4	8.19	987.69	1742.3	99.94	485.46
26	7	4.03	2620	277	124.08	517.5	3.12	499.09	971.78	99.84	512.12
27	7.1	3.66	2379	315	0.48	471.5	3.9	466.06	561.69	82.49	396.18
28	7.1	3.6	2340	322	14.16	427.8	15.99	380.74	691.25	22.72	458.18
29	7.1	5	3250	357	58.32	611.8	27.3	550.1	1046	77.03	522.66
30	7.1	3.55	2308	338	24	577.3	19.11	282.92	1019.3	83.07	556.14
31	7.1	4.17	2711	318	62.64	738.3	16.38	599.52	1118.6	47.97	515.22
32	6.9	3.94	2561	386	14.64	489.9	8.19	629.57	821.75	46.12	434.62
33	7.2	2.94	1911	200	15.6	411.7	30.42	386.99	550.79	17.36	408.58
34	7	3.75	2438	480	10.8	397.9	6.24	275.25	1145.7	45.14	521.42
35	7.2	3.56	2314	354	17.28	473.8	7.8	393.06	724.18	43.58	535.06
36	7.6	3.22	2098	282	19.92	473.8	5.46	408.41	803.41	16.58	480.5
37	7.1	2.78	1807	251	73.44	202.4	9.75	297.5	398.4	59.48	442.68
38	7.1	3.6	2340	205	81.6	395.6	13.65	339.69	579.64	61.23	555.52
39	7.3	2.31	1502	136	45.6	273.7	13.65	169.04	368.54	17.55	391.84
40	7.8	0.27	174	50	7.68	6.9	8.19	53.45	42.5	11.41	65.1
41	7.4	2.98	1937	244	98.64	404.8	7.02	429.47	671.63	7.51	559.24
42	8.3	0.99	607	83	20.4	92	24.96	130.58	147.25	46.51	132.06
43	7.3	2.28	1482	113	15.36	292.1	9.36	235.8	252.96	45.05	499.72
44	7.1	3.46	2249	282	119.04	342.7	16.38	623.95	540.14	11.31	532.58
45	7.4	4.43	2880	208	139.44	719.9	17.16	744.19	1025.3	16.58	516.46
46	7.5	3.42	2223	222	13.44	427.8	24.18	514.97	668.82	32.08	301.32
47	7.3	4.2	2730	300	42.72	625.6	31.59	570.49	933.48	92.63	420.36
48	7.1	6.3	4095	253	74.88	1016.6	56.94	328.8	1674.2	40.07	603.26
49	7.2	4.79	3114	228	28.08	901.6	64.35	274.27	1041.3	58.79	539.4
50	7.4	1.56	1014	111	15.12	142.6	107.64	152.75	163.99	126.95	272.18

Table 1: Geochemical data for the studied aquifer.

Chloride and sodium

Concentrations of sodium and chlorides vary respectively between 6, 9 to 1299, 5 mg/L and 42, 5 to 1742 mg/L. For both Na$^+$ and Cl$^-$, the highest concentration levels characterize the wells situated on the border of the sea (P13, P25 and P48). The spatial distribution of chloride and sodium concentrations illustrate a similar evolution to that salinity: they both increase in the direction of groundwater flow.

Calcium

Ca^{2+} concentrations ranged between 50 and 542 mg/L. The most elevated value was measured in well P25, while lowest level was recorded in well P40. The water of this coastal aquifer is moderately rich in calcium. The mostly analyzed waters are pure, which is in conformity with pH values of these waters. Calcium and pH have an influence on water aggressiveness [4].

Ech	RDJ1	RDJ2	RDJ3	RDJ5	RDJ6	RDJ8	RDJ9	RDJ10	RDJ11	RDJ12	RDJ13	RDJ14	RDJ15	RDJ16	RDJ17	RDJ18	RDJ21	RDJ23	RDJ24	RDJ26
δ^2H	-27,7	-26,7	-27,6	-30,8	-30,1	-32,8	-31,1	-25,1	-31,1	-31,6	-26,7	-24,3	-27,2	-27,0	-25,7	-28,2	-23,4	-31,8	-28,5	-26,6
$\delta^{18}O$	-4,4	-3,6	-4,5	-5,0	-4,6	-5,9	-5,0	-4,0	-4,9	-5,0	-4,4	-4,0	-4,5	-4,1	-4,1	-4,2	-4,2	-5,3	-4,8	-4,1

Table 2: Isotopic data for the studied aquifer.

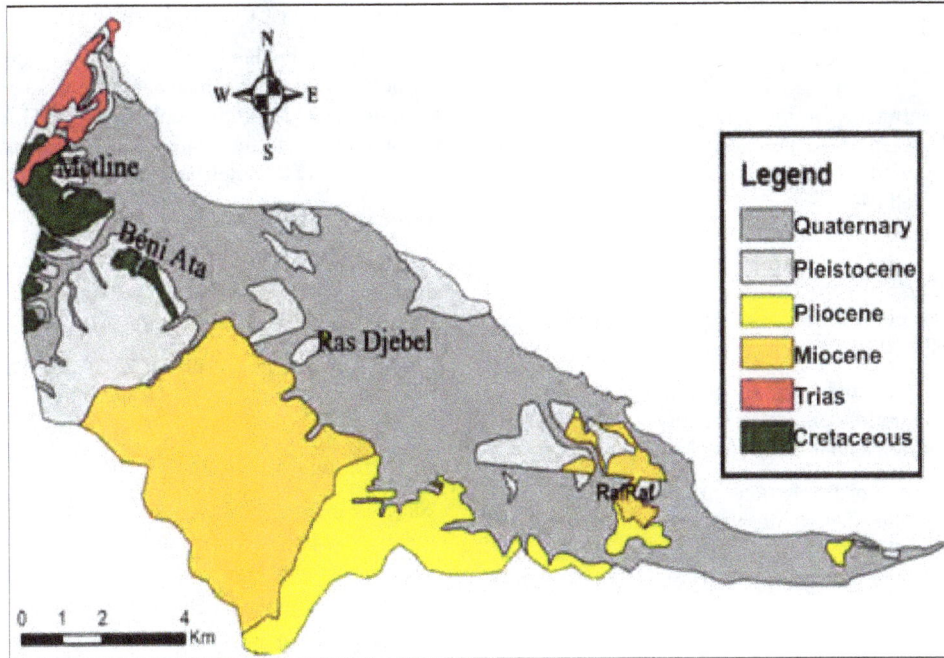

Figure 1: Geological map of RafRaf-Ras Djebel plain.

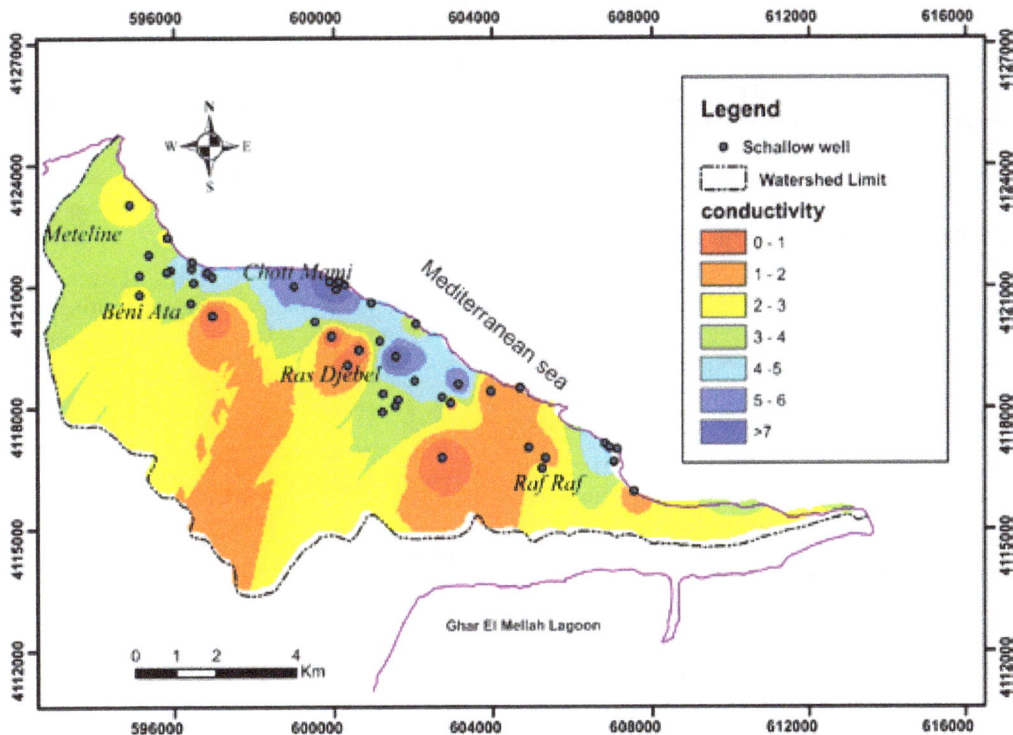

Figure 2: Map of salinity.

*Magnesium

Mg^{2+} concentrations in Ras Djebel-RafRaf aquifer fluctuated between 0, 48 and 207 mg/L. The highest values were recorded for well P25, whereas the lower one was recorded for well P40.

*Potassium

K^+ concentrations are relatively low compared to the concentrations of other cations which values vary between 2, 34 and 107, 64.

*Sulfates

The sulphates concentrations of the water in the Ras Djebel-RafRaf aquifer varied between 53, 45 and 987, 69 mg/L. The highest concentrations were measured for well P25 and P13 while the lowest value is characterized by well P40.

*Nitrates

NO_3^- concentrations varied between 7, 51 and 144, 4 mg/L. The highest concentration was measured for well P6 while the lowest value was characterized by well P41.

Forty two percent of groundwater samples taken during this study, show nitrate concentrations exceeding the maximum European admissible nitrate concentration limit in drinking water (50 mg/L).

The examination of nitrate distribution map (Figure 3) reveals that high nitrate concentrations appear to be related to the fertilizer application and long term flood irrigation practises, highlighting the significant contribution of the return flow of irrigation waters to the degradation of groundwater [5-8].

Saturation Index

Water saturation states for minerals as anhydrite (SI anh), gypsum (SI gyp), aragonite (arg) and calcite (cal) were computed. If the SI is <0, dissolution is considered as the dominant process for the related mineral and when is >0, precipitation of mineral is likely to be occurring in the system [9].

The majority of samples are supersaturated towards aragonite from which the carbonates are precipitated. This supersaturation of water towards these carbonates minerals such as aragonite and calcite shows that the possibility of dissolution of these minerals by the waters is not possible and that cannot contribute to the acquisition of mineralization [10]. In addition, the majority of samples of this groundwater are under saturated towards gypsum.

Piper diagram

The piper diagram [3] (Figure 4) was determined to precisely recognize the different groundwater facies. Nitrate concentration was taken into account when plotting this diagram because of its relative abundance in the groundwater.

The data represented in this groundwater samples are classified into two major groups: $(Ca-Cl-SO_4)$ and $(Na-Cl-NO_3)$ water type.

Relationships between TDS and major elements

Plots of various major elements as a function of TDS value were drawn to determine the contributing elements to groundwater mineralization and the major hydrogeochemical process. Shows that Cl^-, SO_4^{2-} and Na^+ are well correlated with TDS presenting very close values of 1 (respectively: 0, 88; 0, 67 and 0,87). In result chlorides, sulphates and sodium are the main contributors to groundwater mineralization.

However, the contents of calcium and bicarbonate participate significantly is not determining role in the acquisition of mineralization. In addition to TDS, the correlation matrix established between sulphates and Cl^- ion (Figure 5A) shows that the majority of points are situated above the line (1:1). The increase of Cl^- in comparison to sulphates proves that the Cl^- has probably two origins: It is linked to a marine intrusion or dissolution of the halite. This hypothesis is confirmed by the relation between the Cl^- and Na^+ ions (Figure 5B).

The relationship between Ca^{2+} and Cl^- ions (Figure 5C) shows excess of Cl^- against Ca^{2+}, which demonstrates a process of precipitation of gypsum $(CaSO_4, 2H_2O)$ or anhydrite $(CaSO_4)$. Besides, the effect of precipitation is felt in the relationship between the SO_4^{2-} and Ca^{2+} ions (Figure 5D) where the sulphate deficiency is related to the ion exchange.

Isotopic composition of groundwater

Isotope geochemistry techniques are valuable tools in investigating many problems in hydrology and evaluating hydrogeological and hydro chemical controlling mechanisms in any groundwater system [11,12].

They provide significant insights into water origin, recharge circumstances (time and location), water flow directions and residence times [11].

*Stable isotope of water $(\delta^{18}O, \delta^2H)$:

Stable isotopes $(\delta^{18}O, \delta^2H)$ have conservative proprieties and provide information on the groundwater recharge process (Table 2) [13]. They can offer an evaluation of physical process that affects water masses, such as evaporation and mixing [14].

The isotopic compositions of the groundwater samples from the Grombalia shallow aquifer range from -5, 3‰ to -3, 6‰ for $\delta^{18}O$ and from -31‰ to -24, 3‰ for δ^2H.

The correlation diagram of $\delta^{18}O/\delta^2H$ (Figure 6) shows the position of all samples relative to the Global Meteoric Water Line (GMWL: $\delta^2H=8\delta^{18}O+10$) [15] and the Local Meteoric Water Line of the Tunis-Carthage (LMWL: $\delta^2H= 8\delta^{18}O+12, 4$) [16,17].

Nevertheless, these groundwater samples (Figure 6A) can be further divided into two groups:

• The first group, which is placed below the GMWL is probably the consequence of the infiltration of an evaporated component likely deriving from the return flow of irrigation water.

• The second group, placed between the GMWL and the LMWL, shows that the precipitation ensuring the recharge of aquifer from a mixture of oceanic and Mediterranean vapour masses.

The plot of chloride versus $\delta^{18}O$ (Figure 6B) indicates the different processes responsible for the variation of groundwater salinity.

Some water samples show a correlation between these two elements. It is probable that the dissolution mechanism takes precedence over evaporation in the acquisition of the mineralization.

Conclusion

Based on hydrogeological characteristics the aquifer of Ras Djebel Northest Tunisia can be classified into two groups: $Ca-Cl-SO_4$ and $Na-Cl-NO_3$ water type. The acquisition of mineralization mainly through natural mineralization mechanisms such as the ion exchange and the geology of groundwater. Nevertheless, anthropogenic process related to agricultural practices also plays an important role in the salinization of groundwater.

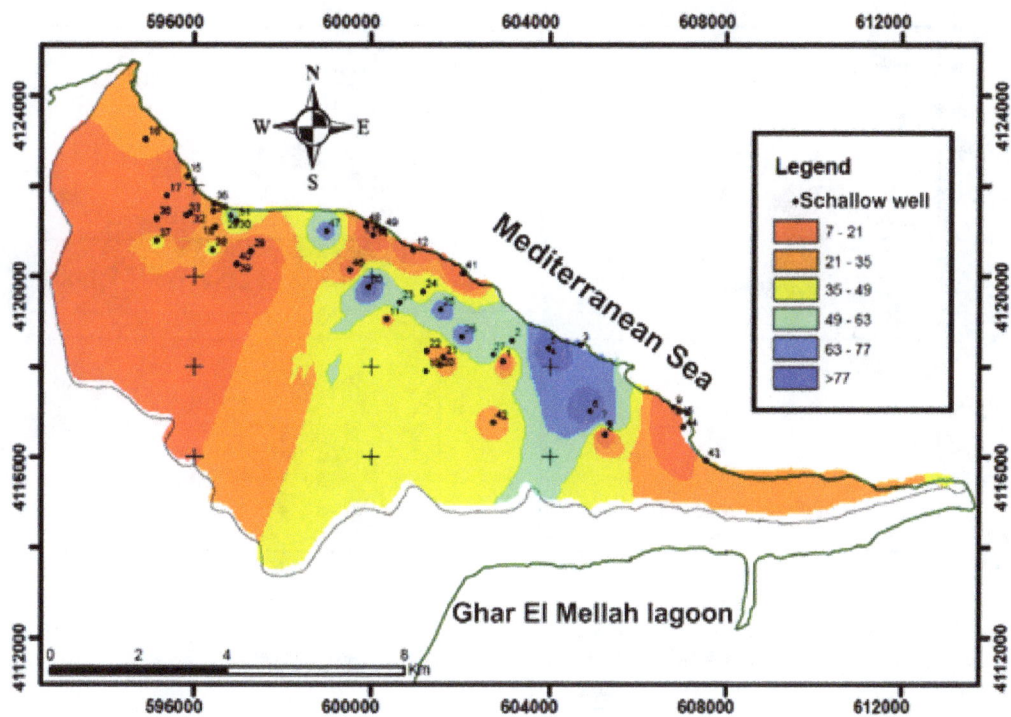

Figure 3: Map of salinity.

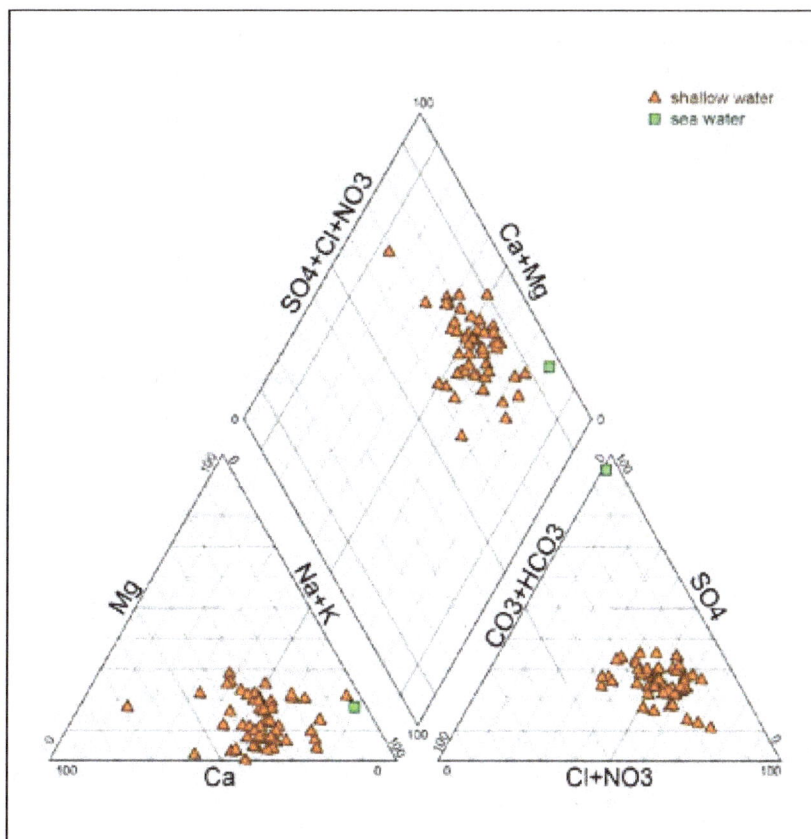

Figure 4: Piper diagram.

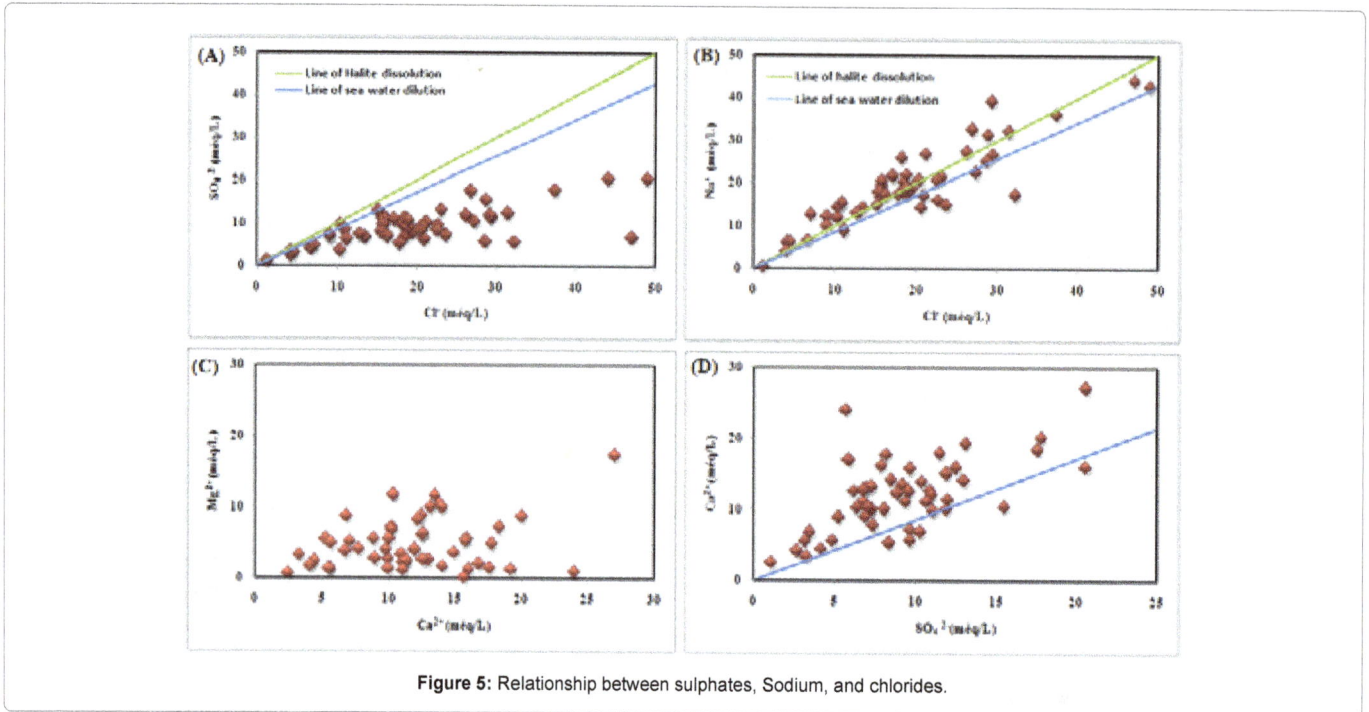

Figure 5: Relationship between sulphates, Sodium, and chlorides.

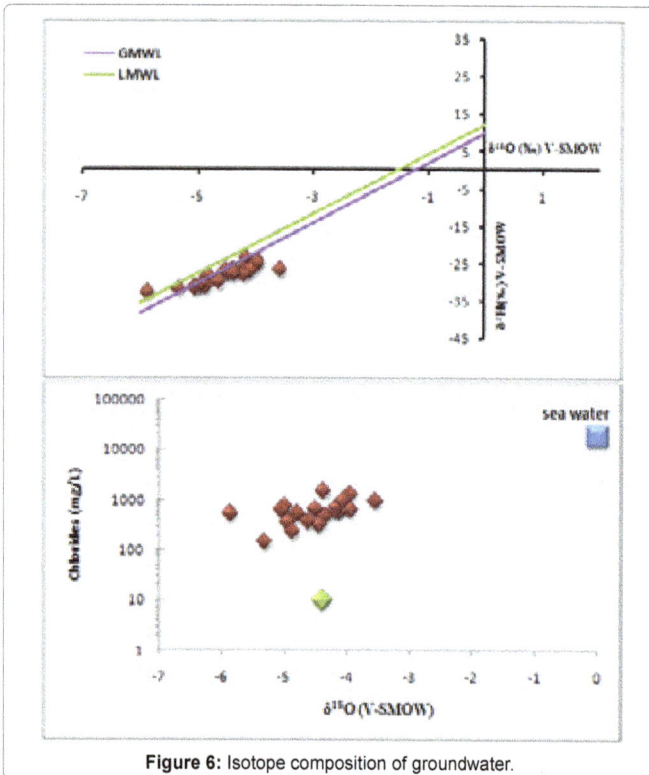

Figure 6: Isotope composition of groundwater.

References

1. Balti M (1986) Notes sur l'exploitation de la nappe phréatique du Ras Djebel Rapport interne DGRE. réf 3/27, p: 19.

2. Burrollet PF (1951) Étude géologique des basins Mio-Pliocène du Nord Est de la Tunisie. Ann Mines et Gèol, n°7, 82p+ annexes.

3. Piper AM (1944) A graphic procedure in the geochemical interpretation of water analyses. Transactions, American Geophysical Union 25: 914-923.

4. Gilbert JR (1990) L'agressivité des eaux de surface en région tropicale, ses conséquences sur le réseau de distribution. TSM, l'eau. 85éme année, N°3, pp: 47-86.

5. Charfi S, Zouari K, Feki S, Mami E (2012) Study of variation of groundwater quality in a coastal aquifer in North eastern Tunisia using multivariate factor analyses. Quaternary International 302: 199-209.

6. Hadas A, Sagiv B, Haruvy N (1999) Agricultural practices, soil fertility management modes and resultant nitrogen leaching rates under semi-arid conditions. Agricultural Water Management 42: 81-95.

7. Oenema O, Boers PCM, Van Erott MM, Fraters B, Vander Meer HG, et al. (1998) Leaching of nitrate from agriculture to groundwater: the effect of policies and measures in the Netherlands. Environmental Pollution 102: 471-478.

8. Rouvier H (1977) Géologie de l'extrême Nord Tunisien: Tectonique et paléogéographique superposes à l'extrémité de la chaine Maghrébine. Thèse de 3éme cycle, Univ Paris VI, 2: 898.

9. Freeze RA, Cherry JA (1979) Groundwater. Prentice Hall Inc., Englewood Cliffs, NJ, USA, p: 604.

10. Ben Hamouda MF (2008) Approche hydrogéochimique et isotopique des systèmes aquifères de CapBon, Cas des nappes de la côte orientale et d'El Haouaria, Tunisie. Thèse en sciences agronomiques de l'INAT.

11. Clark I, Fritz P (1997) Environmental isotopes in Hydrogeology. Lewis Publishers, New York, USA.

12. Tijani MN, Loehnert EP, Uma KO (1996) Origin of saline groundwater's in the Ogoja area, lower Benue trough, Nigeria. Journal of African Earth Sciences 23: 237-252.

13. Gat JR (1971) Comments on the stable isotope method in the regional groundwater investigation. Water Resources Research 7: 980-993.

14. Geyh MA (2000) An Overview of ^{14}C Analysis in the Study of Groundwater. Radiocarbon 42: 99-114.

15. Craig H (1961) Isotopic variation in meteoric water. Science 133: 1702-1703.

16. Singh B, Singh Y, Sekhon GS (1995) Fertilizer-N uses efficiency and nitrate pollution of groundwater in developing countries. Journal of Contaminant Hydrology 20: 167-184.

17. Zouari K, Aranyossy JF, Mamou A, Fontes JC (1985) Étude isotopique et géochimique des mouvements et de l'évolution des solutions de la zone aérée des sols sous climat semi-aride (sud tunisien).

PERMISSIONS

LIST OF CONTRIBUTORS

Khalaf S and Abdalla MG
Irrigation and Hydraulics Department, Faculty of Engineering, El-Mansoura University, Egypt

Imelda N Njogu and Johnson U Kitheka
School of Water Resources Science and Technology, South Eastern Kenya University, Kitui, Kenya

Goutam KS, Tanaya D and Sharanya C
School of Oceanographic Studies, Jadavpur University, Kolkata 700032, West Bengal, India

Anwesha S
Department of Mathematics, Jadavpur University, Kolkata 700032, West Bengal, India

Meenakshi C
Basanti Devi College, Kolkata 700029, West Bengal, India

Yalemsew Adela
Jimma University, Institute of Technology, School of Civil and Environmental Engineering, Ethiopia

Christian Behn
Rostock University, Faculty of Agricultural and Environmental Sciences, Resources Management and Soil Physics, Rostock, Germany

Sadik Ahmed and Ioannis Tsanis
Department of Civil Engineering, McMaster University, Hamilton, Ontario L8S 4L7, Canada

Getahun Kitila and Gizachew Kabite
College of Natural and Computational Science, Wollega University, 395, Ethiopia

Tena Alamirew
Haramaya University, Institute of Technology, 138, Ethiopia

Sarvat Gull, Ahangar MA and Ayaz Mohmood Dar
Department of Civil Engineering, National Institute of Technology, Srinagar, Jammu and Kashmir, India

Yasin Goa Chondie
Areka Agricultural Research Center, Areka, Ethiopia

Agedew Bekele
Awassa Agricultural Research Center, Awassa, Ethiopia

Muhammet Omer Dis, Emmanouil Anagnostou
Civil and Environmental Engineering, University of Connecticut, Storrs, Connecticut, USA

Flamig Zac, Humberto Vergara and Yang Hong
School of Civil Engineering and Environmental Sciences, University of Oklahoma, Norman, Oklahoma, USA

Karoli N. Njau
The Nelson Mandela African Institution of Science and Technology, P.O. Box 447, Arusha, Tanzania

Halima Kiwango
The Nelson Mandela African Institution of Science and Technology, P.O. Box 447, Arusha, Tanzania
Tanzania National Parks, P.O. Box 3134, Arusha, Tanzania

Eric Wolanski
Tropwater, James Cook University, Townsville, Qld. 4810, Australia

Khalaf S, Ahmed AO, Abdalla MG and El Masry AA
Irrigation and Hydraulics Department, Faculty of Engineering, El-Mansoura University, Egypt

Mahasa Pululu S
Department of Geography, Faculty of Natural and Agricultural Sciences, Qwaqwa Campus, University of the Free State, Phuthaditjhaba, South Africa

Palamuleni Lobina G and Ruhiiga Tabukeli M
Department of Geography and Environmental Sciences, School of Environmental and Health Sciences, Mafikeng Campus, North West University, Mmabatho, South Africa

Ward E Sanford
Mail Stop 431, U.S. Geological Survey, Reston, Virginia, 20171, USA

David L Nelms and Jason P Pope
U.S. Geological Survey, Richmond, Virginia, USA

David L Selnick
U.S. Geological Survey, Reston, Virginia, USA

Ayenew Desalegn
Department of Meteorology and Hydrology, Institute of Technology, Arba Minch University, Arba Minch, Ethiopia

Solomon Demissie and Seifu Admassu
School of Civil and Water Resources Engineering, Bahir Dar Institute of Technology, Bahir Dar University, Bahir Dar, Ethiopia

Sudhakar BS
Institute of Science, Nirma University Science & Technology, Ahmedabad-382481, India

Anupam KS
Institute of Engineering & Technology, J.K Lakshmipat Universitsy, Jaipur-302026, India

Akshay OJ
Department of Civil Engineering, Pandit Dindayal Petroleum University, Gandhinagar, India

Wissam Al-Taliby, Ashok Pandit and Howell Heck
Department of Civil Engineering, Florida Institute of Technology, Melbourne, FL 32901, United States of America

Biniyam Yisehak Menna
Department of Meteorology and Hydrology, College of Natural Sciences, Arba Minch University, Arba Minch, Ethiopia

Win Win Zin
Yangon Technological University, Insein Rd, Yangon, Myanmar

Martine Rutten
Delft University of Technology, Stevinweg 1, 2600 GA Delft, The Netherlands

Mzoughi Aroua and Bouhlel Salah
Laboratory of Mineral Resources and Environment, Faculty of Sciences of Tunis, University of Tunis El Manar, Tunis, Tunisia

Ben Hamouda Mohamed Fethi
Hydrology and Isotopic Geochemistry Unit, CNSTN Sidi-Thabet Technology Center, Tunisia

Index

www.ingramcontent.com/pod-product-compliance
Lightning Source LLC
Chambersburg PA
CBHW080702200326
41458CB00013B/4933